Encyclopedia of Photosynthesis

Encyclopedia of Photosynthesis

Edited by **Agatha Wilson**

Callisto Reference

New York

Published by Callisto Reference,
106 Park Avenue, Suite 200,
New York, NY 10016, USA
www.callistoreference.com

Encyclopedia of Photosynthesis
Edited by Agatha Wilson

International Standard Book Number: 978-1-63239-281-7 (Hardback)

Printed in China.

Contents

Preface

This book presents a broad overview on the various aspects of the life giving process of photosynthesis. Photosynthesis is one of the most significant reactions on Earth, and is an essentially interdisciplinary scientific field, with several research groups analyzing and gathering beneficial results from the study of this interesting process. This book discusses the fundamental and applied aspects of photosynthesis. It has been written by specialists in the field of photosynthesis research and associated fields and offers a discourse on distinct and novel topics from photosynthetic inhibitors to relationship between flowering initiation and photosynthesis.

Various studies have approached the subject by analyzing it with a single perspective, but the present book provides diverse methodologies and techniques to address this field. This book contains theories and applications needed for understanding the subject from different perspectives. The aim is to keep the readers informed about the progresses in the field; therefore, the contributions were carefully examined to compile novel researches by specialists from across the globe.

Indeed, the job of the editor is the most crucial and challenging in compiling all chapters into a single book. In the end, I would extend my sincere thanks to the chapter authors for their profound work. I am also thankful for the support provided by my family and colleagues during the compilation of this book.

Editor

Part 1

Fundamental Aspects

1

Photosynthetic Inhibitors

Róbson Ricardo Teixeira, Jorge Luiz Pereira and Wagner Luiz Pereira
Federal University of Viçosa, Viçosa, Minas Gerais State
Brazil

1. Introduction

Life on Earth is dependent on sunlight. In the process known as photosynthesis, plants, algae and certain bacterias are capable of using this source of energy to drive the synthesis of organic compounds. The oxygenic photosynthesis results in the release of molecular oxygen and the removal of carbon dioxide from the atmosphere that is used to synthesize carbohydrates. This process, which involves a complex series of electron transfer reactions, provides the energy and reduced carbon required for the survival of virtually all life on our planet, as well as the molecular oxygen necessary for the survival of oxygen consuming organisms (Nelson, 2011; Nelson & Yocum, 2006; Rutherford & Faller, 2001).

In modern agriculture, farmers continuously face a battle to achieve products in high yields and better quality to feed an ever increasing world population (Stetter & Lieb, 2000). The optimization of agriculture techniques demands, along with other requirements, the application of crop protection agents to control a variety of diseases and pests, among which are weeds. Weeds compete with crops for nutrients, water, and physical space, may harbor insect and disease pests, and are thus capable of greatly undermining both crop quality and yield. In view of the problems caused by weed species, their control is highly desirable.

Among the methods to control weeds, the use of herbicides or weed killers has become the most reliable and least expensive tool for weed control in places where highly mechanized agriculture is practiced. Since the introduction of 2,4-dichlorophenoxyacetic acid (2,4-D) in 1946, several classes of herbicides were developed that are effective for broad-spectrum of weed control (Böger et al, 2002; Cobb, 1992; Ware, 2000).

It is well know that various compounds can interfere with photosynthetic electron transport. This fact has been explored by agrochemical companies to develop an assortment of herbicides to control weeds. Some representative members of commercial photosynthetic inhibitors are diuron (**1**), atrazine (**2**), paraquat (**3**) e diquat (**4**) (Figure 1).

The photosynthetic inhibitors can be divided into two distinct groups, the inhibitors of photosystem II exemplified by diuron (**1**) and atrazine (**2**) and the inhibitors of photosystem I such as paraquat (**3**) and diquat (**4**). It is worth to mention that compounds that inhibit photosystem II account for 30% of the sales in the herbicide market (Draber, 1992).

Although there are a variety of herbicides that can control a broad spectrum of weeds, there is still a necessity for the development of new active ingredients. Several reasons can be mentioned to support this statement. Herbicides should have a favorable combination of properties, such as high specific activity, low application rates, crop tolerance, and low mammalian toxicity. Increasing public concern for environmental pollution derived from agricultural practices also requires that herbicides be rapidly degraded by soilborne microorganisms. Moreover, during the past years, intensive and repeated applications of the same active ingredients cause the selection for and development of herbicide resistance (Devine & Shukla, 2000; Beckie, 2006; Gressel, 2009; Preston, 2004). Starting from 1960s, hundreds of weed biotypes have been reported as surviving herbicide application (Heap, 2011).

Intensive efforts have thus been undertaken to discover new compounds with favorable environmental and safety features to selectively control weeds. In this regard, the photosynthetic system has been target aiming to find new weed killers. In this book chapter, it will be covered the developments concerning the search for new photosynthetic inhibitors during the last ten years.

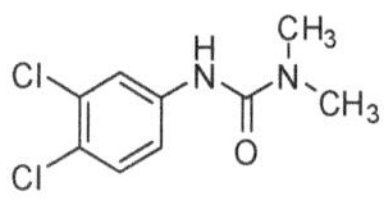

(1)

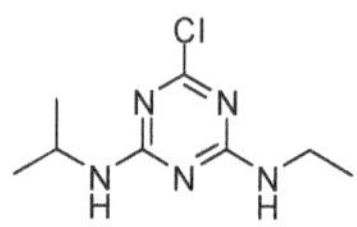

(2)

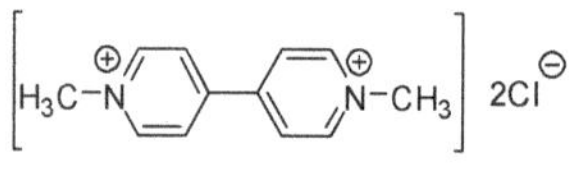

(3)

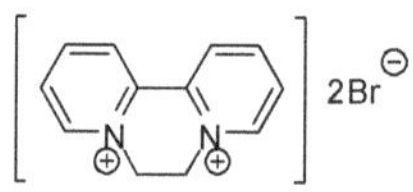

(4)

Fig. 1. Examples of commercial photosynthetic inhibitors.

2. Natural products as source of photosynthetic inhibitors

The strategies used to identify new chemical agents to control weeds can no longer be distinguished from pharmaceutical research and development (Short, 2005). Three major different approaches have been employed. The first one refers to the systematic screening of large numbers of synthetic compounds. Subsequently, lead compounds are optimized. This has been the most widely used strategy by agrochemical companies (Böger et al, 2002; Cobb, 1992; Rüegg et al, 2006; Ware, 2000). The second one is the rational design of specific inhibitors of key metabolic processes (Lein et al, 2004). However, to date such an approach has not been fully successful to produce commercial herbicides.

A third strategy is related to the exploitation of natural products either directly as herbicides (Copping & Duke, 2007) or as leads for the development of new herbicides (Barbosa et al, 2008; Dayan et al, 1999; Dayan et al, 2009; Duke et al, 2000a, 2002; Hütter, 2011; Macías et al, 2007; Pillmoor et al, 1993; Scharader et al, 2010; Strange, 2007; Vyvyan, 2002). This strategy can be considered attractive for several reasons. A vast array of compounds has been isolated from nature and most of them have not been evaluated for agrochemical purposes. Different from what happens to the majority of synthetic agrochemicals, most natural products are water soluble. Moreover, due to natural selection these compounds can present bioactivity in very low concentration. The great variety of chemical structures found in nature can afford chemical agents for weed control that are toxicologically and environmentally benign. Furthermore, the molecular sites where natural products exert their action can be quite different from known molecular targets (Duke et al, 2005; Duke,

Duke et al, 2000b). This is particularly important to overcome the resistance problem. It is worth to mention that the few natural products based herbicides mesotrione, sulcotrione, cinmethylin, bialaphos and gliphosinate (vide infra) act on molecular sites that were not know before they were introduced.

Even though the great variety of natural products has been relatively little explored, several active principles were discovered in this way (Figure 2).

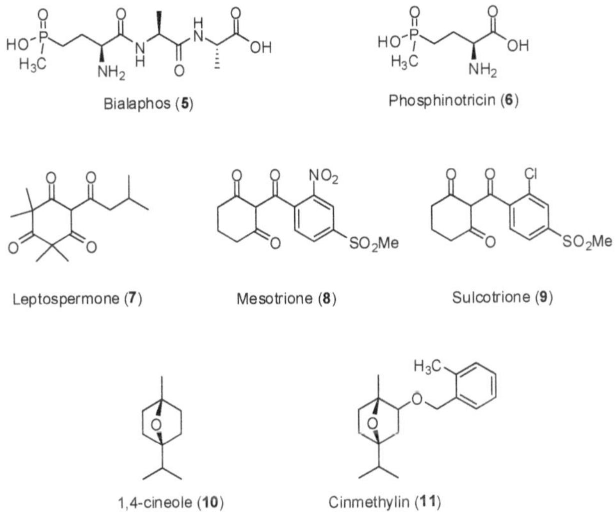

Fig. 2. Structures of compounds **5-11** mentioned in the text.

Bialaphos (**5**) and phosphinotricin (**6**) were originally isolated from various strains of the bacteria *Streptomyces* spp. (Saxena & Pandey, 2001). Bialaphos is a proherbicide that is metabolized into the active ingredient phosphinotricin in the treated plant. Currently, bialaphos is commercialized in Japan with the name of Herbiace®. It should be mentioned that phosphinotricin (**6**) is also produced synthetically as a racemic mixture and commercialized as gluphosinate. Leptospermone (**7**), a major component of the essential oil of *Leptospermun scoparium* (Myrtaceae) (van Klink et al, 1999), was chemically modified to make mesotrione (Mitchell et al, 2001). Mesotrione (**8**) is the active principle of the commercial herbicide Callisto®, which is commercialized by Syngenta and suitable for use in corn fields. Another example is sulcotrione (Chaabane et al, 2005), an herbicide marketed in Europe by Bayer Crop Science under the trade name Mikado®. Sulcotrione (**9**) is used to control a broad range of annual and perennial broadleaf weeds in maize and sugar cane crops. The commercial herbicide cinmethylin (**11**) (Figure 2) is a 2-benzyl ether derivative of the natural product 1,4-cineole (**10**) that was developed to control annual grasses (Romagni et al, 2000a,b; Duke & Oliva, 2004). Moreover, an increasing number of other natural products have been described in the literature as potential leads for the development of chemical agents for weed control among which are coumarins, benzoquinones, flavonoids, terpenoids and lactones (Barbosa et al, 2008). The natural product pool has been explored in the search of new photosynthetic inhibitors and the advances in this regard will be described below.

The diterpene labdane-8α,15-diol (**12**) and its acetyl derivative (**13**) (Figure 3) were isolated from the hexan extract of the stems of *Croton cliatoglanduliferus*. Biological assays carried out with intact spinach chloroplasts showed that these compounds are capable of interfere with ATP synthesis (Morales-Flores et al., 2007).

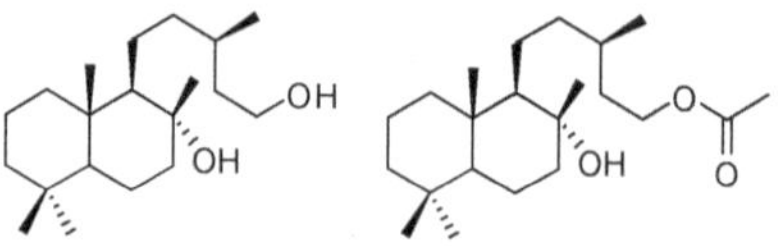

(12) (13)

Fig. 3. Structures of labdane-8α-,15-diol and its acetyl derivative.

It is well established that in the oxygenic photosynthesis there is a transfer of electron from water to $NADP^+$ affording the reduced form NADPH. In this biochemical process, two different reaction centers, known as photosystem II and photosystem I, work concurrently but in series. In the presence of light photosystem II feeds electrons to photosystem I. The electrons are transferred from photosystem II to the photosystem I by intermediate carriers as described by the well known Z scheme of photosynthesis (Figure 4). The net reaction is the transfer of electrons from water to $NADP^+$.

It has been proposed that electron transport is indirectly coupled to phosphorylation (ATP synthesis) through an electrochemical potential of hydrogen ions (protons) build up across the biological membranes involved in the electron transport (Mitchel, 1961). The electrochemical gradient, in turn, is consumed in the formation of ATP from ADP and inorganic phosphate. There is proportionality between changes in pH and changes in hydrogen ion concentrations. The electron transport gives two protons translocated for each electron transferred from photosystem II to photosystem I (Allen, 2003). At pH 8.0 one proton ion is consumed irreversibly in the synthesis of ATP:

$ADP^{-3} + Pi^{-2} \rightarrow ATP^{-4} + {}^{-}OH$.

As can be noticed in the previous equation, ATP formation can be measured in vitro, using intact spinach chloroplast , by determining the basicity of the medium. In the presence of an artificial electron receptor such as methylviologen and under continuous actinic illumination, the pH rise continues to increase linearly with time corresponding to the steady state rate of ATP formation. Under appropriate conditions (Morales-Flores et al., 2007), the rate of ATP formation can be determined by back titration with hydrochloric acid using a microelectrode attached to a potentiometer.

Diterpenes (**12**) and (**13**) (Figure 3) inhibited ATP synthesis coupled to electron transport from water to methylviologen in freshly lysed intact spinach chloroplasts in a concentration dependent manner. The IC_{50} (the concentrations causing 50% inhibition in vitro) were determined being equal to 72 μmol L^{-1} for compound (**12**) and 10 μmol L^{-1} for compound (**13**) (Morales-Flores et al., 2007). As mentioned above, the light-dependent phosphorylation is coupled to electron transport flow. Thus, ATP formation can be inhibited by (a) blockage of the electron transport within thylakoid chain, (b) by dissipation of the H^+ gradient, that is, uncoupling of the ATP synthesis process from the electron transport, and (c) by direct inhibition of the H^+-ATPase complex. Reagents that block electron transport avoid ATP synthesis because the generation of the transmembrane electrochemical gradient is not formed; as previously stated, the driving force for ATP synthesis is dependent upon electron flow. Chemicals that increase proton permeability of thylakoid membranes uncouple phosphorylation from electron flow. Uncoupling agents inhibit ATP synthesis by decreasing the proton gradient but allow electron transport to occur at high rates. In contrast, direct inhibitors of photophosphorylation block both phosphorylation and that portion of electron transport that is a consequence of proton efflux linked to phosphorylation (Veiga et al, 2007a).

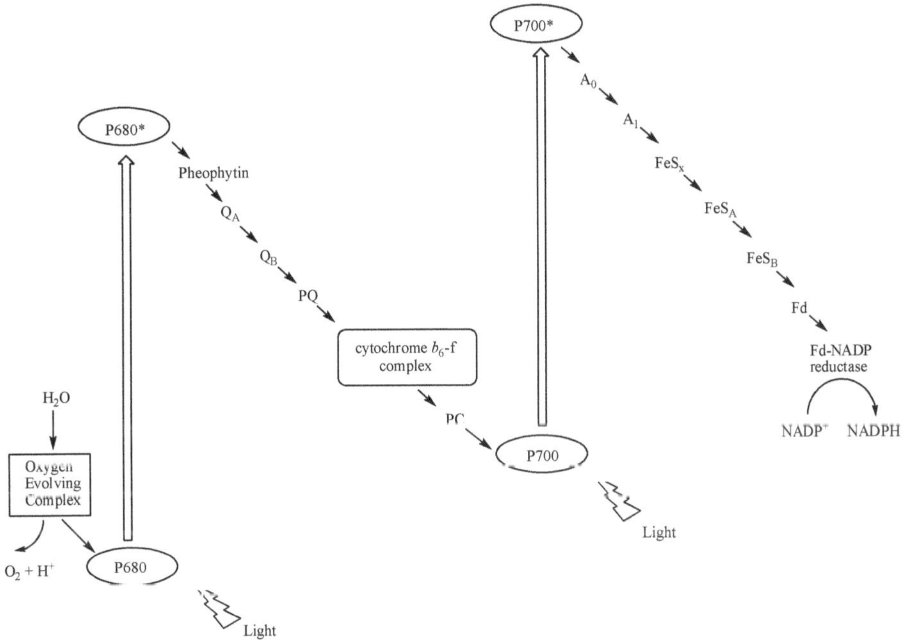

Fig. 4. The Z scheme and the electron transfer processes involved in oxygenic photosynthesis.

In order to elucidate the mechanism of action of (**12**) and (**13**) on photosynthesis, their effect on non-cyclic electron transport from water to methylviologen (basal, phosphorylating, and uncoupled) was investigated. It was found that as concentration of these compounds increased up to 300 μmol L^{-1}, the rates of basal, phosphorilation, and uncoupled electron transport were inhibited being basal and uncoupled electron rates the more affected. The IC_{50} found for compound (**12**) were 200 μmol L^{-1} (basal) and 76 μmol L^{-1} (uncoupled) while for (**13**) IC_{50} for these rates were 31 and 71.5 μmol L^{-1} for basal and uncoupled, respectively.

The partial reactions involved in the electron transport flow can be characterized in great detail by using highly specifc artificial donors and acceptors (Trebst, 2007; Morales-Florest et al, 2007; as cited in Allen & Holmes, 1986). The effect of the natural products (**12**) and (**13**) (Figure 3) on photosystem I and II and their partial reactions were evaluated. It was determined that terpene (**12**) inhibited the partial reaction from water do DCPIP (2,6-dichlorophenol indophenol). Inhibition of this partial reaction indicates that the compounds inhibits the range of reactions of photosystem II electron transport chain from water to Q_B (Figure 4), because oxidezed DCPIP accepts electrons at this level. It was also found that (**12**) inhibits the photosystem II partial reaction from water to sodium silicomolybdate (SiMo).

Inhibition of this partial reaction means that the **(12)** affects the partial reaction from water to quinone Q_A since SiMo is used as electron acceptor at or before Q_A site (Morales-Flores et al, 2007, as cited in Giaquinta & Dilley, 1975). The effect of compound **(12)** on the electron flow from diphenylcarbazide (DPC) to DCPIP was also evaluated. In this case, DCPP donates electrons at P_{680} level and the evaluation is carried out with Tris treated chloroplast. Such a treatment inhibits the oxygen evolving complex activity. Compound **(12)** did not inhibit this last partial reaction (electron flow from DPC to DCPIP). The results found for the activities of compound **(12)** on the partial reactions of photosystem II led to the indication that for this natural product the first site of interaction and inhibition is at the oxygen evolving complex. Further investigations carried out with terpene **(12)** showed that it is able to inhibit the partial reaction from reduced phenylmetasulfate (PMS) and methylviologen. While PMS donates electrons at the P_{680} level, methylviologen accepts electrons at F_x level in photosystem I. Therefore, it was established that the second site of interaction of compound **(12)** is located between P_{700} and F_x of photosystem I electron transport chain (Morales-Flores et al, 2007).

The effect of **(13)** on the partial reactions mentioned above was also evaluated being found the following results. It inhibited the partial reactions from water to DCPIP, from water do SiMo, and from DPC to DCPIP. These results indicate that the compound inhibits the electron flow in photosystem II in the evolving complex system as well as in the path between P_{680} and Q_A or at least at P_{680} site. The labdane **13** also inhibited the partial reaction from reduced PMS to oxidezed methylviologen (Morales-Flores et al, 2007).

The survey of literature revealed that various other terpenes (Figure 5) have been evaluated as potential inhibitors of photosynthesis in a similar manner described for compounds **(12)** and **(13)**.

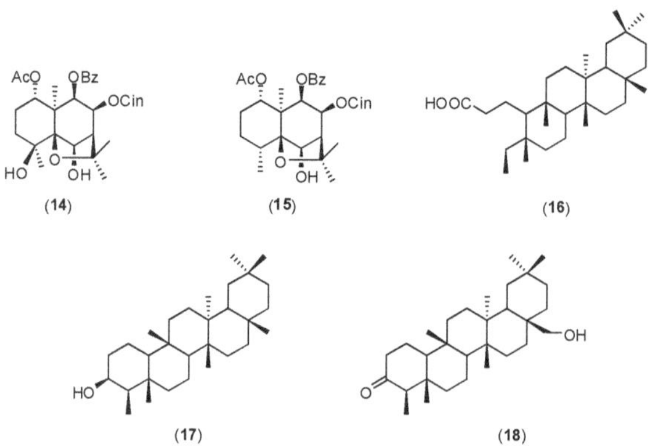

Fig. 5. Structures of photosynthetic inhibitor terpenes **14-18**.

The sesquiterpenoids **(14)** and **(15)** were isolated from the dichoromethane extract of the leaves of *Celastrus vulcanicola* (Torres-Romero et al, 2008). Similar to what happens to other terpenes, both compounds are capable of inhibiting the photosynthetic photophosphorylation in isolated spinach chloroplasts from water to methylviologen in a concentration-dependent manner (IC_{50} = 25 μmol L^{-1} for **14** and IC_{50} = 44 μmol L^{-1} for **15**). At lower concentrations (up to 50 μmol L^{-1}) compound **(14)** inhibited basal and phosphorylating electrons rates (approximately 83% and 79%, respectively); however, at higher concentrations this terpenes reversed both electron transport rates (at 300 μmmol L^{-1}, the electron flow rates were 32% and 68% for basal and phosphorylating, respectively). In addition, this compound also inhibited the uncoupled electron transport rate. Further investigations showed that (**14**) has two targets of interaction: one is located at the oxygen-evolving complex, and the other located at the H^+-ATPase complex. Compound (**15**) behaves likewise to (**14**) albeit being less active (Torres-Romero et al, 2008).

The fractioning of the hexane extract from air-dried leaves of *Maytenus imbricata* led to the isolation of seco-3,4-triterpenoid (**16**). It inhibits the formation of ATP coupled to electron transport from water to methylviologen and its IC_{50} was 148 μmol L^{-1}. Evaluation of the effect of increasing concentration of (**16**) on basal, phosphorylating, and uncoupled electron flow rates showed that the phosphorylating and uncoupled electron flow are inhibited up to 300 μmol L^{-1}; after this concentration both activities are enhanced. Experimental results support the fact that this triterpenoid acts on the oxygen evolving complex since it inhibits the electro flow from water to sodium silicomolybdate (SiMo). Moreover, it behaves as uncoupler activating the Mg^{+2}-ATPase complex (Souza & Silva et al, 2007).

Epifriedelinol (**17**) and canophyllol (**18**) (Figure 5) are friedelane triterpenoids isolated from the stems of *Celastrus vulcanicola*. Their activities on photosynthesis as determined in vitro can be summarized as follows. Epifriedelinol (**17**) and canophyllol (**18**) inhibits the ATP synthesis coupled to electron transport from water to methylviologen as concentration increases presenting IC_{50} = 82 μmol L^{-1} for compound (**17**) and 124 μmol L^{-1} for compound (**18**). The evaluation of their effects on noncyclic electron transport from water to methyl viologen under basal, phosphorylating, and uncoupled conditions showed that epifriedelinol (**17**) did not affect these processes. On the other hand, compound **18** partially inhibited the electron transport rates at different degrees. Both compounds moderately enhanced the light-activated Mg^{+2}-ATPase activity indicating that they affect the variation of pH and thus avoiding the ATP formation. Thus, the friedelane triterpenoids (**17**) and (**18**) has two targest of interaction: one is a decrease of pH variation blocking the ATP formation and the second by interacting and inhibiting the Mg^{+2}-ATPase activity in thylakoids. It was also found that compound (**18**) is capable of inhibiting the electron flow from water to the electron acceptor 2,5-dichloro-1,4-benzoquinone (DCBQ) acting, therefore, as Hill reaction inhibitor (Torres-Romero et al, 2010).

Besides terpenes, the effects of several other natural products (Figure 6) as photosynthetic inhibitors have been investigated.

Demuner and co-workers reported the isolation of 4-methoxy-5-methyl-6-(3-methylbut-2-enyloxy)isobenzofuran-1(*3H*)-one (**19**) from the phytopathogenic fungus *Nymbia alternantherae* (Demuner et al, 2006). This isobenzofuranone acts as Hill reaction inhibitor and uncoupler of photosynthesis, displaying inhibitory activity on ATP synthesis (IC_{50} = 66 μmol L^{-1}).

Siderin (**20**) is a secondary metabolite produce by *Toona ciliate* (Meliaceae). Veiga and co-workers showed that this coumarin inhibited ATP synthesis presenting IC_{50} = 27.0 μmol L^{-1} and did not inhibit photosystem I electron transport. Siderin does inhibit partial reactions of photosystem electron flow frow water to DCPIP, from water do sodium silicomolybdate (SiMo), and partially inhibits electron flow from DPC to DCPIP. All these results support the fact that the site of inhibition of (**20**) is the donor and acceptor sites of photosystem II, between P_{680} and Q_A (Veiga et al, 2007b).

The lactone lasiodiplodin (**21**) was isolated from the ethanolic extract from the fungus *Botryosphaeria rhodina*. As the concentration of (**21**) is raised, ATP synthesis is inhibited

(IC_{50} = 35.6 μmol L^{-1}). The investigation of the phytotoxic effect of this compound on photosynthesis resulted in the identification of three new different sites of interaction and inhibition: one at CF_1 ATPase complex, the second in the oxygen evolving complex, and the third at the electron transfer path between P_{680} and Q_A. These targets are different from that of displayed by synthetic herbicides. This finding corroborates the fact that the exploitation of the natural product pool can afford compounds presenting different molecular targets compared to well known herbicides (Veiga et al, 2007a).

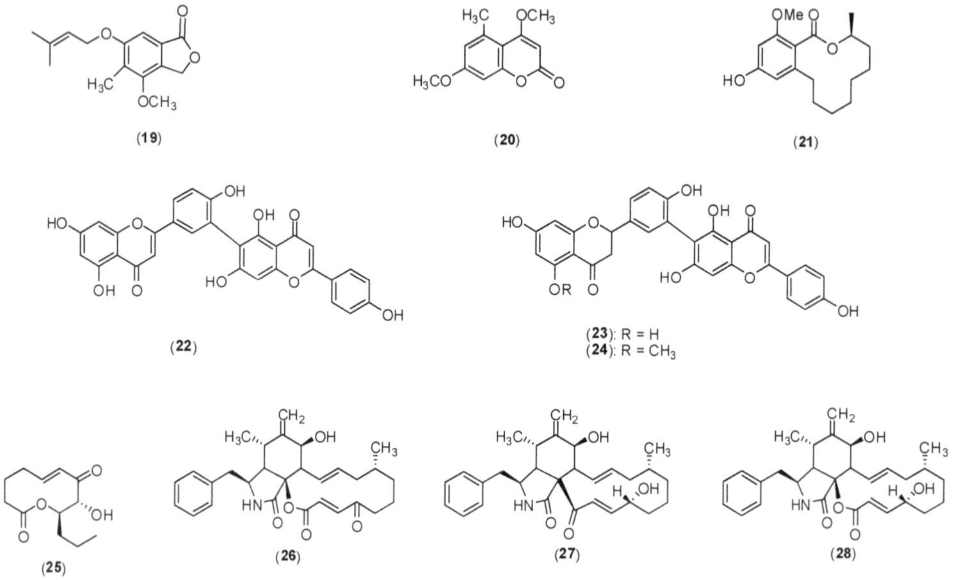

Fig. 6. Structures of compounds **19-28** mentioned in the text.

The bioactivity-guided chemical analysis of *Selaginella lepidophylla* resulted in the isolation of the biflavonoids robustaflavone (**22**), 2,3-dihydrorobustaflavone (**23**), and 2,3-dihydrorobustaflavone-5-methyl ether (**24**) (Figure 6). The in vitro assays revealed that all the isolated flavonoids inhibited the ATP synthesis coupled to electron flow from water to methylviologen in isolated freshly lysed intact spinach chloroplasts. The IC_{50} values for this activity were 44, 39, and 79 μmol L^{-1} for compounds (**22**), (**23**), and (**24**), respectively. Considering the three modes of electron transport (basal, uncoupling and phosphorylating) from water to methylviologen, all of them are partially inhibited by the three flavonoids being compound (**22**) the most active. The mode of action of these compounds were further investigated. Compound (**22**) did not affect the photosystem I. However, it interacts with photosystem II in the span of oxygen evolving complex to P_{680}. The interaction and inhibition target of (**23**) was located at Cytb_6f complex to plastocyanin (PC) (Figure 4). Flavonoid (**24**) had no effect on photosystem I or II. However, it acts as energy transfer inhibitors since increasing concentrations of (**22**), (**23**), and (**24**) inhibited the Mg^{+2}-ATPase activity (Aguilar et al, 2008).

A set of fifteen substances belonging to two groups of organic compounds, nonenolides and cytochalasins, had their phytotoxic effects evaluated against the perennial weed species

Cirsium arvense and *Sonchus arvensis* (Berestetskiy et al, 2008). Among the nonelides evaluate by leaf disc-puncture bioassay, stagonolide A (**25**) displayed the highest phytotoxic effect (necrosis diameter > 6 mm) on leaves of *C. arvense*. Considering the cytochalasins, only cytochalasyn A (**26**) showed high phytotoxic effect on this weed species (necrosis diameter ~3 mm) (Berestetskiy et al, 2008).

The most phytotoxic compound against *S. arvensis* was the cytochalasyn deoxaphomin (**27**) (necrosis diameter ~7 mmm). High phytotoxic effects were also observed for stagonolide A (**25**), cytochalasin A (**26**) and cytochalasin B (**28**) (necrosis diameter ~ 4.5, 5.5 and 4 mm, respectively) (Berestetskiy et al, 2008).

Photometric assays in the range of 450-950 nm showed that stagonolide A (**25**) and cytochalasin B (**28**) presented effects on photosynthesis. Twenty four hours after the observation of the first necrosis on leaf discs, both compounds caused significant decrease of the light absorption at the wave length 450 nm by *C. arvense*. This experimental observation was associated with the reduction of β-carotene or/and chlorophyll *b* content in leaf tissue of *C. arvense* since these pigments have a peak of resonant absorption near the 450 nm. At the wavelength of 530 nm and 550 nm, it was noticed increase in the light absorption for compounds (**25**) and (**28**). However, stagonolide A (**25**) had significantly stronger effect at 550 nm compared to cytochalasin B (**28**). Since cytochromes have a peak absorption in the range of 530-550 nm the substances (**25**) and (**28**) probably increased the concentration of these proteins, and did not affect electron transport. Treatment of leaf discs by stagonolide (**25**) also led to reduction of light absorption in the wavelenghth region between 630-690 nm by *C. arvense leaves* (Berestetsky et al, 2008). The peaks of light absorption in this region are characteristic of chlorophyll intermediates, phytochlorophyllide and chlorophyllide (Berestetskiy et al, 2008, as cited by Duke et al, 1991).

3. Synthetic studies towards the discovery of new photosynthetic inhibitors

In the research and development of new agrochemicals, organic synthesis plays an important role. After the discovery of a lead structure, an optimization process of it is carried out in order to improve the biological activity as well physico-chemical properties of the lead. Within this context, organic chemists utilize a vast array of chemical reactions to prepare derivatives of a lead compound. In this section it will be discussed recent advances concerning synthetic studies aimed to discover new photosynthetic inhibitors. Details about the preparation of the mentioned compounds can be found in the cited references and will not be covered herein since this is not the focus of this book chapter.

Several papers has been published in the last few years dealing with the synthesis of various aromatic nitrogenated compounds and their biological evaluation as photosynthetic inhibitors (Jampilek et al, 2009a,b; Musiol et al, 2007; Musiol et al, 2008; Musiol et al, 2010; Otevrel et al, 2010). The general structures and their associated most active compounds are presented in Figure 7 along with the IC_{50} data. For all of these substances, the in vitro evaluation of their inhibitory activity on the electron transport in isolated intact spinach chloroplasts was determined spectrophotometrically using the artificial electron acceptor 2,6-dichlorophenol indophenol (DCPIP). The rate of photosynthetic electron transport was monitored as photoreduction of DCPIP. The biological assays utilized diuron (**1**) (Figure 1) as positive control which the IC_{50} determined under the conditions used in the experiments was 1.9 $\mu mol\ L^{-1}$.

Quinoline is a structural motif found in various classes of bioactive compound. Musiol and co-workers prepared seventeen quinoline derivatives presenting the general structure (**29**). Some of the synthesized compound could not be biologically evaluated due to poor water solubility. The ten studied quinolines presented low inhibitory activity on photosynthetic electron transport with IC_{50} values ranging from 26 to 487 μmol L^{-1}. Compound (**29a**) was the most active presenting IC_{50} = 26 μmol L^{-1} (Musiol et al, 2007).

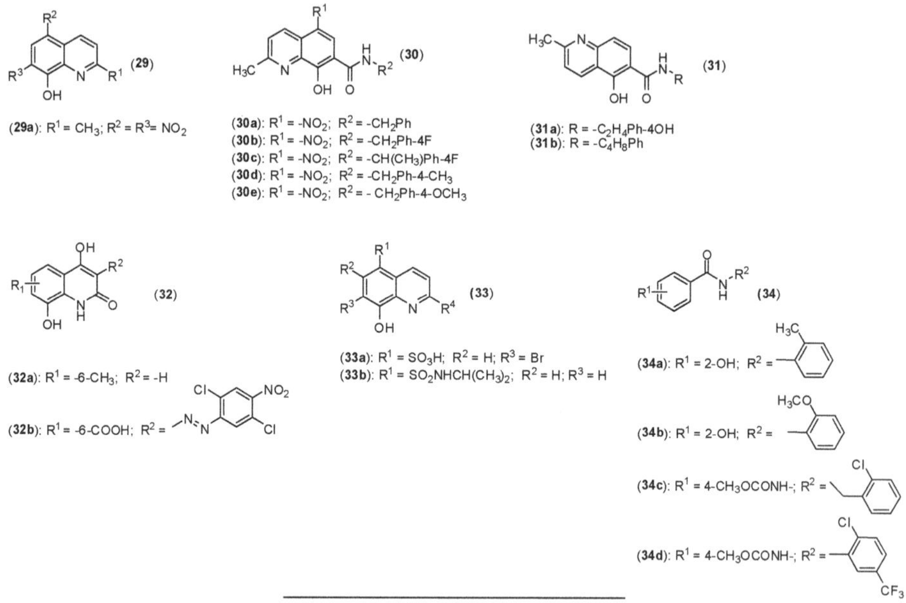

Compound	IC_{50} (μmol L^{-1})
29a	26.0
30a	32.0
30b	95.0
30c	8.7
30d	107.0
30e	137.0
31a	16.0
31b	7.2
32a	157
32b	126
33a	33.5
33b	54.4
34a	2.7
34b	1.6
34c	1.6
34d	1.0
Diuron	1,9

Fig. 7. Structures of aromatic nitrogenated compounds **29-34**.

For the two groups of amides (structures **30** and **31**) based on the quinoline scaffold system, it was found that the for the seventeen compounds of general structure (**30**) the most active ones were those possessing the nitro group (-NO_2) at the R^1 position (**30a-30e**). Compounds lacking this functionality at the specified position were completely inactive. Considering the quinolines (**31**), no simple structure-activity relationship explaining the observed activity was found being the two most active derivatives (**31a**) and (**31b**) (Musiol et al, 2008).
In a series of twelve ring-substituted 4-hydroxy-1*H*-quinolin-2-one derivatives (general structure **32**), all of the compounds displayed very low inhibitory activity (IC_{50} ranging from 126 to 925 μmol L^{-1}) on electron transport in photosynthesis from water to DCPIP (Jampilek et al, 2009a). The best activities were observed with derivatives (**32a**) and (**32b**).
In another series of hydroxilated derivatives, among the fourteen ring-substituted 8-hydroxyquinoline derivatives (**33**), two compounds (**33a, 33b**) showed moderate inhibitory activity (Musiol et al, 2010).

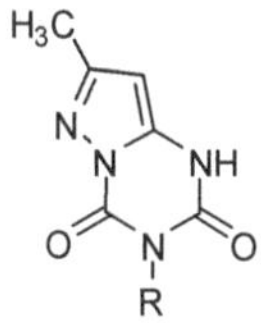

(**35**)

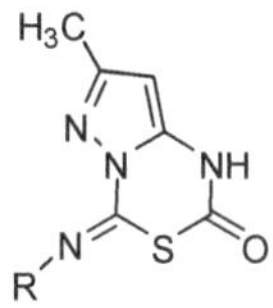

(**36**)

(**35a**): R = phenyl
(**35b**): R = cyclohexyl
(**35c**): R = benzyl
(**35d**): R = ethyl
(**35e**): R = butyl
(**35f**): R = *sec*-butyl

(**36a**): R = cyclohexyl
(**36b**): R = benzyl
(**36c**): R = ethyl
(**36d**): R = butyl
(**36e**): R = butyl

Compound	IC_{50} (μmol L^{-1})
35a	169±86
35b	0.32±0.19
35c	42±8
35d	32±8
35e	6.6±0.8
35f	2.6±0.6
36a	0.21±0.03
36b	18±4
36c	2.9±0.6
36d	0.57±0.11
Diuron	0.27±0.02
Hexazinone	0.11±0.01
Lenacil	0.08±0.02

Fig. 8. Structures of pyrazoles **35** and **36** and the corresponding IC_{50} data.

The investigation of the influence of twelve ring substituted salicylanilides and carbamoylpehnylcarbamates, represented by the general structure (**34**), on photosynthetic apparatus afforded compounds (**34a-34d**) (Figure 7) with biological activities compared do diuron. The remaining evaluated compounds displayed low to moderate activity. In addition, it was also suggested based on experimental results that the site of action of the compounds could be in Q_B, which is the second quinine acceptor on the oxidizing site of photosystem II.

The ability of ten pyrazole derivatives presenting the general structures (**35**) and (**36**) to act as photosystem II inhibitors was evaluated (Vicentini et al, 2004). In the presence of increasing concentrations of the compounds, it was observed that all of them are capable of interfering with the light-driven reduction of ferricyanide by isolated spinach chloroplasts. As can be seen in Figure 8, with the only exception of compound (**35a**), the IC_{50} ranged from 0.2 to 42 μmol L^{-1}. It is important to mention that the effectiveness of the most active compounds was comparable to three commercial herbicides (diuron, hexazinone and lenacil) used as positive controls in the experiments.

The pyrazoles (**35**) and (**36**) were also evaluated in vivo against the blue-green alga *Spirulina platensis* as well as the eukaryotic alga belonging to the genus *Chlorella*. Once again, for some derivatives the observed activities are remarkable and comparable to commercial herbicides used as positive controls (Vicentini et al, 2004).

Even more remarkable biological activity was displayed by another series of pyrazoles presenting the general structures (**37**) and (**38**) (Figura 9). All of these compounds inhibited the Hill reaction from water to ferrycianide in the presence of illuminated chloroplasts, with IC_{50} ranging from 10^{-6} mol L^{-1} to 10^{-4} mol L^{-1}. The efficacy of (**37j**) (IC_{50} = 0.64 μmol L^{-1}) is comparable to commercial herbicides such as diuron, lenacil and hexazinone (IC_{50} = 0.27, 0.08 and 0.11 μmol L^{-1}) (Vicentini et al, 2005).

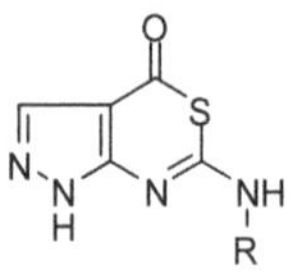

(**37**)

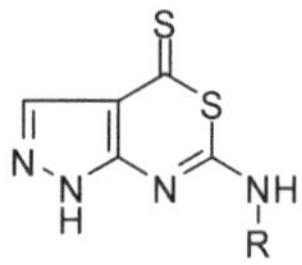

(**38**)

(**37a**): R = 3-trifluoromethylphenyl
(**37b**): R = 4-bromophenyl
(**37c**): R = 4-chlorophenyl
(**37d**): R = 3-chlorophenyl
(**37e**): R = 2-chlorophenyl
(**37f**): R = 4-nitrophenyl
(**37g**): R = ethyl
(**37h**): R = butyl
(**37i**): R = benzyl
(**37j**): R = cyclohexyl

(**38a**): R = 3-trifluoromethylphenyl
(**38b**): R = 4-bromophenyl
(**38c**): R = 4-chlorophenyl
(**38d**): R = 3-chlorophenyl
(**38e**): R = 2-chlorophenyl
(**38f**): R = 4-nitrophenyl
(**38g**): R = ethyl
(**38h**): R = butyl
(**38i**): R = benzyl
(**38j**): R = cyclohexyl

Fig. 9. Structures of pyrazoles **37** and **38** mentioned in the text.

One strategy that could be used to discover novel herbicides is to synthesize analogues of a natural products which is known to present phytotoxic activity. It is also possible to take a structure of a phytotoxic natural product itself and carry out structural modifications on it. Such an approach can result in compounds with improved biological activity as well better physico-chemical properties. The natural product (**39**) (Figure 10) was isolated from *Pterodon polygalaeflorus*. Biological evaluation of (**39**) as potential inhibitor of ATP synthesis in isolated chloroplasts from spinach revealed that this substance is completely inactive. However, its β-lactone derivative (**40**) inhibited this process as concentration increases. For this compound, the IC_{50} was 90 μmol L^{-1}. Similar to other terpenes previously described in this chapter, lactone (**40**) did not affect photosystem I but it did inhibit electron transport through photosystem II by targeting the oxygen evolving complex as well as the redox enzymes of the electron transport chain in the span between P_{680} and Q_A (King-Díaz et al 2006).

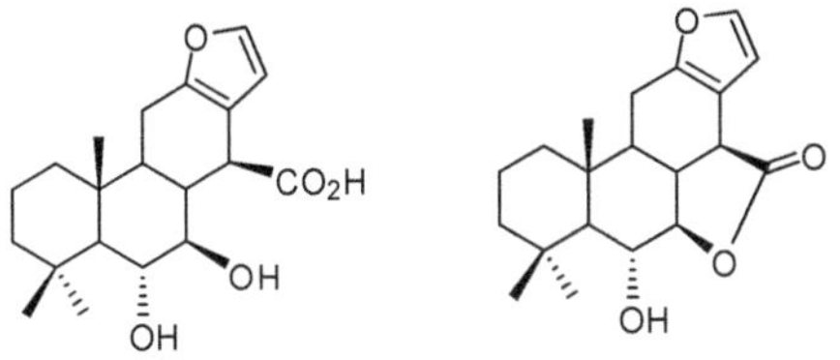

(39) (40)

Fig. 10. Structure of diterpene (**39**) and its synthetic derivative (**40**).

Other derivatives of compound (**39**) have been recently evaluated and present phytotoxic properties on photosynthesis (King-Díaz et al, 2010).
The nostoclides (**41**) correspond to a pair of naturally-occurring lactones produced by a cyanobacterium (*Nostoc* sp.) symbiont of *Peltigera canina* (L.). It has been suggested that these chlorinated compounds may be allelopathic agents since *P. canina* cultures are usually not contaminated with microorganisms (Yang et al, 1993). In view of that, it was decided to investigate the potential phytotoxicity of nostoclide analogues (Barbosa et. al. 2006). In this context, several derivatives (general structures **42** and **43**, Figure 11) were prepared, and their ability to interfere with the ligh-driven reduction of ferricyanide by isolated spinach chloroplasts thylakoid membranes (Hill reaction) was subsequently evaluated (Barbosa et al., 2007; Teixeira et al, 2008).
A number of nostoclide derivatives, at various degrees, exhibited inhibitory properties in the micromolar range against the basal electron flow from water to ferricyanide. As a general trend, the non-brominated derivatives (**43**) presented higher effectiveness than their brominated counterparts. The most active compounds (**43a-43e**) derivatives along with their IC_{50} values are presented in Figure 11.

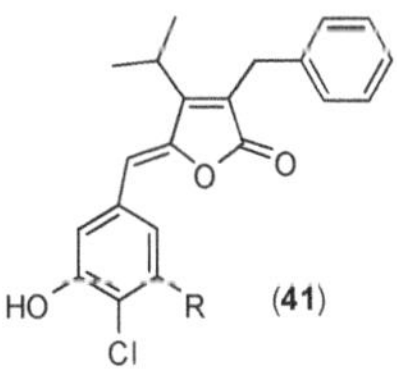

(41)

R = Cl: Nostoclide I
R = H: Nostoclide II

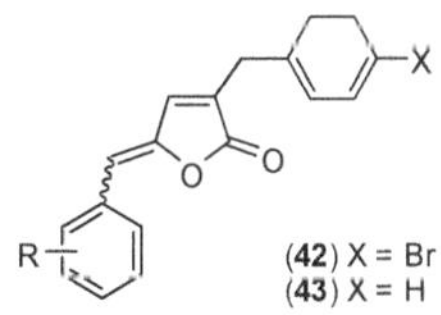

(42) X = Br
(43) X = H

Compound	Benzylidene group	IC_{50} (μmol L^{-1})
43a	Z-4-nitrobenzylidene	1.7±0.7
43b	Z-2-fluorobenzylidene	10.1±3.2
43c	Z-4-trifluoromethylbenzilidene	8.3±2.3
43d	Z-2-trifluoromethylbenzilidene	11.8±4.5
43e	Z-4-ethylbenzylidene	9.1±3.2

Fig. 11. Structure of natural product nostoclides and some synthetic analogues.

More recently, a QSAR (Quantitative Structure Activity Relantionship) investigation was carried out on a series of nostoclide analogues presenting the general structure (**43**) to correlate molecular descriptions with their in vitro biological activity (the ability to interfere with ligh-driven reduction of ferrycianide by isolated spinach chloroplasts thylakoid membranes). The results of this investigation suggested that the degree of inhibition efficiency of this class of compounds is intimately associated with their polarity (Teixeira et al, 2009). Thus, it is likely that new nostoclide analogues with higher polarity could display improved biological activity. At the moment that this book chapter is written, there is no report on the literature concerning the synthesis of nostoclide analogues with higher polarity than the previous ones already prepared.

4. Conclusions

The ongoing need for new agents to control weeds has stimulated the search for new photosynthetic inhibitors. We described in this chapter a variety of compounds presenting this type of activity. The natural products have been explored toward this end resulting in the identification of compounds with various structural motifs. Such an approach has resulted in the discovery of photosynthetic inhibitors with new modes of action. This, in turn, can be helpful in dealing whit resistance a problem to be faced in weed management. It is possible to anticipate that promising inhibitors of photosynthesis will certainly be found by exploring the natural product pool. From nature, it is also possible that more active compounds with low toxicity and improved selectivity will be found. Promising photosynthetic inhibitors has also been revealed by the synthetic studies. One important challenge in the field of weed management is related to selectivity. In other words, chemicals should exert their action only on weeds. As can be noticed in the discussion of the studies published in the literature during the last years, this issue has not been addressed. Future works should also be concerned with this important matter.

5. Acknowledgment

We are greatful to Fundação de Amparo à Pesquisa de Minas Gerais (FAPEMIG) and Fundação Arthur Bernardes (FUNARBE) for financial support.

6. References

Aguilar, M. I.; Romero, M. G.; Chávez, M. I.; King-Díaz, B.; Lotina-Hennsen, B. (2008). Biflavonoids isolated from *Selaginella lepidophylla* inhibit photosynthesis in spinach chloroplasts. *Journal of Agricultural and Food Chemistry*, Vol.56, No.16, (August 2008), pp. 6994-7000, ISSN 0021-8561.

Allen, J. F. (2003). Cyclic, pseudocyclic and noncyclic photophosphorylation: new links in the chain. *TRENDS in Plant Science*, Vol.8, No.1, (January 2003), pp. 15-19, ISSN 1360-1383.

Allen, J. F.; Holmes, N. G. (1986). Electron transport partial reactions. In *Photosyntesis, energy transduction. A practical approach*, M. F. Hipkinns, N. R. Baker (Eds), 103-141, Chap. 5., IRL Press, ISBN 09479467519, Oxford, United Kingdom.

Barbosa, L. C. A.; Demuner, A. J.; de Alvarenga, E. S.; Oliveira, Alberto.; King-Diaz, B.; Lotina-Hennsen, Blas. (2006). Phytogrowth- and photosynthesis-inhibiting properties of nostoclides analogues. *Pest Management Science*, Vol.62, No.3 (March 2006), pp. 214-222, ISSN 1526-498X.

Barbosa, L. C. A.; Rocha, M. E.; Teixeira, R. R.; Maltha, C. R. A.; Forlani, G. (2007). Synthesis of 3-(4-bromobenzyl)-5-(arylmethylene)-*5H*-furan-2-ones and their activity as inhibitors of the photosynthetic electron transport chain. *Journal of Agricultural and Food Chemistry*, Vol.55, No.21 (October 2007), pp. 8562-8569, ISSN 0021-8561.

Barbosa, L. C. A.; Teixeira, R. R.; Montanari, R. M. (2008). Phytotoxic natural products as models for the development of crop protection agents. In: *Current Trends in Phytochemistry*, F. Epifano (Ed.), 21-59, Research Signpost, ISBN 978-81-308-0277-0, Trivandrum, Kerala, India.

Beckie, H. J. (2006). Herbicide-Resistant Weeds: Management Tactics and Practices. *Weed Technology*, Vol.20, No.3, (July 2006), pp. 793-814, ISSN 0890-037X.

Berestestskiy, A.; Dmitriev, A.; Mitina, G.; Lisker, I.; Andolfi, A.; Evidente, A. (2008). Nonelides and cytochalasins with phytotoxic activity against *Cirsium arvense* and *Sonchus arvensis*: a structure-activity relationships study. *Phytochemistry*, Vol.69, No.4, (February 2008), pp. 953-960, ISSN 0031-9422.

Böger, P.; Wakabayashi, K.; Hirai, K. (2002). *Herbicide Classes in Development. Mode of Action, Targets, Genetic Engineering, Chemistry*, Springer-Verlag, ISBN 9783540431473, Berlin, Germany.

Chaabane, H.; Cooper, J. F.; Azouzi, L.; Coste, C. M. (2005). Influence of soil properties on the adsorption-desorption of sulcotrione and its hydrolysis metabolites on various soils. *Journal of Agricultural and Food Chemistry*, Vol.53, No.10, (April 2005), pp. 4091-4095, ISSN 0021-8561.

Cobb, A. (1992). *Herbicides and Plant Physiology*, Chapman and Hall, ISBN 0 412 43860 7, London, United Kingdom.

Copping, L. G.; Duke, S. O. (2007). Natural products that have been used commercially as crop protection agents. *Pest Management Science*, Vol.63, No.6, (June 2007), ISSN 1526-498X.

Dayan, F. E.; Cantrell, C. L.; Duke, S. O. (2009). Natural products in crop protection. *Bioorganic and Medicinal Chemistry*, Vol.17, No.12, (June 2009), pp. 4022-4034, ISSN 0968-0896.

Dayan F.; Romagni, J.; Tellez, M.; Rimando A.; Duke, S. (1999). Managing weeds with natural products. *Pesticide Outlook*, Vol.10, No.5, (October 1999), pp. 185-188, ISSN 0956-1250.

Demuner, A. J.; Barbosa, L. C. A.; Veiga, T. A. M.; Barreto, R. W.; King-Diaz, B.; Lotina-Hennsen, B. (2006). Phytotoxic constituents from *Nimbya alternantherae. Biochemical Systematics and Ecology*, Vol.34, No.11, (November 2006), pp. 790-795, ISSN 0305-1978.

Devine, M. D.; Shukla, A. (2000) Altered target sites as a mechanism for herbicide resistance. *Crop Protection*, Vol.19, No.8-10, (September 2000), pp. 881-889, ISSN 0261-2194

Draber, W. (1992). *Rational Approaches to Structure, Activity, and Ecotoxicology of Agrochemicals.* W. Draber; T. Fujita (Eds), p. 278, CRC, ISBN 0849358590, Boca Raton, Florida, United States.

Duke, S. O.; Dayan, F. E.; Rimando, A. M.; Schrader, K. K.; Aliotta, G.; Oliva, A.; Romagni, J. G. (2002). Chemicals from nature for weed management. *Weed Science*, Vol.50, No.2, (March 2002), pp. 138-151, ISSN 0043-1745.

Duke, S. O.; Dayan, F. E.; Kagan, I. A.; Baerson, S. R. (2005). New herbicide target sites from natural compounds. *ACS Simposium Series 892 (New discoveries in agrochemicals)*, pp. 151-160, ISSN 0097-6156.

Duke, S. O.; Dayan, F. E.; Romagni, J. G.; Rimando, A. M. (2000a). Natural products as sources of herbicides: current status and future trends. *Weed research*, Vol.40, No.1, (February 2000), pp. 99-111, ISSN 0043-1737.

Duke, S. O.; Duke, M. V.; Sherman, T. D.; Nandihalli, U. B. (1991). Spectrophotometric and spectrofluorometric methods in weed science. *Weed Science*, Vol. 39, No.3, pp. 505-13, ISSN 0043-1745.

Duke, S. O.; Romagni, J. G.; Dayan, F. E. (2000b). Natural products as sources of new mechanisms of herbicidal action. *Crop Protection*, Vol.19, No.8-10, (September 2000), pp. 583-589, ISSN 0261-2194.

Duke, S. O.; Oliva, A. (2004). In: *Allelopathy – Chemistry and mode of action of allelochemicals*, F. A. Macías, J. C. G. Galindo, J. M. G. Molinillo, H. G. Cuttler (Eds.), 201-216, CRC Press, Boca Raton, ISBN 0-8493-1964-1.

Giaquinta, R. T.; Dilley, R. A. A partial reaction in photosystem II: reduction of silicomolybdate prior to te site of dichlorophenyldimethylurea inhibition. *Biophysica Acta*, Vol. 387, No.2, pp. 288-305, ISSN 0006-3002.

Gressel, J. Evolving understanding of the evolution of herbicide resistance. (2009). *Pest Management Science*, Vol.65, No.11, (November, 2009), pp. 1164-1173, ISSN 1526-498X.

Heap, I. M. (2011). International Survey of Herbicide Resistant Weeds, In: *WeedScience.com*, May 30th 2011, Available from: www.weedscience.org.

Hüter, O. F. (2011). Use of natural products in the crop protection industry. *Phytochemistry review*, Vol.10, No.2, (June 2011), pp. 185-194, ISSN 1568-7767.

Jampilek, J.; Musiol, R.; Finster, J.; Pesko, M.; Carroll, J.; Kralova, K.; Vejsova, M.; O'Mahony, J.; Coffey, A.; Dohnal, J.; Polanski, J. (2009b). Investigating biological activity spectrum for novel styrylquinazoline analogues. *Molecules*, Vol.14, No.10, (October 2009), pp. 4246-4265, ISSN 1420-3049.

Jampilek, J.; Musiol, R.; Pesko, M.; Kralova, K.; Vejsova, M.; Carroll, J.; Coffey, A.; Finster, J.; Tabak, D.; Niedbala, H.; Kozik, V.; Polanski, J.; Csollei, J.; Dohnal, J. (2009a). Ring-substituted 4-hydroxy-1*H*-quinolin-2-ones: preparation and biological activity. (2009a). *Molecules*, Vol.14, No.3, (March 2009), pp. 1145-1159, ISSN 1420-3049.

King-Díaz, B.; Castelo-Branco, P. A.; dos Santos, F. J. L.; Rubinger, M. M. M.; Ferreira-Alves, D. L.; Veloso, D. P. (2010). Furanoditerpene Ester and thiocarbonyldioxy derivatives inhibit photosynthesis. *Pesticide Biochemistry and Physiology*, Vol.96, No.3, (March, 2010), pp. 119-126, ISSN 0048-3575.

King-Díaz, B.; Pérez-Reyes, A.; Santos, F. J. L.; Ferreira-Alves, D. L.; Veloso, D. P.; Carvajal, S. U.; Lotina-Hennsen, B. (2006). Natural diterpene β-lactone derivative as photosystem II inhibitor on spinach chloroplasts. *Pesticide Biochemistry and Physiology*, Vol.84, No.2, (February, 2006), pp. 109-115, ISSN 0048-3575.

Lein, W.; Börnke, F.; Reindl, A.; Ehrhardt, T.; Stitt, M.; Sonnewald, U. (2004). Target-based discovery of novel herbicides. *Current Opinion in Plant Biology*, Vol.7, No.2, (April 2004), pp. 219-225, ISSN 1369-5266

Macías, F. A.; Molinillo, J. M. G.; Varela, R. M.; Galindo, J. C. G. (2007). Allelopathy - a natural alternative for weed control. *Pest Management Science*, Vol.63, No.4, (April 2007), pp. 327-348, ISSN 1526-498X.

Mitchell, G.; Barlett, D. W.; Fraser, T. E. M.; Hawkes, T. R.; Holt, D. C.; Townson, J. K.; Wichert, R. A. (2001). Mesotrione: a new selective herbicide for use in maize. *Pest Management Science*, Vol.57, No.2, (February 2001), pp. 120-128, ISSN 1526-498X.

Mitchell, P. (1961). Coupling of phosphorylation to electron and proton transfer by a chemiosmotic type of mechanism. *Nature* Vol.19, (July 1961), pp. 144-148, ISSN 0028-0836.

Morales-Flores, F.; Aguilar, M. I.; King-Díaz, B.; de Santiago-Gómez, J.-R.; Lotina-Hennsen, B. (2007). Natural diterpenes from *Croton ciliatoglanduliferus* as photosystem II and photosystem I inhibitors in spinach chloroplasts. *Photosynthesis Research*, Vol.91, No.1 (January 2007), pp. 71-80, ISSN 0166-8595.

Musiol, R.; Jampilek, J.; Nycz, J. E.; Pesko, M.; Carroll, J.; Kralova, K.; Vejsova, M.; O'Mahony, J.; Coffey, A.; Mrozek, A.; Polanski, J. (2010). Investigating the activity spectrum for ring-substituted 8-hydroxyquinolines. *Molecules*, Vol.15, No.1 (January 2010), pp. 288-304, ISSN 1420-3049.

Musiol, R.; Jampilek, J.; Kralova, K.; Richardson, D. R.; Kalinowski, D.; Podeszwa, B.; Finster, J.; Niedbala, H.; Palka, A.; Polanski, J. (2007). Investigating biological activity spectrum for novel quinoline analogues. *Bioorganic and Medicinal Chemistry*, Vol.15, No.3, (February 2007), pp. 1280-1288, ISSN 0968-0896.

Musiol, R.; Tabak, D.; Niedbala, H.; Podeszwa, B.; Jampilek, J.; Kralova, K.; Dohnal, J.; Finster, J.; Mencel, A.; Polanski, J. (2008). Investigating biological activity spectrum of novel quinoline analogues 2: hydroxyquinolinecarboxamides with photosynthesis-inhibiting activity. *Bioorganic and Medicinal Chemistry*, Vol.16, No.8, (April 2008), pp. 4490-4499, ISSN 0968-0896.

Nelson, N. (2011). Photosystems and global effects of oxygenic photosynthesis. *Biochimica et Biophysica Acta*, Vol.1807, No.8, (August 2011), pp. 856-863, ISSN 0005-2728.

Nelson, N. ; Yocum, C. F. Structure and function of photosystems I and II. (2006). *Annual Review of Plant Biology*, Vol.57, (June 2006), 521-565, ISSN 1543-5008.

Otevrel, J. ; Mandelova, Z. ; Pesko, M. ; Guo, J. ; Kralova, K. ; Sersen, F. ; Vejsova, M. ; Kalinowski, D. S. ; Kovacevic, Z. ; Coffey, A. ; Csollei, J. ; Richardson, D. R. ; Jampilek, J. (2010). Investigating the spectrum of biological activity of ring-substituted salicylanilides and carbamoylphenylcarbamates. *Molecules*, Vol.15, No.11, (November 2010), pp. 8122-8142, ISSN 1420-3049.

Pillmoor, J. B.; Wright K.; Terry A. S. (1993). Natural products as a source of agrochemicals and leads for chemical synthesis. *Pesticide Science*, Vol.39, No.2, (June 1993), pp. 131-140 ISSN 0031-613X.

Preston, C. (2004). Herbicide resistance in weeds endowed by enhaced detoxification : complications for management. *Weed Science* Vol.52, No.3 (May 2004), pp. 448-453, ISSN 0043-1745.

Romagni, J. G. ; Allen, S. N. ; Dayan, F. E. (2000a). Allelopathic effects of volatile cineoles on two weedy plant species. *Journal of Chemical Ecology*, Vol.26, No.1, (January 2000), pp 303-313, ISSN 0098-0331.

Romagni, J. G. ; Duke, S. O. ; Dayan, F. E. (2000b). Inhibition of plant asparagine synthetase by monoterpenes cineoles. *Plant Physiology*, Vol.123, No.2, (June, 2000), pp. 725-732, ISSN 0032-0889.

Rüegg, W. T. ; Quadranti M. ; Zoschke, A. Herbicide research and development challenges and opportunities. (2007). *Weed Research*, Vol.47, No.4, (August 2007), pp. 271-275, ISSN 0043-1737.

Rutherford, A. W. ; Faller, P. (2001). The heart of photosynthesis in glorious 3D. *TRENDS in Biochemical Sciences*, Vol.26, No.6, (June 2001), pp. 341-344, ISSN 0968-0004.

Saxena, S. ; Pandey, A. K. (2001). Microbial metabolites as eco-friendly agrochemicals for the next millennium. *Applied Microbiology Biotechnology*, Vol.55, No.4, (May 2001), pp. 395-403, ISSN 0175-7598.

Schrader, K. K. ; Andolfi, A. ; Cantrell, C. L. ; Cimmino, A. ; Duke, S. O. ; Osbrink, W. ; Wedge, D. E ; Evidente, A. (2010). A survey of phytotoxic microbial and plant metabolites as potential products for pest management. *Chemistry and Biodiversity*, Vol.7, No.9, (September 2010), pp. 2261-2280, ISSN 1612-1872.

Short, P. L. (2005). Growing agrochem r & d. *Chemical and Engineering News*, Vol.8, No.38, (September 2005), pp. 19-22, ISSN 0009-2347.

Stetter, J. ; Lieb. F. (2000). Innovation in crop protection : trends in research. *Angewandte Chemie International Edition*, Vol.39, No.10, (May 2000), pp. 1725-1744, ISSN 1433-7851.

Strange, R. N. Phytotoxins produced by microbial plant pathogens. (2007). *Natural Product Reports*, Vol.24, No.1, (January 2007), pp. 127-144, ISSN 0265-0568.

Souza e Silva, S. R. ; Silva, G. D. F. ; Barbosa, L. C. A.; Duarte, L. P. ; King-Diaz, B. ; Archundia-Camacho, F. ; Lotina-Hennsen, B. (2007). Uncoupling and inhibition properties of 3,4-seco-friedelan-3-oic acid isolated from *Maytenus imbricata*. *Pesticide Biochemistry and Physiology*, Vol.87, No.2 (February 2007), pp. 109-114, ISSN 0048-3575.

Teixeira, R. R. ; Barbosa, L. C. A. ; Forlani, G. ; Veloso, D. P. ; Carneiro, J. W. M. (2008). Synthesis of photosynthesis-inhibiting nostoclide analogues. *Journal of*

Agricultural and Food Chemistry, Vol.56, No.7 (April 2008), pp. 2321-2329, ISSN 0021-8561.

Teixeira, R. R. ; Pinheiro, P. F. ; Barbosa, L. C. A. ; Carneiro, J. W. M. ; Forlani, G. (2009). QSAR modeling of photosynthesis-inhibiting nostoclide derivatives. *Pest Management Science*, Vol.66, No.2 (February 2009), pp. 196-202, ISSN 1526-498x.

Torres-Romero, D. ; King-Díaz, B. ; Jiménez, I. A. ; Lotina-Hennsen, B. L. ; Bazzocchi, I. L. (2008). Sesquiterpenes from *Celastrus vulcaniola*. *Journal of Natural Products*, Vol.71, No.8 , (August 2008), pp.1331-1335, ISSN 0163-3864.

Torres-Romero, D. ; King-Díaz, B. ; Strasser, R. J. ; Jiménez, I. A. ; Lotina-Hennsen, B. ; Bazzocchi, I. L. (2010). Friedelane triterpenes from *Celastrus vulcanicola* as photosynthetic inhibitors. *Journal of Agricultural and Food Chemistry*, Vol.58, No.20, (October 2010), pp. 10847-10854, ISSN 0021-8561.

Trebst, A. (2007). Inhibitors in the functional dissection of the photosynthetic electron transport system. *Photosynthesis research*, Vol.92, No.2 (May 2007), pp. 217-224, ISSN 0166-8595.

van Klink, J. W. ; Brophy, J. J. ; Perry, N. B. ; Weavers, R. T. (1999). β-triketones from Myrtaceae : Isoleptospermone from *Leptospermum socparium* and papuanone from *Corymbia dallachiana*. *Journal of Natural Products*, Vol. 62, No.3 (February 1999), pp. 487-489, ISSN 0163-3864.

Veiga, T. A. M. ; González-Vázquez, R. ; Neto, J. O. ; Silva, M. F. G. F. ; King-Díaz, B. ; Lotina-Hennsen, B. (2007b). Siderin from *Toona ciliata* (Meliaceae) as photosystem II inhibitor on spinach thylakoids. *Archives of Biochemistry and Biophysics*, Vol. 465, No.1, (September 2007), pp. 38-43, ISSN 0003-9861.

Veiga, T. A. M. ; Silva, S. C. ; Francisco, A. C.; Filho, E. R. ; Vieira, P. C. ; Fernandes, J. B. ; Silva, M. F. G. ; Müller, M. W. ; Lotina-Hennsen, B. (2007a). Inhibition of photophosphorylation and electron transport chain in thylakoids by lasiodiplodin, a natural product from *Botryosphaeria rhodina*. *Journal of Agricultural and Food Chemistry*, Vol.55, No.10, (May 2007), pp. 4217-4221, ISSN 0021-8561.

Vicentini, C. B. ; Guccione, S. ; Giurato, Laura. Ciaccio, R. ; Mares, D. ; Forlani, G. (2005). Pyrazole derivatives as photosynthetic electron transport inhibitors : new leads and structure-activity relationship. *Journal of Agricultural and Food Chemistry*, Vol.53, No.10, (May 2005), pp. 3848-3855, ISSN 0021-8561.

Vicentini, C. B. ; Mares, D. ; Tartari, A. ; Manfrini, M. ; Forlani, G. (2004). Synthesis of pyrazole derivatives and their evaluation as photosynthetic electron transport inhibitors. *Journal of Agricultural and Food Chemistry*, Vol.52, No.7, (April 2004), pp. 1898-1906, ISSN 0021-8561.

Vyvyan, J. R. (2002). Allelochemicals as leads for new herbicides and agrochemicals. *Tetrahedron*, Vol.58, No.9, (February 2002), pp. 1631-1646, ISSN 0040-4020.

Ware, G. W. (2000). *The Pesticide Book* (5th Edition), Thomson Publications, ISBN 0-913702-63-3, Fresno, California, United States.

Yang, X. ; Shimizu, Y. ; Steiner, J. R. ; Clardy, J. Nostoclide I and II, extracellular metabolites from a symbiotic cianobacterium Nostoc. sp., from the lichen Peltigera canina. (1993). *Tetrahedron*, Vol. 34, No.5 (January 1993), pp. 761-764, ISSN 0040-4039.

2

Photosynthetic Behavior of Microalgae in Response to Environmental Factors

Ben Dhiab Rym
Institut National des Sciences et Technologies de la Mer, Monastir, Tunisia

1. Introduction

Microalgae are prokaryotic or eukaryotic photosynthetic microorganisms that can grow rapidly and live in a wide range of ecosystems extending from terrestrial to aquatic environment. Environmental factors such as temperature, UV-light, irradiance, drought and salinity are known to affect their photosynthesis.

Photosynthesis was the most sensitive process in microalgae, leading to numerous changes in structure and function of the photosynthetic apparatus under various conditions. The photosynthetic response may be involved in the modification of energy collector complexes, antennas, or reaction centers and in the distribution of excitation energy between the two photo systems (PSII and PSI).

Photosynthesis of micro algae is often inhibited by salt stress (Kirst, 1990). Such an inhibition may be explained by a decrease in PSII activity. Indeed, in the green microalgae, salt stress inhibits PSII activity in *Dunaliella tertiolecta* and *Chlamydomonas reinhardtii* that is associated with a state-2 transition (Gilmour et al., 1985, Endo et al., 1995).

In the cyanobacteria, it has been suggested that the decreased PSII activity in salt-stressed cells is associated with the state-2 transition (Schubert et al., 1993; Schubert and Hagemann, 1990). It has been reported that salt stress significantly inhibit the maximal efficiency of PSII photochemistry in *Spirulina platensis* cells and this inhibition is increased with increasing light intensity (Lu et al., 1999; Lu & Zhang, 1999, 2000).

The decreased maximum quantum yield of primary photochemistry (Fv/Fm), interpreted as photodamage (Powles, 1984; Torzillo et al., 1998; Maxwell & Johnson 2000), may be due to an inactivation of PSII reaction centers, an inhibition of electron transport at both donor and acceptor sides of PSII, and a distribution of excitation energy transfer in favor of PSI (Lu & Vonshak, 1999; Lu et al., 1999; Lu & Vonshak, 2002), increasing the cyclic electron flow around PSI (Gilmour et al., 1982; Joset et al., 1996). This decrease is often identified as an adaptive acclimation process of down regulation of PSII, a protective mechanism that helps dissipate excess energy from the photosynthetic apparatus (Allen & Ort, 2001; Tsonev et al., 2003; Hill et al., 2004; Kramer et al., 2004; Hill & Ralph, 2005).

In *Synechocystis*, it has been reported that the decreased PSII activity by salt stress can be explained the fact that salt stress inhibits the repair of photo-damaged PSII by by concealing the synthesis of D1 protein (Allakhverdiev et al., 2002). D1 protein of thylakoid membranes was showed as a sensitive protein to environmental stress conditions: under various unfavorable conditions like drought, nutrition deficiency, heat, chemical stress, ozone fumigation as well as UV-B and visible light stresses can influence the turnover of D1 protein (Giardi et al., 1997). In *Synechococcus* cells, salt stress inactivated both PSII and PSI due to the changes in K/Na ratio (Allakhverdiev et al., 2000).

High temperature stress inhibits the function of the oxidizing side of PSII and oxygen evolution is significantly decreased (Havaux, 1993). A loss of the oxygen evolving complex activity is thought to be due to the release of manganese atoms and the dissociation of the manganese stabilizing 33kDa protein from the PSII reaction center complex (Nash et al., 1985; Yamane et al., 1998).

High temperature stress also results in an inactivation of PSII reaction centers (Bukhov & Carpentier, 1990; Toth et al., 2005). In addition, high temperature results in a shift of the redox equilibrium between the primary acceptor plastoquinone (Q_A) and the secondary acceptor plastoquinone (Q_B) (Pospisil & Tyystjarvi, 1999; Toth et al., 2007). Furthermore, high temperature stress induces a dissociation of the peripheral antenna complex of PSII from its core complex (Armond et al., 1980; Wen et al., 2005).

As response to low temperature, Kenya strain of *Arthrospira platensis* was better acclimated than M2 strain by down-regulating its photosynthetic activity through decreasing antenna size and thus reducing energy flux into the photo-systems, decreasing reaction center density and the performance index, thus decreasing the trapping probability and electron transport beyond Q_A^- unchanged and increasing the energy dissipation flux.

Hence, the Kenya strain minimized potential damage on the acceptor side of PSII as compared to the M2 cells. Acclimation to low temperature was accompanied by an improved mechanism for handling excess energy resulting in an enhanced ability to rapidly repair damaged PSII reaction centers and withstand a high photon flux density stress; which was defined as a cross adaptation phenomenon (Vonshak & Novoplansky, 2008).

Increase in pigment in response to decrease in light intensity has long been considered an important and adaptive response because it increases the cellular efficiency of light-harvesting under light -limiting conditions (Richardson et al., 1983; Falkowski & LaRoche, 1991). This so-called photo-adaptive response can be compared to responses to other environmental factors, including temperature, nutrient availability and chemicals that affect cell metabolism and growth (Laws & Bannister, 1980; Rhee & Cotham, 1981; Verity, 1982; Fabregas et al., 1986; Osborne & Geider, 1986; Geider, 1987; Kana & Gilbert, 1987; Kana et al., 1992). Pigmentation such us Chlorophyll-a, is generally decreased under steady state nutrient limitation or transient starvation, or under lowered temperature, however carotenoids often remain high (Young, 1993) under photo-inhibition. Indeed, carotenoids may act as a screening pigment that blocks excess light, as well as scavenging reactive oxygen species, thereby decreasing the damaging effects of intense illumination at low temperatures (Krause, 1993). Furthermore, at low temperature, photosynthetic capacity decreased due to depress activity of ribulose-1,5-bisphosphate carboxylase (Rubisco) (Li et al., 1984; Raven & Geider, 1988).

Then, results of our works dealing with photosynthetic behaviors of the cyanobacteria *Arthrospira platensis* under various conditions, such as salinity and mixotrophic nutrition mode and those of combined effect of temperature, light intensity and C/N ratio on photo-system II photochemistry in *Cosmarium. sp* isolated from Tunisian geothermal source.

2. Effect of salt concentration on growth, fluorescence, photosynthetic activities and pigment content of the cyanobacteria *Arthrospira platensis*

Arthrospira platensis is commercially produced as a nutrient source for health food, feed and pharmaceutical industries, especially in developing countries. These species are representative of a relatively wide group of filamentous cyanobacteria and have been isolated from various habitats, with low to high ionic strength, in salty and alkaline waters. These cyanobacteria have developed different strategies for their adaptation to extreme environmental conditions, such as salinity, which are commonly encountered during biomass production of outdoor cultures.

Under several conditions, the helical filaments of *Arthrospira platensis* became straight after successive generations. These changes occur after pronounced evaporation, therefore salt stress could be considered as the main cause of this transformation.

In this section we investigated the effect of salt concentration on adapted cells of *Arthrospira platensis*, straight morphone in order to explore salty medium. The physiological behavior was evaluated by studying growth, pigment content, photosynthetic activity, change in the distribution of excitation energy between the two photo-systems (PSI and PSII) and the PSII photochemistry (Ben Dhiab et al., 2007).

The increase in salt concentration from 17mM to 500mM enhanced the growth as expressed by the increase of chlorophyll-a content from 52µg mL^{-1} to 70µg mL^{-1} (Fig.1) and a significant increase in phycobilin and carotenoid per Chlorophyll-a contents.

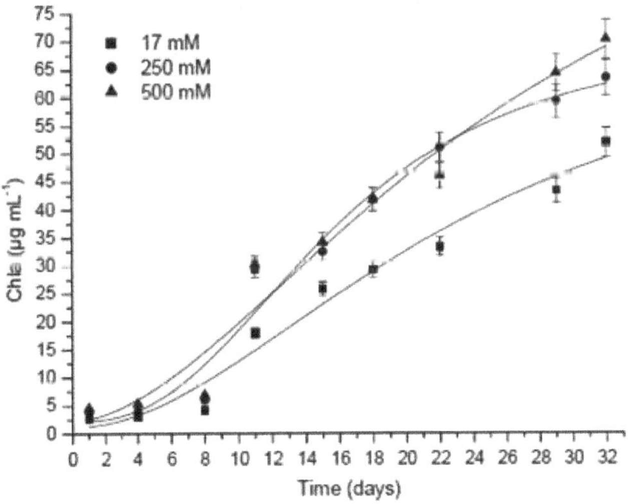

Fig. 1. Chlorophyll-a concentration of *Arthrospira platensis* exposed to different NaCl concentrations. Data are mean ± SEM (n = 3).

The photosynthetic efficiency and the light saturated maximal photosynthetic activity were also increased with the increase of salt concentration; however, dark respiration and compensation points were decreased as shown in (Table 1).

The light state transition regulates the distribution of absorbed excitation energy between the two photo-systems of photosynthesis under varying environmental conditions. In cyanobacteria, there is evidence of the redistribution of energy absorbed by both chlorophyll and phycobilin pigments. Proposed mechanisms differ in the relative involvement of the two pigment types. Changes in the distribution of excitation energy were assessed using 77K fluorescence emission spectroscopy under excitation of both phycobilin at 570 nm and chlorophyll at 440 nm.

Medium NaCl concentration	α	P_c	R_d	P_m
17 mM	5.79 ± 0.97	11.61 ± 4.47	72 ± 4.07	526.32 ± 2.95
250 mM	7.55 ± 0.60	7.62 ± 1.49	43.2 ± 0.00	864.00 ± 6.11
500 mM	6.46 ± 0.72	7.60 ± 3.15	50.4 ± 3.05	623.52 ± 6.11

Table 1. Photosynthetic parameters measured after 15 days of growth of the various cultures (P_m: light saturated maximal photosynthetic activity (μmol O_2 h^{-1} mg^{-1}chl), α: initial slope at the P-I curve (μmol O_2 h^{-1} mg^{-1}chl/μmol photons m^{-2} s^{-1}); Rd: dark respiration (μmol O_2 uptake h^{-1} mg^{-1} chl); Pc: compensation point (light intensity in μmol photons m^{-2} s^{-1} where no net oxygen uptake or evolution was observed)

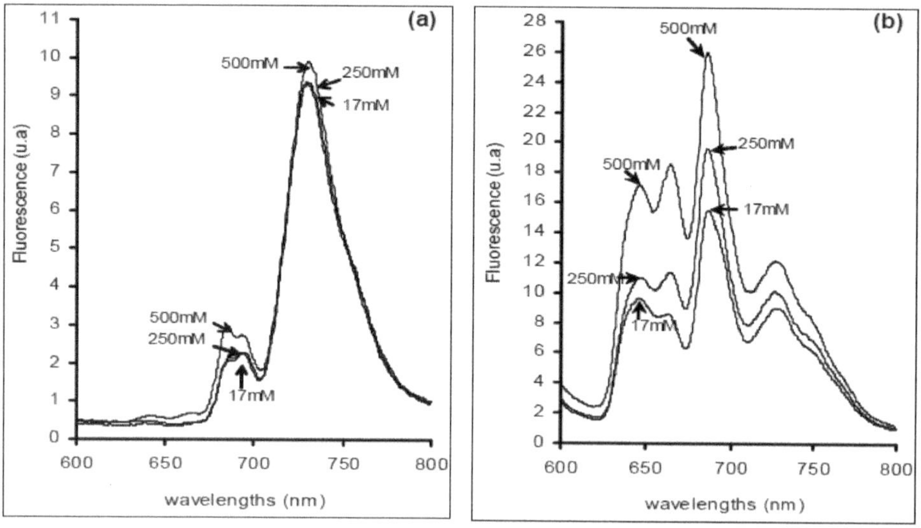

Fig. 2. Fluorescence emission spectra of excitation at 440 nm (a) and of excitation at 570 nm (b) on *Arthrospira platensis* exposed to different NaCl concentration. Spectra are normalized at 800 nm

The ratios between the peak heights at 685 or 695 and 730nm assigned respectively to Photo-system II (PS II) and photo-system I (PS I) (F685/F730) or (F695/F730) were used to obtain qualitative characteristics. These ratios are considered as a relative indicator of the distribution of excitation energy between PSII and PSI and therefore as an indicator of the state of the cells. The increase of NaCl concentration to 500 mM was accompanied with an enhancement of PSII activity per report to PSI under excitation of both chlorophyll and phycobilin (Fig 2).

The ratio between the heights of the peaks at 600 nm and 440 nm, calculated from the emission spectra of PSII at 695 nm (F695,600/F695,440), increased simultaneously with salt concentration of the growth media (Fig 3-a). This effect can be attributed to an increase of the energy transfer between phycobilisomes and PSII, which therefore enhanced the PSII fluorescence emission peak at 695 nm. This may also explain the elevated F695/F730 ratio in salt adapted cells of *Arthrospira platensis*. Additionally, the highest phycobilin/Chlorophyll-a ratio in salt adapted cultures allows the increase of light absorption by phycobilisomes which therefore increases PSII/PSI ratios.

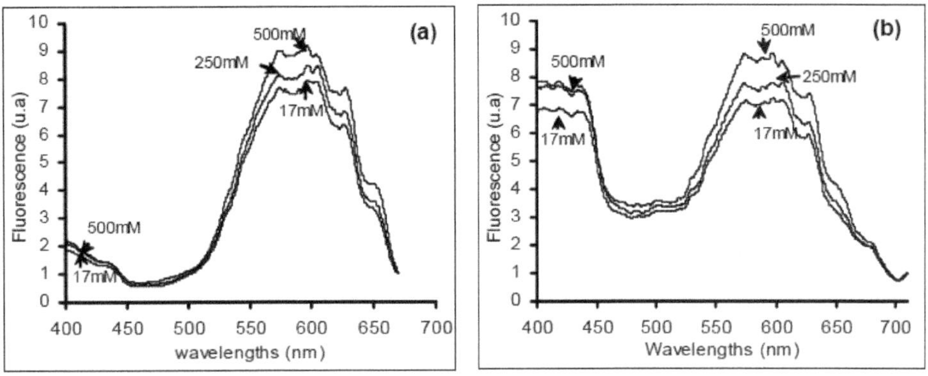

Fig. 3. Excitation spectra of PSII emission at 695 nm normalized at 670 nm (a) and PSI emission at 730 nm normalized at 710 nm (b) on *Arthrospira platensis* exposed to different NaCl concentrations

A direct energy transfer from phycobilisomes to PSI was also observed as shown by the ratio between the heights of the peaks at 600 nm and 440 nm calculated from emission spectra of PSI at 730 nm (F730,600/F730,440) (Fig 3-b). These ratios were close to unity, indicating that the cross section of energy transfer from phycobilisomes and chlorophyll-a to PSI was probably equivalent. These ratios remained relatively unchanged with the increase of NaCl concentration in growth media, resulting in a constant fluorescence emission of PSI. Data of maximal efficiency of PSII photochemistry were not in agreement with the above mentioned results. Indeed, Fv/Fm ratio decreased in salt adapted cultures. However, the trapping flux per PSII reaction center (TR_0/RC) and the probability of electron transport beyond Q_A showed change neither at the donor nor at the acceptor sides of PSII. In plants, Fv/Fm was well defined as an index of the maximal photochemical efficiency of PSII (Bjorkman & Demming, 1987). But this interpretation depended on both F_0 and F_v originating predominantly from sides of PSII (Table 2)

Medium NaCl concentration	Fv/Fm	Ψ_0	TR_0/RC
17 mM	0.46 ± 0.03	0.43 ± 0.08	2.96 ± 0.13
250 mM	0.44 ± 0.03	0.43 ± 0.07	2.83 ± 0.25
500 mM	0.40 ± 0.04	0.41 ± 0.08	2.87 ± 0.18

Table 2. Variable fluorescence parameters measured after 15 days of growth in media with different NaCl concentration; Fv/Fm: maximal efficiency of PSII photochemistry; Ψ_0: probability of electron transport beyond Q_A; TR_0/RC: trapping flux per PSII reaction centre

This assumption was not valid for cyanobacteria (Buchel & Wilhem, 1993; Papageorgiou & Govindjee, 1968; Papageorgiou, 1996; Schreiber et al., 1986), since phycobilin fluorescence interfered with Chlorophyll fluorescence leading to an increase of F_0, and therefore to a decrease of Fv/Fm values. Consequently, PSII contributed only a small proportion of total chlorophyll (Campbell et al., 1998). These data showed that phycobilin content, compared to the chlorophyll content as assessed by (phycobilin/Chlorophyll-a) ratio, increased with salt concentration and therefore affected the Fv/Fm measured values.

Several works studying the effect of salt concentration on physiological behavior of *Arthrospira platensis* showed different results. Vonshak et al., (1988, 1995, 1996), Zeng & Vonshak (1998), Lu et al. (1998, 1999) and Lu & Vonshak (1999, 2002) have shown that an increase in salt concentration led to the decrease of the specific growth rate, photosynthetic efficiency, maximum rate of photosynthesis, phycobilin/Chlorophyll-a ratio and PSII activity. Conversely, dark respiration activity, compensation points and PSI activity were increased. These contradictory results might be attributed to different genetic and environmental factors. Indeed, according to Berry et al. (2003), bioenergetic processes in the cytoplasmic membrane, the thylakoid membrane and the cytoplasm exhibited special adaptation strategies of strains like *Arthrospira platensis* which were mainly realized by using available components from the "tool-box" with different expression levels. In this study, the Compere strain of *Arthrospira platensis* was used, whereas Lu & Vonshak (1999, 2002) and Pogoryelov et al. (2003) have used, respectively, the M2 and Mayse strains.

Changes in the morphology of the trichome (from the helicoidal to the straight form) seem to be accompanied with modifications of physiological behavior of *Arthrospira* in response to the increase of NaCl concentration in growth media. Indeed, according to Jeeji Bai (1985) and Lewin (1980), the comparative behavior of the two morphones (straight and helicoïdal) in pure cultures showed that NaCl addition over and above the basal level inhibits the growth of the helicoidal morphone, while the straight morphone's growth behavior remained unaffected.

Minor variations in culture conditions might induce differences on physiological responses. Indeed, in our study a light intensity of 20 µmol photon m^{-2} s^{-1} was used, which was lower than the one used by Lu and Vonshak (1999, 2002: 50 µmol photon m^{-2} s^{-1}) and Pogoryelov et al. (2003: 60 µmol photon m^{-2} s^{-1}). These experimental light conditions appeared to be appropriated to maintain growth and full activity of the cells under elevated salinity conditions. Indeed, Zeng & Vonshak (1998) have shown that, at a higher light intensity, growing cells showed lower photosynthetic activity after photo- inhibition under salinity stress, compared with cells growing under lower light intensity conditions. In addition, it has been observed that a 12 h salt stress inhibits electron transport at both the donor and acceptor sides of PSII. However, 2 weeks-old adapted cells showed a down-regulation of PSII reaction centers without any inhibition of electron transfer on the donor and acceptor sides of PSII (Lu & Vonshak, 2002). Moreover, the effect of salt stress on PSII function in *Arthrospira* cells might be attributed to a direct interaction of high salt with PS II or more complex interaction through unknown cell components, which remain to be studied further. Indeed, the time course of adaptation to salinity stress induced different results.

3. Combined effect of light intensity and glucose concentration on *Arthrospira platensis* growth and photosynthetic responses

Arthrospira platensis is a photosynthetic cyanobacterium which is able to convert the energy of sunlight into chemical compounds usable by the cell to fix carbon dioxide and release oxygen. This cyanobacterium was also shown to be able of using organic carbon sources in heterotrophic and mixotrophic culture conditions. (Marquez et al., 1993; Chen et al., 1996;

Zhang et al., 1999; Vonshak et al., 2000; Chojnacka & Noworyta, 2004; Lodi et al., 2005; Andrade & Costa 2007). Mixotrophic growth offers a possibility of greatly increasing microalgal cell concentration in batch culture (Richmond, 1988; Marquez et al., 1993, 1995; Zhang et al., 1999). In mixotrophic growth, there are two distinctive processes within the cell, photosynthesis and aerobic respiration. The former is influenced by light intensity and the latter is related to the organic substrate concentration.

The interaction of light and glucose on specific growth rate was found to follow multiplicative growth kinetics (Chojnacka & Noworyta, 2004). The level of light intensity and glucose concentration and their interaction may influence both autotrophic (photosynthesis) and heterotrophic (oxidative metabolism of glucose) processes and therefore influence cell growth.

In this section, we expose primarily the combined effects of light intensity and glucose concentration on maximal biomass concentration, maximum specific growth rate, maximum net photosynthetic rate, and dark respiration rate and secondly comparative analysis on growth and photosynthetic responses between photoautotrophic and mixotrophic cultures (Ben Dhiab et al., 2010).

The effect of light intensity and glucose concentration on growth and photosynthesis was investigated using designs of response surface modelling (RSM) as shown in Table 3.

Experiment	Factors		Responses			
	Light (μmol photons m^{-2} s^{-1})	Glucose (g L^{-1})	Maximal Biomass X_{max} (g L^{-1})	Maximal specific growth rate (μ) (day^{-1})	Net Photosynthesis (P_n) (μmol O_2 mg Chl^{-1} h^{-1})	Dark respiration (R_d) (μmol uptake O_2 mg Chl^{-1} h^{-1})
1	50	0.5	0.51	0.28	43.77	45.85
2	100	0.5	0.44	0.24	48.56	41.60
3	150	0.5	0.73	0.43	84.32	46.76
4	50	1.5	0.73	0.33	46.91	45.85
5	100	1.5	0.80	0.19	47.61	50.87
6	150	1.5	0.83	0.49	135.79	46.16
7	50	2.5	0.85	0.35	49.21	50.74
8	100	2.5	0.91	0.37	67.78	63.27
9	150	2.5	1.33	0.49	132.7	65.00
10	100	1.5	0.86	0.23	67.83	52.93
11	100	1.5	0.84	0.23	65.29	39.94

Table 3. Maximal specific growth rate (μ_{max}), maximal biomass concentration (X_{max}), net photosynthetic rate (P_n) and dark respiration rate (R_d) for various culture conditions of light intensity and glucose concentration.

Analysis of the results was performed by MODDE. 7.0. The effect of each factor and their interactions was obtained by ANOVA with confidence interval of 90%.

Growth was characterized by two responses: maximum biomass concentration (X_{max}) and maximum specific growth rate (μ_{max}), whereas photosynthesis was evaluated by maximum net photosynthetic rate (P_n) and dark respiration rate (R_d).

The results showed that *Arthrospira platensis* grew in the presence of organic substrate (glucose) in the light. Independently of light intensity and glucose concentration, rates of the instantaneous relative growth, net photosynthesis, and dark respiration demonstrated two different phases (Fig. 4).

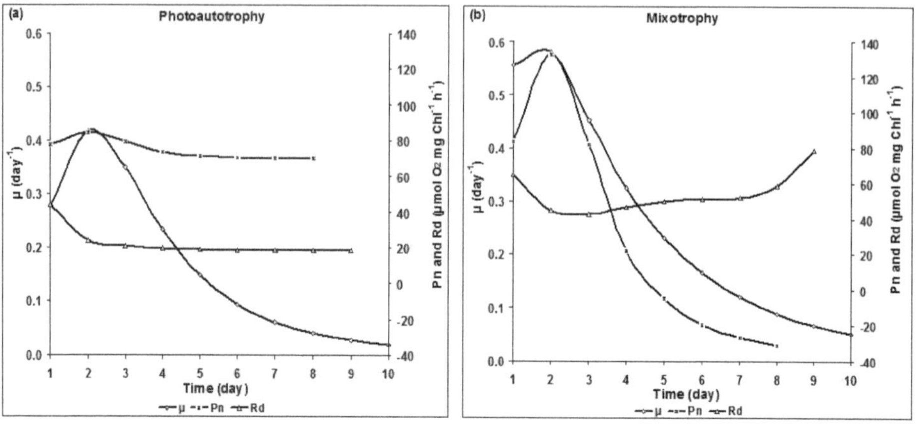

Fig. 4. Rates of the instantaneous relative growth, net photosynthesis, and dark respiration under photoautotrophic (a) and mixotrophic (b) conditions at a light intensity of 150 μmol photons $m^{-2}s^{-1}$

The first phase occurred during the first 3 to 4 days. It was characterized by the highest rates of instantaneous relative growth and net photosynthesis, as well as the preponderance of photosynthetic activity even in mixotrophic cultures, as seen by the increase in pH until the third day. This result might be supported by the data reported by Yang et al. (2000) who found that light was the major source for ATP production in the early phase of mixotrophic cultivation.

The second phase occurred from the fourth day onwards. It was characterized by the decrease of the instantaneous relative growth rates. In photoautotrophic cultures, rates of maximal net photosynthesis and dark respiration were maintained at constant values which are lower than those observed during the first phase. However, in mixotrophic cultures, net photosynthetic rate was reduced to negative values and dark respiration rate increased. Thus, we suggest that metabolic activity was based essentially on photoheterotrophy. This suggestion is supported by the decrease of pH during this phase, due to the release of carbon dioxide, caused by the heterotrophic component of mixotrophic metabolism as reported by Hase et al. (2000).

These results are in contrast with those observed by Chen & Zhang (1997) who showed that heterotrophic metabolism dominated in the first phase, then decreased subsequently as the glucose was consumed, and later photosynthesis became predominant. The same comment was signaled by Andrade & Costa (2007). Indeed, these suggestions were based only on the glucose consumption and phycocyanin content. Our results showed effectively that the total glucose in the medium was consumed during the first 3 days, considered as the photoautotrophic phase and characterized by the highest net photosynthetic rate. Thus, we hypothesise that glucose was stored as reserve carbohydrate (glycogen) to be metabolized in the second phase considered as heterotrophic. Indeed, Pelroy et al. (1972) showed in

Synechocystis sp. that exogenous glucose is mainly stored as glycogen under illumination before being metabolized for the maintenance of cells (Yang et al., 2000). Our hypothesis is supported by results of Martinez and Orus (1991) who noted that respiratory rate was noticeably enhanced in mixotrophic cultures, reflecting the increasing rate of glucose metabolism after the induction of glucose uptake ability. Moreover, in the second phase and additional to the decrease of the net photosynthetic rate, we noted a reduction of the photochemical efficiency of photosystem II accompanied by lower electron transport rate. Therefore, organic carbon sources reduced the photosynthetic efficiency in this phase, and the enhancement of biomass of *Arthrospira platensis* implied that organic sources had more pronounced effects on respiration than on photosynthesis. These conclusions are in agreement with results obtained with *Phaeodactylum tricornutum* under mixotrophic culture (Liu et al., 2009).

All the above results confirmed that, in the second phase, *Arthrospira platensis* might use and metabolize glucose and then shift to the heterotrophic nutrition mode. Comparison between mixotrophic and autotrophic cultures showed that the former were characterized by the highest values of the instantaneous relative growth rates and maximal biomass concentration. These results are consistent with those obtained by Marquez et al. (1993, 1995), Vonshak et al. (2000), Zhang et al. (1999), and Andrade and Costa (2007). Marquez et al. (1993) and Hata et al. (2000) suggested that in mixotrophic culture, a simultaneous uptake of organic compounds and CO_2 takes place as carbon sources for cell synthesis, and it is then expected that CO_2 will be released via respiration and will be rapidly trapped and reused under sufficient light intensity. Thus, mixotrophic cells acquire the energy by catabolizing organic compounds via respiration and converting light energy into chemical energy via photosynthesis (Hata et al., 2000).

Effectively, both photosynthetic and dark respiration rates were the highest in mixotrophic cultures, as observed by Vonshak et al. (2000) in *Arthrsopira platensis*, Kang et al. (2004) in *Synechococcus sp.*, Yu et al. (2009) in *Nostoc flagelliforme*, and Orus et al. (1991) in *Chlorella vulgaris*. The biomass gain recorded in mixotrophic cultures was achieved in the second phase characterized by the nutritive salt reduction and the decrease of the available photons for cells as a consequence of shading caused by the increase in cell density. Such conditions seem to stimulate heterotrophic growth which gives the possibility to increase biomass concentration (Chen et al., 1996; Chen & Zhang, 1997). Martinez et al. (1997) indicated in *Chlorella pyrenoidosa* that light contributes the energy needed for growth and cell maintenance, while glucose is used for the formation of biomass.

Furthermore, the results of Yu et al. (2009) demonstrated that, during the first 4 days, the cell concentration in mixotrophic culture of *Nostoc flagelliforme* was lower than the sum of those in photoautotrophic and in heterotrophic culture. However, from the fifth day, the cell concentration in mixotrophic culture surpassed the sum of those obtained from the other two trophic modes. The reason for this phenomenon requires further investigation.

Growth and photosynthetic activity of mixotrophic cultures were evaluated with respect to light intensity and glucose concentration using response surface methodology as shown in Fig 5. The experimental design showed that polynomial models dependent on light intensity and glucose concentration could describe relatively accurately the maximal biomass concentration, maximum specific growth rate, and maximum net photosynthetic rate.

The results showed clearly that all these responses were influenced by both factors. Furthermore, the interaction of both factors showed that at low light intensity, glucose had a low effect on maximum biomass concentration and maximum net photosynthetic rate.

However, at the highest light intensity (150 µmol photons $m^{-2}s^{-1}$), the effect of glucose was positive and the responses were more sensitive to it. The effect of this organic carbon substrate might be attributed to the protective role of glucose or to the shift in light intensity at which photo-inhibition occurs, as explained by Chojnacka & Marquez-Rocha (2004). Thus, cells growing heterotrophically might use part of the O_2 produced by cells growing photoautotrophically, decreasing dissolved oxygen concentration; this can help reduce photooxidative damage. As has been observed in the response surface plot, the maximum biomass concentration (1.33 gL^{-1}) was obtained at the highest light intensity (150 µmol photons $m^{-2}s^{-1}$) and glucose concentration (2.5 gL^{-1}). The same conditions improved maximum specific growth rate (0.49 day^{-1}) and maximum net photosynthetic rate (139.89 µmol µmol O_2 mg Chl^{-1} h^{-1}). Conditions favouring high biomass production and maximum net photosynthetic rate were, however, not optimal. Indeed, the optimal conditions of light intensity and glucose concentration are not achieved in the experimental range used in this

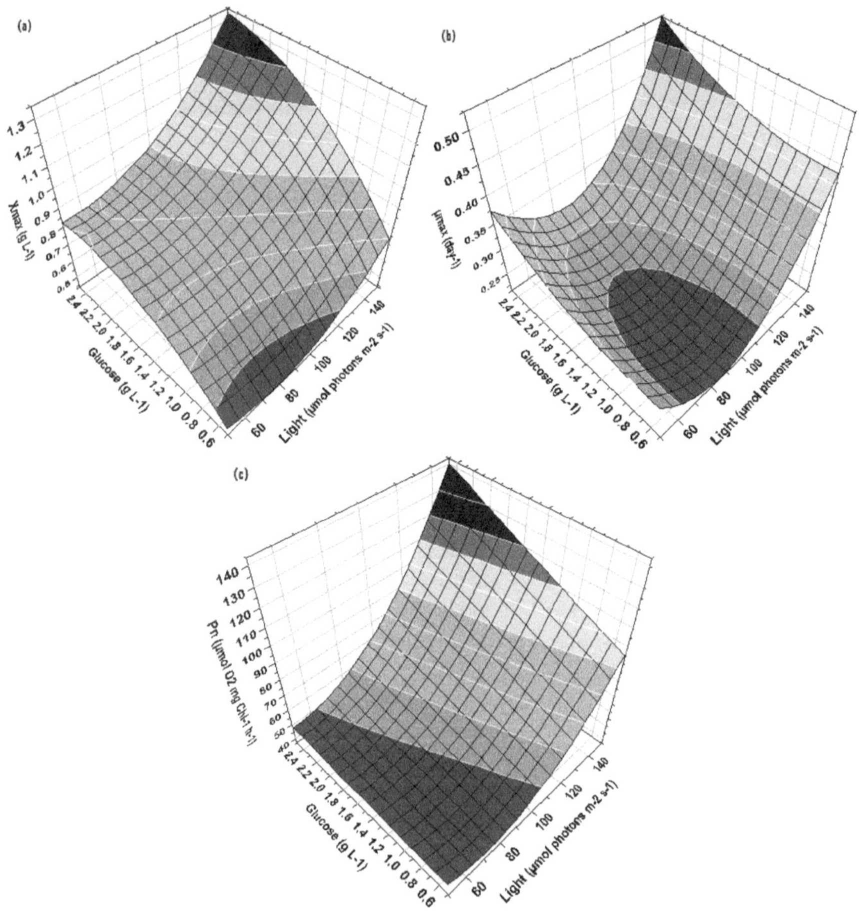

Fig. 5. Response surface plot vs light intensity and glucose concentration for maximal biomass concentration (a), maximal specific growth rate (b) and net photosynthesis (c).

study. Therefore, further studies that extend the experimental range of light intensity and glucose concentration might be required to reveal optimal conditions that maximize growth and photosynthesis of mixotrophic cultures. For the same species, Zhang et al. (1999), using a number of mathematical models, found that the optimal initial glucose concentration and light intensity were 2.5 g L^{-1} and 48 μmol photons m^{-2} s^{-1} (4 klx), respectively. However, Chojnacka (2003) determined optimal growth parameters to be 2.5 gL^{-1} glucose concentration and 126 μmol photons $m^{-2}s^{-1}$ (10.5 klx) light intensity. This difference in the optimal light intensity, as commented by Chojnacka & Marquez-Rocha (2004), could be due to the different methods of light intensity measurement and distribution of cells inside the culture vessels. Considering all the findings drawn from the experimental design, it is also recommended that data from batch cultures should be further examined to develop much more accurate models.

4. Combined effect of temperature, light intensity and C/N on photo system II in *Cosmarium. sp* isolated from Tunisian geothermal using polyphasic rise of fluorescence transients

Hot spring microalgae are exposed to daily and seasonal fluctuations in temperature and light, which are the major factors determining photosynthetic and growth rates. These factors intensities may disturb the balance between the absorption of energy through photosynthesis and the ability to utilize this energy. Such conditions may cause higher excitation pressure on PSII that result in damage to the photosynthetic apparatus (Huner et al., 1998; Yamamoto, 2001; Wilson et al., 2006).

Within the photosynthetic apparatus, photo-system II (PSII) is the most sensitive component of the electron transport chain (Cajanek et al., 1998). Damage to PSII is often the first manifestation of stress. Among partial reactions of PSII, the oxygen evolving complex (OEC) is particularly heat sensitive (Georgieva et al., 2000).

Numerous studies have shown that high temperature stress has various effects on PSII function.

Several major regulatory mechanisms may be involved in the protection of the PSII apparatus from photo-damage under stress conditions: photochemical processes related to electron transport, non photochemical processes by which excess energy is dissipated as heat fluorescence or transferred to other systems, and modification in D1 protein turnover under excess energy stress (Prasil et al., 1992; Campbell et al., 1998; Huner et al., 1998; Melis, 1999; Adams et al., 2001; Yamamoto, 2001; Tsonev & Hikosaka, 2003).

Polyphasic rise in Chlorophyll-a fluorescence (OJIP test) has been used as a tool to evaluate modifications in PSII photochemistry in a wide range of studies, not only in higher plants (Tomek et al., 2001; Ban Dar & Leu, 2003; Force et al., 2003; Zhu et al., 2005; Strauss et al., 2006) but also in algae (Hill et al., 2004; Hill & Ralph, 2005; Kruskopf & Flynn, 2006) and cyanobacteria (Strasser et al., 1995, 2004; Lu & Vonshak, 1999; Lu et al., 1999; Qiu et al., 2004; Lazar, 2006). This tool has been used in a variety of studies, including structure and function of the photosynthetic apparatus, characterization of vitality and physiological condition, and selection of species tolerant to stress conditions (Hermans et al., 2003, Goncalves & Santos, 2005).

Fluorescence induction kinetics (Kautsky curve) of all photosynthetic organisms show a polyphasic rise between initial (F_0) and maximum (F_m) fluorescence yields (Schreiber & Neubauer, 1987; Strasser et al., 1995; Srivastava et al., 1999). These phases were designed as O, J, I and P, and can be visualized using a logarithmic time scale. By monitoring fluorescence transients and quenching, it is possible to obtain information on the absorption, transfer and dissipation of energy by PSII.

In this section we study the response of photo-system II under combined environmental factors of temperature, light intensity and Carbon/Nitrogen ratio in *Cosmarium sp* isolated from Tunisian hot spring using Polyphasic rise of fluorescence transients.

Analysis of fluorescence transients in *Cosmarium sp* might be provide information about changes taking place in the structure, conformation and functional of the photosynthetic apparatus, especially in PSII under these environmental factors.

The increase of temperature from 20°C to 60°C causes a significant decrease in the minimal (F_0) and maximal (F_m) fluorescence (Fig 6). A decrease in the fluorescence yield in *Cosmarium* cells can be attributed to an inhibition of electron flow at oxidizing site of PSII (Lu & Vonshak, 2002). The decrease in F_m and fluorescence at J, I, P may be due to two reasons, first by inhibition of electron transport at the donor side of the PSII which results in the accumulation of $P680^+$ (Govindjee, 1995; Neubauer & Schreiber., 1987) and second due to a decrease in the pool size of Q_A^-.

The same results were observed in *Synechocystis* cells exposed to Sb at 10 mg L^{-1}. In fact F_m was decreased and the shape of J–I–P phase became flat with the increase of Sb concentration, indicating that PSII were partially inactivated and could not be closed, and the reduction of PQ (non-photochemical phase) was inhibited (Zhang et al., 2010). In *Spirulina platensis*, Zhao et al. (2008) showed that F_m decreased but F_0 increased significantly with increasing temperature. A loss of the JI-phase at higher temperatures suggests that heat stress resulted in the destruction of the oxygen-evolving complex which may be due to a loss of the manganese cluster activity.

The reduction in the J and P steps observed in the Kenya strain of *Arthrospira platensis* grown at low temperature reflects a bigger decline in the concentration of Q_A, Q_B^{2-}, and plastoquinol (PQH_2) and therefore a decrease in the accumulation of reduced Q_A (Govindjee, 2004; Zhu et al., 2005; Lazar 2006) as well as a decrease in the concentration of active reaction centers (Govindjee, 2004; Zhu et al., 2005; Lazar, 2006).

At the donor side, the quenching effect of the variable fluorescence yield at J, I and P was also ascribed to the deterioration of the water splitting system (Strasser, 1997).

Moreover, the efficiency of the water-splitting complex on the donor side of PSII as expressed from (F_v/F_0) (Schreiber et al., 1994) is the most sensitive component in the photosynthetic electron transport chain. In this study the increase of temperature from 20°C to 60°C decreased this ratio from 2.71 to 0.14. This decrease results from photosynthetic electron transport impairment as indicated by Pereira et al. (2000). The same results were also observed in barely under salt concentration (Kalaji et al., 2011) and in *Synechocystis sp.* exposed to Sb (Zhang et al., 2010).

Area over the fluorescence induction curve between F_0 and F_m is proportional to the pool size of the electron acceptor Q_A on the reducing side of PSII. If the electron transfer from reaction center to quinone pool is blocked, this area will be dramatically reduced (Strasser et al., 2000). As compared to the initial inoculums, the area over the fluorescence curve was decreased with increasing temperature at 60°C to reach ratios of 0.28 and 0.65. However these ratios were augmented to 3.5 and 4.24 times in the lowest temperature at 20°C. This decrease in area over the fluorescence curve suggests that high temperature inhibits the electron transfer rates at the donor side of PSII. This inhibition becomes most considerable under the interaction of the highest temperature and light intensity (60°C/130µmol photons m^{-2} s^{-1}) which lead to the blockage of the electron transfer from reaction center to quinone

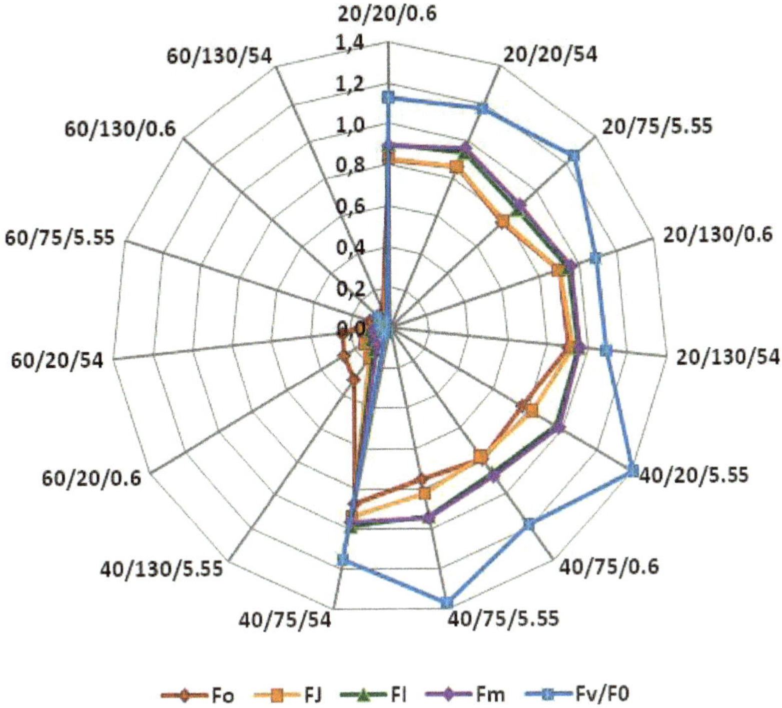

Fig. 6. Spider plot of parameter of fluorescence intensities responses of the OJIP steps and F_v/F_0

pool. These results are in agreement with those of Mehta et al. (2010) which showed an inhibition in the electron transfer rates at the donor side of PS II in *Triticum aestivum* leaves, treated with 0.5 M NaCl.

To localize the action of high temperature in electron transport chain on acceptor side of PSII, the kinetics of relative variable fluorescence (V_j), was calculated. V_j is equivalent to (F_j-F_0/F_m-F_0). F_j is the fluorescence at J step at 2 ms, relative variable fluorescence (V_j) at 2 ms for unconnected PS II units, equals to the fraction of closed RCs at J step expressed as proportion of the total number of the RCs that can be closed (Force et al., 2003). Efficiency with which a trapped exciton can move an electron in to the electron transport chain further than Q_A^- ($\Psi_0 = ET_0/TR_0$) was also measured.

Increasing temperature from 20°C to 60°C provoke an increase of V_j from 85% to 147% and a decrease of Ψ_0 from 112% to 62%. Thus, revealed a loss of Q_A^- reoxidation capacity and an inhibition of electron transport at the acceptor side of PSII and also beyond Q_A^-. The same results were showed in wheat leaves (Mehta et al., 2010), *Spirulina platensis* (Lu & Vonshak, 2002) and *Synechocystis sp.* (Zhang et al., 2010) treated respectively with 0.5 M NaCl, 0.8M NaCl and antimony potassium tartrate (Sb).

Results of specific energy fluxes demonstrated that an increase of temperature from 20°C to 60°C provoked the raise of the effective antenna size per reaction center (ABS/RC) (from 33% to 74.5%), the trapping energy flux per reaction center (from 78% to 493%) and dissipated energy flux per reaction center (from 13% to 936%).

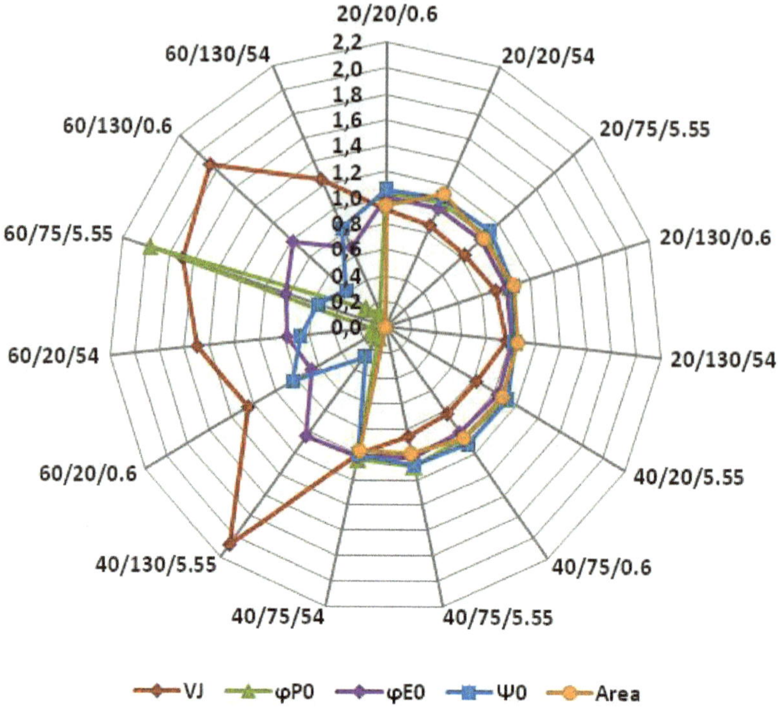

Fig. 7. Spider plot of parameter of quantum efficiencies (φ_{P0}, φ_{E0}, Ψ_0), area, and relative variable fluorescence intensity at the J-step (V_j).

The ratio of ABS/RC demonstrates average antenna size and expresses the total absorption of PS II antenna chlorophylls divided by the number of active (in the sense of Q_A reducing) reaction centers. Therefore, the antenna of inactivated reaction centers are mathematically added to the antenna of the active reaction centers. Consequently, the increase of this ratio was justified by an inactivation of some active reaction centers.

TR_0/RC ratio represents the maximal rate by which an exciton is trapped by the RC resulting in the reduction of Q_A. So it refers only to the active (Q_A to Q_A^-) centers (Force et al. 2003). The increase in this ratio indicates that all the Q_A has been reduced but it is not able to oxidize back due to stress, as a result the reoxidation of Q_A^- is inhibited so that Q_A cannot transfer electrons efficiently to Q_B.

DI_0/RC corresponds to the ratio of the total dissipation of untrapped excitation energy from all RCs with respect to the number of active RCs. Dissipation may occurs as heat, fluorescence and energy transfer to other systems. It is influenced by the ratios of active/inactive RCs. The ratio of total dissipation to the amount of active RCs increased (DI_0/RC) due to the high dissipation of the active RCs.

These ratios conclusively describe that the number of inactive centers have increased due to high temperature in *Cosmarium sp*. These results were confirmed by the decrease in the concentration of active PSII reaction centers (RC/CSm) from 315% to 5% at 60°C, resulted in a similar decrease in maximum quantum yield for primary photochemistry (φ_{P0}) and in the quantum yield of electron transport (φ_{E0}) respectively from 239% to 50% and from

264% to 33%. Furthermore the increase of temperature from 20°C to 60°C provokes the decrease in the performance index (PI_{ABS}) from 2361% to 122% which indicates the sample vitality.

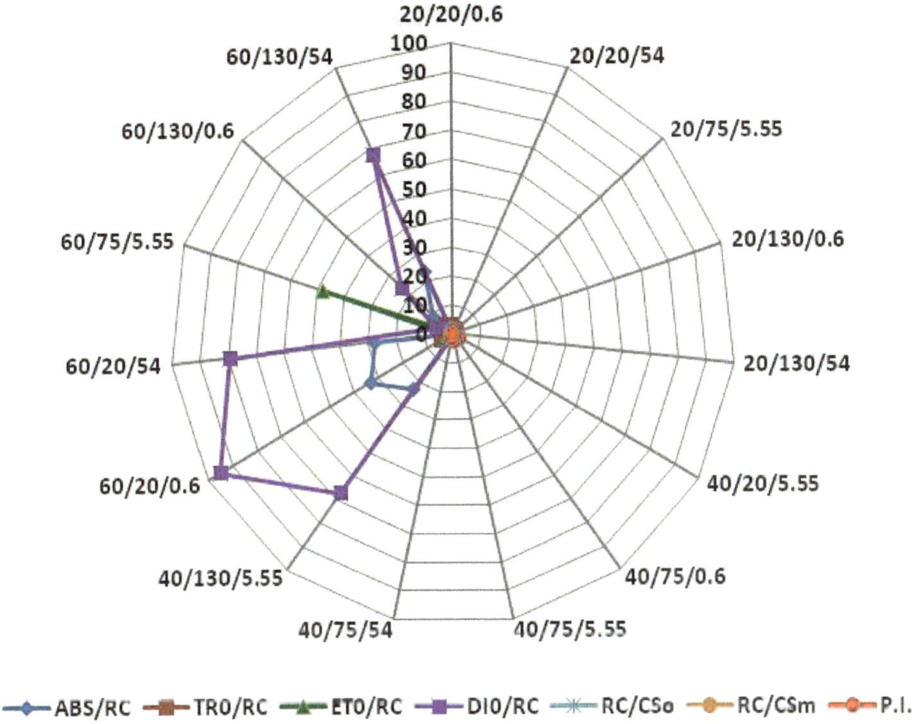

Fig. 8. Spider plot of specific energy fluxes parameters

All these results were in line with those observed by Mathur et al. (2010) and Mehta et al. (2010) in wheat leaves of *Triticum aestivum* as a response to high salt stress.

According to Zhao et al. (2008), heat stress in *Spirulina platensis* resulted in a decrease in RC/ABS, suggesting further a decrease in the total number of active reaction center. It should be pointed out that the expression RC/ABS refers to Q_A reducing reaction centers of PSII (Strasser & Strasser, 1995; Strasser et al., 2000). The decrease in RC/ABS in heat-stressed *Spirulina* cells also indicates a decrease in the total number of Q_A reducing reaction centers of PSII. These results further indicate that heat stress caused a shift of the equilibrium towards Q_A, making the electrons tend to stay longer on Q_A, suggesting that heat stress induced an increase in the Q_B non- reducing reaction centers of PSII.

Dissipation in this context refers to the loss of absorbed energy through heat, fluorescence and energy transfer to other systems (Strasser et al., 2000), and is represented by the equation DI_0/RC= (ABS/RC)-(TR_0/RC). Therefore, dissipation can be thought of as the absorption of photons in excess of what can be trapped by the reaction center. In illuminated *Monstera* leaves, the excitation energy in the antenna of the active reaction centers was in excess of that required for trapping and some energy was dissipated.

The energy pipeline model as an effect of elevated temperature was deduced using biolyzer HP3 software. This model gives information about the efficiency of flow of energy from antennae to the electron transport chain components through the RC of PSII. The area or the width of the arrows for each of the parameters, ABS/CS_0, TR_0/CS_0, ET_0/CS_0 and DI_0/CS_0, indicate the efficiency of light absorption, trapping, electron transport and dissipation per cross-section of PSII, respectively, in *Cosmarium sp.* treated in 20°C and 60°C.
The effect of elevated temperature on *Cosmarium* cells was represented phenomenologically per excited cross-section (CS) area (Fig. 9).

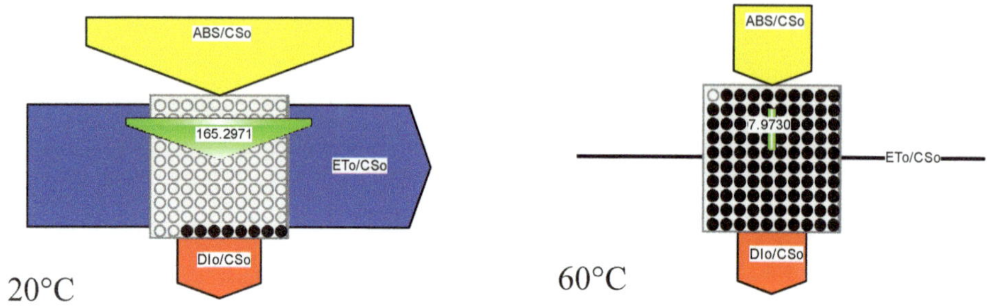

Fig. 9. Phenomenological fluxes of cell suspension pipeline model of *Cosmarium* sp. treated at two temperature conditions (20°C and 60°C).

Results showed a decrease in ABS/CS_0, ET_0/CS_0 under elevated temperature (60°C). Indeed, ABS/CS_0 describes the number of photons absorbed by antenna molecules of active and inactive PSII RCs over the excited cross-section of the tested sample and is represented by the dark-adapted F_0. The ABS/CS_0 can be substituted as an approximation by the fluorescence intensity, F_0. A decrease in ABS/CS_0 at high temperature indicates a decrease in the energy absorbed per excited cross-section. ET_0/CS_0 represents electron transport in a PSII cross-section and indicates the rate of reoxidation of reduced Q_A via electron transport over a cross-section of active RCs (Force et al., 2003). A decrease in this ratio indicates inactivation of RC complexes and the Oxygen Evolving Complex (OEC) and also suggests that the donor side of PSII has been affected.
DI_0/CS_0, represents the total dissipation measured over the cross-section of the sample that contains active and inactive RCs. A decrease in the density of active RCs (indicated as open circles) and an increase in the density of inactive RCs (indicated as filled circles) was observed in response to elevated temperature. Dissipation occurs as heat, fluorescence and energy transfer to other systems. An increase in energy dissipation at high temperature indicates that energy available for photochemistry is reduced under stress conditions (Strasser, 1987; Strasser et al., 1996, 2000; Kruger et al., 1997; Force et al., 2003).

5. Conclusion

In conclusion, our findings showed that morphological change of *Arthrospira platensis* trichome from helicoidal to the straight morphone was accompanied with physiological modifications as response to salt stress. Indeed, the increase of NaCl concentration in growth media to 500mM enhances the growth, the photosynthetic efficiency and PSII

activity as compared to 17mM cultures. The elevated PSII activity might be attributed to the increase of phycobilin content which led to an increase of light absorption by phycobilisomes relative to that of chlorophyll-a and therefore an increase of the energy transfer between phycobilisomes and PSII. Furthermore, the fluorescence emission spectra of phycobilin excitation showed that the energy in excess might be directly dissipated by fluorescence of phycobilin in order to protect the PSII against the photo inhibition.
The straight form seems to be more tolerant to salt; it maintains full activity under 500 mM NaCl medium whereas the optimal growth medium for the helicoidal morphone is 17 mM NaCl. Therefore, a better understanding of salt stress on PSII may help optimize the productivity of the microalgal cultures grown outdoors.
In order to increase and optimize the biomass productivity, this strain was performed under two trophic modes (autotrophic and mixotrophic) using factorial design. As revealed by the kinetics of maximal net photosynthetic, dark respiration rates and instantaneous growth rates, mixotrophic cultures showed two phases. The first was distinguished by the preponderance of the photoautotrophic mode while the second was based mainly on photoheterotrophy. The synergistic effect of photosynthesis and glucose oxidation enhanced the growth rate and the biomass concentration of *Arthrospira platensis* under mixotrophic mode as compared to autotrophic mode.
The combined effect of temperature, light intensity and C/N ratio on *Cosmarium sp* demonstrate that *Cosmarium* cells isolated from hot spring water maintained their maximum photochemical efficiency under 20°C. Nevertheless this specie was acclimated and withstood under 60°C by down-regulating electron transport at both donor and acceptor sides of PSII. Otherwise, the conversion of some active reaction centers of PSII into inactive form provides a protective mechanism for quenching excessive energy. Indeed the excitation energy in excess would be dissipated as heat and fluorescence or transferred to other systems via electron transport pathways such respiration or state transition between PSII and PSI suggested to be a part of the acclimation mechanisms to environmental stress.
Ultimately to withstand and grow under harsh environments or extreme conditions, several adaptive mechanisms might be developed. These mechanisms were accompanied with their ability to adjust their photosynthetic apparatus in order to acclimate to the prevailing environmental conditions.

6. References

Adams, WW., Dommig Adams, B; Rosenstiel, TN; Ebbert ,V; Brightwell, AK; Barker, DH., Zarter, C. (2001). Photosynthesis, xanthophylls, and D1 phosphorylation under winter stress., In: PS2001 *Proceedings: 12th International congress on photosynthesis.* CSIRO publishing, Melbourne, Australia.

Allakhverdiev, S.I., Nishiyama, Y., Miyairi, S., Yamamoto, H., Inagaki, N., Kanesaki ,Y., Murata, N. (2002). Salt stress inhibits the repair of photodamaged photosystem II by suppressing the transcription and translation of psbA genes in *Synechocystis. Plant Physiol,* 130: 1443-1453.

Allen, D. J. & Ort, DR. (2001). Impacts of chilling temperatures on photosynthesis in warm-climate plants. *Trends Plant Sci,* 6: 36– 42.

Andrade, MR & Costa, JAV. (2007). Mixotrophic cultivation of microalga *Spirulina platensis* apparatus, *Biochim. Biophys. Acta,* 601: 433–442.

Armond, P.A., Bjorkman, O., Staehelin, L.A. (1980). Dissociation of supramolecular complexes in chloroplast membranes. A manifestation of heat damage to the photosynthetic apparatus. *Biochim Biophys Acta,* 601 (3):433-43.

Ban Dar, H. & Leu, K.L. (2003). A possible origin of the middle phase of polyphasic chlorophyll fluorescence transient. Funct. *Plant Biol.* 30:571–6.

Ben Dhiab, R., Ben Ouada, H., Boussetta, H., Frank, F., El Abed, A., Brouers, M. (2007). Growth, Fluorescence, Photosynthetic O_2 production and pigment content of salt adapted cultures of *Arthrospira platensis. J Appl Phycol.* 19: 293-30.

Ben Dhiab, R., Ghenim, N., Trabelsi, L., Yahia, A., Challouf, R., Ghozzi, K., Ammar, J., Omrane, H., Ben Ouada, H. (2011). Modeling growth and photosynthetic response in *Arthrospira platensis* as function of light intensity and glucose concentration using factorial design. *Journal of Applied Phycology,* 22 (6): 745-752

Berry, S., Bolychevtseva, YV., Rögner, M., Karapetyan, NV. (2003). Photosynthetic and respiratory electron transport in the alkaliphilic cyanobacterium *Arthrospira (Spirulina) platensis. Photosynth Res* 78:67–76

Björkman, O. & Demming, B. (1987). Photon yield of O_2 evolution and chlorophyll fluorescence characteristics at 77 K among vascular plants of diverse origins. *Planta* 170: 489–504

Buchel, C. & Wilhem, C. (1993). In vivo analysis of slow chlorophyll fluorescence induction kinetics in algae: progress, problems and perspectives. *Photochem Photobiol* 58: 137–148

Bukhov, N.G. & Carpentier, R. (2000). Heterogeneity of photosystem II reaction centers as influenced by heat treatment of barley leaves. *Physiologia Plantarum,* 110: 279–285

Cajanek, M., Stroch, M., Lachetova, I., Kalina, J., Spunda, Y. (1998). Characterization of the photosystem II inactivation of heat stressed barley leaves as monitored by the various parameters of chlorophyll a fluorescence and delayed fluorescence. Journal of *Photochemistry and Photobiology,* 47, 39–45

Campbell, D., Hurry, V., Clarke, A., Gustafson, P., Oquist, G. (1998). Chlorophyll fluorescence analysis of cyanobacterial photosynthesis and acclimation. *Microbiol Mol Biol Rev,* 62: 667–683

Chen, F. & Zhang, Y. (1997). High cell density mixotrophic culture of *Spirulina platensis* on glucose for phycocyanin production using a fed-batch system. *Enzyme Microb Technol,* 20: 221-224

Chen, F., Zhang, Y., Guo, S. (1996). Growth and phycocyanin formation of *Spirulina platensis* in photoheterotrophic culture. *Biotechnol Lett,* 18: 603-608

Chojnacka, K. & Marquez-Rocha, FJ. (2004.) Kinetic and stoichiometric relationships of the energy and carbon metabolism in the culture of microalgae. *Biotechnology,* 3: 21-34

Chojnacka, K. & Noworyta, A. (2004). Evaluation of *Spirulina sp.* Growth in photoautotrophic, heterotrophic and mixotrophic cultures. *Enzyme Microb Technol,* 34: 461-465

Chojnacka, K. (2003). Heavy metal ions removal by microalgae *Spirulina sp.* In the processes of biosorption and bioaccumulation. Dissertation, Wroclaw University of Technology Poland.

Endo, T., Schreiber, U., Asada, K. (1995). Suppression of quantum yield of photosyntem II by hyperosmotic stress in *Chlamydomonas reinhardtii, Plant Cell Physiol.* 36: 1253–1258

Fabregas, J., Herrero, C., Cabezas, B., Liano, R., Abalde, J. (1986). Response of the marine microalga *Dunaliella tertiolecta* to nutrient concentration and salinity variations in batch cultures. *Journal of Plant Physiology,* 125: 475-4-84

Falkowski, PG. & LaRoche, J. (1991). Acclimation to spectral irradiance in algae. *Journal of Phycology*, 27:8-14

Force, L., Gritchley, C., Van Rensen, JJS. (2003). New fluorescence parameters for monitoring photosynthesis in plants 1. The effect of illumination on the fluorescence parameters of the JIP test. *Photosynth Res*, 78: 17-33

Geider, RJ. (1987). Light and temperature dependence of carbon to chlorophyll a ratio in microalgae and cyanobacteria: implications of physiology and growth of phytoplankton. *New Phycologist*, 106: 1-34

Georgieva K., Tsonev T., Velikova V., Yordanov I. (2000). Photosynthetic activity during high temperature of pea plants. *Journal of Plant Physiology*, 157: 169–176

Giardi, MT., Masojidek, J., Godde, D. (1997). Effects of abiotic stresses on the turnover of the D1 reaction center II protein. *Physiol. Plant.* 101, 635-642

Gilmour, DJ., Hipkins, MF., Boney, AD. (1982). The effect of salt stress on the primary processes of photosynthesis in *Dunalielia tertiolecta. Plant Sci. Lett.* 26, 325-330

Gilmour, DJ., Hipkins, MF., Webber, AN., Baker, NR., Boney, AD. (1985). The effect of ionic stress on photosynthesis in *Dunaliella tertiolecta, Planta* 163: 250–256

Goncalves, JFC. & Santos, UM. (2005). Fluorescence technique as a tool for selecting tolerant species to environments of high irradiance. *Brazil. J. Plant Physiol*, 17: 307–13

Govindjee (1995). Sixty-three years since Kautsky: chlorophyll a fluorescence. Austr. J. Plant Physiol. 22, 131–160

Govindjee (2004). Chlorophyll a fluorescence: a bit of basics and history. In Papageorgiou, GC. & Govindjee (eds.) Chlorophyll a fluorescence: A signature of photosynthesis. Springer, Dordrecht, the Netherlands, pp. 321-362.

Hase, R., Oikaw,a O., Sasao, C., Morita, M., Watanabe, Y. (2000). Photosynthetic production of microalgal biomass in a raceway system under greenhouse conditions in Sendai city. *J Biosci Bioeng*, 89: 157-163

Hata, J.I, Hua, Q., Yang, C., Shimizu, K., Taya, M. (2000). Characterization of energy conversion based on metabolic flux analysis in mixotrophic liverwort cells, *Marchantia polymorpha. Biochem Eng J*, 6: 65-74

Havaux, M. (1993). Characterization of thermal damage to the photosynthetic electron transport system in potato leaves, *Plant Sci*, 94: 19–33

Hermans, C., Smeyers, M., Rodriguez, RM., Eyletters, M., Strasser, RJ. & Delhaye, JP. (2003). Quality assessment of urban trees: a comparative study of physiological characterization, airborne imaging and on site fluorescence monitoring by the OJIP-test. *J. Plant Physiol*, 160: 81–90.

Hill, R. & Ralph, PJ. (2005). Diel and seasonal changes in fluorescence rise kinetics of three scleractinian corals. *Funct. Plant Biol.* 32: 549–59

Hill, R., Larkum, AWD., Frankart, C., Kuhl, M., Ralph, PJ. (2004). Loss of functional photosystem II reaction centers in zooxanthellae of corals exposed to bleaching conditions: using fluorescence rise kinetics. *Photosynth. Res*, 82: 59–7.

Huner, NPA., Oquist, G., Sarhan, F. (1998). Energy balance and acclimation to light and cold. *Trends Plant Sci*, 3: 224–30

Jeeji Bai, N. (1985). Competitive exclusion or morphological transformation? A case study with *Spirulina fusiformis. Arch Hydrobiol Suppl* 71, *Algal Stud*, 191: 38–39

Joset, F., Jeanjean, R., Hagemann, M. (1996). Dynamics of the response of cyanobacteria to salt stress: deciphering the molecular events. *Physiol. Plant*, 96: 738-744

Kalaji, HM., Govindjee, Bosa, K., Koscielniak, J., Golaszewska, KZ. (2011). Effects of salt stress on photosystem II efficiency and CO_2 assimilation of two Syrian barley landraces. Environmental and Experimental Botany 73: 64-72

Kana, T., Gilbert, PM. (1987). Effect on irradiance up to 2000 μE $m^{-2}s^{-2}$ on marine *Synechococcus* WH 7803-I. Growth: pigmentation and cell composition. *Deep-Sea Res,* 34: 479–516

Kana, TM., Feiwel, NL., Flynn, LC. (1992). Nitrogen starvation in marine *Synechococcus* strains: clonal differences in phycobiliprotein breakdown and energy coupling. *Marine Ecology Progress Series,* 88: 75-82.

Kang, R., Wang, J., Shi, D., Cong, W., Cai, Z., Ouyang, F. (2004). Interaction between organic and inorganic carbon sources during mixotrophic cultivation of *Synechococcus sp. Biotechnol Lett,* 26: 1429-1432

Kirst, GO. (1990). Salinity tolerance of eukaryotic marine algae, *Annu. Rev. Plant Physiol. Plant Mol. Biol.* 41: 21–53.

Kramer, DM., Avenson, TJ., Kanazawa, A., Cruz, JA., Ivanov, B., Edwards, GE. (2004). Therelationship between photosynthetic electron transport and its regulation. In Papageorgiou, GC. & Govindjee (eds.) Chlorophyll a Fluorescence: A Signature of Photosynthesis. Springer, Dordrecht, the Netherlands, pp. 251–78.

Krauss, N., Hinrichs, W., Witt, I., Fromme, P., Protzkow, W. (1993). 3-Dimensional structure of the system-I of photosynthesis at 6 angstrom resolution. *Nature,* 361: 326-331

Krüger, GHJ., Tsimilli-Michael, M., Strasser, RJ. (1997). Light stress provokes plastic and elastic modifications in structure and function of photo system II in camellia leaves. *Physiologia Plantarum,* 101: 265-277

Kruskopf, M. & Flynn, KJ. (2006). Chlorophyll content and fluorescence responses cannot be used to gauge reliably phytoplankton biomass, nutrient status or growth rate. *New Phytol,* 169: 525–36

Laws, EA. & Bannister, (1980). Nutrient- and light-limited growth of *Thalassiosira fluviatilis* in continuous culture with implications for phytoplankton growth in the ocean. *Limnol. Oceanog,* 25 (3): 457-473

Lazar, D. (2006). The polyphasic chlorophyll a fluorescence rise measured under high intensity of exciting light. *Funct. Plant Biol,* 33: 9-30

Lewin, RA. (1980.) Uncoiled variants of *Spirulina platensis* (Cyanophyceae: Oscillatoriaceae). *Arch Hydrobiol* Suppl 60, *Algal Stud,* 26: 48–52

Li, WKW., Smith, JC., Platt, T. (1984). Temperature response of photosynthetic capacity and carboxylase activity in Arctic marine phytoplankton. *Marine Ecology Progress Series,* 17: 237-243.

Liu, X., Duan, S., Li, A., Xu, N., Cai, Z., Hu, Z. (2009). Effect of organic carbon sources on growth, photosynthesis, and respiration of *Phaeodactylum tricornutum. J Appl Phycol,* 21: 239-246

Lodi, A., Binaghi, L., De Faveri, D., Carvalho, JCM., Converti, A. (2005). Fed-batchmixotrophic cultivation of *Arthrospira (Spirulina) platensis* (Cyanophycea) with carbon source pulse feeding. *Ann Microbiol,* 55: 181-185

Lu, C. & Zhang, J. (1999). Effects of salt stress on PSII function and photoinhibition in cyanobacterium *Spirulina platensis, J. Plant Physiol,* 155: 740–745.

Lu, C. & Zhang, J. (2000). Role of light in the response of PSII photochemistry in the cyanobacterial *Spirulina platensis* to salt stress, *J. Exp. Bot,* 51: 911–917.

Lu, CM., Torzillo, G., Vonshak, A. (1999). Kinetic response of photosystem II photochemistry in cyanobacterium *Spirulina platensis* to high salinity is characterized by two distinct phases. *Aust J Plant Physiol*, 26: 283–292

Lu, CM. & Vonshak, A. (1999). Characterization of PSII photochemistry in salt adapted cells of cyanobacterium *Spirulina platensis*. *New Phytol*, 141: 231–239

Lu, CM. & Vonshak, A. (2002). Effect of salinity on photo system II function in cyanobacterial *Spirulina platensis* cells. *Physiol Plant*, 114: 405–413

Lu, CM., Zhang, J., Vonshak, A. (1998). Inhibition of quantum yield of PSII electron transport *in Spirulina platensis* by osmotic stress may be explained mainly by an increase in the proportion of the Q_B non reducing PSII reaction centres. *Aust J Plant Physiol*, 25: 689–694

Marquez, FJ., Nishio, N., Nagai, S., Sasaki, K. (1995). Enhancement of biomass and pigment production during growth of *Spirulina platensis* mixotrophic culture. *J Chem Tech Biotech*, 62: 159-164

Marquez, FJ., Sasaki, K., Kakizono, T., Nishio, N., Nagai, S. (1993). Growth characteristics of *Spirulina platensis* in mixotrophic and heterotrophic conditions. *J Ferment Bioeng*, 76: 408-410

Martinez, F. & Orus, MI. (1991). Interactions between glucose and inorganic carbon metabolism in *Chlorella vulgaris* strain UAM 101. *Plant Physiol*, 95: 1150-1155

Martinez, F., Camacho, F., Jiménez, JM., Espinola, JB. (1997). Influence of light intensity on the kinetic and yield parameters of *Chlorella pyrenoidosa* mixotrophic growth. *Process Biochem*, 32: 93-98

Mathur, S., Jajoo, A., Mehta, P., Bharti, S. (2010). Analysis of elevated temperature-induced inhibition of Photosystem II by using chlorophyll a fluorescence induction kinetics in wheat leaves (*Triticum astivum*). *Plant Biol*, doi:1111/j-1438-8677-2009.00319.X

Maxwell, K. & Johnson, GN. (2000). Chlorophyll fluorescence- a practical guide. *J Exp. Bot.* 51: 659–68

Mehta, P., Jajoo, A., Mathur, S., Bharti, S. (2010). Chlorophyll a fluorescence study effects of high salt stress on photosystem II in wheat leaves. *Plant Physiol. Biochem*, 48: 16-20

Melis, A. (1999). Photosystem II damage and repair cycle in chloroplasts : what modulates the rate of photodamage in vivo ? *Trends Plant Sci*, 4: 130-135

Nash, D., Miyao, M., Murata, N. (1985). Heat inactivation of oxygen evolution in photosystem II particles and its acceleration by chloride depletion and exogenous manganese, *Biochim. Biophys. Acta*, 807 127–133

Neubauer, C. & Schreiber, U. (1987). The polyphasic rise of chlorophyll fluorescence upon onset of strong continuous illumination: 1. Saturation characteristics and partial control by the Photo system II acceptor side. *Z Naturforsch*, C 42: 1246-1254

Orus, MI., Marco, E., Martinez, F. (1991). Suitability of *Chlorella vulgaris* UAM 101 for heterotrophic biomass production. *Bioresour Technol*, 38: 179-184

Osborne, BA. & Geider, RJ. (1986). Effect of nitrate limitation on photosynthesis of the diatom *Phaeodactylum tricornutum Bohlin* (Bacillarophyceae). *Plant, Cell and Environment* 9: 617-625

Papageorgiou, G. & Govindjee (1968). Light induced changes in the fluorescence yield of chlorophyll a in vivo *Anacystis nidulans*. *Biophys J*, 8: 1299–1315

Papageorgiou, GC. (1996). The photosynthesis of cyanobacteria (blue bacteria) from the perspective of signal analysis of chlorophyll a fluorescence. *J Sci Ind Res* 55: 596–617

Pelroy, RA., Rippka, R., Stanier, RY. (1972). Metabolism of glucose by unicellular blue-green algae. *Archiv für Mikrobiologie,* 87: 303-322

Pereira, WE., de Siqueira, DI., Martinez, CA., Puiatti, M. (2000). Gas exchange andchlorophyll fluorescence in four citrus rootstocks under aluminium stress. *J. Plant Physiol,* 157: 513-520

Pogoryelov, D., Sudhir, PR., Kovàcs, L., Combos, Z., Brown, I., Garab, G. (2003). Sodium dependency of the photosynthetic electron transport in the alkaliphilic cyanobacterium *Arthrospira platensis. J Bioenerg Biomembranes,* 35: 427–437

Pospisil, P. & Tyystjarvi, E. (1999). Molecular mechanism of high-temperature-induced inhibition of acceptor side of photosystem II, *Photosynth. Res,* 62: 55–66

Powles, SB. (1984). Photoinhibition of photosynthesis induced by visible light. *Annu. Rev. Plant Physiol,* 35: 15–44

Prasil, O., Adir, N., Ohad, I. (1992). Dynamics of photosystem II: mechanism of photoinhibition and recovery process. *Topics in photosynthesis,* 11: 295-348

Qiu, B., Zhang, A., Zhou, W., Wei, J., Dong, H., Liu, Z. (2004). Effects of potassium on the photosynthetic recovery of the terrestrial cyanobacterium, *Nostoc flagelliforme* (Cyanophyceae) during rehydration. *J. Phycol,* 40: 323–32

Raven , JA. & Geider, RJ. (1988). Temperature and algal growth. *New Phycologist,* 110: 441-461.

Rhee, GY. & Gotham, IJ. (1981). The effect of environmental factors on phytoplankton growth : light and the interactions of light with nitrate limitation. *Limnol. Oceanogr,* 26 (4): 649-659

Richardson, K., Beardall, J., Raven, JA. (1983). Adaptation of unicellular algae to irradiance: an analysis of strategies. *New Phytologist,* 93: 157-191

Richmond, A. (1988). *Spirulina,* In: Borowitzka, MA., Borowitzka, LJ. (eds.) Microalga biotechnology. Cambridge, Cambrige University Press, pp 85-121

Schreiber, U. & Bilger, W. (1987). Rapid assessment of stress effects on plant leaves by chlorophyll fluorescence measurements. In: Tenhunen GD. (eds.), Plant response to stress. Springer Verlag, New York, pp. 27-53

Schreiber, U. & Neubauer, C. (1987). The polyphasic rise of chlorophyll fluorescence upon onset of strong continuous illumination. 2. Partial control by the photosystem II donor side and possible ways of interpretation. *Z. Naturforsch,* C 42: 1255-1264

Schreiber, U., Bilger, W., Neubauer, C. (1994). Chlorophyll fluorescence as a non invasive indicator for rapid assessment of in vivo photosynthesis. In: Schulze, ED., Caldwell, MM. (eds.) Ecophysiology of photosynthesis. Springer, Berlin Heidelberg New York, pp 49-70

Schreiber, U., Schliwa, U., Bilger, W. (1986). Continuous recording of photochemical and non photochemical chlorophyll fluorescence quenching with a new type of modulation fluorometer. *Photosynth Res,* 10: 51–62

Schubert, H., Fluda, S., Hagemann, M. (1993). Effects of adaptation to different salt concentrations on photosynthesis and pigmentation of the cyanobacterium *Synechocystis sp.* PCC 6803. *J. Plant Physiol,* 142 (3): 291-295

Schubert, H. & Hagemann, M. (1990). Salt effects on 77K fluorescence and photosynthesis in the cyanobacterium *Synechocystis sp.* PCC 6803, *FEMS Microbiol. Lett,* 71: 169–172.

Srivastava, A., Strasser, RJ., Govindjee (1999). Greening of peas: parallel measurements of 77K emission spectra, OJIP chlorophyll a fluorescence transient, period four oscillation of the initial fluorescence level, delayed light emission, and P700. *Photosynthetica,* 37: 365-392

Strasser, BJ. (1997). Donor side capacity of PSII probed by chlorophyll a fluorecence transients. *Photosynth, Res*, 52: 147-155

Strasser, BJ., Eggenberg, P., Strasser, BJ. (1996). How to work without stress but with fluorescence. *Bulletin de la Societé Royale des Sciences de Liege*, 65: 330-349

Strasser, BJ. & Strasser, RJ. (1995). Measuring fast fluorescence transients to address environmental questions: the JIP test. In: Mathis, P. (eds.) Photosynthesis: from light to biosphere. Kluwer, Dordrecht, pp 977-980

Strasser, RJ. (1987). Energy pipeline model of the photosynthetic apparatus. In: Biggins, J. (eds.), Progress in Photosynthesis Research. Matinus Nijhoff Publishers, Dordrecht, The Netherlands, pp 717–720

Strasser, RJ., Srivastava, A., Govindjee (1995). Polyphasic chlorophyll a fluorescence transient in plants and cyanobacteria. *Photochem Photobiol*, 61: 32-42

Strasser, RJ., Srivastava, A., Tsimilli-Michael, M. (2000). The fluorecence transient as a tool to characterize and screen photosynthetic samples. In Yunus, M., Pathre, U., Mohanty, P. (eds.), Probing Photosynthesis Mechanisms, regulation and Adaptation. Taylor and Francis, London, UK, pp 445-483.

Strasser, RJ., Tsimilli-Michael, M., Srivastava, A. (2004). Analysis of the chlorophyll a fluorescence transient. In Papageorgiou, GC. & Govindjee (eds.) Chlorophyll a Fluorescence: A Signature of Photosynthesis. Springer, Dordrecht, the Netherlands, pp. 321–62.

Tomek, P., Lazar, D., Ilik, P., Naus, J. (2001). On the intermediate steps between the O and P steps in chlorophyll a fluorescence rise measured at different intensities of exciting light. *Aust J Plant Physiol*, 28: 1151–1160

Torzillo, G., Bernardini, P., Masojidek, J. (1998). On-line monitoring of chlorophyll fluorescence to assess the extent of photoinhibition of photosynthesis induced by high oxygen concentration and low temperature and its effect on the productivity of outdoor cultures of *Spirulina plantensis* (cyanobacteria). *J. Phycol*, 34: 504–10

Toth, SZ. & Strasser, RJ. (2005). The specific rate of QA reduction and photosystem II heterogeneity. In: Van Est, A. & Bruce, D. (eds.) Photosynthesis: fundamental aspects to global perspectives. Allen Press Inc, Lawrence, pp 198–200

Toth, SZ., Schansker, G., Garab, G., Strasser, R.J. (2007). Photosynthetic electron transport activity in heat-treated barley leaves: the role of internal alternative electron donors to photosystem II, *Biochim. Biophys. Acta*, 1767: 295–305

Tsonev, T., Velikova, V., Georgieva, K., Hyde, PF., Jones, HG. (2003). Low temperature enhances photosynthetic downregulation in French bean (*Phaseolus vulgaris* L.) plants. *Ann. Bot*, 91:343–52.

Tsonev, TD. & Hikosaka, K. (2003). Contribution of photosynthetic electron transport, heat dissipation, and recovery of photoinactivated photosystem II to photoprotection at different temperatures in *Chenopodium album* leaves. *Plant Cell Physiol*, 44: 828-835

Verity, PG. (1982). Effects of temperature, irradiance and daylength on the marine diatom *Leptocylindrus danicus* Cleve. 4. Growth. *Journal of Experimental Marine Biology and Ecology* 60: 209-222

Vonshak, A. & Novoplansky, N. (2008). Acclimation to low temperature of two *Arthrospiraplatensis* (Cyanobacteria) strains involves down regulation of PSII and improved resistance to photoinhibition. *J. Phycol*, 44: 1071–1079

Vonshak, A., Chanawonge, L., Bunnag, B., Tanticharoen, M. (1995). Physiological characterization of *Spirulina platensis* isolates: Response to light and salinity. *Plant Physiol,* 14: 161–166

Vonshak, A., Cheung, SM., Chen, F. (2000). Mixotrophic growth modifies the response of *Spirulina (Arthrospira) platensis* (cyanobacteria) cells to light. *J Phycol,* 36: 675-679

Vonshak, A., Guy, R., Guy, M. (1988). The response of the filamentous cyanobacterium *Spirulina platensis* to salt stress. *Arch Microbiol,* 150: 417–420

Vonshak, A., Kancharaksa, N., Bunnag, B., Tanticharoen, M. (1996). Role of light and photosynthesis on the acclimation process of the cyanobacterium *Spirulina platensis* to salinity stress. *J Appl Phycol,* 8: 19–124

Wen, XG., Gong, HM., Lu, C. (2005). Heat stress induces an inhibition of excitation energy transfer from phycobilisomes to photosystem II but not to photosystem I in a cyanobacterium *Spirulina platensis, Plant Physiol. Biochem,* 43: 389–395.

Wilson, KE., Ivanov, AG., Oquist, G., Grodzinski, B., Sarhan, F., Huner, NPA. (2006). Energy balance, organellar redox status, and acclimation to environmental stress. *Can. J. Bot,* 84: 1355-1370

Yamamoto, Y. (2001). Quality control of photosystem II. *Plant Cell Physiol,* 42: 121-128

Yamane, Y., Kashino, Y., Koike H., Satoh, K. (1998). Effects of high temperature on the photosynthetic systems in spinach oxygen-evolving activities, fluorescence characteristics and the denaturation process, *Photosynth. Res,* 57: 51–59

Yang, C., Hua, Q., Shimizu, K. (2000). Energetics and carbon metabolism during growth of microalgal cells under photoautotrophic, mixotrophic and cyclic light-autotrophic/dark-heterotrophic conditions. *Biochem Eng J,* 6: 87-102

Young, AJ. (1993). Factors that affect the carotenoid composition of higher plants and algae. In Carotenoids in Photosynthesis, Young, AJ. & Britton, G. (eds.) London: Chapman and Hall, pp. 161-205.

Yu, H., Jia, S., Dai, Y. (2009). Growth characteristics of the cyanobacterium Nostoc flagelliforme in photoautotrophic, mixotrophic and heterotrophic cultivation. *J Appl Phycol,* 21: 127-133

Zeng, MT. & Vonshak, A. (1998). Adaptation of *Spirulina platensis* to salinity stress. *Comp Biochem Physiol,* Part A 120: 113-118

Zhang, D., Pan, X., Mu, G., Wang, J. (2010). Toxic effects of aantimony on photosystem II of *Synechocystis sp.* As probed by in vivo chlorophyll fluorescence. *J. Appl Phycol,* 22: 479-488

Zhang, XW., Zhang, YM., Chen, F. (1999). Application of mathematical models to the determination optimal glucose concentration and light intensity for mixotrophic culture of *Spirulina platensis. Process Biochem,* 34: 477-481

Zhao, B., Wang, J., Gong, H., Wen, X., Ren, H., Lu, C. (2008). Effects of heat stress on PSII photochemistry in a cyanobacterium *Spirulina platensis. Plant Science,* 175: 556-564

Zhu, XG., Govindjee, Baker, NR., DeSturler, E., Ort, DR., Long, SP. (2005). Chlorophyll a fluorescence induction kinetics in leaves predicted from a model describing each discrete step of excitation energy and electron transfer associated with photosystem II. *Planta,* 223: 114-133.

3

Crassulacean Acid Metabolism in Epiphytic Orchids: Current Knowledge, Future Perspectives

Gilberto Barbante Kerbauy, Cassia Ayumi Takahashi,
Alejandra Matiz Lopez, Aline Tiemi Matsumura, Leonardo Hamachi,
Lucas Macedo Félix, Paula Natália Pereira,
Luciano Freschi and Helenice Mercier
São Paulo University,
Brazil

1. Introduction

1.1 Crassulacean Acid Metabolism (CAM)

Crassulacean Acid Metabolism (CAM) is one of three photosynthetic assimilation pathways of atmospheric CO_2, together with the photosynthetic pathways C_3 and C_4 (Silvera et al., 2010a). The CAM is characterized by the temporal separation between CO_2 fixation and its assimilation into organic compounds. In CAM plants, CO_2 is fixed during the dark period through the action of the enzyme phosphoenolpyruvate carboxylase (PEPC), which uses CO_2 for carboxylation of phosphoenolpyruvate (PEP), giving rise to oxaloacetate (OAA). The OAA formed is converted into malate by the action of malate dehydrogenase (MDH). Then, this organic acid is transported to the vacuole along with H^+ ions, causing the typical nocturnal acidification of CAM plants. During the light period, the decarboxylation of malate and refixation of the CO_2 by the enzyme ribulose bisphosphate carboxylase oxygenase (RUBISCO - C_3 cycle) takes place in the cytosol, causing a decrease of acidity in the tissues (Herrera, 2009; Luttge, 2004; Silvera et al., 2010b) (Figure 1).

The CAM pathway can be separated into four phases (Luttge, 2004; Osmond, 1978; Silvera et al., 2010b). Phase I is characterized by the opening of stomata during the night, the uptake and subsequent fixation of atmospheric CO_2 by PEPC in the cytosol and the formation of organic acids, such as malate. Phase II consists of fixing CO_2 by the enzyme RUBISCO and PEPC concurrently, a phase characterized essentially by the decrease in the activity of PEPC and the start of the activity of RUBISCO. Phase III consists of the reduction of stomatal opening, efflux of organic acids from the vacuole and subsequent decarboxylation of these acids. Finally, phase IV comprises the depletion in the stock of organic acids associated with an increase of stomata conductance.

Due to nighttime fixation of atmospheric CO_2, CAM plants exhibit greater water use efficiency (EUA) when compared with the photosynthetic pathways C_3 and C_4 (Herrera, 2009), given that CAM plants use 50 to 100 g of water per gram of CO_2 fixed, while C_3 plants use 400 to 500 g (Drennam & Nobel, 2000). The ratio of transpiration is 3- to 10-fold lower in CAM plants than in C_3 (Kluge & Ting, 1978). Besides the EUA, another advantage of CAM comprises mechanisms to minimize the damage caused by reactive oxygen species (ROS) (Sunagawa et al., 2010).

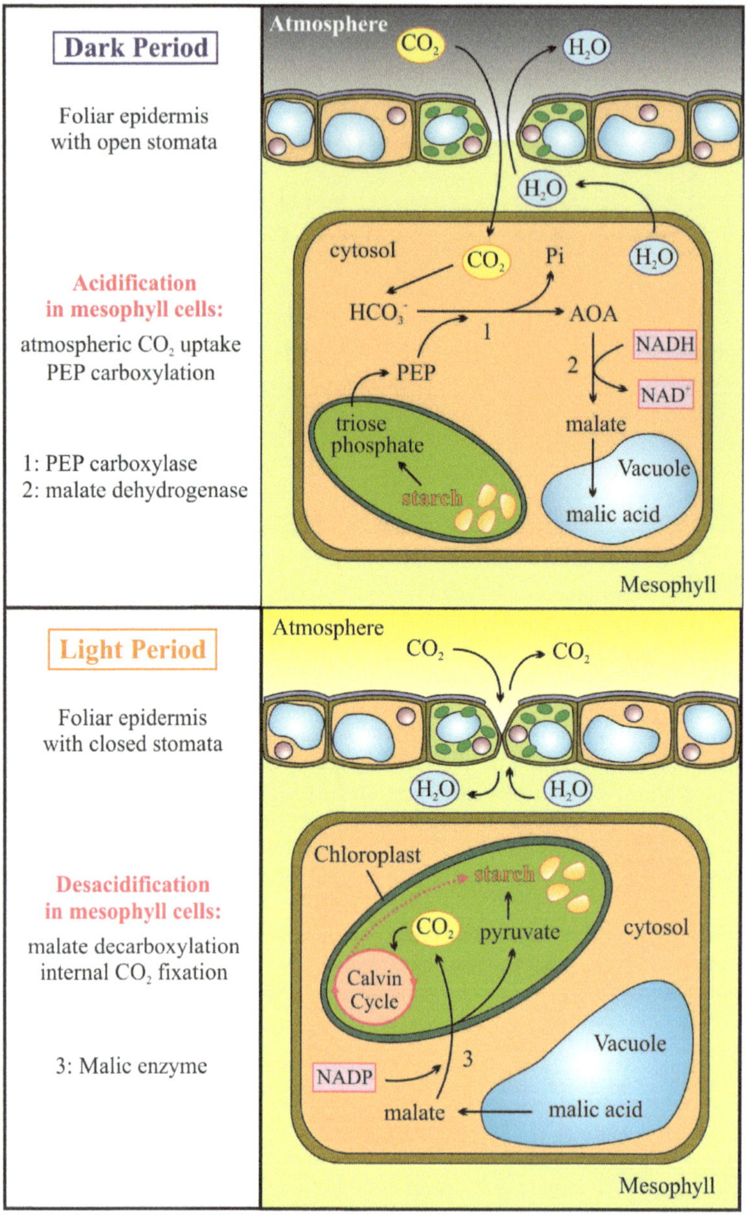

Fig. 1. Temporal separation of CO_2 uptake and fixation typical of the crassulacean acid metabolism (CAM). Atmospheric CO_2 is taken up and fixed during the night (upper panel). Nighttime-accumulated organic acids are decarboxylated during the day, and the internally released CO_2 is refixed (lower panel). The nighttime stomatal opening allows a reduction of water loss by transpiration, which is an important adaptive trait of CAM pathway. PEP: phosphoenolpyruvate, AOA: oxaloacetate.

Regarding its occurrence, CAM species are distributed in semiarid, tropical and subtropical environments, including epiphytic species in the humid tropics (Silvera et al., 2010a, 2010b). The more representative plant families in number of CAM species are Aizoaceae, Asclepiadaceae, Asteraceae, Bromeliaceae, Cactaceae, Crassulaceae, Euphorbiaceae, Portulacaceae and Orchidaceae. In the Orchidaceae, at least half of all species are classified as CAM (Winter & Smith, 1996).

The CAM pathway can operate in different modes: obligate CAM, facultative CAM, CAM-cycling and CAM-idling (Figure 2). Obligate CAM species exhibit high accumulation of organic acids at night and nocturnal CO_2 fixation even under optimal environmental conditions (Kluge & Ting, 1978). On the other hand, facultative CAM species, also known as C_3-CAM, are plants capable of performing C_3 photosynthesis under favorable growth conditions and switching to CAM mode when challenged by environmental constraints, such as water limitation or excessive light incidence. In both constitutive and facultative CAM species, CAM is more strongly expressed in mature tissues (Gehrig et al., 2005), while young plants or young tissues of a mature plant tend to perform exclusively C_3 photosynthesis (Avadhani et al., 1971). CAM-cycling consists of diurnal CO_2 fixation and accumulation of organic acids; however, the stomata remain closed at night, and the CO_2 necessary for the nighttime formation of the organic acids is exclusively obtained from the respiratory activity of the plant tissues. Finally, plants in CAM-idling mode exhibit only a small accumulation of organic acids, and the stomata remain closed both day and night; this nocturnal accumulation of organic acids is also provided by the recycling of respiratory CO_2. Overall, CAM-cycling is considered a weak CAM, while CAM-idling is currently believed to consist of a strong CAM mode (Luttge, 2004). Facultative C_3-CAM species, such as *Guzmania monostachia* (Bromeliaceae) and *Talinum triangulare* (Portulacaceae), can be induced to CAM by various environmental factors, such as drought stress (Freschi et al., 2010; Herrera et al., 1991), photoperiod (Brulfert et al., 1988), salinity (Winter & Von Willert, 1972), nitrogen deficiency (Ota, 1988), phosphorus deficiency (Paul & Cockburn, 1990) and high photosynthetic photons flux (Maxwell, 2002). The CAM induction by environmental factors is usually fast and reversible, a conspicuous example of plasticity. Exemplifying the plasticity of C_3-CAM species, plants of facultative C_3-CAM bromeliad *G. monostachia* have recently been shown to be clearly induced to a CAM-idling mode of photosynthesis when maintained under drought stress during a seven-day period. Interestingly, these same plants returned to a typical C_3 condition after a subsequent period of seven days of rehydration (Freschi et al., 2010).

Despite the above-mentioned favorable adaptive traits, the CAM pathway exhibits some disadvantages related to biomass productivity and photorespiration. The biomass productivity in grams of dry weight/m/day in CAM plants is between 1.5 and 1.8, while in C_3 plants the productivity is 50-200 (Black, 1973). The energy costs for the assimilation of each CO_2 molecule is significantly higher in CAM plants than in C_3 species. The estimated stoichiometry of ATP: NADPH: CO_2 of C_3 plants is 3: 2: 1, while for CAM plants it is 4.8: 3.2: 1 (Winter et al., 1978). Regarding photorespiration, it was verified that the vigorous assimilation of the CO_2 provided by decarboxylation of organic acids after stomata closure can result in increases of up to 40% in the O_2 concentration inside the leaves during phase III (Spalding et al., 1979). Hence, the photorespiration in CAM plants occurs not only in phase IV of CAM, when the stock of malate was exhausted and the stomata are open for CO_2 absorption to occur, but also in phase III, where the CO_2 concentration is counterbalanced by the O_2 concentration (Luttge, 2011).

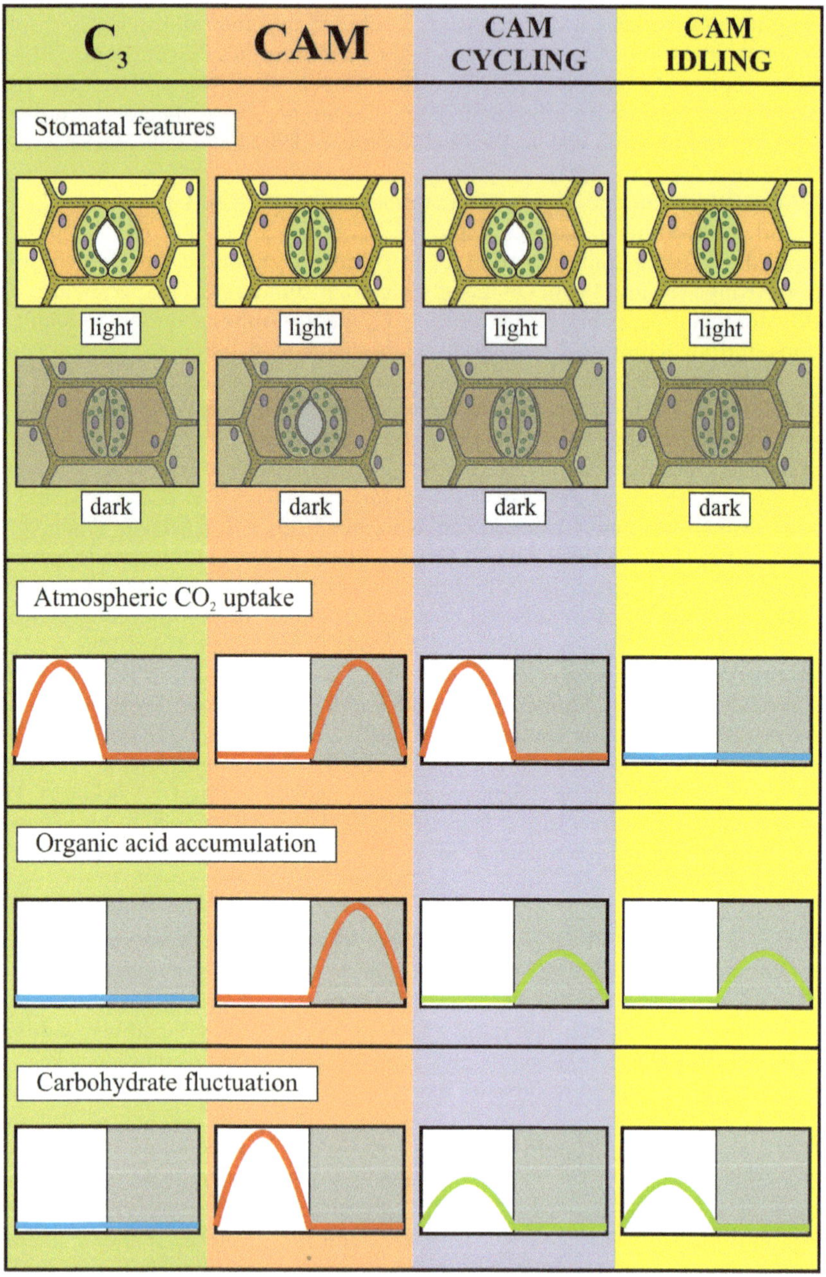

Fig. 2. Different modes of CAM in relation to C_3, constitutive CAM, CAM-cycling and CAM-idling. Differences related to the opening of stomata, atmospheric CO_2 uptake, flux of organic acids and flux of carbohydrates. The gray box represents the dark period, while the white box represents the light period.

Despite the decrease of biomass productivity and photorespiration problems, the fixation of CO_2 via CAM evidently guarantees a maximum carbon gain combined with a minimum loss of water supplies in the plant tissues. In xeric habitats, some plants are only able to survive in low precipitation periods mainly due to the presence of CAM photosynthesis. In this sense, the epiphytic habitat can be considered an excellent example of xeric habitat, where some epiphytic plants, including orchids, often face situations of limited water supply.

2. Crassulacean acid metabolism in the epiphytic habitat

Epiphytes are plants that spend much or all of their lives attached to other plants (Benzing, 1990). They are responsible for much of the biotic diversity that makes humid tropical forests the most complex of all the world's ecosystems (Gentry & Dodson, 1987). Canopy-based species constitute virtually one third of the total vascular flora in some pluvial neotropical forests (Benzing 1990). The majority of epiphytes, including Araceae, Gesneriaceae, Cactaceae and pteridophytes, can be found in the parts of the canopy which have greater quantities of moisture and nutrients than are available to their more xeric counterparts growing in arid zone canopies (Gentry & Dodson, 1987). Their occurrence in drier sites is less common and usually involves fewer taxa but not necessarily lower abundance. Foremost species of these taxa are the few stress-tolerant bromeliads (Bromeliaceae) and orchids (Orchidaceae), which are commonly classified as the extreme epiphytes (Benzing, 1973). The Orchidaceae has been more successful than any other lineage in colonizing tree crowns. About two out of three epiphytes are orchids; at least 70% of the family is canopy-adapted (Benzing, 1990).

Aerial substrata can impose severe environmental stresses on the survival of epiphytic vegetation (Benzing, 1987). The adaptations that facilitate extreme epiphytism are numerous and involve many morphological or anatomical characteristics of the plant body as well all stages of the life cycle. Their success of survival in drier sites of the canopy is associated mainly with refined water-balance mechanisms, singular nutritional modes and efficient reproductive systems (Benzing, 1990). Some of the adaptive traits of epiphytes can be summarized as high water use efficiency, crassulacean acid metabolism, low surface to volume ratio, aerial seed dispersion, mycoheterotrophy (Orchidaceae), holdfast structures, slow growth rate and efficient mineral use (Benzing, 1973; Dodson, 2003; Yoder et al., 2010).

The availability of nutrients and water in the forest canopy is sporadic and totally dependent on the rainy period. Of the natural water supplies available to epiphytes, precipitation intercepted by the canopy and reaching the ground by flowing down the trunks or by falling through the foliage usually contains the most abundant quantities of nutrient solutes (Benzing, 1973). Nevertheless, a liter of stem-flow or fall-through rarely contains more than a few milligrams of any specific mineral nutrient (McColl, 1970; Sollins & Drewry, 1970). Therefore, most extreme epiphytes need to take and accumulate water and all required nutrients during brief intervals (few hours or minutes) when the shoot or root systems are in contact with rainfall and leachates. The extreme epiphytes often have some characteristics that enable plants to absorb and accumulate water and minerals faster and with greater efficiency. In species of the Orchidaceae, the roots of many orchids are covered with a special structure called velamen, which acts as a sponge, allowing the root to immobilize a temporary but highly accessible reservoir of moisture and minerals (Benzing &

Ott, 1981). Moreover, the velamen slows root transpiration, provides mechanical protection and assists in the attachment of orchids to the bark of host trees (Ackerman, 1983; Benzing et al., 1982). Epiphytic orchids have considerable succulence in their shoot organs (leaves and pseudobulbs), which are important reservoir structures for storing water and nutrients (Benzing, 1990).

The vegetative growth is strongly influenced by water supply, the shortage of which can be considered the most severe environmental stress in the epiphytic habitat. The extreme epiphytes need to adjust water-balance mechanisms in all plant tissues rapidly and constantly, via appropriate stomatal and photosynthetic responses, to avoid irreversible drought injuries. The drought endurance observed in the majority of epiphytes is provided by a strong CAM photosynthetic behavior, which promotes a very favorable water economy (Benzing & Ott, 1981).

A high number of epiphytic species perform CAM photosynthesis (Benzing, 1989; Luttge, 2004). Among orchids species it is likely that at least half could perform this type of photosynthetic pathway owing to the high number of epiphytic species, of which the Epidendroideae subfamily is the richest in epiphyte CAM species. This subfamily is believed to have radiated at the beginning of the Tertiary (Ramirez et al., 2007), a period marked by climatic changes such as soil aridification and declining CO_2 concentrations (Pearson & Palmer, 2000), favoring the survival of the epiphyte species displaying CAM photosynthesis. Nevertheless, CAM does not seem to be related to phylogenetic relationships among the taxa. Silvera et al. (2009) suggest that among the Orchidaceae, CAM arose independently at least 10 times from a C_3 ancestor. Indeed, the enzymatic machinery to perform CAM is present in all plants, including those performing C_3 photosynthesis exclusively, and the differences between both photosynthetic pathways are mainly associated with the regulation of such machinery (Silvera, 2010a).

The existence of a significant correlation between photosynthetic pathways and epiphytism has already been observed by Silvera et al. (2009). The phylogenetic analyses showed that C_3 photosynthesis is the ancestral state and that CAM has evolved multiple independent origins, indicating the great evolutionary flexibility of CAM in Orchidaceae (Silvera et al., 2009). Moreover, when using maximum likelihood to trace epiphytism as a character state across the orchid phylogeny, the authors have also verified that the terrestrial habit is the ancestral state within tropical orchids and, similar to CAM, the epiphytic habit is derived. Throughout evolutionary time, the CAM divergence observed by $\delta^{13}C$ analysis is consistently accompanied by divergence in epiphytism, demonstrating a functional relationship between these traits. Correlated divergence between the photosynthetic pathway and epiphytism is likely an important factor contributing to the burst of speciation that occurred in diverse epiphytic orchid clades (Silvera et al., 2009).

Besides epiphytism, it is currently accepted that CAM is also strongly linked with a certain degree of succulence, as commonly observed in members of the Crassulaceae (Kluge et al., 1993) and Orchidaceae (Silvera et al., 2005), in which the leaf thickness provides a higher storage capacity for organic acid accumulation (Ting, 1985). Although succulence and CAM usually coincide, some epiphytes are an exception to this rule (Benzing, 1990) since there are clear demonstrations that some epiphytic orchid species with thin leaves performed CAM, while some species with thick leaves performed C_3 metabolism (Silvera et al., 2005). However, Zotz et al. (1997) have clearly demonstrated the correlation between chlorenchyma thickness and CAM in the leaves of different orchids. In their study, they have shown that species with strong CAM had thicker chlorenchyma, highlighting that the importance of the relationship between leaf thickness and CAM does not reside in the thickness of the leaf, but rather in the thickness of the chlorenchyma, as in most cases the succulence is due to the presence of a thick hydrenchyma, which does not contribute to CAM (Winter et al., 1983) in terms of organic acid storage capacity.

Most of the knowledge of the distribution of CAM orchids comes from studies in New Guinea, Australia (Earnshaw et al., 1987; Winter et al., 1983), Panama (Silvera et al., 2005, 2009, 2010; Zotz & Ziegler, 1997; Zotz, 2004), Mexico (Mooney et al., 1989) and Costa Rica (Silvera, 2009). These surveys reveal an increase in the number of epiphyte CAM orchid species following the forest precipitation frequency, rising from 25% in wet tropical rainforests and moist tropical forests to 100% in dry forests (Figure 3A). Among the Orchidaceae an evolutionary correlation was found between the photosynthetic pathways and epiphytism, following a direct relation between the epiphytic orchid habit and the presence of CAM, which is more evident in regions ranging from 0 to 500 m when compared to higher regions (Silvera et al., 2009). The same authors attribute this relation to the canopy height of the forests at these altitudes, favoring epiphytism and, consequently, the CAM pathway. Indeed, even in the same location, the presence of CAM epiphytes increases with canopy height, ranging from 7% in the forest understory to 50% in exposed canopy sites (Zotz & Ziegler, 1997). In regions above 2000 m, the tree height is lower and, by consequence, so is the canopy. Therefore, the epiphyte habitat in these regions is reduced (Figure 3B).

It is worth mentioning that the main technique to determine the photosynthetic pathway in the surveys of orchids was based on the quantification of the stable isotope ^{13}C (expressed in $\delta^{13}C$‰) on plants leaves. The typical range of $\delta^{13}C$‰ for C_3 plants is between -33‰ to -22.1‰, while for strong CAM plants it ranges from -22‰ to -12‰ (Elheringer & Osmond, 1989). Several studies done in different taxa have shown a bimodal pattern of the frequency distribution of $\delta^{13}C$‰ values among orchid species, following their photosynthetic pathway (Holtum et al., 2005; Motomura et al., 2008b; Pierce et al., 2002; Silvera et al., 2005, 2009, 2010a, 2010b); this behavior is characterized by a cluster formed by a high number of C_3 species (indicated by values near -28‰) and a smaller cluster formed by CAM species (with values near -16‰); between the clusters are intermediate values that could represent weak CAM or facultative CAM species (usually in response to a stress, such as drought) (Silvera et al., 2009). In fact, it was found that although some orchid species presented $\delta^{13}C$‰ values typical of a C_3 plant, they were capable of nighttime carbon fixation, reflected by an increase in tissue titratable acidity (Silvera et al., 2005). Therefore, the characterization of the photosynthetic pathways by isotopic measurement alone tends to underestimate the number of species capable of performing some degree of CAM (Pierce et al., 2002; Silvera et al., 2005, 2009, 2010a, 2010b; Winter et al., 2008).

Nowadays, it is thought that approximately 40% of the tropical orchid species could exhibit some form of CAM (strong, weak or facultative) (Silvera et al., 2010a), but there is insufficient knowledge about the photosynthetic pathways in other non-leaf organs, such as pseudobulbs or roots. Also, studies about the capacity of a photosynthetic shift between C_3 and CAM triggered by environmental conditions are very rare in the literature.

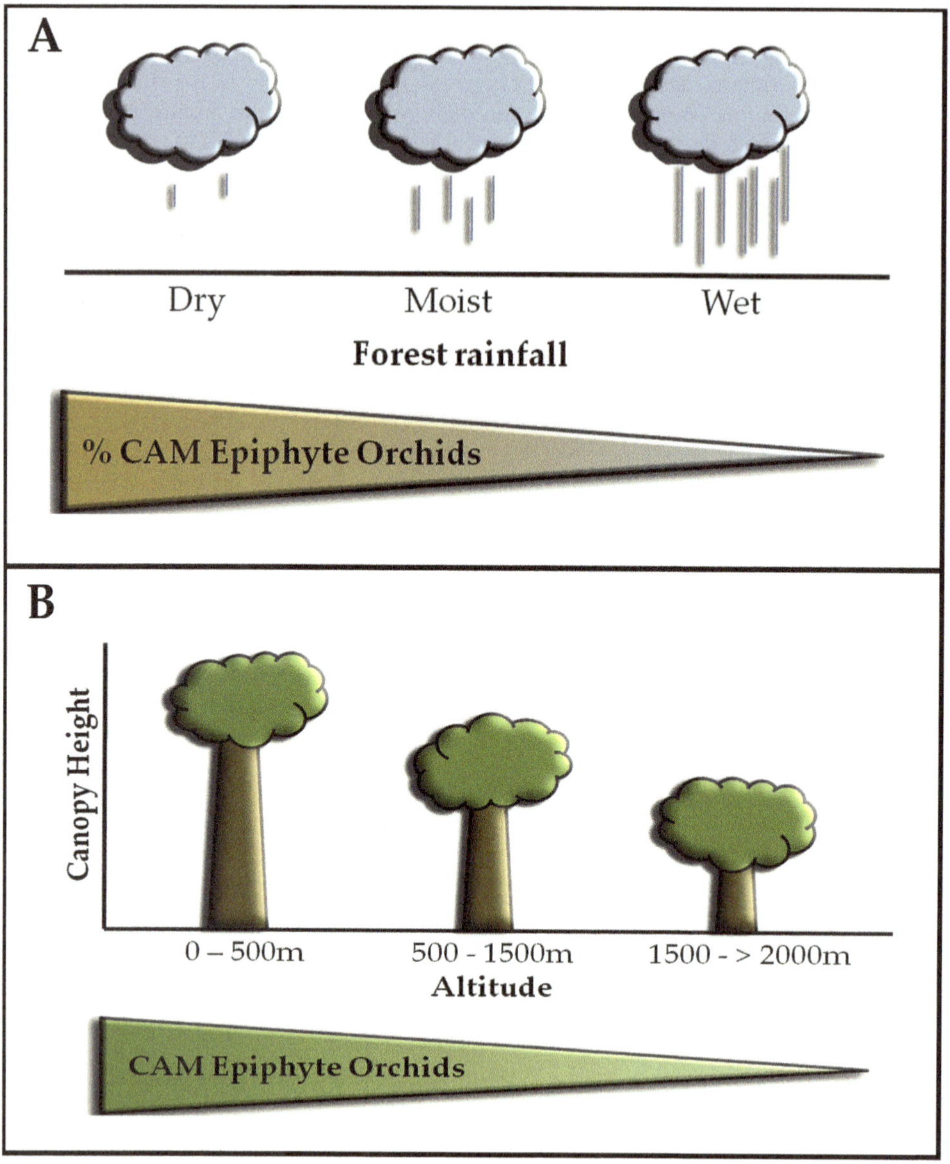

Fig. 3. The occurrence of CAM epiphytic plants according to (A) forest rainfall and (B) variation of altitude and canopy height.

3. Crassulacean acid metabolism in orchid leaves

CO_2 fixation in orchid leaves has been widely studied in the past few years. Most of these studies suggest that thin-leaved orchids present C_3 photosynthesis pathway (Calvin-Benson cycle), while thick-leaved succulent orchids photosynthesize basically via CAM pathway

(Avadhani et al., 1982). Goh et al. (1977) and Winter et al. (1983) observed that the majority of orchid species and hybrids that possess succulent leaves presented nocturnal stomatal opening and acidity rhythms typical of CAM. Orchids like *Cattleya, Laelia, Brassavola* and *Sophronitis* have thick leaves, and all of those are reported as CAM plants (Avadhani & Arditti, 1981, Avadhani et al., 1982). On the other hand, C_3 orchids like *Arundina graminifolia* and *Oncidium* Goldiana are characterized by thin leaves (Hew & Yong, 2004). When under water stress, daytime CO_2 uptake is greatly reduced by thin or thick orchid leaves. In some CAM orchids submitted to an extended water stress, the nighttime CO_2 uptake can be severely curtailed due to the closure of stomata even at night. These plants under severe drought for a long period of time may convert to a CAM-idling mode, in which organic acids fluctuate without exogenous CO_2 uptake (Fu & Hew, 1982).

In addition to succulence and water storage capacity, there are other leaf features observed by Cushman (2001) and Cushman & Borland (2002) in *Epidedrum secundum,* such as thick cuticles, large and vacuolated cells with capacity to store organic acids and reduced stomata size and frequency. The stomata of some Orchidaceae species occur only on the lower epidermis of the leaves, and occasionally they can be located within hyperstomatic chambers, as seen in *Arachnis* cv. Maggie Oei, *Aranda* cv. Deborah, *Arundina graminifolia, Bromheadia finlaysoniana, Cattleya bowringiana* X *C. forbesii* and *Spathoglottis plicata* (Orchidaceae) (Goh et al., 1977). Orchid leaf stomata may vary in shape, size and distribution (Goh et al., 1977). In addition, as reported previously by Withner et al. (1974), there are many species in which upper leaf stomata are not seen, contributing to water maintenance. It is commonly accepted that the stomatal rhythm in CAM plants is due to internal CO_2 concentration (Kluge, 1982). In CAM orchids, normally thick-leaved, stomatal opening/closing seems to be regulated by dark fixation of CO_2 in the mesophyll cells, which reduces CO_2 in the internal atmosphere, promoting stomatal opening at night (Goh et al., 1977).

The capacity of the leaves of CAM plants to accumulate acids at night has been shown to increase as they unfold, till the leaves are completely expanded reaching their maturity; however, it decreases in the senescence stage (Ranson & Thomas, 1960). Goh et al. (1984) observed that in *Arachnis* Maggie Oei there was a reduction in acidity fluctuation in young and old leaves by half when compared to mature green leaves and an even smaller fluctuation in yellow senescent leaves, ranging from 20% to 30%. It is worth noting that in some CAM plants the PEPC activity is much higher in mature leaves than in young ones (Amagasa, 1982; Lerman et al., 1974; Nishida, 1978). In full-grown leaves of *Arachnis* Maggie Oei, the stomatal rhythms as well as the CO_2 exchange pattern are consistent with CAM as shown by acidity fluctuations. The stomata open in late afternoon and acidity increases as CO_2 is absorbed. During daytime, de-acidification occurs and stomata are closed (Goh et al., 1977).

In a study of eighteen *Cymbidium* species, Motomura et al. (2008b) verified that there are different CAM intensities, ranging from weak to strong. They found that three strong CAM *Cymbidium* species have thicker leaves than other species: >1.0 mm and <0.7 mm, respectively. In contrast, weak CAM species displayed thin leaves, like C_3 species. Their studies corroborate preview data that emphasize the existence of a tendency for less negative $\delta^{13}C$ values as the leaf thickness increases, while in thinner leaves a wide range of $\delta^{13}C$ values can be found (Earnshaw et al., 1987; Silvera et al., 2005; Zotz & Ziegler 1997; Winter et al., 1983).

Moreira et al. (2009), in a comparative study regarding photosynthetic and structural features in leaves of *Dichaea cogniauxiana* and *Epidendrum secundum*, noted that diurnal titratable acidity fluctuations indicate the presence of CAM in the second species. Moreover, several morphological features in *E. secundum* leaves are typical of plants with this photosynthetic pathway, including the occurrence of a thick mesophyll with few stomata and a wide cuticle. On the other hand, in *D. cogniauxiana*, a C_3 species, there was a neglegible diurnal acidity variation and the leaves were less succulent and covered by a thin cuticle.

In order to investigate whether the major compounds which are produced during the dark $^{14}CO_2$ fixation in orchids are similar to those in other plants, Knauft & Arditti (1969) undertook a study of *Cattleya* leaves. A high dark CO_2 uptake and a prominent diurnal acidity rhythm were reported for *Cattleya* orchids, which have succulent leaves (Borriss, 1967; Nuerenbergk, 1963), while in thin-leaved orchids such as *Cymbidium*, they did not observe organic acid production from CO_2 dark fixation (Hatch, et al., 1967). They also noticed that three organic acids arose during CO_2 dark fixation: malate, citrate and a third unidentified one whose relative amounts varied based on temperature change (from 18°C to 29°C). The major acids at 18°C were malate and citrate, while at 29°C the predominant acid was malate. The unidentified acid increased from 13% to 30% with rising temperature (Knauft & Arditti, 1969).

4. Crassulacean acid metabolism in orchid pseudobulbs

Pseudobulbs are commonly described as enlarged internodes provided with a thick cuticle, epidermis devoid of stomata with gross cell walls resulting in an impervious organ, fundamental parenchyma, vascular bundles and water storage cells that determine a succulent aspect. Although devoid of stomata, the pseudobulbs have chloroplasts in part of their cells indicating, therefore, a certain capacity of photosynthetic activity (Milaneze-Gutierre & da Silva, 2004; Oliveira &, Sajo 2001). The pseudobulb is a characteristic of epiphytic orchids and secondary terrestrial orchids (Hew et al., 1998; Kozhevnikiva & Vingranova, 1999, Stancato et al., 2001; Zimmerman, 1990). This organ has been frequently studied as a storage organ capable of storing minerals (Zimmerman, 1990), water (Stancato et al., 2001; Zimmerman, 1990) and carbohydrates (Hew et al., 1998; Stancato et al., 2001; Zimmerman, 1990). Although there is considerable information about its storage function, there are few studies focusing on pseudobulb photosynthesis, especially regarding the occurrence of the CAM pathway.

Aschan & Pfanz (2003) revised stem photosynthesis, and distinguished it in four categories: two characterized by internal CO_2 refixation (wood photosynthesis and corticular bark photosynthesis) and two characterized by net photosynthesis (stem photosynthesis and CAM in stem succulent plants).

The hermetic feature of most orchid pseudobulbs do not allow them to fix carbon from the air. The only possibility is recycling the respiratory CO_2 generated by the voluminous underlying parenchyma (Ng & Hew, 2000). Hew et al. (1998) detected the presence of chlorophylls and PEPC and Ribulose-1,5-biphosphate carboxilase/oxigenase (Rubisco) activities in pseudobulbs and other non-foliar organs of *Oncidium* Goldiana and suggested that the photosynthesis in non-foliar organs was regenerative. Nevertheless, in the same study they observed that the $\delta^{13}C$ values of pseudobulbs were not characteristic of a CAM plant.

Winter et al. (1983) studied stems and leaves of 82 species from the Orchidaceae and discovered that 53 species exhibited $\delta^{13}C$ values compatible with dark CO_2 fixation. Species , possessing pseudobulbs and having $\delta^{13}C$ values of a CAM plant in leaves presented had similar $\delta^{13}C$ values in pseudobulbs as well. Curiously, the majority of pseudobulbs examined exhibit an enrichment in ^{13}C by up to 3% compared to corresponding leaves, suggesting that there is less discrimination against this isotope in these organs. This could not been explained at the time since, according to the authors, virtually nothing was known about their gas exchange. Another interesting fact was that the leafless species *Bulbophyllum minutissimum*, which have pseudobulbs with thick chlorenchyma and stomata in a depression, had $\delta^{13}C$ values characteristic of CAM. In this case, the pseudobulbs may play a considerable role in performing the CAM pathway and, therefore, contribute to the carbon acquisition in this species.

One of the most interesting studies about the role of pseudobulbs for net CO_2 uptake in light and dark period was done by Ando & Ogawa (1987). They measured CO_2 exchange of shoots of *Laelia anceps* and noted that pseudobulbs alone were apparently not able to assimilate carbon due to their absence of stomata, unlike the leaves. However, when the entire shoot was illuminated during the light period, the leaf assimilated carbon during both the light and dark periods. Notwithstanding, when the pseudobulbs were submitted to dark conditions during the light period, the leaf only assimilated carbon at night revealing a possible influence of the pseudobulbs in the light and dark CO_2 uptake by the leaf. The authors hypothesized that the organic acid, fixed during the night by the leaf, is transported to the pseudobulbs and decarboxylated in this organ during the day. This transport is up-regulated by the exposure of the pseudobulbs to light, causing a decline in the amount of organic acid in the leaf and, therefore, stimulating the net carbon uptake during the day.

The above-mentioned studies clearly raise questions about the occurrence and functionality of the CAM in pseudobulbs of orchids. In fact, current data on this issue remain remarkably scarce and more studies are needed to clarify the mechanisms of gas exchange, discrimination of carbon isotopes, types of organic acids produced or stored in pseudobulbs and translocation of those compounds between leaves and pseudobulbs.

5. Crassulacean acid metabolism in orchid roots

5.1 Structure and function of velamen on aerial roots

A typical orchid root has a cortex with chloroplasts in cortical cells and an exodermis enveloped by the velamen, a multilayered epidermis originating at the root tip. The numbers of cell layers depends on the species and growing conditions (Benzing et al., 1982; Dycus & Knudson, 1957). There is some disagreement about the importance of velamen in aerial roots. Dycus & Knudson (1957), in a study involving the species *Laelia purpurata*, *Cattleya labiata* and *Vanda* hybrid, observed no active phosphorus absorption on aerial roots even after immersion in a solution of this nutrient, except after mechanical injury of the velamen or after roots entered a solid substrate. They proposed that aerial roots which were not able to absorb the phosphorus had lost the ability to retain water and nutrients. However, more recent studies of several species proved that salts in solution can be taken and translocated by aerial roots, an interpretation supported by Benzing et al. (1982). Based on morphological and anatomical observations with the terrestrial orchid *Sobralia macrantha*, the authors proposed that velamen act as a sponge, due to the presence of an internal dead space, which can be a temporary and accessible source of water and minerals. The presence of many mitochondria and external passage cells with well-developed membranes support the role of roots in nutrient accumulation. Between rainstorms, velamen and exodermis can act as a boundary layer that slows roots transpiration (Benzing, 1990; Benzing et al., 1983). The velamen contains living cells on the root tip, which remain green, and dead cells, which can mask the green color of cortex when dry, reflecting light. After absorbing water, the velamen becomes transparent or translucent (Aschan & Pfanz, 2003; Dycus & Knudson, 1957; Goh et al., 1983). According to Benzing et al. (1983), velamentous roots are a basic but important adaptation of Orchidaceae to conquer an epiphytic environment.

5.2 Photosynthesis on aerial roots and presence of CAM

The capacity of green orchid aerial roots to photosynthesize is well known; however, in leafy orchids, roots have a secondary role in photosynthesis, unlike in shootless plants (Aschan & Pfanz, 2003; Benzing et al., 1983; Dycus & Knudson, 1957; Goh et al., 1983). Goh et al. (1983) observed that diurnal CO_2 exchange pattern and acidity fluctuations in the aerial roots of the CAM hybrids *Arachnis* Maggie Oei and *Aranda* Deborah were typical of CAM, but based on $^{14}CO_2$ incorporation studies, the C_3 pathway seemed to operate during the day and might represent the major pathway of CO_2 fixation. The authors concluded that these roots were not completely autotrophic and were still dependent on the leaves.

In a more recent study, Motomura et al. (2008a) identified a low degree of CAM expression in aerial roots of two *Phalaenopsis* species, but there were variations into CAM expression between aerial roots of the same plant. Different regions of aerial roots also seem to have different intensities of CAM expression, as observed by Martin et al. (2010) in 12 epiphytic orchids. They separated aerial roots in green root tips and white portions. Although 11 taxa performed CAM in leaves, three taxa performed CAM in the white portion of the roots (one of them from *Phalaenopsis* genus) and only one taxon performed CAM in the green portion. There was no correlation between the presence and intensity of CAM in leaves and roots from the same plant. They concluded that roots of the CAM species analyzed were too shaded to perform CAM, as proven by high chlorophyll amounts and low chlorophyll a/b ratios.

Apparently, the presence of CAM in roots is not related with its presence in leaves. C_3 root photosynthesis can be found in CAM orchids, but CAM in roots was not found in C_3 orchids. Moreira et al. (2009) worked with leaves and roots of the C_3 *Dichaea cogniauxiana* and the CAM *Epidendrum secundum* and observed that roots of both species had ratios of $^{13}CO_2/^{12}CO_2$ and titratable acid fluctuation typical of a C_3 plant. Gehrig et al. (1998) analyzed the photosynthetic behavior and isolated the different forms of PEPC in leaves, stem and aerial roots of the obligate CAM *Vanilla planifolia*. They demonstrated that aerial roots have low malate accumulation at night and expressed only Ppc V2, the "housekeeping" isoform of PEPC, unlike leaves and stem, which had substantial malate accumulation and expressed the Ppc V1, the isoform related to CO_2 fixation in CAM. They found that loss of CO_2 in the light was low for aerial roots, so it was proposed that during the day, respiratory CO_2 was partially refixed by C_3 photosynthesis.

5.3 Photosynthesis in aerial roots of shootless orchids

In leafy orchids, no expression of CAM or expression of weak CAM is expected in the roots, even in plants classified as CAM since the roots do not possess a mechanism to control water loss like stomata in leaves, so carbon uptake at night would not be advantageous in terms of water conservation. Moreover, photosynthesis in roots plays only a secondary function since leaves are the main source of photoassimilates. However, this is not the case of shootless orchids, where roots assume most or full responsibility for carbon gain for the whole plant, besides mineral uptake and assimilation. Compared to roots of leafy plants, they usually have a thinner velamen layer and larger, chloroplast-containing parenchyma cells in the cortex (Benzing et al., 1983; Benzing & Ott, 1981; Winter et al., 1985).

The presence of CAM in roots appeared more common in this type of orchid compared to leafy ones, as demonstrated by Benzing & Ott (1981). In a study using 12 orchid taxa, although all leafy plants performed CAM, small quantities of CO_2 were assimilated by their roots during day and night. The only species which exhibited CAM rhythms in roots were shootless taxa and one leafy taxon with a small number of well-developed leaves. Similarly, Benzing et al. (1983) observed that, compared to leaves, roots of *Epidendrum radicans* and *Phalaenopsis amabilis* had much weaker diurnal fluctuations in titratable acid, which was not the case for the shootless orchid *Polyradicion lindenii*, which showed more intense dark acidification than roots of the two leafy plants.

In a study of the leafless orchid *Campylocentrum tyrridion*, Winter et al. (1985) observed a nocturnal increase in titratable acidity content and a daily cycle of CO_2 uptake that resembled classical CAM. Cockburn et al. (1985) found similar results in the shootless orchid *Chiloschista usneoides*. The major acid produced at night was malic acid, which indicates CO_2 uptake via PEPC. They observed that despite the absence of stomata, leakage of CO_2 was low, indicating a balance between CO_2 carboxylation and fixation, so the CO_2 concentration in roots and in the atmosphere is the same. Considering that classical CAM plants maintain the balance of CO_2 by controlling stomata aperture, the authors proposed the term astomatal CAM to this variant.

In all these cases, the functional significance of the CAM for aerial roots is not related to a water conserving mechanism but most probably to the role of CAM as a CO_2 concentrating mechanism. CO_2 uptake is limited by velamen and exodermis; therefore, by performing C_3 photosynthesis the low partial pressure of CO_2 can lead to damage of the photosynthetic apparatus in high light intensities. Another advantage of CAM could be the recycling of CO_2 produced by respiration of the plant tissues and of the endocellular fungi since this kind of association is very common in Orchidaceae (Benzing et al., 1983, Winter et al., 1985).

6. Crassulacean acid metabolism in orchid floral organs

One of the first studies on photosynthesis in orchid flowers was in 1968 by Dueker & Arditti, in which they studied the occurrence and contribution of photosynthesis in two different varieties of green *Cymbidium* flowers, the green color of which is due to chlorophyll present in the floral parts.

Even though green leaves are considered the main sources of photosynthate production, Studies support the notion that reproductive organs, such as greenish flowers, can be photosynthetically active (Dueker et al., 1968; Weiss et al., 1988). Among plants, some have green flowers, while others have photosynthetic parts associated with the inflorescence. These structures may contribute positively to total carbon gain and the energy costs of reproduction (Antlfinger & Wendel, 1997; Bazzaz et al., 1979; Marcelis & Hofman-Eijer, 1995; Reekie & Bazzaz, 1987).

The most important advantage of photosynthesizing flowers is their position relative to light. Sepals or bracts are normally shaped in the outermost layer protecting the reproductive parts throughout the floral bud stage. This provision takes full advantage of radiant light energy and may result in a higher carbon fixation, which means that photoassimilates needed for inflorescence growth can be supplemented at least in part by their own photosynthesis (Antlfinger & Wendel, 1997; Khoo et al., 1997). Despite this, it is noteworthy that these organs are one of the main plant sinks, requiring the input of photoassimilates from the leaves and/or pseudobulbs (Yong & Hew, 1995)

Previous work demonstrated that based on chlorophyll content green parts of reproductive structures have up to three times higher photosynthetic assimilatory capacity than green leaves of the same plant species (Heilmeier & Whale 1987; Luthra et al., 1983; Smillie, 1992; Werk & Ehleringer, 1983; Williams et al., 1985). Surveys of floral orchid organs demonstrated some capacity for $^{14}CO_2$ fixation in the light. These studies also indicated that the fixation rates vary among flower parts, plant varieties and stage of floral development (Antlfinger & Wendel, 1997; Dueker et al., 1968; Khoo et al., 1997). Since the amount of CO_2 fixation is highest in the sepals, lower in the petals (Dueker et al., 1968; Khoo et al., 1997) and lowest in the ovary, despite their similarity in appearance between the petals and sepals, these results suggest certain metabolic differences between them (Dueker et al., 1968).

In order to regulate floral gas exchange, stomata like those in leaves can be found in the epidermal layer of petals, but their density is normally lower than that found in leaves (He et al., 1998; Hew et al., 1980). Hew et al. (1980) observed that stomata in orchid flowers are not functional since they could not respond to light intensity, CO_2 and abscisic acid (ABA). Nevertheless, Goh in 1983 indicated the capacity for photosynthesis in orchid flowers, measuring gas exchange, acidity fluctuation and compounds synthesized when providing $^{14}CO_2$ in three succulent-leaf orchids, *Arachnis, Aranda, Dendrobium,* and one thin-leaf *Oncidium* Goldiana. They verified that orchids with succulent leaves performed night CO_2 fixation and accumulation of acidity, and radioactive malate was formed when $^{14}CO_2$ was provided characterizing CAM, while *Oncidium* Goldiana exhibited C_3 pattern. Thus, it appears that flowers of at least some succulent orchids are capable of both C_3 and CAM photosynthesis. Dueker & Arditti (1968) also observed a capacity for CO_2 fixation at night in *Cymbidium* flowers.

Given that most existing information on the CO_2 exchange and carbon gain of vascular epiphytes is on leaves, data on the carbon gain in flowers are limited (Zotz & Hietz, 2001). Several studies have characterized the responses of flowers from CAM orchids to environmental conditions such as different light irradiances (Khoo et al., 1997; He et al., 1998; He & Teo, 2007) and temperatures (He et al., 1998) through the contents of pigments, acidity and certain photosynthetic parameters, such as PSII and Fv / Fm. Nevertheless, none of these surveys have, in fact, characterized CAM metabolism in flowers. Since this is an understudied field, more attention is clearly needed to determine whether different organs of a given plant can perform CAM.

7. Conclusion

Crassulacean acid metabolism is characterized by nocturnal CO_2 fixation and organic acid accumulation in the vacuole by the decarboxylation of PEP catalyzed by the enzyme PEPC. The decarboxylation of the organic acids occurs during the day, and the CO_2 released is reattached by the enzyme RUBISCO after the closing of the stomata. By closing stomata during most of the daytime, CAM plants exhibit greater efficiency in water use. Thus, CAM plants can inhabit semiarid and arid environments, including deserts, exposed rock outcrops and epiphytic habitats. Despite these advantages, the CAM pathway is also accompanied, however, by lower biomass productivity, higher energy expenditure and the occurence of photorespiration during phases III and IV.

The main technique used to distinguish C_3, C_4 and CAM plants is based on the quantification of stable isotopes of ^{13}C in plant leaves. However, this technique has a certain degree of variability among individuals of the same species or between plant material derived from the same plant, as in shaded leaves and leaves exposed to light of the same plant. In fact, for several orchid species, values of $\delta^{13}C$‰ typical of C_3 plants were found, despite the fact that they clearly exhibited the ability to fix some of the carbon at night, as reflected in the increased nocturnal acidity in the tissues. This illustrates why the use of only isotopic measurements to determine the type of photosynthetic metabolism tends to underestimate the number of species capable of expressing CAM photosynthesis. Therefore, other parameters, such as day/night fluctuations in titratable acidity, activity of enzymes of CAM and diurnal patterns of gas exchange, are needed to determine the photosynthetic pathway.

Other problems that researchers need to face when determining the photosynthetic pathway of orchid plants is the fact that many experiments are planned using only some parts of the plant, normally leaves, and other organs such as roots, pseudobulbs and flowers are often not included in the study analysis. Moreover, few studies give attention to the fact that some orchids have plasticity in switching between C_3 and CAM photosynthesis in response of changes in environmental conditions.

Despite the existence of technical difficulties in studying the CAM features, there is no doubt that CAM plants have sparked the curiosity of many researchers around the world for decades. Nowadays, it is well known that the majority of CAM plants have been found living in the epiphytic habitat along with many species of Orchidaceae. The proportion of CAM epiphytic orchid flora is completely associated with the degree of water availability in the ecosystems. The occurrence of CAM orchid species increases from wet tropical rainforest and moist tropical forests to dry forests, and steadily declines with increasing altitude, which is entirely related with the increase of mean annual precipitation, and, within a single site, the percentage of CAM epiphytic orchids increases with canopy height. The scarcity of water is arguably the most important and severe abiotic stress in the epiphytic habitat. The epiphytic orchids need to adjust water-balance mechanisms in all plant tissues rapidly and constantly, via appropriate stomatal and photosynthetic responses, to avoid irreversible drought injuries and maintain water storage. The drought endurance observed in the majority of epiphytic orchids is provided by a strong CAM photosynthetic behavior, which promotes a very favorable water economy.

The CAM features also appear to be linked with succulence in orchid plant tissues. The existence of a thicker chlorenchyma tissue, which was detected in some strong CAM orchids, can be important to increase the capacity of organic acid storage. Interestingly, the thickness of the chlorenchyma is not entirely associated with leaf thickness. In the majority of cases the succulence is due to the presence of a thick hydrenchyma, which does not contribute to CAM in terms of night-produced organic acid storage capacity. Therefore, there are epiphytic orchid species with thin leaves performing CAM, while some species with thick leaves display typical C_3 photosynthesis.

Most of studies on water relation and CAM photosynthesis in vascular epiphytes have only looked at the leaf tissues. According to Zotz (1999), focusing on leaves alone may lead to a skewed picture of plant functioning. Orchids have organs other than leaves that also exhibit considerable succulence (e.g. pseudobulbs) which are important reservoir structures for storing water and nutrients. The translocation of water between organs may be an important mechanism to maintain near-constant water content in leaves even during times of drought, while allowing substantial fluctuations in the water content of stems or roots, indicating that non leaf organs may have great importance and influence in the water relations and, therefore, photosynthetic activity in many orchids.

Despite the fact that the leaf is considered the main site where photosynthesis occurs, other organs are able to perform the photosynthetic functions, as long they have all essential biochemical requisites, such as chlorophyll and functional chloroplasts. In most cases, the leaf assumes a secondary function, like respiratory CO_2 refixation, and assimilates are not exported to sink organs. This carbon recycling enables the plant to reduce water loss, which is extremely important to plant survival in the case of epiphytes. The organs of epiphytic orchids also have many other adaptations which enable water economy. Leaves can be succulent and have stomata only in lower epidermis or, in some cases, inside hyperstomatic chambers, which provides a more stable environment around stomata. In roots, the existence of velamen is associated with epiphytic habitats and provides water absorption and conservation, mechanical protection and attachment to substrate. Pseudobulbs can appear in some orchid species as a variation of stem that provides drought tolerance.

The CAM expression can occur in organs like stems, roots and/or flowers. In stems or pseudobulbs, it seems to be associated with the presence of CAM in leaves. In flowers, it was proposed that both C_3 and CAM genetic information can be expressed in CAM orchids. Photosynthesis in this organ is less intense and can contribute to reproductive costs. In roots, however, no correlation was found and C_3 photosynthesis appears to be more common in roots, even in CAM species. However, when leaves are not present on the plant body, stems or pseudobulbs and/or roots might have an important and more relevant role in CAM expression since they are the unique source for photosynthetic assimilates. The lack of some plant organs, like leaves, enables more water economy and investment in reproductive structures.

It is also relevant to mention the importance of uncovering the interaction between different organs of the plant. According to Zotz & Hietz (2001), it is still premature to link the physiology of a single organ with the entire individual, or the physiology of one individual to the entire community. There are rare studies which seek integration between organs in the physiological studies, as exemplified by the work of Ando & Ogawa (1987), who analyzed the influence of light on both leaves and pseudobulbs. It is important to consider that orchid organs other than leaves (e.g. roots, pseudobulbs and flowers) might also have great importance to the survival of these plants in the dry epiphytic habitat since these organs might perform different physiological functions, including, perhaps, distinct modes of CAM photosynthesis. Certainly, a more integrated view is needed in CAM studies of vascular epiphytes to allow a better understanding of the functional importance of each organ to the whole plant.

8. References

Ackerman, J.D. (1983). On the evidence for a primitively epiphytic habit in orchids. *Systematic Botany*, Vol.8, No.4, (October-December 1983), pp. 474-477, ISSN 0363-6445

Amagasa, T. (1982). Change in properties of phosphoenolpyruvate carboxylase in *Kalanchoe daigremontiana* with leaf age. *Plant & Cell Physiology*, Vol.23, No.8, (September 1982), pp. 1471-1474, ISSN 0032-0781

Ando, T. & Ogawa, M. (1987). Photosynthesis of leaf blades in *Laelia anceps* Lindl. is influenced by irradiation of pseudobulb. *Photosynthetica*, Vol.21, No.4, pp. 588-590, ISSN 0300-3604

Antlfinger, A.E. & Wendel, L.F. (1997). Reproductive effort and floral photosynthesis in *Spiranthes cernua* (Orchidaceae). *American Journal of Botany*, Vol.84, No.6, (June 1997), pp. 769–780, ISSN 0002-9122

Aschan, G. & Pfanz, H. (2003). Non-foliar photosynthesis – a strategy of additional carbon acquisition. *Flora*, Vol.198, No.2, pp. 81-97, ISSN 0367-2530

Avadhani, P.N. & Arditti, J. (1981). Carbon fixation in orchids. In: *Proceedings of World Orchid Symposium*, pp. 79-85, Harbour Press, Sydney.

Avadhani, P.N.; Goh, C.J.; Rao, A.N. & Arditti, J. (1982). Carbon fixation in orchids. In: *Orchid Biology, Reviews and Perspectives II*, J. Arditti, (Ed.), Vol.123, No.3, (January 1973), pp. 173-193, Cornell University Press, Ithaca, New York

Avadhani, P.N.; Osmond, C.B. & Tan, K.K. (1971). Crassulacean acid metabolism and the C_4 pathway of photosynthesis in succulent plants, In: *Photosynthesis and photorespiration*, M.D. Hatch, C.B. Osmond & R.O. Slatyer, (Eds.), pp. 288–293, Wiley-Interscience, ISBN 0471359009, New York, USA.

Bazzaz, F.A. & Carlson, R.W. (1979). Photosynthetic contribution of flowers and seeds to reproductive effort of an annual colonizer. *New Phytologist*, Vol.82, No.1, (January 1979), pp. 223–232, ISSN 0028-646X

Benzing, D.H. & Ott, D.W. (1981). Vegetative reduction in epiphytic Bromeliaceae and Orchidaceae: its origin and significance. *Biotropica*, Vol.13, No.2, (June 1981), pp. 131-140, ISSN 0006-3606

Benzing, D.H. (1973). Mineral nutrition and related phenomena in Bromeliaceae and Orchidaceae. *The Quarterly Review of Biology*, Vol.48, No.2, (June, 1973), pp. 277-290, ISSN 0033-5770

Benzing, D.H. (1987). Vascular epiphytism: taxonomic participation and adaptive diversity. *Annals of the Missouri Botanical Garden*, Vol.74, No.2, pp. 183-204, ISSN 0026-6493

Benzing, D.H. (1989). The evolution of epiphytism. In *Vascular Plants as Epiphytes: evolution and Ecophysiology*, U. Luttge, (Ed.), Vol.76, pp. 15–41, Springer-Verlag, ISBN 3540507965, Berlin, Germany

Benzing, D.H. (1990). *Vascular epiphytes: general biology and related biota*, Cambridge University Press, ISBN 0-521-26630-0, Cambridge, England

Benzing, D.H.; Friedman, W.E.; Peterson, G. & Renfrow, A. (1983). Shootlessness, velamentous roots, and the pre-eminence of Orchidaceae in the epiphytic biotope. *American Journal of Botany*, Vol.70, No.1, (January 1973), pp. 121-133, ISSN 0002-9122

Benzing, D.H.; Ott, D.W. & Friedman, W.E. (1982). Roots of *Sobralia macrantha* (Orchidaceae): structure and function of the velamen-exodermis complex. *American Journal of Botany*, Vol.69, No.4, (April 1982), pp. 608-614, ISSN 0002-9122

Black, J. R. & Clanton, C. (1973). Photosynthetic carbon fixation in relation to net CO_2 uptake. *Annual Review of Plant Physiology*, Vol.24, No.24, (June 1973), pp. 253- 286, ISSN 0066-4294

Borriss, H. (1967). Kohlenstoff-Assimilation and diurnaled Saiurerhythmus epiphytischer Orchideen. *Orchidee*, Vol.7, pp. 396, ISSN 0758-2560

Brulfert, J.; Kluge, M.; Gluçlu, S. & Queiroz, O. (1988). Interaction of photoperiod and drought as CAM inducing factors in *Kalanchoe blossfeldiana* Poelln. Cv. Tom Thumb. *Journal of Plant Physiology* Vol.133, No.2, (September 1988), pp. 222-227, ISSN 0176-1617

Cockburn, W.; Goh, C.J. & Avadhani, P.N. (1985). Photosynthetic carbon assimilation in a shootless orchid, *Chiloschista usneoides* (Don) LDL. A variant on Crassulacean acid metabolism. *Plant Physiology*, Vol.77, No.1, (January 1985), pp. 83-86, ISSN 0032-0889

Cushman, J.C. & Borland, A.M. (2002). Induction of Crassulacean acid metabolism by water limitation. *Plant Cell & Environment*, Vol.25, No.2, (February 2002), pp. 295–310, ISSN 0140-7791

Cushman, J.C. (2001). Crassulacean acid metabolism. A plastic photosynthetic adaptation to arid environments. *Plant Physiology*, Vol.127, No.4, (December 2001), pp. 1439–1448, ISSN 0032-0889

Dodson, C.H. (2003). Why are there so many orchid species? *Lankesteriana*, Vol.3, No.2, (May 2003), pp. 99-103, ISSN 14093871

Drennan, P. M. & Nobel, P. S. (2000). Responses of CAM species to increasing atmospheric CO_2 concentrations. *Plant, Cell & Environment*, Vol.23, No.8, (August 2000), pp. 767-781, ISSN 0140-7791

Dueker, J. & Arditti, J. (1968). Photosynthetic $^{14}CO_2$ fixation by green Cymbidium (Orchidaceae) flowers. *Plant Physiology*, Vol.43, No.1, (January 1968), pp. 130-132, ISSN 0032-0889

Dycus, A.M. & Knudson, L. (1957). The role of the velamen of the aerial roots of orchids. *Botanical Gazette*, Vol.119, No.2, (December 1957), pp. 78-87, ISSN 0006-8071

Earnshaw, M.J.; Winterm K;, Ziegler, H.; Stichler, W.; Cruttwell, N.E.G.; Kerenga, K.; Cribb, P.J.; Wood, J.; Croft, J.R.; Carver, K.A. & Gunn, T.C. (1987). Altitudinal changes in the incidence of Crassulacean acid metabolism in vascular epiphytes and related life forms in Papua New Guinea. *Oecologia*, Vol.73, No.4, (October 1987), pp. 566–572, ISSN 0029-8549

Ehleringer, J.R. & Osmond, B.O. (1989). Stable isotopes, In: *Plant physiological ecology*, R.W. Pearcy, J. Ehleringer, H.A. Mooney & P.W. Rundel, (Eds), pp. 281–300, Chapman & Hall, ISBN 0713126906, London, England

Freschi, L.; Takahashi, C.A.; Cambuí, C.A.; Semprebom, M.T.R.; Cruz, A.B.; Mioto, P.T.; Versieux, L.M.; Calvente, A.; Latansio-Aidar, S.R.; Aidar, M.P.M. & Mercier, H. (2010). Specific leaf areas of the tank bromeliad *Guzmania monostachia* perform distinct functions in response to water shortage. *Journal of Plant Physiology*, Vol.167, No.7, (May 2010), pp. 526-533, ISSN 0176-1617

Fu, C.F. & Hew, C.S. (1982). Crassulacean acid metabolism in orchids under water stress. *Botanical Gazette*, Vol. 143, No. 3, (September 1982), pp. 294-297, ISSN 0006-8071

Gehrig, H.; Faist, K. & Kluge, M. (1998). Identification of phosphoenolpyruvate carboxylase isoforms in leaf, stem and roots of the obligate CAM plant *Vanilla planifolia* Salib. (Orchidaceae): a physiological and molecular approach. *Plant Molecular Biology*, Vol.38, No.6, (December 1998), pp. 1215-1223, ISSN 0167-4412

Gehrig, H.H.; Wood, J.A.; Cushman, M.A.; Virgo, A.; Cushman, J.C. & Winter, K. (2005). Large gene family of phosphoenolpyruvate carboxylase in the Crassulacean acid metabolism plant *Kalanchoe pinnata* (Crassulaceae) characterized by partial cDNA sequence analysis. *Functional Plant Biology*, Vol.32, No.5, (May 2005), pp. 467–472, ISSN 1445-4408

Gentry, A. & Dodson, C.H. (1987). Diversity and biogeography of Neotropical vascular epiphytes. *Annals of the Missouri Botanical Gardens*, Vol.74, No.2, pp. 205-233, ISSN 0026-6493

Goh, C. J.; Wara-Aswapati, O. & Avadhani, P.N. (1984). Crassulacean acid metabolism in young orchid leaves. *New Phytologist*, Vol.96, No.4, (April 1984), pp. 519-526, ISSN 0028-646X

Goh, C.J. (1983). Rhythms of acidity and CO_2 production in orchid flowers. *New Phytologist*, Vol.93, No.1, (January 1983), pp. 25–32, ISSN 0028-646X

Goh, C.J.; Arditti, J. & Avadhani, P.N. (1983). Carbon fixation in orchid aerial roots. *New Phytologist*, Vol.95, No.3, (November 1983), pp. 367-374, ISSN 0028-646X

Goh, C.J.; Avadhani, P.N.; Loh, C.S.; Hanegraaf, C. & Arditti, J. (1977). Diurnal stomatal and acidity rhythms in orchid leaves. *New Phytologist*, Vol.78, No.2, (March 1977), pp. 365-372, ISSN 0028-646X

Hatch, M.D.; Slack, C.R. & Johnson, H.S. (1967). Further studies of a new pathway of photosynthesis. Carbon dioxide fixation in sugar cane and its occurrence in other plant species. *Biochemical Journal*, Vol.102, No.2, (February 1967), pp. 417, ISSN 0264-6021

He, J. & Teo, L. C. D. (2007). Susceptibility of green leaves and green flowers petals of CAM orchid *Dendrobium* cv. Burana Jade to high irradiance under natural and tropical conditions. *Photosynthetica*. Vol.45, No.2, (June 2007), pp. 214-221, ISSN 0300-3604

He, J.; Khoo, G. H. & Hew, C. S. (1998). Susceptibility of CAM Dendrobium leaves and flowers to high light and high temperature under natural tropical conditions. *Environmental and Experimental Botany*. Vol.40, No.3, (December 1998), pp. 255-264, ISSN 0098-8472

Heilmeier, H. & Whale, D. M. (1987). Carbon dioxide assimilation in the flowerhead of Arctium. *Oecologia*, Vol.73, No.1, (August 1978), pp. 109–115, ISSN 0029-8549

Herrera, A. (2009). Crassulacean acid metabolism and fitness under water deficit stress: if not for carbon gain, what is facultative CAM good for? *Annals of Botany*, Vol.103, No.4, (February 2009), pp. 645- 653, ISSN 0305-7364

Herrera, A.; Delgado, J. & Paraguatey, I. (1991). Occurrence of Crasulacean acid metabolism in *Talinum triangulare* (Portulacaceae). *Journal of Experimental Botany*, Vol.42, No.4, (April 1991), pp. 493- 499, ISSN 0022-0957

Hew, C. S.; Koh, K. T. & Khoo, G. H. (1998). Patern of photoassimilate portioning in pseudobulbous and rhizomatous terrestrial orchids. *Enviromental and Experimental Botany*, Vol.40, No.2, (October 2008), pp. 93-104, ISSN 0098-8472

Hew, C.S. & Yong, W.H. (2004). *The physiology of tropical orchids in relation to the industry*, World Scientific Publishing Co. Pte. Ltd., ISBN 9789812388018, Singapore.

Hew, C.S.; Lee, G.L. & Wong, S.C. (1980). Occurrence of non-functional stomata in the flowers of tropical orchids. Annals of Botany, Vol.46, No.2, (August 1980), pp.195–201, ISSN 0305-7364

Holtum, J.A.M. & Winter, K. (2005). Carbon isotope composition of canopy leaves in a tropical forest in Panama throughout a seasonal cycle. *Trees-Structure and Function*, Vol.19, No.5, (May 2005), pp. 545–551, ISSN 0931-1890

Khoo, G.H.; He, J. & Hew, C.S. (1997). Photosynthetic utilization of radiant energy by CAM *Dendrobium* flowers. *Photosynthetica*, Vol.34, No.3, (February 1997), pp. 367–376, ISSN 0300-3604

Kluge, M. & Ting, I.P. (1978). *Crassulacean acid metabolism. Analysis of an ecological adaptation*, Springer-Verlag, ISBN 3540089799, Berlin, Heidelberg, New York

Kluge, M. (1982). Crassulacean acid metabolism (CAM), In: *Photosynthesis: Development, Carbon Metabolism and Plant Productivity*, Govindjee, (Ed.), Vol.2, pp. 231-262, Academic Press, ISBN 0122943023, New York, USA

Kluge, M.; Brulfert, J.; Lipp, J.; Ravelomanana, D. & Ziegler, H. (1993). Acomparative study of $\delta^{13}C$ analysis of Crassulacean acid metabolism (CAM) in *Kalanchoë* (Crassulaceae) species of Africa and Madagascar. *Botanica Acta*, Vol.106, No.4, pp. 320–324, ISSN 0932-8629

Knauft, R.L. & Arditti, J. (1969). Partial identification of dark $^{14}CO_2$ fixation products in leaves of *Cattleya* (Orchidaceae). New Phytologist, Vol.68, No.3, (July 1969), pp. 657–661, ISSN 0028-646X

Kozhevnikova, A. D. & Vinogranova, T. N. (1999). Pseudobulb structure in some boreal terrestrial orchids. *Systematic and Geography of Plants*, Vol. 68, (June, 1999), pp. 59-65, ISSN 1374-7886

Lerman, J. C.; Deleens, E.; Nato, A. & Moyse, A. (1974). Variation in the carbon isotope composition of a plant with Crassulacean acid metabolism. *Plant Physiology*, Vol.53, No.4, (April 1974), pp. 581-584, ISSN 0032-0889

Lüttge, U. (2004). Ecophysiology of Crassulacean acid metabolism (CAM). *Annals of Botany*, Vol.93, No.6, (June 2004), pp. 629-652, ISSN 0305-7364

Lüttge, U. (2011). Photorespiration in phase III of Crassulacean acid metabolism: evolutionary and ecophysiological implications. *Progress in Botany*, Vol.72, No.5, (October 2010), pp. 371-384, ISSN 978364213145

Luthra, Y.P.; Sheoran, I.S. & Singh, R. (1983). Photosynthetic rates and enzyme activities of leaves, developing seeds and pod-wall of pigeon pea (*Cajanus cajan* L.). *Photosynthetica*, Vol.17, No.2, (April 1983), pp. 210–215, ISSN 0300-3604

Marcelis, L.F.M. & Hofman-Eijer, L.R.B. (1995). The contribution of fruit photosynthesis to the carbon requirement of carbon requirement of cucumber fruits as affected by irradiance, temperature and ontogeny. *Physiologia Plantarum*, Vol.93, No.3, (April 1995), pp. 476–483, ISSN 0031-9317

Martin, C.E.; Mas, E.J.; Lu, C. & Ong, B.L. (2010). The photosynthetic pathway of the roots of twelve epiphytic orchids with CAM leaves. *Photosynthetica*, Vol.48, No.1, (March 2010), pp. 42-50, ISSN 0300-3604

Maxwell, K. (2002). Resistance is useful: diurnal patterns of photosynthesis in C_3 and Crassulacean acid metabolism epiphytic bromeliads. *Functional Plant Biology*, Vol.29, No.6, (June 2002), pp. 679-687, ISSN 1445-4408

Mccoll, J.G. (1970). Properties of some natural waters in a tropical wet forest of Costa Rica. *Bioscience*, Vol.20, No.20, (November 1970), pp. 1096-1100, ISSN 0006-3568

Milaneze-Gutierre, M. A. & da Silva, C. I. (2004). Caracterização morfo-anatômica dos órgãos vegetativos de Cattleya walkeriana Garndner (Orchidaceae). *Acta Scientarum Biological Sciences*, Vol.26, No.1, (June 2008), pp. 91-100, ISSN 1679-9283

Mooney, H.A.; Bullock, S.H. & Ehleringer, J.R. (1989). Carbon isotope ratios of plants of a tropical forest in Mexico. *Functional Ecology*, Vol.3, No.2, pp. 137-142, ISSN 0269-8463

Moreira, A.S.F.P.; Lemos, J.P.; Zotzb, G. & Isaias, R.M.S. (2009). Anatomy and photosynthetic parameters of roots and leaves of two shade-adapted orchids, *Dichaea cogniauxiana* Shltr. and *Epidendrum secundum* Jacq. *Flora*, Vol.204, No.8, (April 2009), pp. 604-611, ISSN 0367-2530

Motomura, H.; Ueno, O.; Kagawa, A. & Yukawa, T. (2008a). Carbon isotope ratios and the variation in the diurnal pattern of malate accumulation in aerial roots of CAM species of *Phalaenopsis* (Orchidaceae). *Photosynthetica*, Vol.46, No.4, (August 2008), pp. 531-536, ISSN 1573-9058

Motomura, H.; Yukawa, T.; Ueno, O. & Kagawa, A. (2008b). The occurrence of Crassulacean acid metabolism in *Cymbidium* (Orchidaceae) and its ecological and evolutionary implications. *Journal of Plant Research*, Vol.121, No.2, (February 2008), pp. 163–177, ISSN 0918-9440

Ng, C. K. Y. & Hew, C. S. (2000). Orchid pseudobulbs - "false" bulbs with a genuine importance in orchid growth and survival! *Scientia Horticulturae.* Vol. 83, (June 2000), pp.165-172, ISSN 0304-4238

Nishida, K. (1978). Effect of leaf age on light and dark $^{14}CO_2$ fixation in a CAM plant, *Bryophyllum calycinum*. *Plant & Cell Physiology*, Vol.19, No.19, (September 1978), pp. 935-941, ISSN 0032-0781

Nuerenbergk, E. L. (1963). On the CO_2 metabolism of orchids and its ecological aspect. *Proceeding of 4th Worth Orchid Conference*, pp. 158

Oliveira, V. C. & Sajo, M. G. (2001). Morfo-anatomia caulinar de nove espécies de orchidaceae. *Acta Botânica Brasílica*. Vol.15, No.2, (June 2001), pp. 177-188, ISSN 1677-941X

Osmond, C.B. (1978). Crassulacean acid metabolism - curiosity in context. *Annual Review of Plant Physiology*, Vol.29, No.29, (June 1978), pp. 379-414, ISSN 0066-4294

Ota, K. (1988). Stimulation of CAM photosynthesis in *Kalanchoe blossfeldiana* by transferring to nitrogen-deficient conditions. *Plant Physiology*, Vol.87, No.2, (June 1988), pp. 454-457, ISSN 1532-0889

Paul, M.J; Cockburn, W. (1990). The stimulation of CAM activity in *Mesembryanthemum crystallinum* in nitrate- and phosphate-deficient conditions. *New Phytologist*, Vol.114, No.3, (March 1990), pp. 391- 398, ISSN 0028-646X

Pearson, P.N. & Palmer, M.R. (2000). Atmospheric carbon dioxide concentrations over the past 60 million years. *Nature*, Vol.406, No.6797, (August 2002), pp. 695–699, ISSN 0028-0836

Pierce, S., Winter, K. & Griffiths, H. (2002). Carbon isotope ratio and the extent of daily CAM use by Bromeliaceae. *New Phytologist*, Vol.156, No.1, (October 2002), pp. 75–83, ISSN 0028-646X

Ramırez, S.R.; Gravendeel, B.; Singer, R.B.; Marshall, C.R. & Pierce, N.E. (2007). Dating the origin of the Orchidaceae from a fossil orchid with its pollinator. *Nature*, Vol.448, No.7157, (August 2007), pp. 1042–1045, ISSN 0028-0836

Ranson, S. L. & Thomas, M. (1960). Crassulacean acid metabolism. *Annual Review of Plant Physiology*, Vol.11, No.1, (June 1960), pp. 81-110, ISSN 0066-4294

Reekie, E.G. & Bazzaz, F.A. (1987). Reproductive efforts in plants. I. Carbon allocation to reproduction. *American Naturalist*, Vol.129, No.6, (June 1987), pp. 876–896, ISSN 0003-0147

Silvera, K.; Neubig, K.M.; Whitten, M.; Williams, N.H.; Winter, K. & Cushman, J.C. (2010a). Evolution along the Crassulacean acid metabolism continuum. *Functional Plant Biology*, Vol.37, No.11, (October 2010), pp. 995-1010, ISSN 1445-4408

Silvera, K.; Santiago, L.S. & Winter, K. (2005). Distribution of Crassulacean acid metabolism in orchids of Panama: evidence of selection for weak and strong modes. Functional of Plant Biology, Vol.32, No.5, (May 2005), pp. 397–407, ISSN 1445-4408

Silvera, K.; Santiago, L.S. & Winter, K. (2010b). The incidence of Crassulacean acid metabolism in the Orchidaceae derived from carbon isotope ratios: a checklist o the flora of Panama and Costa Rica. *Botanical Journal of the Linnean Society*, Vol.163, No.2, (June 2010), pp. 194-222, ISSN 1095-8339

Silvera, K.; Santiago, L.S.; Cushman, J.C. & Winter, K. (2009). Crassulacean acid metabolism and epiphytism linked to adaptive radiations in the Orchidaceae. *Plant Physiology*, Vol.149, No.4, (April 2009), pp. 1838–1847, ISSN 0032-0889

Smillie, R.M. (1992). Calvin cycle activity in fruit and the effect of heat stress. *Scientific Horticulture*, Vol.51, No.1-2, (July 1992), pp. 83–95, ISSN 0304-4238

Sollings, P. & Drewry, G. (1970). Electrical conductivity and flow rate of water through the forest canopy, In: *A Tropical Rain Forest*, H.T. Odum, (Ed.), Chap. H, pp. 137-155, U.S. Atomic Energy Commission, Washington, D.C.

Spalding, M.H.; Stumpf, D.K.; Ku, M.S.B.; Burris, R.H. & Edwards, G.E. (1979). Crassulacean acid metabolism and diurnal variation of internal CO_2 and O_2-concentrations in Sedum praealtum DC. *Australian Journal of Plant Physiology*, Vol.6, pp. 557-567, ISSN 0310-7841

Stancato, G. C.; Mazzafera P. & Buckeridge, M. S. (2001). Effect of a drought period on the mobilization of non-structural carbohydrate, photosynthetic efficiency and water status in an epiphytic orchid. *Plant Physiology and Biochemistry*, Vol.39, No.11, (November 2001), pp. 1009-1016, ISSN 0981-9428

Sunagawa, H.; Cushman, J.C. & Agarie, S. (2010). Crassulacean acid metabolism may alleviate production of reactive oxygen species in a facultative CAM plant, the common ice plant *Mesembryanthemum crystallinum* L. *Plant Production Science*, Vol.13, No.3, (June 2010), pp. 256-260, ISSN 1349-1008

Ting, I.P. (1985). Crassulacean acid metabolism. *Annual Review of Plant Physiology*, Vol.36, No.1, (June 1985), pp. 595–622, ISSN 0066-4294

Weiss, D.; Schönfeld, M. & Halevy, A.H. (1988). Photosynthetic activities in the *Petunia* corolla. *Plant Physiology*, Vol.87, No.3, (March 1988), pp. 666–670, ISSN 0032-0889

Werk, K.S. & Ehleringer, J.R. (1983). Photosynthesis by flowers in *Encelia farinosa* and *Encelia californica* (Asteraceae). *Oecologia*, Vol.57, No.3, (March 1983), pp. 311–315, ISSN 0029-8549

Williams, K.; Koch, G.W.; Mooney, H.A. (1985). The carbon balance of flowers of *Diplacus aurantiacus* (Scrophulariaceae). *Oecologia*, Vol.66, No.4, (July 1985), pp. 530–535, ISSN 0029-8549

Winter, K. & Smith, J.A.C. (1996). An introduction to Crassulacean acid metabolism. Biochemical principles and ecological diversity, In: *Crassulacean acid metabolism. Biochemistry, ecophysiology and evolution*, K. Winter & J.A.C. Smith, (Eds.), pp. 1-13, Springer-Verlag, ISBN 3540581049, Berlin, Germany.

Winter, K. & Von Willert, D.J. (1972). NaCl-induzierter Crassulaceanaurestoffwechsel bei *Mesembryanthemum crystallinum. Zeitschrift fur Pflanzenphysiologie*, Vol.67, pp. 166-170, ISSN 0044-328X

Winter, K.; Garcia, M. & Holtum, J.A.M. (2008). On the nature of facultative and constitutive CAM: environmental and developmental control of CAM expression during early growth of Clusia, Kalanchoe, and Opuntia. *Journal of Experimental Botany*, Vol.59, No.7, (February 2008), pp. 1829–1840, ISSN 0022-0957

Winter, K.; Lüttge, U.; Winter, E. & Troughton, J.H. (1978). Seasonal shift from C_3 photosynthesis to Crassulacean acid metabolism in *Mesembryanthemum crystallinum* growing in its natural environment. *Oecologia*, Vol.34, No.2, (January 1978), pp. 225-237, ISSN 1432-1939

Winter, K.; Medina, E.; Garcia, V.; Mayoral, M.L. & Muniz, R. (1985). Crassulacean acid metabolism in roots of a leafless orchid, *Campylocentrum tyrridion* Garay & Dunsterv. *Journal of Plant Physiology*, Vol.118, No.1, pp. 73-78, ISSN 0176-1617

Winter, K.; Wallace, B.J.; Stocker, G.C. & Roksandic, Z. (1983). Crassulacean acid metabolism in Australian vascular epiphytes and some related species. *Oecologia*, Vol.57, No.1-2, (March 1983), pp. 129–141, ISSN 0029-8549

Withner, C.L.; Nelson, P.K. & Wejksnora, P.J. (1974). The anatomy of orchids, In: *The Orchids-Scientific Studies*, C.L. Withner, (Ed.), pp. 267, John Wiley and Sons, Inc., ISBN 0471957151, New York, USA

Yoder, J. A.; Imfeld, S.M.; Heydinger, D.J.; Hart, C.E.; Collier, M.H.; Gribbins, K.M. & Zettler, L.W. (2010). Comparative water balance profiles of Orchidaceae seeds for epiphytic and terrestrial taxa endemic to North America. *Plant Ecology*, Vol.211, No.1, (November 2010), pp. 7-17, ISSN 1385-0237

Yong, J.W. & Hew, C.S. (1995). Partitioning of ^{14}C assimilates between sources and sinks during different growth stages in the sympodial thin-leaved orchid Oncidium Goldiana. *International Journal of Plant Sciences A*, Vol.156, No.2, (March 1995), pp. 188-196, ISSN 1058-5893

Zimmerman, J. K. (1990). Role of pseudobulbs in growth and flowering of *Catasetum viridifolium* (Orchidaceae). *American Journal of Botany*. Vol.77, No.4, (June 1990), pp. 533-542, ISSN 0002-9122

Zotz, G. & Hietz, P. (2001). The ecophysiology of vascular epiphytes: current knowledge, open questions. *Journal Experimental Botany*, Vol.52, No.364, (November 2001), pp. 2067–2078, ISSN 0022-0957

Zotz, G. & Ziegler, H. (1997). The occurrence of Crassulacean acid metabolism among vascular epiphytes from Central Panama. *New Phytologist*, Vol.137, No.2, (October 1997), pp. 223–229, ISSN 0028-646X

Zotz, G. (1999). What are backshoots good for? Seasonal changes in mineral, carbohydrate and water content of different organs of the epiphytic orchid *Dimerandra emarginata*. *Annals of Botany*, Vol. 84, No.6, (December 1999), pp. 791-798, ISSN 0305-7364

Zotz, G. (2004). How prevalent is Crassulacean acid metabolism among vascular epiphytes? *Oecologia*, Vol.138, No.2, (October 2004), pp. 184–192, ISSN 0029-8549

4

Characterization of Chestnut Behavior with Photosynthetic Traits

José Gomes-Laranjo et al.*
Centre for the Research and Technology of Agro-Environmental and Biological Sciences (CITAB), Universidade de Trás-os-Montes e Alto Douro (UTAD), Vila Real Portugal

1. Introduction

European chestnut (*Castanea sativa* Mill.) covers in total 2.53 million hectares, of which 2.2 million hectares are forests and the remaining parts, are orchards. In Europe it is growing in an area comprised by 27° N and 53° N latitude, from sea level in seaside regions to 2000 m above sea level (a.s.l.) in the south of Europe. According to Fernández-López et al. (2005), chestnut species is characterized by the existence of some differentiation among extreme populations, which can be supposed due to its long-range distribution across the Mediterranean region, through varying climatic conditions. As reported by Heiniger (1992), chestnut is a good indicator of warm regions with oceanic climate.

In spite of *C. sativa* Mill. be characterized as a mesophilic species, nowadays the chestnut shows many growth limitations, which partially can be ascribed to the climate changes, since they influence abiotic and biotic factors and consequently photosynthetic productivity. From abiotic factors, water and heat stress have been the most important limitations, inducing less growth, less vigor increasing susceptibility to the biotic factors such as ink and blight diseases.

When compared with other hardwood species, the trees can absorb the same amount of carbon, but due to the fast-growing chestnut can store more carbon in less time. So, this agro forest system has been identified as an important way to slow climate change.

Photosynthesis, according to (Givnish, 1988), provides green plants with almost all of their chemical energy, being central to their activity to compete and reproduce. So understanding

* Lia-T. Dinis[1], Luís Martins[2], Ester Portela[1], Teresa Pinto[1], Marta Ciordia Ara[3], Isabel Feito Díaz[3], Juan Majada[3], Francisco Peixoto[4], S. Pereira Lorenzo[5], A.M. Ramos Cabrer[5], Changhe Zhang[1], Afonso Martins[1] and Rita Costa[6]

[1]*Centre for the Research and Technology of Agro-Environmental and Biological Sciences (CITAB), Universidade de Trás-os-Montes e Alto Douro (UTAD), Vila Real, Portugal*

[2]*Department of Forest & Landscape, Universidade de Trás-os-Montes e Alto Douro (UTAD) Apartado, Vila Real, Portugal*

[3]*(SERIDA), Consejería de Medio Rural y Pesca, Villaviciosa, Asturias, Spain*

[4]*CECAV, University of Trás-os-Montes e Alto Douro, Vila Real, Portugal*

[5]*University of Santiago de Compostela, Department of Crop Production, Campus de Lugo, Spain*

[6]*INRB, I.P. L-INIA Quinta do Marquês, Oeiras, Portugal*

photosynthetic performances of each species, its varieties or even its cultivars is crucial to understand and advise the future.

This chapter will present an overview about photosynthetic studies done in chestnut species emphasizing the abiotic stresses (drought and heat stress) and biotic stresses (ink disease) issues, contributing to understand the impact of climate change in chestnut.

2. Ecological chestnut behaviour according to its European distribution

European chestnut is the only species of genus *Castanea* in Europe showing an outstanding evolutionary history from its likely origin in North-east Turkey and Caucasus region in respect to other European forest tree species during last 9,000 years before present (YBP) (Mattioni et al., 2008). Palynological studies support also two main fast expansion periods, about 5,000 YBP due to glacial Pleistocene Epoch and 2,000 YBP, during the Roman Empire. Actually, in Europe, *C. sativa* Mill. is commonly found between the Canary Islands, the most southern point (27° N latitude) and the most northern point defined by a line passing in the south of the United Kingdom, northern Germany, Poland and Ukraine (52° N latitude). In terms of altitude, chestnut is quite widespread, since it grows at sea level in coast regions until 400 m and 2000 m a.s.l. in the inner regions of the European continent. The lowest elevations are recommended for the highest latitudes and vice versa (Bounous, 2002).

The wide distribution of European chestnut, and the higher phenotypic plasticity observed in populations coming from arid regions in comparison to those that are from more wet areas, suggest the substantial adaptive variation existing among populations (Fernández-López et al., 2005). In a study comparing progenies from several European climatic contrasting locals growing under the same climatic conditions (EU funded Cascade project), populations from Greece started growth earlier followed by southern latitude progenies (south Italy and south Spain) while the plants from north Spain and North Italy initiated later. Height growth of the northern populations was higher than growth of the southern plants. The southern ones also showed an earlier growth cessation, budbreak, and a longer juvenile period than those from more north latitudes (Fernández-López et al, 2005). The importance of budbreak is due to the sensibility of the young leaves to latest frost during spring times, besides other factors such as drought tolerance.

The European chestnut presents a fair effective number of alleles, decreasing diversity from northern to the most southern populations (Eriksson et al., 2005b). Nevertheless, in Iberian Peninsula there were detected more than 350 genotypes in 574 accessions (Pereira-Lorenzo et al., 2010). According to these authors, in Iberian Peninsula the two main variability origins are located in North and Centre, being the most southern ecotypes (Andalusia and Canary Islands) assigned to both of these zones, which might suggest a colonization process. This colonization by varieties from the north part of Iberian Peninsula, since there are quite different edaphoclimatic regions, suggests that this species has potential to adapt themselves to new climatic conditions and by this way to the new context of climate changes. This ability of long-term species to respond and to adapt to environmental changes though natural selection is due mainly to their high intrapopulation genetic diversity (Martin et al., 2010), demonstrating genomic SSRs significantly higher levels of diversity in terms of number of alleles, expected heterozygosity and level of polymorphism among populations. While previous studies using isoenzymes showed limited diversity, the new methodologies using microsatellites have demonstrated high number of alleles and gene diversity, even in European cultivated stands. More than a hundred of chestnut ecotypes from Portugal and Spain were characterized (Costa et al., 2009; Dinis et al., 2011a; Martín et al., 2007, Pimentel- Pereira, et al., 2007; Pereira-Lorenzo et al., 2006), showing significant adaptive differences to the different edaphoclimatic conditions: (Dinis et al., 2011a). In the same way, in Spain almost 50 varieties spreading from the northern to the southern and Canary Islands were studied, showing the chemical data of fruits low correlations with environmental parameters indicating that these differences could be mainly ascribed to genetic differences (Pereira-Lorenzo et al., 2006). In Spain this diversity will be certainly higher, since only in Andalusia (located in south of Spain), there were detected 43 ecotypes, showing them quite different morphological traits, namely at the level of leaves, spines on the burs, catkins, fruit shape and size (Martín et al., 2007). So, Man has also influenced the structural and genetic characteristics of chestnut populations through cultural practices (propagation, grafted plants, silvicultural or agronomical practices, etc), leading to widely heterogeneous stand typologies all over Europe: a) naturalized forests with coppices exploited for timber, b) managed coppice, an agro-forest system cultivated for fruit and timber production and c) orchards, a crop for fruit production (Mattioni et al., 2008), perfectly demonstrating the multipurpose character of this species. In spite of human influence, three main gene pools across Europe were identified (Villani et al., 2010): a) the north-eastern Turkish pool, b) the Greek gene pool and c) the Mediterranean Turkish gene pool, from which the European seems to be originated. In all these situations, they have been considered the natural hybridations and introgressions. These are due to the introduction in Europe during first half of the 20th Century, of Asiatic species a cause of the resistant ink disease breeding programs. Although climate and the grazing regime are influential, the degree of silvicultural management appears to be the most important factor determining floristic composition of forests and their long-term sustainability (Konstantinidis et al., 2008).

Because of its different domestication levels, chestnut has undergone natural and artificial selection pressure, which have shaped the actual genetic and phenotypic traits, and so expected capacity of adaptation to environmental stress factors. Particularly evident until now, there is the capacity to drought tolerance adaptation, which underlines a higher phenotypic plasticity in populations from arid regions as compared to those from humid areas (Villani et al., 2010). Concerning rainfall, the minimum for chestnut is 800 mm, but it can growth well in rainy areas (2000 mm) like Galicia, located in northwestern Spain (Pereira-Lorenzo et al., 2006).

Plants from this species are moderately thermophilic and well adapted to ecosystems with a annual mean temperature ranging between 8 and 15 °C and monthly mean temperatures between May and October over 10 °C. Temperature is also important for a good pollination (July), with an adequate range between 25 °C and 30 °C (Bounous, 2002).

The capability to take in water and nutrients and to make an efficient use of these resources is an important feature of species adapted to the Mediterranean environment (Zhangh et al., 2011). C_3 plants discriminate during photosynthesis against ^{13}C (1.1% of total carbon in atmosphere), the heavy stable isotope of C (Lauteri et al., 2004), being its depletion very affected by environmental conditions. So, the carbon isotope discrimination (Δ), a complex physiological trait involved in acclimation and adaptive processes, estimates the amount of ^{13}C in plants, which by its way is dependent on the plant response. Climatic conditions from each local, temperature and precipitation, can be conveniently characterized by the xerothermic index, which takes in account each growth season (Eriksson et al., 2005a), the biggest values meaning driest and warmest conditions. Across European chestnut populations displays significant Δ variability (Lauteri et al., 2009). In general, chestnut growth decreases with the increase in xerothermic index, in opposition to the Δ values. Then highest Δ values were measured in progenies from the warmest locals (e.g. Greece and south of Spain). On the other way, the highest variability on Δ values presented by ecotypes from Mediterranean locals in response to increase in temperature indicates a higher adaptation potential against climate changes than that of those from wet locals (Lauteri et al., 2004). The same is also true for drought tolerance, meaning that Mediterranean ecotypes might have more tolerance to drought than other ones from contrasting locals. The highest Δ values correspond to the lowest water use efficiency (WUE) (Lauteri et al., 2004). These authors demonstrated that under the same climatic conditions, progenies from ecotypes originating from xeric locals (*e.g.* Mediterranean locals) can present higher *A* and g_s rates but surprisingly lower WUE than those from mesic (wet) locals.

Under field conditions, on Trás-os-Montes (located in the Northeast of Portugal) chestnut trees have maximal photosynthesis rate (*A*), 7.6 $\mu molCO_2.m^{-2}.s^{-1}$, in September (Figure 1), when temperature reaches 25 °C. This month corresponds to an important phase of bur growth. In relation to maximal leaf transpiration (*E*), it was measured during July (29 °C, 4.8 $mmolH_2O.m^{-2}.s^{-1}$), which can be positive in terms of leaf cooling and consequently in *A*. So, the worst *A* measured in July comparatively to September, might be due to heat stress, relieving the importance of leaf water content in this month, and indirectly the importance of xylem growth to allow leaf water uptake.

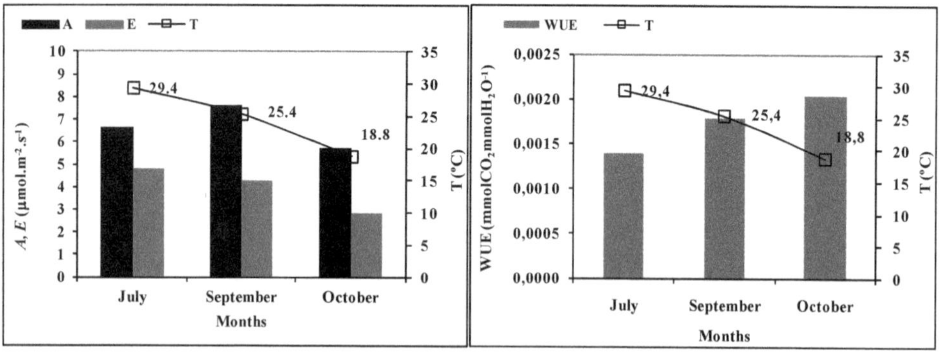

Fig. 1. Variation of photosynthesis (*A*), leaf transpiration (*E*), water use efficiency (WUE) and environmental temperature (T) during July, September and October, in adult chestnut trees (n=8852) with near 40 years old. Measurements were done in Trás-os-Montes region, located in the Northeast of Portugal, during 6 successive years.

The highest production of xylem cells is observed between mid-May and mid-June and phloem ones from mid-June until mid-July. Wood and phloem production mainly terminated in the middle of August while differentiation of xylem cells lasted until mid-

October (Cufar et al., 2011). The highest production in relation to water use efficiency (WUE) increases from July (0.0014 $mmolCO_2.mmolH_2O^{-1}$) to October (Gomes-Laranjo et al., 2007), following the standard curve above referred.

In conclusion, populations from wet regions seem to have less phenotypic plasticity than those from dry and warm regions, indicating that the first ones might also be more vulnerable to possible climate changes (Lauteri et al., 2009; Villani et al., 2010).

3. Ecophysiological characterization of chestnut

3.1 Characterization of Portuguese varieties by gas exchange traits, as a function of temperature and radiation

Near 85% of Portuguese chestnut area (around 35,000 ha) is located in Trás-os-Montes, a region located in the Northeast of Portugal (41°30′N, 6°59′E), 9-13°C being the mean annual temperature and 600-1200 mm the amount of annual rainfall (Figure 2).

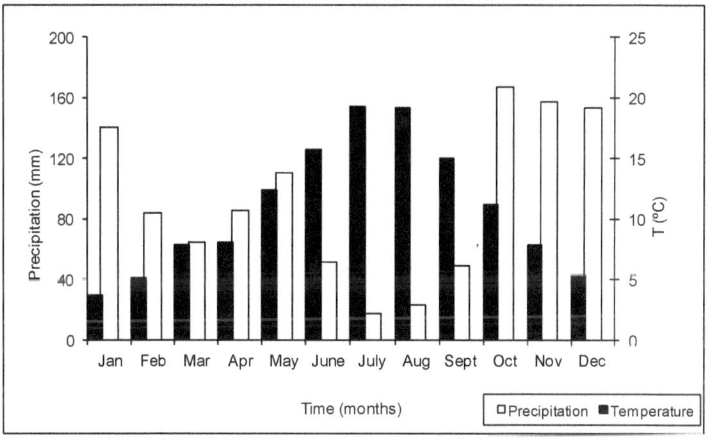

Fig. 2. Month values of precipitation amount and daily mean temperature in Trás-os-Montes region. Values represent the mean between 1988 and 1994.

Chestnut is a dim-light species. In a study done in Trás-os-Montes Region, including a sample constituted by 13 varieties (see Figure 6), half of the maximal A (A_{100}) was obtained when PPFD level reached near 400 $\mu mol.m^{-2}.s^{-1}$, instead 75% of A_{100} was measured under PPFD of 850 $\mu mol.m^{-2}.s^{-1}$, corresponding this radiation to half of maximal PPFD. These results indicate that trees might have better adaptation to the shade north-facing slopes, rather than south-facing ones (Gomes-Laranjo et al., 2008a). South-facing orchards have higher mean temperatures than north-facing ones. In the former orchards, there is higher evapotranspiration but lower water availability, which induces less growing trees, and so less vigour, predisposing them to biotic stresses such ink disease (induced by the soil born oomycete *Phytophothora cinnamomi*). Maximal CO_2 assimilation (A_{100}), was obtained at about 24 °C (Figure 3), while half of maximal A was measured around 11-12 °C and 37-38 °C of ambient temperatures.

This relatively low tolerance to hot temperatures was attributed to the level of thylakoid fatty acid saturation, so that the most tolerant cultivars (*e.g.* Aveleira) have a higher level of saturation than the worst ones (*e.g.* Judia) (Gomes-Laranjo et al., 2006).

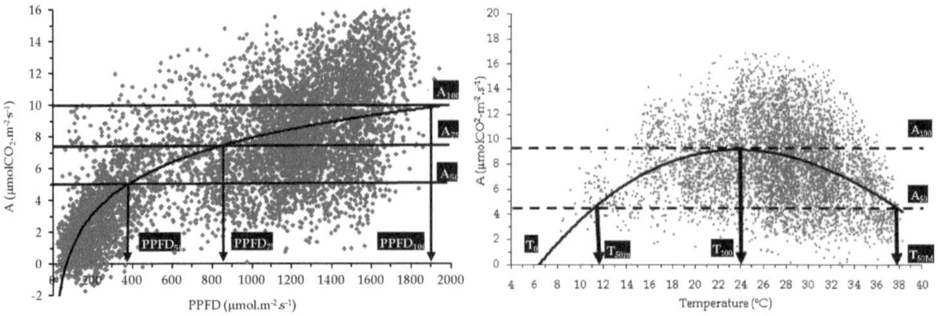

Fig. 3. Study of photosynthesis rate (A) as a function of radiation (PPFD, left) and temperature (T, right). For PPFD, A was modulated by logarithmic curve $A = 3.0108\ln(PPFD) - 12.87$ ($R^2 = 0.541$), instead of temperature, there was selected a second polynomial curve $A = -0.0002T^3 - 0.0432T^2 + 1.6543T - 8.9259$ ($R^2 = 0.121$). Measurements (n=8852) were taken in adult trees (40 years old) from 13 Portuguese varieties, during 4 years, between July and October, with an infra-red gas analyser (model LCA-2, Analytical Development Co., Hoddesdon, UK). For PPFD curve, there was selected the measurements done between 16 °C and 32°C, the temperature curve was determined under plenty PPFD (>900 µmol.m^{-2}.s^{-1}).

Leaf transpiration is a physical process of evaporation that causes the water content of the leaves to drop. This produces a gradient of water potential from the root to the leaves. In consequence a flow of water from the soil to the roots, stems, leaves and evaporation from here to the atmosphere through stomata, happens. This water movement has two main functions: a) promote the mineral upward to the plant, and b) promote the regulation of leaf temperature due to high specific heat value of water (Epstein & Bloom, 2005). So, a higher E is better for leaf temperature regulation. Contrarily to A, dim-light did not significantly

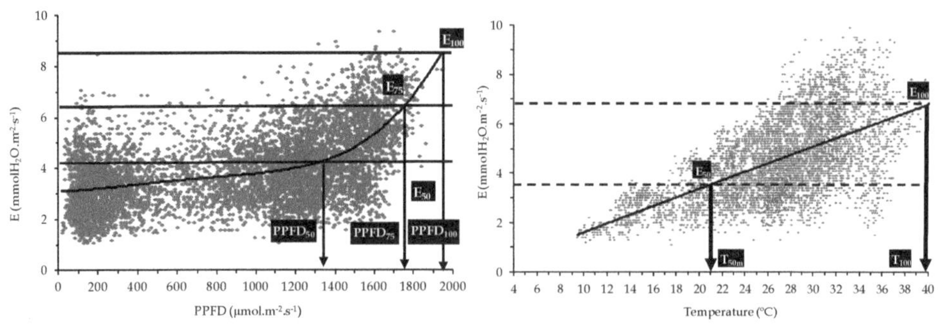

Fig. 4. Study of transpiration (E) rates in European chestnut trees as a function of radiation (PPFD, left) and temperature (T, right). For PPFD the polynomial equation is $E = 1E^{-12}PPFD^4 - 3E^{-9}PPFD^3 - 3E^{-6}PPFD^2 + 8E^{-5}PPFD + 3.109$ ($R^2 = 0.308$) and for temperature, E can be represented by $E = -0,0002T^2 + 0.1813T - 0.1967$ ($R^2 = 0.360$).
For other conditions see Figure 3.

affect *E*, half rate of it being around 1300 $\mu mol.m^{-2}.s^{-1}$ (Figure 4). Although, above this PPFD level, this strong influences *E*. On the other way, there is visible a strong increase in *E* as a response to the temperature rise (Figure 4), from 1.5 to 3.5 and to 7 $mmolH_2O.m^{-2}.s^{-1}$ when temperature increases from 10 to 21 and to 40 °C, respectively. On the other way it must be believed that a higher diversity of *E* was observed under highest temperatures than under lowest ones. This might indicate that significant differences among varieties in this parameter could be ascribed.

The water use efficiency (WUE) firstly considered as water used per unit plant material generated (Briggs & Shantz, 1914), varying between 0.001 and 0.005 or less, clearly indicates that transpiration is a wasteful process. This is the result between two conflicting requirements: the need for exposure of moist, green cells to light and that for open pathways to allow CO_2 to diffuse to these cells. In what concerns the influence of PPFD, the strongest increase in WUE was obtained under dim-light conditions (WUE_{75}~600 $\mu mol.m^{-2}.s^{-1}$, Figure 5). These findings are partially consistent with the influence of T on WUE, where the highest efficiency, 0.003 $\mu mol\ CO_2.mmol\ H_2O^{-1}$, was found under lowest temperatures (14°C). The half value of WUE appears when T reaches 30 °C, a typical temperature measured during summer times in the South of Europe.

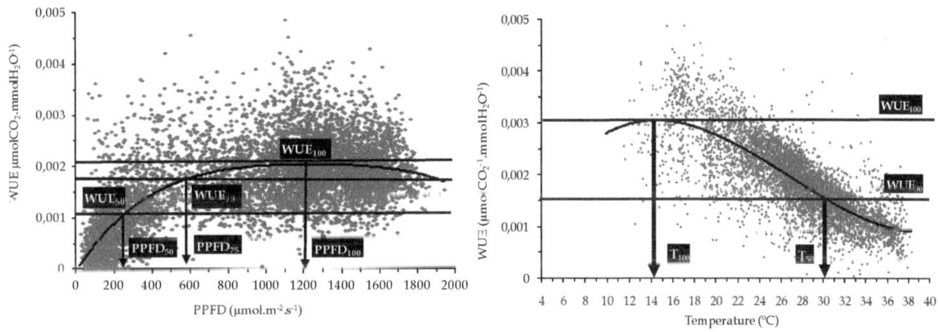

Fig. 5. Characterization of water use efficiency variation (WUE) as a function of the abiotic factors: radiation (PPFD, *left*) and temperature (T, *right*). Temperature influence can be modulated by the equation $WUE = 0.0002T^3 - 0.0126T^2 + 0.2098T + 2.3388$ ($R^2 = 0.627$), while for PPFD the respective equation is $WUE = -6E^{-16}PPFD^4 + 3E^{-12}x^3 - 6E^{-09}x^2 + 6E^{-06}x - 5E^{-05}$ ($R^2 = 0.466$).

In relation to Portuguese varieties, optimal temperature (T_{100}) changes between 22 °C ('Lada') and 29 °C ('Boaventura'), 7°C being the wide of the interval. Concerning the heat tolerance, T_{50} interval range is only 3°C, between 35°C ('Judia' and 'Negral') and 38°C ('Aveleira'). Results suggest that globally, the varieties from Valpaços (780 m a.s.l.) have lower T_{100} than those from Vinhais (780 m a.s.l.), while the varieties from Vila Real (730 m a.s.l.) presented high T_{50} (Figure 6). The infection of trees with *Phytophthora cinnamomi* Rands (the oomycete that causes ink disease) induces higher heat tolerance as the shift on T_{100} and T_{50}, presented by infected 'Judia' trees, indicates.

The state of the lipids plays an important function in the temperature response at the thylakoid membranes, since it might function as a buffer, in a determined range of temperatures. Fatty acids of the thylakoid glicerolipids must be sufficiently unsaturated in order to allow a convenient mobility of the electron transfer carrier plastoquinone, but at the same time must preserve its integrity, namely in relation to the above cited impermeability of the protons (Blackwell et al. 1994). In chestnut chloroplasts the most significant fatty acids are α–linolenic acid (18:3) (40-50%) and palmitic acid (16:0) (20-30%). Comparing the balance of the lipid composition in three varieties, Aveleira (Vinhais), Longal (Valpaços) and Judia (Valpaços), the ratio of unsaturated/saturated fatty acid was 1.86, 2.27 and 2.40, the former being also the variety with lowest T_{50}, followed by Longal, being Aveleira the most tolerant to heat.

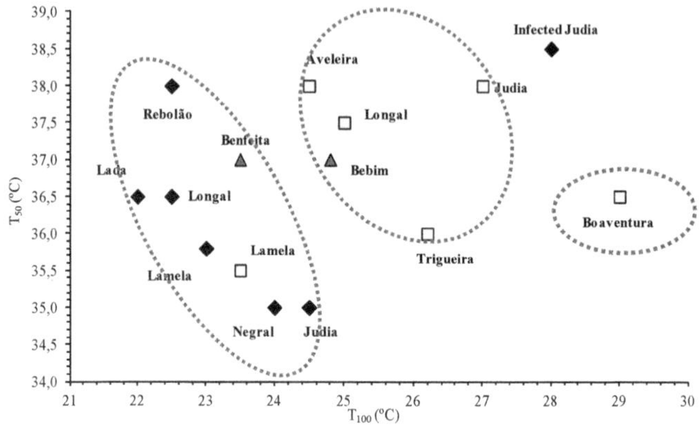

Fig. 6. Distribution of Portuguese varieties according to their optimal temperature (T_{100}) and temperature for half (T_{50}) of maximal rate for *A*. Varieties in different origins, Valpaços, Vila Real and Vinhais, are indicated by lozenge, triangle and square, respectively. The label 'Judia infected' correspond to infected trees with *Phytophthora cinnamomi* Rands.

3.2 Spanish varieties

In Spain, there are 124,053 hectares of pure *C. sativa* Mill. forests and orchards and 55,416 hectares of chestnut mixed with other broadleaves (mainly oak, holm oak, cork and beech) according to the Third National Forestry Inventory published in 2006 (www.marm.es). However, this does not include the area formed by the Canary Islands, with 2000 hectares, or the estimated 5000 hectares in the provinces of Zamora, Salamanca, Ávila and Málaga (Pereira-Lorenzo et al., 2009). It is found from sea level in some parts of the northern coast of the Iberian Peninsula up to 1500 m in the Sierra Nevada (Granada, Southeastern Spain; 37° N, 03° W).

Four models of management for chestnut can be distinguished in Spain: i) orchards for nut production based on grafted trees with selected cultivars; ii) coppice stands for timber production; iii) high forests for timber production; iv) grafted orchards for nut and timber production with cultivars like 'Garrida', 'Loura' and 'Parede' which, in humid areas, are pruned in a way which favours the development of long trunks that are sold at high prices (Pereira-Lorenzo et al., 2009). Chestnut coppice distribution is largely dependent on the selvicultural techniques applied historically in the region. As a result, coppice exploitation for timber is more common in Asturias than in the rest of Spain.

The identification and evaluation of the genetic resources of grafted chestnuts in Spain began by examining morphological characters and isoenzymes enabling the discrimination of these materials. Pereira-Lorenzo et al. (2006) evaluated 701 accessions corresponding to 168 cultivars; 31 from Andalusia (12 cultivars), 293 from Asturias (65), 25 from Castilla-Leon (9), 4 from Extremadura (2) and 348 from Galicia (80). The main morphological characters examined showed considerable variation in relation to geographic area. Furthermore, the isoenzyme data demonstrated greater variability in the North, a finding related to the historic importance of chestnut cultivation in this area. Examination of the F-statistics, however, showed that there is in fact only limited genetic differentiation, but high heterozygosity, linked to grafting, between areas. The same methodology was applied to the Canary Islands and a further 38 cultivars were characterized (Pereira-Lorenzo et al., 2007).
More recently, Pereira-Lorenzo et al. (2011) have used microsatellites to confirm this high degree of conservation of this genetic variability in the Iberian Peninsula, finding that diversity in chestnut orchards was mostly explained by hybridization (up to 77%), with mutation only accounting for 6%. Previous results (Pereira-Lorenzo et al., 2010) showed that there are two main origins of variability in the Iberian Peninsula, one in the North and the other in the Central Iberian Peninsula. This division roughly corresponds to contrasting climatic conditions in Spain; i.e., mesic areas (moderately humid environment) and xeric (drier environment) areas, respectively. The Northern group is characterized by a wetter and milder climate than the Southern, which has lower monthly rainfall in summer (Ps) and both higher average annual T and higher average T of the hottest month (Allué-Andrade, 1990).
Water availability is known to be one of the most limiting factors in relation to photosynthesis and plant productivity (Boyer, 1996). Furthermore, predicted climate warming effects in the Iberian Peninsula (Ramírez-Valiente et al., 2009) suggest considerable changes in this respect. To address these issues, an ecological characterization of chestnut has been conducted: the first part of the investigation focused on water relations and growth traits (Ciordia et al., under review); the second concentrated on water relations and physiological traits (Ciordia, 2009) and the results are presented in the remainder of this section.
One and a half-year-old open pollinated seedlings of 10 half-sib progenies of the main Spanish fruit cultivars of *C. sativa* Mill. were chosen from each of the two main gene pools (Pereira-Lorenzo et al., 2010): Northern (specifically from Asturias and Galicia) and Central Iberian Peninsula (specifically from the Canary Islands and Andalusia in Southern Spain). Chestnut seedlings were either well-irrigated throughout the experiment, or were subjected to mild drought stress followed by suspension of irrigation, with the aim of determining the vulnerability to drought of each progeny and of assessing the effect of the treatment on the phenotypic variation of the plants. Significant variations in most adaptive traits studied, such as juvenile growth, biomass allocation, leaf morphology and water relations, were observed.
In terms of photosynthetic and water traits, our results show that, in general and irrespective of geographic provenance, water stress modified physiological response: significantly reducing water conductance (K), gas exchange measurements (CO_2 assimilation rate, *A*; transpiration rate, *E*; stomatal conductance to CO_2, g_s) with the exception of internal CO_2 concentration (C_i), and the maximal apparent efficiency of Photosystem II, estimated as Fv/Fm, whilst intrinsic water use-efficiency (WUE_i) and initial fluorescence (Fo) increased and carbon isotope composition ($\delta^{13}C$) became less negative.

As expected, given that *C. sativa* Mill. is considered to be an anisohydric species, lower values of water potential, both at predawn (Ψ_{wpd}) and at midday (Ψ_{wmd}), were obtained in plantlets subjected to water stress. However, the recovery of $\square_{wpd}$ compared to that at midday was not a general strategy for all chestnut progenies under water stress. Adjustment in water potential, together with variation in *E*, led to variation in K, which was found to decrease significantly when the plants were subjected to water stress. This reduction was most marked in the materials originating from Andalusia and the Canary Islands.

CO_2 assimilation rate (*A*) was also found to decrease when soil water availability is low and, in this sense, a negative lineal correlation between *A* and Ψ_{wpd} and Ψ_{wmd} occurs. Furthermore, in some genotypes where Ψ_{wpd} and Ψ_{wmd} became very low and were in fact very similar, indicating that the plants were at critical matric potential (Ψ_{mcrit}), the reduction in photosynthetic efficiency was drastic with values below 6 µmol CO_2 $m^{-2}.s^{-1}$, a decrease mainly associated with a reduction in g_s. Stomatal closure is generally the first response to drought and it is the dominant limitation to *A* at mild and moderate drought (Flexas & Medrano, 2002). In the Spanish chestnut progenies studied, as in other plant species, an adaptive mechanism in response to water stress can be observed, which tends to increase WUE_i. This is because the relationship between reduction in carbon assimilation and g_s is not a linear one and results in photosynthetic efficiency being affected less, and less quickly, than water loss (Chaves et al., 2003).

Provided that materials collected from different habitats are sampled under uniform garden conditions, variation in $\delta^{13}C$ among geographic origins can be employed to infer genetic diversity in WUE_i. In general, studies on genetic diversity in $\delta^{13}C$ for forest tree species from both mesic and xeric environments suggest the existence of significant, and often considerable, variation in WUE_i (e.g. Cregg et al., 2000; Li & Wang, 2003). In our study WUE_i was seen to increase under water limitation as demonstrated not only by the doubling of the A/E coefficient in the experimental group, but also by changes in $\delta^{13}C$. This parameter was found to be more negative under control than experimental conditions. Family genetic variation ranged from 2.1‰ under drought conditions to 1.6‰ under control conditions. This is in line with findings from a common garden trial with 6 provenances of *Castanea* where the range of variation of carbon isotope discrimination (Δ) in leaf dry matter was also about 1.6‰ (Lauteri et al., 1997), which suggests a maximum between population differences in WUE_i of over 16% (Ferrio et al., 2005, as cited in Farquhar & Richards, 1984). Understanding genetic variation in WUE_i through $\delta^{13}C$ in relation to geographic and/or climatic gradients is essential in evaluating adaptation patterns in drought-prone areas. However, in our study, we have not found there to be significant differences between the geographic areas of Northern Spain (humid area) compared to Andalusia and the Canary Islands (dry area). In fact, progenies from the latter cultivated under water stress conditions reached more negative values than those from the Northern group. This apparent contradiction can be explained if we take into account that stomatal opening and closing not only regulates transpiration but also plays an important role in thermal regulation.

The shape of the leaf also contributes to thermal regulation. According to Nicotra et al. (2008) more lobated leaves have a finer boundary layer leading to more efficient heat exchange and reducing differences in T between leaf and air. They also suggest that there is a relationship between degree of leaf lobation (ILB), carbon fixing and water loss and speculate that this is probably the effect of evolutionary convergence towards adaptation to drought and heat as indicated by the fact that more lobated leaves have a lower specific leaf area (SLA) due to their higher proportion of veins and thus have an *A* per unit of mass equivalent to less lobated leaves.

The results from our experiment on the whole support the findings of Nicotra et al. (2008). We found that ILB is a plastic parameter which varies in relation to water availability, increasing under stress conditions, and that SLA decreases as water availability drops. According to Fischer & Turner (1978), lower values of SLA may be associated with higher resistance to drought. Contrary to the findings of Abrams (1994) who found that xeric genotypes generally produced leaves with lower SLA than those from the mesic region, in our study, SLA did not vary significantly between the two geographic groups. ILB could be considered a favourable selection characteristic in water limited environments, obviously to a large degree linked to summer rainfall which in turn is one of the environmental factors which distinguishes the two groups in this study.

However, we did not find increase in ILB to be associated with increases in carbon fixation: there was a trend, stronger for the Northern group, for C_i to increase under water stress simultaneous with reduction in carbon assimilation. This result may be explained by alterations in the photosystem II (PSII), which may also be the cause of reductions in the rate of *A* and *E*. In the chestnut progenies, the maximal apparent efficiency of PSII is maintained within a fixed range for healthy, well watered plants (Björkman & Demming, 1987). Any stress, such as drought, may be reflected by a drop in this relation (Öquist, 1987), and this was indeed seen in our trial, principally in relation to material from the Northern pool. We also noted a significant increase in initial fluorescence (F_o), providing support for the idea that PSII reaction centres are damaged rather than a protective temporary inhibition in response to excess luminosity. Under water limitations, the proportion of absorbed light used by the plant in the course of photosynthesis and photorespiration is less than it would be in a turgid plant. In this situation, plants can reduce light absorption or divert the light through alternative processes such as thermic dissipation which can account for up to 75% of absorbed light (Niyogi, 1999). In such conditions, leaves experience a transitory reduction in PSII photochemical efficiency or suffer photoinhibition, and, in a worst case scenario, photooxidative destruction. Thus photosynthesis regulation can be seen to be a dynamic process regulated by dissipation of thermal energy (Chaves et al., 2003).

In summary, we found water stress to have a great effect on growth as well as most of the morphological and physiological traits studied in *C. sativa* Mill. juvenile progenies of the main fruit Spanish cultivars, all of which demonstrated its capacity to respond to drought. In spite of the large phenotypic variation observed between progenies, when analysing the data, significant differences were also found between the two gene pools (Northern and Southern) for most morphological and growth traits, as well as for K. This corresponds to the genetic differentiation between these groups found by Pereira-Lorenzo et al. (2010) and Martin et al. (2010) using molecular markers and detected previously by Mattioni et al. (2008). The increased vigour of progenies from the Northern Iberian Peninsula can therefore be attributed to differences in the gene pool of this group compared to that of plants from the rest of Spain.

Moreover, a significant geographical trend related to monthly rainfall in summer (Ps) was found under both experimental and control treatments for juvenile growth and certain leaf morphology traits. According to previous studies (Villani et al., 1994; Lauteri et al., 1997, 2004), genetic, morphological and physiological differentiation of wild chestnut populations spread along a Turkish transect stretching from the Black Sea coast to the Mediterranean area were associated with macroclimatic variables, particularly with precipitation (Pigliucci et al., 1990). However, Conedera et al. (2009) suggested studying the ecological plasticity of the chestnut in more depth in view of possible climate change, as a combination of increased temperatures and a significant reduction in summer precipitation, correlated to the geomorphologic site conditions, has been associated with damage in chestnut stands in the Southern Alps, thus leading the same authors to hypothesize that this species does not have an effective mechanism to protect against over-transpiration in extremely hot and dry weather.

The variability demonstrated in our study may be the result of domestication processes at the local level, i.e., the materials used in different localities are usually the ones which are best adapted to the local conditions even though they may not be the most productive. That said, both origins of chestnut genetic variability offer valuable experimental material for further studies of characteristics and processes involved in divergence and speciation as well as for the safeguarding and development of genetic resources for use in breeding and afforestation programs.

3.3 Hybrid clones

In an attempt to reduce the impact of the ink disease, many strategies have been taken into consideration, from characterizing soil suppressive features to examining orchard management related factors (Martins et al., 2010; Martins et al., 2005; Portela & Pinto, 2005) as well as assessing the impact of oomycete attack on the water relations of chestnut (Gomes-Laranjo et al., 2004; Maurela et al., 2001). In their studies, Robin et al., (2006) observed near 20% of tolerant trees in several populations of *C. sativa* Mill., suggesting that resistant genes could already exist within the species. In the work on genetic resources of *C. sativa* Mill. species, Aravanopoulos (2005) and Eriksson et al. (2005a) also found significant differences in tolerance against *Phytophtora cambivora* (Petri) Buisman infection in England, where the populations were the most sensitive, as opposed to those from Galicia.

Nonetheless, tree breeding in Europe has also focused on adding resistant genes to the gene pool of chestnut. This has been achieved through hybridization with the more resistant Asiatic *C. crenata* Sieb. and Zucc. species. These plants (*C. sativa* Mill. × *C. crenata* Sieb. and Zucc. or the reciprocals), known to be highly tolerant, are largely used as a clonal rootstock in European orchards (Cortizo et al., 1996; Fernandes, 1966; Gomes, 1982; Salesses et al., 1993). Portuguese production of hybrid clones started in the early 1950s by Fernandes (1952), and led to an important collection of clones, COLUTAD (Abreu et al., 1999), being one of the most popular and promising discoveries. Some of these clones showed good features for wood production while others are more suitable as rootstock for *C. sativa* Mill. varieties, thus making these clones first class material for new plantations, as proposed by the National Centre of Forest Seeds (CENASEF).

Japanese chestnut (*C. crenata* Sieb. and Zucc.) and Chinese chestnut (*C. mollissima* Blume) and *C. Sativa* Mill. species are thought to share the same origin, located in eastern Asia (Lang et al., 2007). According to these authors, *C. crenata* Sieb. and Zucc. is the most basal; in relation to Chinese species, it is positioned in a monophyletic clade; the North American and European species are its sister group. *C. crenata* Sieb. and Zucc. also presents the most divergent haplotype in relation to the other *Castanea* species, as a result of the unique climate of the island (Lang et al., 2007). According to Bounous & Beccaro (2002), *C. crenata* Sieb. and Zucc. grows in volcanic soils on the South of Japan, where summers register high levels of precipitation. Due to its origin, Japanese chestnut grows vigorously in humid and

warm climates. The earlier bud burst of these hybrids makes them more vulnerable to frost damage than *C. sativa* Mill., as well as less tolerant to drought.
In a study done with the Euro-Japanese clones growing in CENASEF, they showed 1.5 to 2 times significantly higher in g_s, E and A values than the values measured in Portuguese variety 'Judia'. In fact the former's maximal photosynthetic rate was obtained between 32.5 and 35 °C, whereas for the latter the optimal T was 26.5 °C, explaining by this way the differences between both genotypes.
In terms of leaf water potential, the A_{100} occur between -0.5 and -1.1 MPa for all plants, whereas the values inducing a reduction of 50% in A_{100} are considerably different within hybrid clones, ranging between -2.8 MPa and -1.8 MPa. These values are higher than the -1.2 MPa determined for *C. sativa* Mill. plants.
On the other study to compare the effect of wild (*C. sativa* Mill.) and hybrid (*C.sativa* Mill. x *C. crenata* Sieb. and Zucc.- Ca90) rootstock on Judia, there was observed an increase on heat tolerance when there is used Ca90 rootstock.

3.4 Effect of the air temperature on photosynthetic capacity

Temperature controls the developmental rate of many organisms. The total amount of heat required, between the lower and upper thresholds, for an organism to develop from one point to another in its life cycle is calculated in units called degree-days (°D). This represents the physiological time that provides a common reference for the development of organisms (Cesaraccio et al., 2001; Zalom et al., 1983).
Analysis of Figure 7, shows a strict relation between daily mean A, mean T and their dependence on place's altitude in Trás-os-Montes Region (Portugal). In fact, as the altitude increases, T decreases but A also increases. The maximal value (11.5 μmol $CO_2.m^{-2}.s^{-1}$) was measured at 23-25 °C, corresponding to the daily mean values observed at 850 and 1050 m a.s.l. places. The minimal A value (65% of the maximal rate) was measured in the warmest local, 450 m a.s.l. being 29.15 °C of daily mean T (Figure 7). In the lowest altitude local the amount of heat was 2587 °D and in the highest altitude place that was 1879°D (in the period between March and October in 2007). Other important difference observed in tree response to altitude, and consequently to the T is the daily variation of A (Figure 7). In the highest altitudes, it stays in the maximal range until the middle of the afternoon, while in the lowest altitudes it starts to decrease in the morning, indicating loss on productivity. On the other side, A is much higher in the highest altitude than in the lowest one.

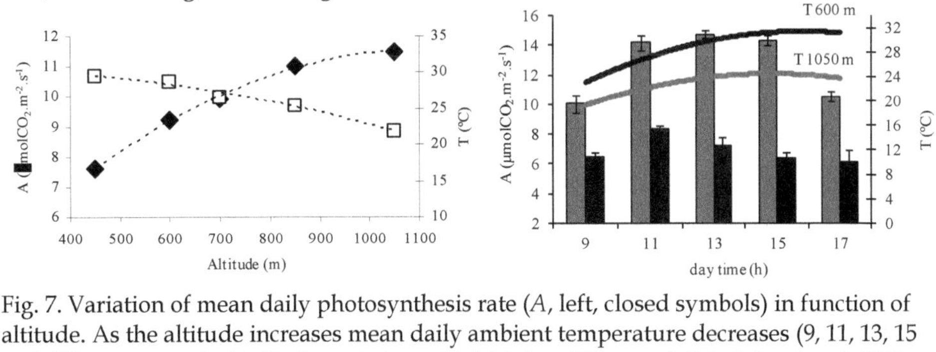

Fig. 7. Variation of mean daily photosynthesis rate (A, left, closed symbols) in function of altitude. As the altitude increases mean daily ambient temperature decreases (9, 11, 13, 15 and 17 h; open symbols). Daily variation of A (right) at 450 m a.s.l. (black bars) and 1050 m a.s.l. (gray bars).

Trees exhibit maximal *A* when Ψ_w is situated in the range of -0.8 to -1.3 MPa; *A* was inhibited by 10% at -1.6 MPa (Gomes-Laranjo et al., 2008a). Under dry soil conditions water potential can decrease to -2.0 MPa, which can induce a reduction in *A* of about 50%. The influence of a pair of factors, water stress and thermoinhibition, may constitute part of the explanation for the Fv/Fm diminution during morning in the warmest localities (450, 600 and 700 m a.s.l.), and the increase for the highest altitude orchards (1050 m a.s.l.). It is important to say that this parameter reflects the maximal efficiency of PSII, and consequently adequate internal conditions to have a high *A*. These conditions are strongly dependent on T, since physical state of thylakoid membranes (where photosynthetic electron chain takes place) is strongly T dependent. Adaptation of membranes to T has been reported by Burkey et al., (1997), since an augmentation in unsaturation level was detected as a function of the altitude increase (Table 1). Variations on Chla/b and Chl/Car ratios, which reflect sun and shade adaptation, also indicate adaptation of plants to the environmental conditions, because Chla/b ratio inversely increases in function of altitude increment and Chl/Car has a similar variation against altitude (Gomes-Laranjo et al., 2008b). These two variations are consistent with the known acquired tolerance to warm and sunny conditions, since Chla is the main pigment present in the photosystem I, which is located in exposed thylakoid membranes, and carotenoids have the chlorophyll protection function against photoinhibition.

Altitude	Chltot	Chla/b	Chl/Car	Total fatty acid		PI	UI	Malonic aldehyde x10^{-4} (mM)	
(s.l.m)	mg.cm^{-2}			Saturated (%)	insaturated (%)			Control	ADP-Fe
1050	121.5 b	3.12 c	4.8 a	27.0	73.0	111.5	171.3	1.35	3.88
900	145.9 a	3.10 c	5.0 a	32.4	67.6	97.5	154.2	1.67	4.57
700	99.1 c	3.30 b	4.4 b	38.2	61.8	119.2	156.3	2.00	4.91
600	143.9 a	3.40 b	4.6 b	33.1	66.9	47.5	146.1	1.90	4.53
450	80.9 d	3.60 a	3.9 c	43.9	56.1	79.6	121.2	3.12	5.36

Table 1. Variation of photosynthetic pigment content (n=10), fatty acid composition (n=3) and malonic aldehyde (n=3) in leaves from adult trees (var. 'Judia'). PI represents the peroxidation index and UI the unsaturation index.

Leaf pigment content was higher on trees growing in lower altitude than that in higher altitude regions, since thermoinhibition speeds light saturation of the photosynthetic process (Boardman, 1977). Moreover, increase in Chla/b suggests higher proportion of stacking thylakoid membranes, which in turn might induce higher *A*, if any stress factor imposes (Anderson et al., 1988). Additionally, thiobarbituric acid reactive substances presented as malonic aldehyde show a decrease in peroxidation activity as the altitude increases, which could indicate a better response to oxidative stress with the altitude.

3.5 Photosynthetic apparatus characterization of leaves from different sides of chestnut canopy

Chestnut (*C. sativa* Mill.) is a large deciduous tree, reaching a height of 40 m and 6 m to 7 m diameter of canopy (Bounous, 2002). In such canopies, it is possible to identify a deep heterogeneity in the light availability around the crown (north, east, south and west regions) besides an enormous internal canopy region (Figure 8).

According to the measurements, leaves located in the north side only have less than 20-% of the south side level of PPFD, making the south part of the canopy hotter than the opposite part. Concerning east and west sides, they have between 74% and 67% of the south PPFD (Table 2). These facts induce a cascade of occurrences at many levels of biological organization (Boardman, 1977; Osmond & Chow, 1988). Leaf area of the south, east, west and north of tree top was different. The highest leaf area was in the northern canopy. The shadow leaves were larger, thinner and adapted to responding more effectively to the less light available and its diffused nature.

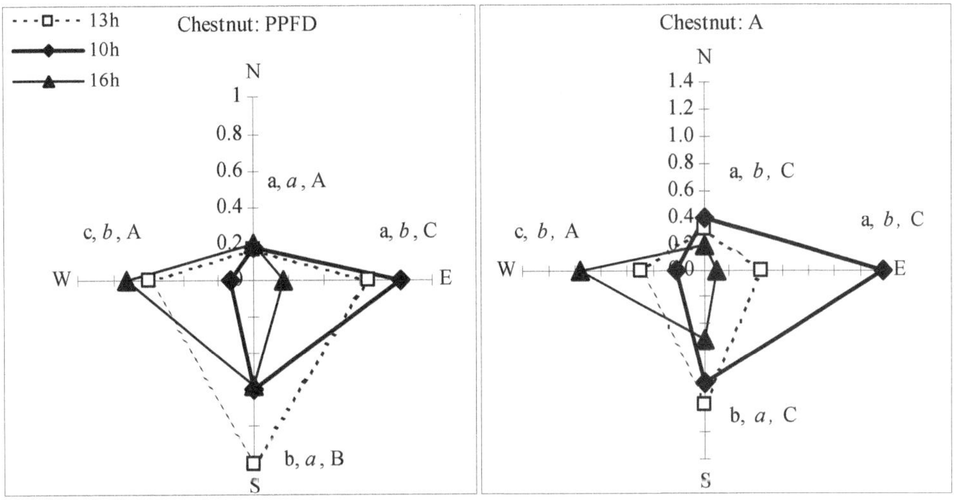

Fig. 8. Study of the radiation (PPFD) (*left*) and photosynthesis (A) (*right*) variation at 10, 13 and 16 h in north, east, south and west sides of 'Judia' crown. In each side, letters represent the result of the comparison between 10 h (normal letters) 13 h (italic letters) and 16 h (caps letters) hours. The values with common letters are not significantly different, according to Fisher test, 5%.

This species has typical mesomorphic leaves with a dorsiventrally flattened lamina. Concerning the lamina internal anatomy, in the upper epidermis there are two layers, and in the lower epidermis only a single layer (Figure 9). Both epidermises have cells with straight anticlinal walls. However, the cells were more regular in the upper than in the lower surface. In the glabrous adaxial surface there is a thicker cuticle layer than that in the abaxial surface, which is coated by a dense trichome layer (depending on variety, it can be almost absent in north leaves).

Differences in light saturation point between south and north leaves were found (Figure 10), showing the former leaves highest PPFD saturation point, as there is known for other sun and shade broadleaves (Lichtenthaler, 1985; Osmond & Chow, 1988). Additionally, north leaves presented higher *A* than south leaves below 400 μmol m^{-2} s^{-1} radiation intensity. The former also presented lowest compensation point

Parameters	North		East		South		West	
PPFD (µmol m^{-2} s^{-1})	137	*d*	673	*b*	912	*a*	611	*c*
LT (°C)	24.4	*b*	25.1	*a*	25.1	*a*	24.8	*ab*
Histology								
Thikness (µm)	188.7	*a*	194.6	*c*	239.2	a	208.7	*b*
Upper epidermis (µm)	22.0	*c*	23.8	*b,c*	29.1	a	24.4	*b*
Lower epidermis (µm)	15.2	*a*	15.0	*a*	16.2	a	15.4	*a*
Palisade mesophyll (µm)	82.4	*c*	87.8	*b*	110.8	a	86.8	*b*
Spongy mesophyll (µm)	73.4	*b*	66.6	*c*	82.2	a	83.6	*b*
Pal/Spongy	1.19	*b*	1.38	*a*	1.39	a	1.07	*c*
Transmitance (%)	5.44	*b*	5.25	*a*	5.24	a	5.36	*a*
Photosynthetic pigments								
Total Chlorophyll (mg.cm^{-2})	63.3	*b*	63.9	*b*	67.6	*ab*	67.0	*a*
Chl*a* / Chl*b*	2.91	*b*	3.09	*a*	3.14	*a*	3.08	*a*
Carotenoides (mg.cm^{-2})	12.1	*a*	13.0	*a*	14.2	*a*	14.5	*a*
Chl/Car	5.23	*a*	4.92	*b*	4.76	*c*	4.62	*c*
Gas Exchanges								
g_s (mmol m^{-2} s^{-1})	169	*b*	190	*a*	188	*a*	183	*a*
A (µmol(CO_2) m^{-2} s^{-1})	1.47	*c*	3.14	*b*	3.89	*a*	3.02	*b*
E (mmol(H_2O) m^{-2} s^{-1})	3.26	*c*	3.54	*b*	3.74	*a*	3.56	*b*
WUE (µmol(CO_2) mmol(H_2O)$^{-1}$)	0.49	*c*	0.84	*b*	1.05	*a*	0.75	*b*
A/PPFD (µmol(CO_2) mmol^{-1})	0.0064	*c*	0.0037	*b*	0.0044	*a*	0.0048	*b*
Ci (ppm)	292	*a*	281	*b*	268	*d*	275	*c*

Means followed by the same letter are not significant different at P<0.05 (Fisher test).

Table 2. Comparison of the mean values of photosynthetic active radiation (PPFD), leaf temperature (LT) and leaf histology, contents of photosynthetic pigments and gas exchanges among canopy sides (n = 225).

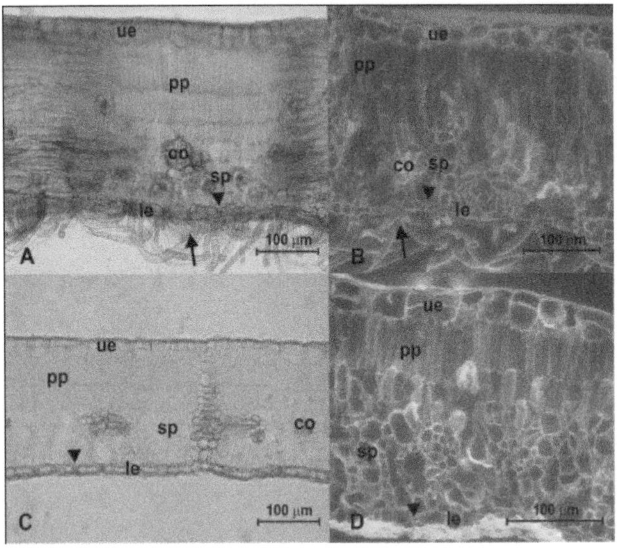

Fig. 9. Optical (A, C) and scanning electron microscopic (B, D) micrographs of the south (A, B) and north (C, D) of leaf cross-section from 'Martaínha' variety. Legend: upper epidermis (ue), lower epidermis (le), palisade parenchyma cells (pp), spongy parenchyma (sp), calcium oxalate crystals (co), stomata (arrowheads), trichomes (arrows).

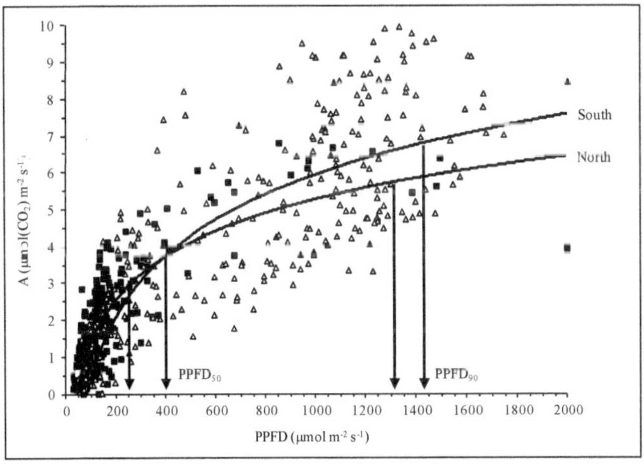

Fig. 10. Study of the correlation between PPFD intensity and photosynthesis rate in north and south sides. Data were obtained in July, August and September. Arrows represent the PPFD value for 90% and 50% of maximal *A* in south and north sides. Logarithmic equation analysis was used to determine the equation of the best-fitting line. The values of r^2 were 0.64 (north) and 0.68 (south), respectively.

Changes in thylakoid membrane surface potential, induced by the electron transfer chain, were studied. South side chloroplast presented highest thylakoid membrane potential (Figure 11). This acclimation to high T is normally associated with a greater degree of saturation of fatty acids in membrane lipids which turns membranes less fluid (Gomes-Laranjo et al., 2006). These authors also demonstrated that chloroplasts from north side showed approximately 5% more of surface thylakoids than those from south chloroplasts.

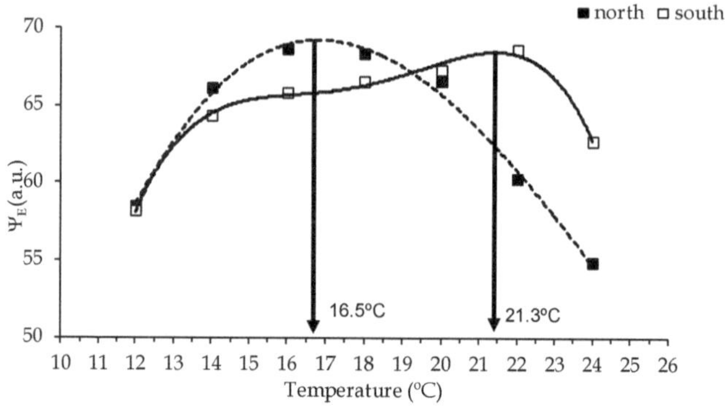

Fig. 11. Influence of temperature in thylakoid membrane potential of 'Judia' chloroplasts from leaves collected in south and north-side of canopy. For membrane potential measurements there was used the molecular probe 9-amino-6-chloro-methoxyacridine (Gomes-Laranjo et al., 2006).

4. Impact of different orchard management strategies on gas exchange rates

4.1 Soil management

Producers have submitted their orchards to intense management practices, such as soil tillage, fertilization and irrigation to increase productivity. In a study done between 2003 and 2007, in Trás-os-Montes region (Portugal), the effect of tillage and irrigation on the predawn leaf water potential (Ψ_{wpd}) (Figure 12) and gas exchanges rates (Figure 13) were demonstrated.

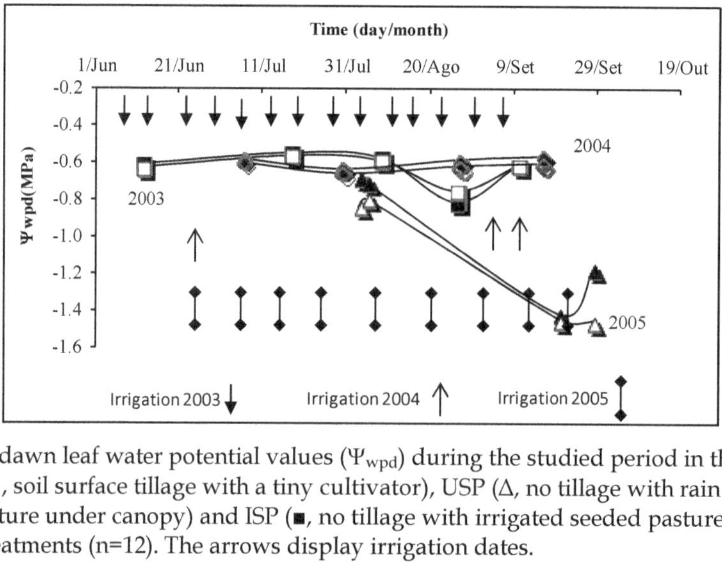

Fig. 12. Predawn leaf water potential values (Ψ_{wpd}) during the studied period in three years, for TTC (▲, soil surface tillage with a tiny cultivator), USP (Δ, no tillage with rain feed seeded pasture under canopy) and ISP (■, no tillage with irrigated seeded pasture under canopy) treatments (n=12). The arrows display irrigation dates.

As shown in Figure 12, in 2003 and 2004, no significant differences in Ψ_{wpd} were found among treatments, with a level around -0.6 MPa. These results showed a complete recovery of the water level during the night, suggesting the availability of enough water for chestnut requirements in the soil, even without irrigation, and favourable conditions for chestnut growth. This observation provides an evidence that trees can absorb water according to its availability across the root zone, which is consistent with earlier observations above-mentioned. In relation to photosynthetic rate (*A*) (Figure 13), no significant differences in cultural practices were detected in 2003 and 2004, following the same pattern as Ψ_{wpd}. Nevertheless, significant differences were detected between the measurements at different T. Independent on the measurement date, highest values of *A*, in the range of 9-9.5 µmol $CO_2.m^{-2}.s^{-1}$, were measured when T reached 28-29 °C, decreasing to 8-8.5 µmol $CO_2.m^{-2}.s^{-1}$ at 33 °C and to 4-5 µmol $CO_2.m^{-2}.s^{-1}$, in August, when T rose up to 37.5 °C. These data are consistent with above lines and previous studies (Gomes-Laranjo et al., 2006) as also with the values obtained for August 2005, showing that photosynthetic rate of *C. sativa* Mill. seems more dependent on atmospheric T than on soil water, under the climatic conditions of the studied region.

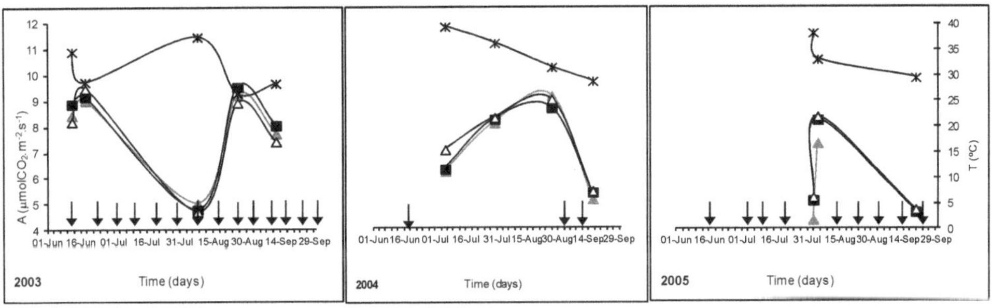

Fig. 13. Obtained results for photosynthesis rate (*A*) and atmospheric temperature (T) on TTC (▲, soil surface tillage with a tine cultivator), USP (Δ, no tillage with rain feed seeded pasture under canopy) and ISP (■, no tillage with irrigated seeded pasture under canopy) treatments in 2003, 2004 and 2005 from June to September. When present, each arrow indicates irrigation in the ISP treatment.

Nevertheless, related to 2005, a decrease in *A* was observed from 5th August to 20th September, from 8.4 to 4.8 µmol $CO_2.m^{-2}.s^{-1}$, in spite of the maintenance of atmospheric T at 30-32 °C, which might be due to the water stress as the very low Ψ_{wpd} (-1.6 MPa) suggested.

Fertilization with nitrogen, phosphorus and potassium is common practice among producers; however applications of magnesium (Mg) and boron (B) are hardly carried out in chestnuts. Both nutrients, Mg and B greatly affect vegetative growth, namely the size and shape of the leaves and obviously LAI and photosynthetic activity. Portela et al. (1999) showed that trees with pronounced yellowing due to Mg deficiency may reduce LAI to values less than 50%, and nut size and productivity were also greatly decreased (Portela et al, 2003). Although B deficiency in chestnuts has been identified and B fertilization was carried out in some orchards with unequivocal positive effects on nut production, only recently post-treatment evaluation was carried out (Portela et al., 2011).

4.2 Boron applications

The trees suffering B deficiency show death of shoot tips and the terminal buds systematically fail; many branches are leafless; leaves are small and narrow, sometimes malformed and with irregular shape; the internodes are shortened and a rosetting arrangement of younger leaves can be observed (Portela et al., 2011). In 2008, fruit setting and productivity were increased fourfold with B applications. Epstein and Bloom (2005) emphasized the role of B in increasing the translocation rate of the sugars through the phloem from the photosynthetic tissues to the actively growing regions and also the developing fruits.

Soil B application in a non-irrigated orchard (around 15 years old) located at 600 m a.s.l in Trás-os-Montes (corresponding to an amount of 2500 °D, in the period of March-October), can increase the water potential in chestnut. This conclusion was drawn from 3 years of B experiment (2006-2009), where in October the leaf water potential (Ψ_w) in fertilized trees with B was -2.14 MPa, while in control plants the value shifted to -2.24 MPa. As a consequence, a cascade of benefits was obtained. In the gas exchange rates (Table 3, unpublished data), an increase in g_s, *E*, *A* were observed as so in chlorophyll content. Photosynthesis rate increased by 52% in 2008 (under 27°C) and by 16% in 2009 (under 31°C). It is noteworthy that values from 2009 were obtained under heat stress (optimal T for chestnut is around 27 °C) inducing a reduction of about 50% in *A*.

Year / Treatment	T (°C)	PPFD ($\mu mol.m^{-2}.s^{-1}$)	gs ($mmol.m^{-2}.s^{-1}$)	E ($mmol.m^{-2}.s^{-1}$)	A ($\mu molCO_2.m^{-2}.s^{-1}$)	WUE ($\mu molCO_2.mmolH_2O$)	A/Ci ($\mu molCO_2.\mu bar$)	A/gs ($\mu molCO_2.mmol$)	Ψ_w (MPa)
2007									
B0	28,09	1180,00	127,79 a	2,57 b	11,77 a	5,24 a	0,08 a	0,11 a	-1,24 a
B1	29,58	1180,80	129,28 a	2,89 a	12,19 a	4,59 b	0,09 a	0,11 a	-1,26 a
2008									
B0	27,60	1305,00	84,00 b	2,27 b	8,29 b	3,66 b	0,04 b	0,10 a	
B1	27,38	1305,00	158,00 a	3,13 a	12,65 a	4,08 a	0,06 a	0,08 b	
2009									
B0	31,50	1392,00	33,02 b	1,18 b	6,73 b	7,15 a	0,38 a	0,25 a	-2,44 a
B1	31,65	1392,00	48,42 a	1,65 a	7,79 a	5,11 b	0,14 b	0,18 b	-2,14 b

Table 3. Effect of boron fertilization on gas exchange rates in chestnut trees.

Boron application to the soil might have created better conditions for development of the root system and therefore a better acquisition of water from the soil. The role of B in root development is well supported by Dell and Huang (1997).

4.3 Silicon applications

Silicon (Si) is regarded as a beneficial element of higher plants (Epstein, 2001). It may increase plant resistance against biotic, e.g., diseases and pests (Hou & Han, 2010; Osunacanizalez et al., 1991) and abiotic stresses, e.g., water deficit (Gong et al., 2003), salt (Saqib et al., 2008) and heavy metals (Neumann & zur Nieden, 2001). Our recent work has shown that foliar application of Si significantly increased chestnut plant g_s and *A* (Figure 14, unpublished data).

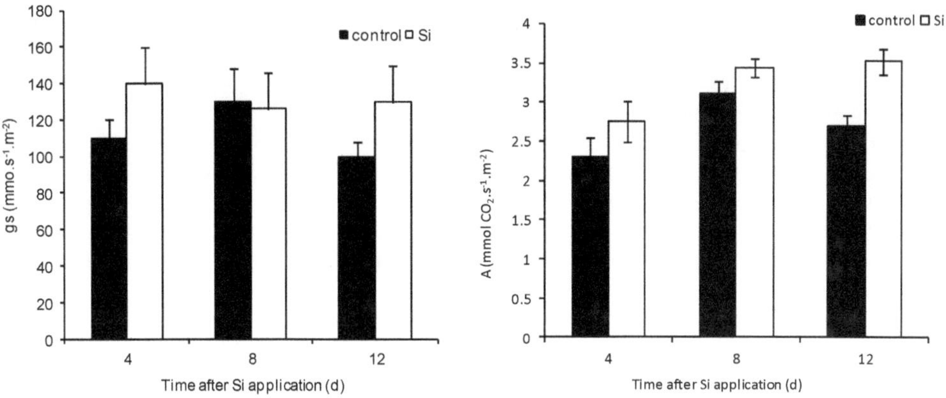

Fig. 14. The effect of silicon (Si) treatment on chestnut (*C. sativa* Mill. M.) plantlet g_s and *A*.

5. Understanding mechanisms of ink disease tolerance by analysing alterations in secondary metabolite synthesis

The oomycete pathogen *Phytophthora cinnamomi* Rands. is generally found in areas with acid to neutral soils containing low amounts of nutrients (mainly low content of phosphorus and calcium), low organic matter, and fewer micro-organisms and also presenting poor drainage, where average annual rainfall is greater than 500 mm. The severity of the symptoms is related to climatic and soil characteristics, which have been shown to be greater in south-facing stands (Martins et al., 2005). The metabolism seems to be directly affected by the water stress provoked by the presence of the oomycete. The most important cause could be the blockage of the xylem and the consequent lack of water in the leaves (Bryant et al., 1983). Infection frequently occurs as a response to several biochemical stimuli exuded from host plants, which are mainly near root injuries. These stimulating compounds are sugar enriched, which attract and promote zoospore germination in plant tissues. A recent study, where resistant and sensitive plantlets were infected, shows an increase of soluble sugars in resistant ones (Dinis et al., 2011b). It is known that the increase in water stress induces accumulation of soluble sugars in leaves (Quick et al., 1992), and consequently an increase in osmotic strength, which in most of times occurs at expenses of starch content diminution (Gomes-Laranjo et al., 2004a). On the other side, there is expected an inhibition in the phloem loading, due to the lack of water, increasing by this way, the starch synthesis in chloroplasts. It is well known that the water stress can also alter carbon assimilation and the partitioning between sucrose and starch (Quick et al., 1992), resulting in an increase in sucrose concentration (Chaves, 1991). Soluble sugar, such as sucrose and glucose, either acts as substrate for cellular respiration or as osmolyte to maintain cell homeostasis. Fructose is not related to osmoprotection but it is related to secondary metabolite synthesis, namely with the synthesis of erythrose-4-P, which acts as a substrate for lignin and phenolic compound syntheses (Rosa et al., 2009).

Dinis et al. (2011b), have observed a strong increase in phenols content (50%) in resistant plantlets with increase of soluble sugars, constituting a part of the defence mechanism against the hyphae invasion. Resistance to ink disease is attributed essentially to phenolic content, because most of them have fungicidal activity. Gallic acid is the most predominant phenolic acid in chestnut (De Vasconcelos et al., 2009). Non-structural carbohydrates tend to accumulate and more phenolic compounds are produced (Bryant et al., 1983). Some of those secondary metabolites, like the polyphenols, function as plant antioxidants due to their free radical scavenging property.

Concerning the protein content opposing results were found. Gomes-Laranjo et al. (2004b) has shown an increase in infected plants with *P. cinnamomi Rands.*, as many other authors also found in plants under water stress (Bacelar et al., 2006; Rosa et al., 2009; De Vasconcelos et al., 2009). However, Dinis et al., (2011b) obtained a decrease in soluble proteins content in the first answered period after infection. This could be a consequence of an increase in protein hydrolysis to provide an increase in amino acids under leaf drought conditions (Yadav et al., 1999). One of the most important responses of plants to drought is an overproduction of different types of compatible solutes (Ashraf & Harris, 2004; Serraj & Sinclair, 2002) namely total free amino acids, proline and soluble sugars. Among several biochemical indexes of water stress, proline accumulation and the decline in protein synthesis in plants have been widely reported. Other authors reported that the accumulation of certain cytosolutes, particularly proline and glycine betaine (a quaternary amine) could avoid the negative effects of the cell osmotic potential decrease, without interfering with protein synthesis (Raggi, 1994). Proline synthesis has been associated with protein hydrolysis induced by water deficit (Bacelar et al., 2006). In the second period, the soluble protein content increased (Dinis et al., 2011b), probably due to defense enzyme activity against the oomycete invasion (Ricardi el al., 1998; Tabaeizadeh, 1998).

In conclusion, leaf water deficit, caused by xylem blockage, due to the invasion of the oomycete hyphae, triggers phenol synthesis and phloem loading, resulting in the damage at the physiological and biochemical levels, which ultimately leading to plant death.

6. Genetic modifications targeting improved chestnut tolerance against abiotic stresses

Abiotic stresses adversely affect growth and productivity and trigger a series of morphological, physiological, biochemical and molecular changes in plants. Drought, extreme T, and saline soils are the most common abiotic stresses that plants have to face. Globally, approximately 22% of the agricultural land is saline (FAO, 2004), and areas under drought have been already expanding and this is expected to increase further (Burke et al., 2006). Gene expression profiles of either drought- or salt-stressed barley plants indicated that, although various genes were differentially regulated in response to different stresses, they possibly induce a similar defense response (Ozturk et al., 2002).

Progress in breeding for drought tolerance has consequently been limited. Molecular biology, however, provides new tools that promise better understanding of the mechanisms of drought stress and drought tolerance. Drought tolerance is a complex trait, and breeding for tolerance has been hampered by interactions between genotype and environment. From the conventional plant breeding point of view, several characteristics and processes have been considered important in drought tolerance improvement.

Similarly, many physiological and morphological (phenotypic) characters are considered important in adaptation to drought stress. Osmotic adjustment, in which the plant increases the concentration of organic molecules in the cell water solution to bind water, is one example of the mechanism that alleviates some of the detrimental effects of drought. A thicker layer of waxy material at the plant surface and more extensive and deeper rooting are the others. Root development plays a major role in a plant's response to water availability. Root development is restricted in acid soils, because of aluminium toxicity. Phosphorus is also highly fixed in acid soils and this too adversely affects root development. Therefore, improving aluminium tolerance and phosphorus uptake indirectly improves drought tolerance. Similarly, physiological and biochemical traits that might enhance drought tolerance have been proposed but only a few of these mechanisms have been demonstrated to be casually related to the expression of tolerance under field conditions.

It has been reported that photosynthesis and several other related physiological traits differ significantly between drought-tolerant and susceptible genotypes. Some crops are naturally more drought tolerant than others, and are obviously better suited to drought environments (Gomes-Laranjo et al., 2006).

Gomes-Laranjo et al. (2006) studied the T effect in three chestnut (*C. sativa* Mill.) varieties' behaviour, 'Judia', 'Longal' and 'Aveleira' and they found differences among them indicating, the least adaptability of 'Judia' to high T (T100, 23.5 °C) and also 'Longal', showing a shift of one degree increase (T100, 24.5 °C) in the optimal T for photosynthesis, and 'Aveleira' which exhibits quite different behaviour (T100, 26 °C and T50, 38.5 °C). When different clones of the 'Judia' variety were studied in terms of heat tolerance, some intra-clonal variety was observed, clones with higher tolerance to heat were found and that seems to be associated with a "memory" heat, since they followed a pattern of behaviour similar to the climatic conditions of the place of origin (Dinis et al., 2011c, d). These genotypes may be interesting for selective breeding for heat resistance. Adaptive traits have been evaluated in progenies from local varieties (Ciordia et al., 2011) between the most genetically differentiated areas in Spain (Pereira-Lorenzo et al., 2010, 2011), the North and the Centre-South. Progenies showed significant differences in growth and morphological aspects, as a strategy for adaptation to water stress, quite common in central and southern Spain.

Identification of areas of the genome that have a major influence on drought tolerance, namely Quantitaive Trait Loci (QTL), could allow marker assisted selection (MAS) to be used to identify those plants from a population that are likely to be better adapted to drought. These areas of the genome are invariably numerous and large, and it is a further step to identify the genes underlying the QTL and assess their contributions to drought tolerance. In addition to accounting for variation in drought tolerance directly, these QTL will also largely determine root morphology and development, and may well govern expression of a whole range of other associated genes. Once the major QTLs have been identified, they might be transferred among plants using linked molecular markers associated with them.

In the mapping project undertaken for the European chestnut (*C. sativa* Mill.) a family obtained from controlled crosses between two trees with contrasting phenotypes, with respect to the efficiency of water use was used: the female parent ('Bursa') belonging to type adapted to drought in the Western part of Turkey, and the male parent ('Hopa') adapted to flooding in Eastern Turkey. Two genetic linkage maps were constructed for the first time for European chestnut, based on this plant material and different markers: RAPD, ISSR and isozyme markers for identification of genomic regions (QTL) related to water use efficiency (Casasoli et al, 2004). QTLs were detected for bud flush, growth and carbon isotope discrimination using female and male parents originated from two Turkish chestnut populations adapted to a drought and humid environment, respectively (Casasoli et al.,

2004). Phenology showed the higher proportion of stable QTLs. Phenotypic correlations and co-localizations among QTLs for different adaptive traits was related with the genetic adaptation of the female parent to drought. Homeologous genomic regions identified between oak and chestnut allowed to compare QTL positions (Casasoli et al., 2006). Co-location of the QTL controlling the timing of bud burst was significant between the two species.

To date, 517 markers were mapped in *C. sativa* Mill. linkage map, covering 80% of their genome, and 12 linkage groups that were aligned to obtain a consensus between the male and female parents. In order to obtain orthologous markers for comparative mapping, microsatellite markers developed for species of *Quercus* and *C. sativa* Mill. were tested in both species. Nineteen loci, 15 of which from *Quercus* and four from *C. sativa* Mill., were integrated into the two maps previously established, allowing the first comparative mapping between the linkage groups of the two species (Barreneche et al., 2004). The same loci were also used to perform the alignment of the maps obtained for the European chestnut with inter map x-specific *C. mollissima* Blume x *C. dentata* (Marsh.) Borkh. (Sisco et al., 2005).

The candidate-gene approach is a powerful and robust method. Compared to the genome wide mapping strategy, the chances of finding markers linked to putative QTL are maximized, since the selection of candidate-gene markers is based on known relationships between biochemistry, physiology and the agronomic character under study.

Genetic association between allelic variants and trait differences on a population scale is a powerful, and relatively recent approach to identifying genes or alleles that contribute to variation in adaptive traits (Long & Langley 1999; see Neale & Savolainen, 2004 for conifers). Regarding forest trees, progress on identification of drought-related genes and development of expressional studies are relatively recent (Chang et al. 1996; Dubos & Plomion, 2003; Watkinson et al., 2003). The molecular basis of dehydration tolerance in trees is extremely complex and a wide variety of expressional candidate genes have been suggested (González-Martínez et al., 2006).

Being multigenic as well as a quantitative trait, it is a challenge to understand the molecular basis of abiotic stress tolerance and to manipulate it as compared to biotic stresses (Amudha & Balasubramani, 2010). Stress-induced gene expression can be broadly categorized into three groups: (1) genes encoding proteins with known enzymatic or structural functions, (2) proteins with as yet unknown functions, and (3) regulatory proteins. Intuitively, genetic engineering would be a faster way to insert beneficial genes than through conventional or molecular breeding. Also, it would be the only option when genes of interest originate from cross barrier species, distant relatives, or from non-plant sources (Bhatnagar-Mathur et al., 2008).

Initial attempts to develop transgenics (mainly tobacco) for abiotic stress tolerance involved "single action genes" i.e., genes responsible for modification of a single metabolite. However, that approach has overlooked the fact that abiotic stress tolerance is likely to involve many genes at a time, and that single-gene tolerance is unlikely to be sustainable (Bhatnagar-Mathur et al., 2008).

Therefore, a second "wave" of transformation attempts to transform plants with the third category of stress-induced genes, namely, regulatory proteins has emerged. Through these proteins, many genes involved in stress response can be simultaneously regulated by a single gene encoding stress inducible transcription factor (Kasuga et al., 1999), thus offering possibility of enhancing tolerance towards multiple stresses including drought, salinity, and freezing. Regarding the genetic transformation of chestnut, apart from the success obtained for American chestnut (Polin et al., 2006), where we can find reports of transgenic plants with resistance genes to *Cryphonectria parasitica* in field trials (Powel et al., 2005), the studies attempted for European chestnut didn't show significant results of stable transformation (Seabra & Pais, 1998, Corredoira et al., 2004).

7. Conclusion

A great deal of work has been done to assess the biodiversity in chestnut species, part of them with the aim to understand species' potentialities facing to climate changes.

Chestnut is quite spread in the world, mostly of those regions being located in the North Hemisphere. The cradle of chestnut, is attributed to a region in east of China, from where it spread for many regions, Europe being one of the most important areas. Nowadays, European chestnut occupies areas in the south part of Europe, mainly corresponding to the Mediterranean countries, ranging between 27º N and 53º N, where climate change impacts can be more significant. In fact, it is consensual that the main consequences of those, might be an increase in T and a decrease in the water availability. The large spread of chestnut in Europe during thousands of years has induced long-term adaptations and by this way a certain specialization of genotypes according to the local edaphoclimatic characteristics of each one. According to elegant studies performed in the Cascade project (Eriksson et al., 2005b), chestnut presents a substantial adaptive variation among populations. They demonstrated that Mediterranean ecotypes might have more tolerance to drought than other ones from wet locals. Additionally, Pereira-Lorenzo et al. (2010) verified that the south Iberian areas of chestnut were colonized by genotypes from the north Iberian region concluding by this way that this species has potential to adapt itself to new climatic conditions and eventually to the new context of climate changes. Anyway, chestnut is not very much heat tolerant as outlined in the present chapter. Air temperatures in the range of 22 to 30 ºC are optimal for it growth. Thus for this species, the summer T is one of the most decisive factors.

Chestnut is a dim-light species presenting significant adaptive degree even inside same canopy, where shady and sunny leaves where also characterized. Chestnut is also an anisohydric species, with considerable buffer capacity during summer times on leaf water Potential, being -0.8 MPa is the typical value of predawn leaf water potential without water stress. This buffer capacity is attributed to deep root system which allows plant to absorb water from the deep soil layers. As this species is less heat tolerant, transpiration plays a decisive role in leaf cooling and by this way in promoting the best photosynthetic rates for such edaphoclimatic conditions. In consequence, chestnut needs great quantity of water, perfectly supporting the popular dictate for chestnut "Chestnut wants to boil in July and drink in August". So, increase of water absorption capacity, must be integrated into the new management strategies. The impacts of boron and silicon applications were studied with promising results. The above referred abiotic stresses, which induce loss of plant vigor, sensitize plant to pests and diseases. The most common, the root rot ink disease attacks the fragile chestnuts, blocking xylem and stopping water uptake, triggering phenol synthesis.

Besides agronomic characteristics, heat and drought stress are two important traits that must continue to be in account for future studies. Morphological and ecophysiological studies in strict connection with genetic studies aiming to identify tolerant genotypes, including the necessary variety breeding programs, must continue in future. The employ of

QTL's, which allows the identification of determined genome areas, could allow marker assisted selection (MAS) to be used to identify tolerant plants from a population that are likely to be better adapted to those stresses.

7. References

Abrams, M.D. (1994). Genotypic and phenotypic variation as stress adaptations in temperate tree species: a review of several case studies. *Tree Physiology*, Vol.14, No.7-8-9, pp. 833-842. ISSN: 0829-318X.

Abreu, C.G.; Carvalho, L.; Gaspar, M.J.; Gomes, A.L.; Colaço, J. & Cardoso, A.O. (1999). Assessment of resistance to chestnut ink disease. *Acta Hort. (ISHS)*, Vol. 494, pp. 363-368. ISSN: 0567-7572.

Allué-Andrade, J.L. (1990). Atlas fitoclimático de España-Taxonomías. *Colección Monografías* INIA, No. 69. ISBN 84-7498-362-2, Madrid.

Amudha, J. & Balasubramani , G. (2011). Recent molecular advances to combat abiotic stress tolerance in crop plants. *Biotechnology and Molecular Biology Review* Vol. 6, No. 2, pp. 31-58. ISSN 1538-2273.

Anderson, J.M.; Chow, W.S. & Goodchild, D.J. (1988). Thylakoid membrane organisation in sun/shade acclimation. In: Ecology of photosynthesis in sun and shade, Evans, J. R. & Caemmerer, Sv IIIWWA (eds.), pp. 11-26., CSIRO, ISBN: 0643048235, Melbourne.

Aravanopoulos, E.A. (2005). Phenotypic Variation and Population Relationships of Chestnut (*Castanea sativa* Mill.) in Greece, Revealed by Multivariate Analysis of Leaf Morphometrics. *Acta Hort. (ISHS)*, Vol. 693, pp. 233-240. ISSN: 0567-7572.

Ashraf, M. & Harris, P.J.C. (2004). Potential biochemical indicators of salinity tolerance in plants. *Plant Sci.* Vol. 166, pp. 3-16. ISSN: 0168-9452.

Ashraf, M.; Rahmatullah, A.R.; Afzal, M.; Tahir, M.A.; Kanwal, S. & Maqsood, M.A. (2009) Potassium and silicon improve yield and juice quality in sugarcane (*Saccharum officinarum* l.) under salt stress. *J Agron Crop Sci* Vol. 195, pp.284-291. ISSN: 0931-2250.

Bacelar, E.A.; Santos, D.L.; Moutinho-Pereira, J.M.; Gonçalves, B.C.; Ferreira, H.F. & Correia, C.M. (2006). Immediate responses and adaptive strategies of three olive cultivars under contrasting water availability regimes: Changes on structure and chemical composition of foliage and oxidative damage. *Plant Sci.* Vol. 170, pp. 596-605. ISSN: 0168-9452.

Barreneche, T.; Casasoli, M.; Russel, K.; Akkak, A.; Meddour, H.; Plomion, C. & Villani, F.K. (2004) Comparative mapping between *Quercus* and *Castanea* using simple-sequence repeats (SSRs).*Theor Appl. Genet* Vol. 108, pp. 558–566. ISSN: 0040-5752.

Bhatnagar-Mathur, P.; Vadez, V.K. & Sharma, K. (2008). Transgenic approaches for abiotic stress tolerance in plants: retrospect and prospects. *Plant Cell Reports* Vol. 27, pp. 411–424. ISSN: 0721-7714.

Björkman, O. & Demming, B. (1987). Photon yield of evolution and chlorophyll fluorescence characteristics at 77 K among vascular plants of diverse origins. *Planta,* Vol.170, pp. 489-504. ISSN: 0032-0935.

Blackwell, M.; Gibas, C.; Gygax, S.; Roman, D. & Wagner, B. (1994). The plastoquinone diffusion coefficient in chloroplasts and its mechanistic implications. *Biochim Biophys Acta*, Vol. 1183, pp.533-543. ISSN: 0006-3002.

Boardman, N.K. (1977) Comparative photosynthesis of sun and shade plants. *Ann. Rev. Plant Physiology*, Vol. 28, pp. 355-377. ISSN: 0066-4294.

Bounous, G. (2002). Il castagno. Coltura, ambiente es utilizzazione in Italia e nel mondo. Edagricole, Italy.

Bounous, G. & Beccaro, G. (2002) Chestnut culture: Directions for establishing new orchards. FAO-CIHEAM, *Nucis-Newletter*, Vol. 11, pp. 30-34. ISSN: 1020-0797

Boyer, J.S. (1996). Advances in drought tolerance in plants. *Advances in Agronomy*, Vol.56, pp. 187–218. ISSN: 0065-2113.

Briggs, L. & Shantz, H. (1914). Relative water requirments of plants. *International Journal of Agricultural Research*, Vol. 3, pp. 1-63. ISSN: 2228-7973.

Bryant, J.P. & Chapin, K.D.R. (1983). Carbon/nutrien balance of boreal plants in relation to vertebrate herbivore. *Oikos*, Vol. 40:357-68. ISSN: 0030-1299.

Burke, E.J.; Brown, S.J. & Christidis, N. (2006). Modeling the recent evolution of global drought and projections for the twenty-first century with the Hadley centre climate model. *J Hydrometeor*, Vol. 7, pp. 1113–1125. ISSN: 1525-755X.

Burkey, K.O.; Wilson, R.F. & Wells, R. (1997) Effects of canopy shade on the lipid composition of soybean leaves. *Physiologia Plantarum*, Vol. 101, pp. 591-598. ISSN: 0031-9317.

Casasoli, M.; Derory, J.; Morera-Dutrey, C.; Brendel, O.; Porth, I.; Guehl, J-M.; Villani, F. & Kremer, A. (2006). Comparison of QTLs for adaptive traits between oak and chestnut based on an EST consensus map. *Genetics*, Vol. 172, pp. 533-546. ISSN: 0016-6731.

Casasoli, M.; Pot, D.; Plomion, C.; Monteverdi, M.C.; Barreneche, T.; Lauteri, M. & Villani, F. (2004). Identification of QTLs affecting adaptive traits in *Castanea sativa* Mill. *Plant Cell Environ.*, Vol. 27, pp. 1088-1101. ISSN: 0140-7791.

Cesaraccio, C.; Spano, D.; Duce, P. & Snyder, R.L. (2001). An improved model for determining degree-day values from daily temperature data. *Biometeorology*, Vol. 45, No. 4, pp. 161-169. ISSN: 0020-7128.

Chang, S.J.; Puryear, J.D.; Dias, M.A.D.L.; Funkhouser, E.A.; Newton, R. J. & Cairney, J. (1996). Gene expression under water deficit in loblolly pine (*Pinus taeda*): isolation and characterization of cDNA clones. *Physiol. Plant*, Vol. 97, pp. 139–148. ISSN: 0031-9317.

Chaves, M.M. (1991). Effects of water deficits on carbon assimilation. *Journal of Experimental Botany*, 42:1-16. ISSN: 0022-0957.

Chaves, M.M.; Maroco, J.P. & Pereira, J.S. (2003). Understanding plant responses to drought-from genes to the whole plant. *Functional Plant Biology*, Vol.30, pp. 239-264. ISSN: 1445-4408.

Ciordia, M. (2009). *Variabilidad genetica de castaño (Castanea sativa* Mill.*) en España mediante marcadores moleculares, caracteres morfológicos y adaptativos*. Phd thesis. Escola Politécnica Superior. Universidad de Santiago de Compostela.

Ciordia, M.; Feito, I.; Pereira-Lorenzo, S.; Fernández, A. & Majada, J. (2011). Adaptive diversity in *Castanea sativa* Mill. half-sib progenies in response to drought stress. *Environmental and Experimental Botany* (under review).

Conedera, M.; Barthold, F.; Torriani, D. & Pezzatti, G.B. (2009). Drought sensitivity of *Castanea sativa* Mill.: the study case of summer 2003 in the Southern Alps. *Acta Hort. (ISHS)*, No.866, pp. 297-302, ISSN 0567-7572.

Corredoira, E.; Montenegro, D.; San-José, M.C.; Vieitez, A.M. & Ballester, A. (2004). Agrobacterium-mediated transformation of European chestnut embryogenic cultures. *Plant Cell Rep.*, Vol. 23, No. 5, pp. 311-318. ISSN: 0721-7714.

Cortizo, E.V.; Madriñan, M.L.V. & Madriñán, F.J.V. (1996). El Castaño. Edilesa, ISBN: 84-8012-134-3, Léon.

Costa, R.; Ferreira-Cardoso, J.; Pimentel-Pereira, M.; Borges, O. & Gomes-Laranjo, J. (2009). Variedades portuguesas de castanha. In: Castanheiros. Técnicas e Práticas, Gomes-Laranjo, J.; Peixoto, F. & Ferreira-Cardoso, J. (eds) pp. 193-212, Pulido Consulting-Indústria Criativa & Universidade Trás-os-Montes e Alto Douro, ISSN: 978-989-96241-0-8, Vila Real.

Cregg, B.M.; Olivas-García, J.M. & Hennessey, T.C. (2000). Provenance variation in carbon isotope discrimination of mature ponderosa pine trees at two locations in the Great Plains. *Canadian Journal of Forest Research-Revue Canadienne de Recherche Forestiere*, Vol.30, No.3, (March 2000), pp. 428-439. ISSN: 0045-5067.

Cufar, K.; Cherubini, M.; Gricar, J.; Prislan, P.; Spina, S. & Romagnoli, M. (2011). Xylem and phloem formation in chestnut (*Castanea sativa* Mill.) during the 2008 growing season. *Dendrochronologia*, Vol. 29, pp. 127-134. ISSN: 1125-7865.

De Vasconcelos, M.C.B.M.; Bennett, R.N.; Rosa, E.A.S. & Ferreira-Cardoso, J. (2009). Industrial processing effects on chestnut fruits (*Castanea sativa* Mill.). 2. Crude protein, free amino acids and phenolic phytochemicals. *Int J Food Sci Tech.*, Vol. 44, pp. 2613–2619. ISSN: 0950-5423.

Dell, B., and Huang, L. (1997). Physiological response of plants to low boron. *Plant Soil*, Vol 193, pp. 103–120. ISSN 0032-079X.

Dinis, L.T.; Peixoto, F.; Pinto, T.; Costa, R.; Bennett, R.N. & Gomes-Laranjo; J. (2011a). Study of morphological and phenological diversity in chestnut trees (Judia' variety) as a function of temperature sum. *Environmental and Experimental Botany*, Vol. 70, pp. 110-120. ISSN: 0098-8472.

Dinis, L-T.; Peixoto, F.; Zhang, C.; Martins, L.; Costa, R. & Gomes-Laranjo, J. (2011b). Physiological and biochemical changes in resistant and sensitive chestnut (*Castanea*) plantlets after inoculation with *Phytophthora cinnamomi* Rands.. *J. Physiological and Molecular Plant Pathology*, Vol. 75, pp. 146-156. ISSN: 0885-5765.

Dinis, L-T.; Ferreira-Cardoso, J., Peixoto, F., Costa, R. & Gomes-Laranjo, J. (2011c). Study of morphological and chemical diversity in chestnut trees (var. 'Judia') as a function of temperature sum. *Cyta* (DOI: 10.1080/19476337.2010.512394).

Dinis, L.-T., Peixoto, F., Ferreira-Cardoso, J., Morais, J.J.L. Borges, A.D.S., Nunes, F.N., Coutinho, J.F., Costa, R., Gomes-Laranjo J. (2011d). Chemical composition and technological properties in chestnut trees (var. 'Judia') as a function of edaphoclimatic conditions. *Journal of the Science of Food and Agriculture* (accepted).

Dubos, C. & Plomion, C. (2003). Identification of water-deficit responsive genes in maritime pine (*Pinus pinaster* Ait.) roots. *Plant Mol. Biol.*, Vol. 51, pp. 249–262. ISSN: 0167-4412.

Epstein, E. (2001). Silicon in plants: Facts *vs.* concepts. In: *Silicon in Agriculture*, Datnoff, L.E.; Snyder, G.H. & Korndorfer, G.H. (eds). Elsevier Science B.V. ISSN: 0167-4412, Amsterdam.

Epstein, E. & Bloom, A.J. (2005). Mineral nutrition of plants: *Principles and perspectives*, Sinauer Associates, Inc. Publishers, ISSN: 0878931724, Sunderland.

Eriksson, G.; Jonson, A.; Laureti, M. & Pliura, A. (2005a). Genetic Variation in Drought Response of *Castanea sativa* Mill. Seedlings. *Acta Hort. (ISHS)*, Vol. 693, pp. 247-254. ISSN: 0567-7572.

Eriksson, G.; Pliura, A.; Fernandez-Lopez, J.; Zas, R.; Silva, R.B.; Villani, F.; Bucci, G.; Casasoli, M.; Cherubini, M.; Lauteri, M.; Mattioni, C.; Monteverdi, C.; Sansotta, A.; Garrod, G.; Mavrogiannis, M.; Scarpa, R.; Spalato, F.; Aravanoupoulos, P.; Alizoti, E.; Drouzas, A.; Robin, C.; Barreneche, T.; Kremer, A.; Rornane, F.; Grandjanny, M.; Grossman, A.; Botta, R.; Marinoni, D.; Akkak, A.; Diamandis, S.; Perlerou, H.; Vannini, A.; Vettraino, A.M.; Russell, K. & Buck E. (2005b) Management of genetic resources of the multi-purpose tree species *Castanea sativa* Mill. *Acta Hort. (ISHS)*, Vol. 693, pp. 373-386. ISSN: 0567-7572.

Fernandes, C.T. (1952). Doenças do Castanheiro. Parasitas do Género *Phytophthora* de Bary. Direcção Geral dos Serviços Florestais e Aquícolas, Lisboa.

Fernandes, C.T. (1966). A «doença da tinta» dos Castanheiros. Direcção Geral dos Florestais e Aquícolas, Alcobaça.

Fernández-López, J.; Zas, R.; Diaz, R.; Aravanopoulos, F.A.; Alizoti, P.G.; Botta, R.; Mellano, M.G.; Villani, F.; Cherbini, M. & Eriksson, G. (2005). Geographic Variability among Extreme European Wild Chestnut Populations. *Acta Hort. (ISHS)*, Vol. 693, pp. 181-186. ISSN: 0567-7572.

Ferrio, J.P.; Resco, V.; Williams, D.G.; Serrano, L. & Voltas, J. (2005). Stable isotopes in arid and semi-arid forest systems, *Investigación Agraria: Sistemas y Recursos Forestales*, Vol.14, No.3, pp. 371-382. ISSN: 1131-7965.

Fischer, R.A. & Turner, N.C. (1978). Plant productivity in the arid and semiarid zones. *Annual Review of Plant Physiology*, Vol.29, pp. 277-317. ISSN: 0066-4294.

Flexas, J. & Medrano, H. (2002). Drought-inhibition of photosynthesis in C_3 plants: stomatal and non-stomatal limitations revisited. *Annals of Botany*, Vol. 89, No.2, pp. 183-189. ISSN: 0305 7364.

Givnish, T.J. (1988). Adaptation to sun and shade: A whole-plant perspective. In: Ecology of photosynthesis in sun and shade, Evans, J. R. & Caemmerer, Sv IIIWWA (eds.), pp. 63-92, CSIRO, ISBN: 0643048235, Melbourne.

Gomes, A.L. (1982). Revisão crítica sobre a cultura do castanheiro em Portugal. University of Trás-os-Montes and Alto Douro, Vila Real.

Gomes-Laranjo, J.; Coutinho, J.P.; Galhano, V. & Ferreira-Cardoso J.V. (2008b). Differences in photosynthetic apparatus of leaves from different sides of the chestnut canopy. *Photosynthetica*, Vol. 46, pp. 63-72. ISSN: 0300-3604.

Gomes-Laranjo, J.; Coutinho, J.P. & Peixoto F. (2008a). Ecophysiological characterization of *C. sativa* Mill. *Acta Hort. (ISHS)*, Vol. 784, pp. 99-105. ISSN: 0567-7572.

Gomes-Laranjo, J.; Coutinho, J.P.; Peixoto, F.; Araujo-Alves, J. (2007). Ecologia do castanheiro. In: Castanheiros, Gomes-Laranjo, J., Ferreira-Cardoso, J; Portela, E. & Abreu, C. (eds.), pp. 109-149, University of Trás-os-Montes and Alto Douro, ISBN: 978-972-669-844-9, Vila Real.

Gomes-Laranjo, J.; Peixoto, F.; Sang, H.; Kraayenhof, R. & Torres-Pereira, J. (2004a). A comparative study on membrane potentials of chestnut (*C. sativa* Mill. *cv* 'Aveleira') and spinach thylakoids. *Biochimica et Biophysica Acta-Bioenergetics*, Vol. 1658, pp. 254-254. ISSN: 0005-2728.

Gomes-Laranjo, J.; Araújo-Alves, J.P.L.; Ferreira-Cardoso, J.V.; Pimentel-Pereira, M.J.; Abreu, C.G. & Torres-Pereira, J.M.G. (2004b). Effect of chestnut ink disease on photosynthetic performance. *J Phytopathology*, Vol. 152, pp. 138-44. ISSN: 0931-1785.

Gomes-Laranjo, J.; Peixoto, F.; Sang, H. & Torres-Pereira, J. (2006). Study of the temperature effect in three chestnut (*Castanea sativa* Mill.) cultivars' behaviour. *Journal of Plant Physiology*, Vol. 163, pp. 945-955. ISSN: 0176-1617.

Gong, H.J.; Chen, K.M.; Chen, G.C.; Wang, S.M. & Zhang, C.L. (2003). Effects of silicon on growth of wheat under drought. *J Plant Nutr.*, Vol. 26, pp. 1055-1063. ISSN: 0190-4167.

González-Martínez, S.; Ersoz, E.; Brown, G.R. ;Wheeler, W. & Neale, D. (2006) -DNA Sequence Variation and Selection of Tag Single-Nucleotide Polymorphisms at Candidate Genes for Drought-Stress Response in *Pinus taeda* L. *Genetics*, Vol. 172, No. 3, pp.1915–1926. ISSN: 0016-6731.

Heiniger, U.C. & Conedera, M. (1992). Chestnut forests and chestnut cultivation in Switzerland. In: *Proceedings of the International Chestnut Conference*, Double, W.L. (Ed.), pp. 175-178, West Virginia University, Morgantown.

Hemmelgarn, M. (2004). Chestnuts: a healty feast. *Journal of the American Chestnut Foundation*, Vol. 18, No. 1, pp. 35-37. ISSN: 1931-4906

Hou, M.L. & Han, Y.Q. (2010). Silicon-mediated rice plant resistance to the asiatic rice borer (*Lepidoptera: Crambidae*): Effects of silicon amendment and rice varietal resistance. *J Econ Entomol.*, 103:1412-1419. ISSN: 0022-0493.

Kasuga, M.; Liu, Q.; Miura, S.; Yamaguchi-Shinozaki, K.; Shinozaki, K. (1999). Improving plant drought, salt, and freezing tolerance by gene transfer of a single stress inducible transcription factor. *Nat Biotechnol.*, Vol. 17, pp. 287–291. ISSN: 1087-0156.

Konstantinidis, P., Tsiourlis, G., Xofis, P. & Buckley, G. (2008). Taxonomy and ecology of *Castanea sativa* Mill. forests in Greece. *Plant Ecology*, Vol. 195, pp. 235-256. ISSN: 1385-0237.

Lang, P.; Dane, F.; Kubisiak, T.L.; Huang, H. (2007). Molecular evidence for an Asian origin and a unique westward migration of species in the genus *Castanea* via Europe to North America. *Mol. Phylogenet Evol.*, Vol. 43, No. 1, pp. 49-59. ISSN: 1055-7903.

Lauteri, M., Monteverdi, C., & Scarascia-Mugnozza, G. (2009). Preservation of chestnut (*Castanea sativa* Mill.) genetic resources and adaptative potential in relation to environmental changes. *Acta Hort. (ISHS)*, Vol. 866, pp. 677-692. ISSN: 0567-7572.

Lauteri, M.; Pliura, A.; Monteverdi, M.C.; Brugnoli, E. & Villani, F. (2004). Genetic variation in carbon isotope discrimination in six European populations of *Castanea sativa* Mill. originating from contrasting localities. *Journal of Evolutionary Biology*, Vol.17, pp. 1286-1296. ISSN: 1420-9101.

Lauteri, M.; Scartazza, M.; Guido, M.C. & Brugnoli, E. (1997). Genetic variation in photosynthetic capacity, carbon isotope discrimination and mesophyll conductance in provenences of castanea sativa Mill. adapted to different environments. *Functional Ecology*, Vol.11, No.6, pp. 675-683. ISSN: 0269-8463.

Li, C.Y. & Wang, K.Y. (2003). Differences in drought responses of three contrasting *Eucalyptus microtheca* F. Muell. populations. *Forest Ecology and Management*, Vol.179, No.1, (July 2003), pp. 377-385. ISSN: 0378-1127.

Lichtenthaler, H.K. (1985). Differences in morphology and chemical composition of leaves grown at different light intensities and qualities. In: *Control of leaf ground*, Baker, N.

R.; Davies, W. J. & Ong, C.O., Vol. 27, pp 201-219, S.E.B. Seminar Series, Cambridge University Press, Cambridge.

Long, A.D. & Langley, C.H. (1999). The power of association studies to detect the contribution of candidate genetic loci to variation in complex traits. *Genome Res.*, Vol. 9, pp. 720-731. ISSN: 1088-9051.

Martín, M.; Moral, A.; Martín, L.; Alvarez, J. (2007). The Genetic Resources of European Sweet Chestnut (*Castanea sativa* Mill. Miller) in Andalusia, Spain. *Genetic Resources and Crop Evolution*, Vol. 54, pp. 379-387. ISSN: 0925-9864.

Martin, M.A.; Mattioni, C.; Cherubini, M.; Taurchini, D. & Villani, F. (2010). Genetic diversity in European chestnut populations by means of genomic and genic microsatellite markers. *Tree Genetics and Genome*, Vol.6, No.5, pp. 735-744. ISSN: 1614-2942.

Martins, A., Raimundo, F.; Borges, O.; Linhares, I.; Sousa, V.; Coutinho, J.P.; Gomes-Laranjo, J.; Madeira, M. (2010). Effects of soil management practices and irrigation on plant water relations and productivity of chestnut stands under Mediterranean conditions. *Plant Soil*, Vol., 327, pp. 57-70. ISSN: 0032-079X.

Martins, L.M., Macedo, F.W., Marques, C.P., Abreu, C.G. (2005). Assesment of Chestnut Ink Disease Spread by Geostatistical. *Acta Hort. (ISHS)*, Vol. 693, pp. 621-625. ISSN: 0567-7572.

Mattioni, C.; Cherubini, M.; Micheli, E.; Villani, F. & Bucci, G. (2008). Role of domestication in shaping *Castanea sativa* Mill. genetic variation. *Tree Genetics and Genome*, Vol.4, pp. 563-574. ISSN: 1614-2942.

Maurela, M.; Robina, C.; Capdeviellea, X.; Loustaub, D. & Desprez-Loustaua, M.-L. (2001). Effects of variable root damage caused Effects of variable root damage caused by *Phytophthora cinnamomi* Rands. on water relations of chestnut saplings. *Ann. For. Sci.*, Vol. 58, pp. 639–651. ISSN: 1286-4560.

Ministerio de Medio Ambiente, Dirección General de Conservación de la Naturaleza. (2000). *Tercer Inventario Forestal (IFN3)*, 2-VI-2011, Available from: http://www.marm.es/es//biodiversidad/servicios/banco-de-datos biodiversidad/informacion-disponible/ifn3.aspx.

Neale, D.B. & Savolainen, O. (2004). Association genetics of complex traits in conifers. *Trends Plant Sci.*, Vol. 9, No. 7, pp. 325-30. ISSN: 1360-1385.

Neumann, D. & Nieden, U. (2001). Silicon and heavy metal tolerance of higher plants. *Phytochemistry*, Vol. 56, pp. 685-692. ISSN: 0031-9422.

Nicotra, A.B.; Cosgrove, M.J.; Cowling, A.; Schlichting, C.D. & Jones, C.S. (2008). Leaf shape linked to photosynthetic rates and temperature optima in South African *Pelargonium* species. *Oecologia,*Vol.154, No.4, pp. 625-635. ISSN: 0029-8549.

Niyogi, K.K. (1999). Photoprotection revisited: genetic and molecular approaches. *Annual Review of Plant Physiology and Plant Molecular Biology*, Vol.50, pp. 333-359. ISSN: 1040-2519.

Öquist, G. (1987). Environmental stress and photosynthesis, In: *Progress in Photosynthesis Research*, J. Biggins, (Ed.), Vol.4, pp. 1-10, Martinus Nijhoff Publishers. ISBN-10: 9024734495, Dordrecht.

Osmond, C.B. & Chow, W.S. (1988). Introduction: Ecology of photosynthesis in the sun and shade: summary and prognostications. In Ecology of photosynthesis in sun and shade, Evans, J. R. & Caemmerer, Sv IIIWWA (eds.), pp. 1-10, CSIRO, ISBN: 0643048235, Melbourne.

Osunacanizalez, F.J.; Dedatta, S.K. & Bonman, J.M. (1991). Nitrogen form and silicon nutrition effects on resistance to blast disease of rice. *Plant and Soil*, Vol. 135, No. 2, pp. 223-231.

Ozturk, Z.N.; Talame, V.; Deyholos, M.; Michalowski, C.B.; Galbraith, D.W.; Gozukirmizi, N.; Tuberosa, R.; Bohnert, H.J. (2002). Monitoring large-scale changes in transcript abundance in drought- and saltstressed barley. *Plant Mol Biol.*, Vol. 48, pp. 551-573. ISSN: 0735-9640.

Pereira-Lorenzo. S.; Ramos-Cabre.r A.M.; Díaz-Hernández. M.B.; Ciordia-Ara. M. & Ríos-Mesa. D. (2006). Chemical composition of chestnut cultivars from Spain. *Scientia Horticulturae*, Vol. 107, pp. 306-314. ISSN: 0304-4238.

Pereira-Lorenzo, S.; Diaz-Hernández, M.B. & Ramos-Cabrer, A.M. (2009). Spain, In: *Following chestnut footprints*, Avanzato, D. (Ed.), pp. 134-142, ISBN 978-90-6605-632-9, Leuven.

Pereira-Lorenzo, S.; Díaz-Hernández, M.B. & Ramos-Cabrer, A.M. (2006). Use of highly discriminating morphological characters and isoenzymes in the study of Spanish chestnut cultivars. *Journal of the American Society for the Horticultural Science*, Vol.131, No.6, pp. 770-779. ISSN: 0003-1062.

Pereira-Lorenzo, S.; Lourenço, R.M.; Ramos-Cabrer, A.M.; Ciordia, M.; Marqués, C.A.; Borges, O. & Barreneche, T., (2011). Chestnut cultivar diversification process in the Iberian Peninsula, Canary Islands and Azores. *Genome*, Vol.54, pp. 301-315. ISSN: 0831-2796.

Pereira-Lorenzo, S.; Lourenço, R.M.; Ramos-Cabrer, A.M.; Marques, C.A.; Serra da Silva, M.; Manzano, G. & Barreneche, T. (2010). Variation in grafted European chestnut and hybrids by microsatellites reveals two main origins in the Iberian Peninsula. *Tree Genetics and Genome*, Vol.6, No.5, pp. 701-715, ISSN 1614-2950.

Pereira-Lorenzo, S.; Ríos-Mesa, D.; González-Díaz, A.J. & Ramos-Cabrer, A.M. (2007). *Los castañeros de Canarias*, CCBAT - Cabildo de Tenerife, ISBN 84-87340-80-6, CAP - Cabildo de La Palma.

Pigliucci, M.; Villani, F. & Benedettelli, S. (1990). Geographic and climatic factors associated with the spatial structure of gene frequencies in *Castanea sativa* Mill. from Turkey. *Journal of Genetics*, Vol.69, No.3, pp. 141-149. ISSN: 0022-1333.

Pimentel- Pereira, M.; Gomes-Laranjo, J.; Pereira-Lorenzo, S. (2007). Análise dos caracteres morfométricos de variedades portuguesas. In: Castanheiros, Gomes-Laranjo, J., Ferreira-Cardoso, J; Portela, E. & Abreu, C. (eds.), pp. 95-108, University of Trás-os-Montes and Alto Douro, ISBN: 978-972-669-844-9, Vila Real.

Polin, L. D.; Liang, H.F.; Ronald, E. M.; Nishii, D. L. , Newhouse, A. E., Nairn, C. J., Powell, W. A. & Maynard, C. A. (2006). Plant Cell. *Tissue and Organ Culture*, Vol. 84, No. 1, pp. 69-79. ISSN: 0167-6857.

Portela, E.; Ferreira-Cardoso, J. & Louzada, J. (2011). Boron application on a chestnut orchard. Effect on yield and quality of nuts. *Journal of Plant Nutrition*, Vol. 34, pp. 1245-1253. ISSN: 0190-4167.

Portela, E.; Roboredo, M. & Louzada, J. 2003. Assessment and description of magnesium deficiencies in chestnut groves. *Journal of Plant Nutrition*, Vol. 26, pp. 503-523. ISSN: 0190-4167.

Portela, E.; Ferreira-Cardoso, J.; Roboredo, M. & Pimentel-Pereira, M. (1999). Influence of magnesium deficiency on chestnut (*Castanea sativa* Mill.) yield and nut quality. In: D. Anaç and Martin-Prével (eds.), Improved Crop Quality by Nutrient Management, pp 153-157, Kluwer Academic Publishers, ISBN 0-7923-5850-3. Dordrecht.

Portela, E. & Pinto, R. (2005). Site Factors and Management Pratices Associated with the Incidence of Blight in Chestnut Orchards in Northeastern Portugal. *Acta Hort. (ISHS)*, Vol. 693, pp. 575-580. ISSN: 0567-7572.

Powel, W.A., Merkel, S.A., Liang, H. & Maynard, C.A. (2005). Blight resistance technology: transgenic approaches. In: Steiner, K.C. & Carlson, J.E. (eds), Proceedings of the Conference on Restoration of American Chestnut to Forest Lands, ISSN: 0013-87464–6. Washington.

Quick, W.P.; Wendler, R.; David, M.M.; Rodrigues, M.L.; Passarinho, J.A. (1992). The effect of water stress on photosynthetic carbon metabolism in four species grown under field conditions. *Plant Cell Environ.*, Vol.15, pp. 25-35. ISSN: 0140-7791.

Raggi, V. (1994).Changes in free amino acids and osmotic adjustment in leaves of water stressed bean. *Plant Physiolology*, 91, pp. 427-34. ISSN: 0032-0889.

Zhang, C., Gomes-Laranjo, J., Correia, C.M., Moutinho-Pereira, J., Carvalho Gonçalves, B.M., Bacelar, E., Peixoto, F.P. e Galhano, V. (2011). Response, tolerance and adaptation to abiotic stress of olive, grapevine and chestnut in the Mediterranean region: Role of abscisic acid, nitric oxide and MicroRNAs. In: 3- Plants and Environment. Série Abiotic Stress, Lazinica, A. ed., vol.. Intech, Rijeka, pp. 179-206.

Ramírez-Valiente, J.A.; Valladares, F.; Gil, L. & Aranda, I. (2009). Population differences in juvenile survival under increasing drought are mediated by seed size in cork oak (*Quercus suber*, L.). *Forest Ecology and Management*, Vol. 257, pp. 1676-1683. ISSN: 0378-1127.

Ricardi, F.; Gazeau, P. & Zivy, M. (1998). Protein changes in response to progressive water deficit in maize. *Plant Phisiology*, Vol. 117, pp. 1253-63. ISSN: 0032-0889.

Robin, C.; Morel, O.; Vettraino, A.-M.; Perlerou, C.; Diamandis, S. & Vannini, A. (2006). Genetic variation in susceptibility to *Phytophthora Cambivora* in European chestnut (*Castanea sativa* Mill.). *Forest Ecology and Management*, Vol. 226, pp. 199-207. ISSN: 0378-1127.

Rosa, M., Prado, C., Podazza, G., Interdonato, R., González, J.A. & Hilal, M., (2009). Soluble sugars: Metabolism, sensing and abiotic stress: A complex network in the life of plants. *Plant Signaling & Behavio.*, Vol. 4, pp. 388 - 93. ISSN: 1559-2316.

Salesses, G.; Chapa, J. & Chazerans, P. (1993). Screening and breeding for ink disease resistance. In: *International Congress on Chestnut*, Antognozi, E. (Ed.), pp. 545-549, University of Perugia, Spoleto.

Saqib, M.; Zorb, C. & Schubert, S. (2008). Silicon-mediated improvement in the salt resistance of wheat (*Triticum aestivum*) results from increased sodium exclusion and resistance to oxidative stress. *Funct Plant Biol.*, Vol. 35, pp. 633-639. ISSN: 1445-4408.

Seabra Costa, R. & Pais, M.S. (1998). Genetic Transformation of European Chestnut (*C. sativa* Mill.). *Plant Cell Reports*, Vol. 17, pp. 177-182. ISSN: 0721-7714.

Serraj, R. & Sinclair, T.R. (2002). Osmolyte accumulation: can it really help increase crop yield under drought conditions? *Plant Cell Environ.*, Vol. 25, pp. 333–41. ISSN: 0140-7791.

Singh, K.; Singh, R.; Singh, J.P.; Singh, Y. & Singh, K.K. (2006). Effect of level and time of silicon application on growth, yield and its uptake by rice (*Oryza sativa* Mill.). *Indian J Agr Sci.*, Vol. 76, pp. 410-413. ISSN: 0019-5022.

Sisco, P.H.; Kubisiak, T.L.; Casasoli, M.; Barreneche, T.; Kremer, A.; Clark, C.; Sederoff, R.R.; Hebard, F.V.; Villani, F. (2005). An improoved genetic map for *Castanea mollissima* Blume/*Castanea dentata* and its relationship to the genetic map of *Castanea sativa* Mill.. *Acta Hort. (ISHS)*, Vol. 693, pp. 491-495. ISSN: 0567-7572.

Tabaeizadeh, Z. (1998). Drought-induced responses in plant cells. *International Review of Cytology*, Vol. 182, pp. 193-247. ISSN: 0074-7696.

Villani, F.; Mattioni, C.; Cherubini, M.; Lauteri, M. & Martin, M. (2010). An integrated approach in assess the genetic and adaptative variation in *Castanea sativa* Mill. *Acta Hort. (ISHS)*, Vol. 866, pp. 91-95. ISSN: 0567-7572.

Villani, F.; Pigliucci, M. & Cherubini, M. (1994). Evolution of *Castanea sativa* Mill. in Turkey and Europe. *Genetic Research*, Vol.63, pp. 109-116. ISSN: 0016-6723.

Watkinson, J. I.; Sioson, A.; Vasquez-Robinet, C.; Shukla, M. & Kumar, D. (2003). Photosynthetic acclimation is reflected in specific patterns of gene expression in drought-stressed loblolly pine. *Plant Physiology* 133: 1702–1716. ISSN: 0032-0889.

Yadav, V.K.; Gupta,V. & Nyflam, Y. (1999). Hormonal regulation of nitrate in gram (*Cicer arietinum*) genotypes under drought. *Indian J Agr Sci.*, Vol. 69, pp. 592-595. ISSN: 0019-5022.

Zalom, F.G.; Goodell, P.B.; Wilson, L.T.; Barnett, W.W.& Bentley; W.J. (1983). Degree days: The calculation and use of heat units in pest management. University of California. Division of Agriculture and Natural Resources Leaflet 21373.

5

Interaction Between Flowering Initiation and Photosynthesis

Giedrė Samuolienė and Pavelas Duchovskis
Institute of Horticulture, Lithuanian Research Centre for Agriculture and Forestry, Kaunas
Lithuania

1. Introduction

The vast majority of the biological processes are dependent on solar radiation. Photosynthesis is the main process which intermediate between light and plant development. Plants utilize solar radiation as a source of energy for photosynthesis, drives water and nutrient transport (Ballaré and Casal, 2000). Besides they use it as environmental cue to modulate a wide range of physiological responses from germination to fruiting (Suetsugu and Wada, 2003), modulates several metabolic pathways affecting cell metabolism, also is the basis to plant structure and molecule production, thus part of what is produced by photosynthesis is used in photomorphogenesis (Quail, 2007). That composes a complex development program called photomorphogenesis.

From many developmental processes that define plant form and function, flowering is of exceptional interest. A lot of horticulturally important plants are depended upon flowering. Much effort is being put into regulating the timing of flowering. Depending on particular species sensitivity to photoperiod, the transition of apex to the reproductive stage is affected by the duration of light (Leavy and Dean, 1998; Nocker, 2001). Many flowering-time studies are based on *Arabidopsis thaliana* model because classic events, the daylenght sensing mechanisms can be light mediated (Bernier et al., 1993). Over the years physiological studies have led to four separate but herewith interdependent models for the control of flowering: photoperiodic induction, non-photoperiodic (autonomous/vernalization), induction by gibberellins and by carbohydrates. The floral transition in biennial photoperiod-sensitive and cold-required plants is associated with an increased content of carbohydrates in apical meristems (Blazquez, Weigel 2000). According to Corbesier et al. (1998), the concentration of sucrose increases dramatically in phloem exudates upon photoinduction in both short and long day plants, even when the photoinductive treatment and accumulation of non-structural carbohydrates limits photosynthesis by feedback regulation (Araya et al., 2006; Araya et al., 2010; Paul and Driscoll, 1997; Paul and Foyer, 2001;). Moreover, sucrose may function as long-distance signalling molecule during floral induction (Bernier et al., 1993; Leavy and Dean, 1998). Meanwhile, accumulation of glucose has been shown to suppress expression of photosynthetic genes and induce leaf senescence, via the signalling hexokinase pathway (Dai et al., 1999). Araya et al. (2006) states, that repression of photosynthesis occurs mainly in leaves that accumulates starch. Though starch per se is not metabolically active, hexokinase is a sensor for sugar repression of photosynthesis (Araya et al., 2010). According to Paul and Foyer (2001), starch synthesis is promoted when sucrose synthesis is restricted and in many plant species leaf starch serves as a transient sink to accommodate excess photosynthate that cannot be converted to sucrose and explored. Thus several mechanisms for the carbohydrate repression of photosynthesis have been discussed: accumulation of non-structural carbohydrates in leaves represses photosynthesis through sucrose synthesis and accumulation of sugar phosphates in the cytosol (Chen et al., 2005); through starch accumulation which causes the deformation of chloroplasts (Nakano et al., 2000); soluble sugars suppress the expression of photosynthetic genes (Paul and Foyer, 2001). Araya et al. (2006) describes carbohydrate repression of photosynthesis in relation to leaf developmental stages, as the metabolic roles of carbohydrates dramatically change depending on the leaf age. The photosynthetic system is constructed by importing carbohydrates to young sink leaves, while mature source leaves with high photosynthetic activities export photosynthates to sink organs. As *Apiaceae* plants form below-ground storage organs, the translocation of assimilates and the distinct distribution of biomass between leaves and storage organs is a sensitive indicator for changes in environment conditions. Thus the transport and distribution of non-structural carbohydrates between plant organs is important for triggering the complete sequence of photosynthesis action and floral evocation. The distinct distribution of biomass between the leaves and below-ground storage organ, forming a clear model of assimilate partitioning between source and sink, can be a sensitive indicator for changes in environmental conditions. Moreover, it can play a key role as 'cross-talk' of response pathway with other flowering induction pathways. Therefore it is important to understand the biological regularities of plant vegetative growth and generative development. The photo and thermo induction is needed for biennial plants for the formation of inflorescence stem and flowers. The biological background of juvenile period and insensibility to the influence of photo- and thermo induction in the physiology of plant ontogenesis is not explored in detail yet. However still, the participation of photosynthetic system and its primary metabolites in flower initiation processes is not clearly understood. It is sill unclear how photosynthetic pigments, non-structural carbohydrates and other materials distribute and interact during photo and thermo induction and during other flowering initiation processes. Thus, it is very important to understand the mechanisms of plant morphogenesis, to control the growth and development processes and theirs' ratio on the purpose to optimize the formation of productivity elements during different ontogenesis stages.

Despite on light duration (photoperiod), light quality (spectral composition) and quantity (flux density) also make influence on plant morphogenetic processes (Hohewoning et al., 2010; Matsuda et al., 2004). The genetic investigations showed that response to light is the outcome of various photoreceptors information acting through complex interacting signal network (Carvalho et al., 2011; Folta and and Childers, 2008; Leavy and Dean, 1998). Therefore this problem is being tried to solve by artificial lighting with solid-state light-emitting diodes (LEDs). As pant physiology and development is regulated not only via photosynthesis (chlorophylls, carotenoids) but also through specific photomorphogenetic photoreceptors (phytochromes, cryptochromes and phototropin) (Ballaré and Casal, 2000; Carvalho et al., 2011). Thus the results from LEDs lighting suggest that the precise selection of particular spectral components enables to modulate the photomorphogenetic responses of the plant (Folta and and Childers, 2008; Goins et al., 1997; Hohewoning et al., 2010; Matsuda et al., 2004). The light signal transduction pathways involve a complex system, which include photoreceptors absorbing particular wavelengths. Chlorophyll *b* absorb blue (absorption maximum at 430, 455 nm) and red (640nm) light, carotenoids absorb near 450 nm, and chlorophyll *a* absorb near 660 nm and triggers photosynthetic process (Bell et al., 2000; Carvalho et al., 2011), but these pigments reflect green light (Carvalho et al., 2011). The photomorphogenetic response depends on red (600 - 700 nm) and far-red (700 - 800 nm) absorbing phytochromes and blue/UV-A (320 - 400 nm) absorbing cryptochromes action (Bradburne et al., 1989; Bell et al., 2000; Carvalho et al., 2011). As it is known that blue light plays important roles in photomorphogenesis, chlorophyll biosynthesis, maturation of chloroplast, photosynthesis or stomatal opening (Hogewoning et al., 2010). Whereas red light is important for elongation processes, changes in plant anatomy, and development of photosynthetic apparatus of plants (Goins et al. 1997;). Thus blue and red LEDs have been used for studies in many areas of photobiological research such as photosynthesis, chlorophyll synthesis and morphogenesis with model plants such as *Arabidopsis*. However, very few studies have been carried out on the effects of LEDs on physiological responses or morphogenesis of *Apiaceae* plants. Besides, there is no much data about the effects of different solid-state light-emitting diodes spectral composition (blue, red, far-red, green, yellow, UV-A) on the growth and development of edible carrot, carbohydrate and chlorophyll contents and distribution between plant organs. Thus the manipulation of photosynthetically active and morphogenic light allows regulating allocation and use of photosynthate within the developing plant.

2. Materials and methods

2.1 Growth conditions and plant material

Edible carrot (*Daucus sativus* (Hoffm.) Röhl.) and common caraway (*Carum carvi* L.) were initially grown in vegetative tumbler, 54x34x15 cm in size, placed in a greenhouse until particular developmental level needed for special experiment (16 hour photoperiod and 21/16°C day/night temperature were maintained). Peat (pH ≈ 6) was used as a substrate.

2.2 Flowering initiation under controlled environment

Carrots with 5 and 9 leaves in rosette, common caraway with 9 leaves in rosette were grown in a phytotron chambers with different photo and thermo periods for 120 days: 0hr and +4°C; 8hr and +4°C; 16hr and +4°C; 8hr and +21/17°C; 16hr and +21/17°C. Then evocation, flower initiation and differentiation processes were investigated under illumination with the photoperiod of 16-hr and +21/16±2°C day/night temperatures.

2.3 The control of morphogenesis and photophysiological processes by light-emitting diodes (LED)

To induce vernalization processes carrots with 9 leaves in rosette were moved from the greenhouse to the phytotron chambers under low temperature (+4°C) treatment for 120 days. Then evocation, flower initiation and differentiation processes were investigated under illumination with the photoperiod of 16 hour and +21/17±2°C day/night temperatures maintained for one month. The originally designed (Tamulaitis et al., 2005) light emitting diode based lighting units, consisting of commercially available wavelengths: blue (445 nm, LuxeonTM type LXHL-LR5C, *Lumileds Lighting*, USA), red (638 nm, delivered by AlGaInP LEDs Luxeon™ type LXHL-MD1D, *Lumileds Lighting*, USA), red (669 nm, L670-66-60, *Epitex*, Japan), and far red (731 nm, L735-05-AU, *Epitex*, Japan) LEDs were used. Illumination with different spectra was generated by LED-based illuminator (Table 1). The lighting treatment started at vegetative stage. The analyses were performed before after four weeks LED treatment.

Light spectral components	Photon flux densities, μmol $m^{-2}s^{-1}$
445, 638, 669, 731 nm (Basal)	167,2
445, 638, 669 nm	167,9
445, 638, 731 nm	160,1
638, 669, 731 nm	169,5
455, 638 nm	160,4
638 nm	150,0

Table 1. The composition of light spectral components (LED)

2.4 Determination of photosynthetic pigments

About 0.2 g of fresh leaf tissue was ground with 0.5 g $CaCO_3$, washed with pure acetone, and filtered through cellulose filter. Sample was diluted till 50 ml with 100% of acetone. Chlorophyll *a*, *b* and carotenoids (carot.) were measured by spectrophotometric method of Wetshtein (Gavrilenko and Zigalova, 2003). The spectrum of photosynthetic pigments was measured at 440.5 nm, 662 nm and 644 nm respectively.

2.5 Determination of non–structural carbohydrates

Fructose (Fru) glucose (Glu) and sucrose (Suc) were measured by high performance liquid chromatography (HPLC) method. About 1 g of fresh plant tissue (leaves, zone of apical meristems or root-crop) was ground and diluted with +70° C 4 ml double distilled water. The extraction was carried out for 24 h. The samples were filtered using cellulose acetate (pore diameter 0.25 μm) syringe filters. The analyses were performed on Shimadzu HPLC (Japan) chromatograph with refractive index detector (RID 10A), oven temperature was maintained at +80° C. Separation of carbohydrates was performed on Shodex SC-1011 column (300 x 4.6 mm) (Japan), mobile phase - double distilled water. The sensitivity of the HPLC method was established using a method validation protocol (ICH, 2005).

2.6 Physiological indices

Flowering initiation stages were described according to Duchovskis (2000).

The net assimilation rate (NAR) of a plant was defined as its growth rate per unit leaf area (LA) for any given time period (day). It can be calculated as:

NAR (g cm^{-2} d^{-1})=(1/LA)(dW/dt), where LA is leaf area (cm^2) and dW/dt is the change in plant dry mass per unit time.

The leaf area was measured by "WinDias" leaf area meter (Delta-T Devices Lts, UK). Leaves and root-crops were dried in a drying oven at 105° C for 24 h for the determination of dry mass.

2.7 Statistical analysis

The analysis were performed in seven (biometrical measurements) or five (analytical measurements) replications and data analysis was processed using one-way analysis of variance *Anova*, the Duncan 's LSD test to trial mean at the confidence level $p = 0.05$. The standard deviation of mean to express values of NAR was used.

3. Results and discussion

3.1 Flowering and physiological signals

As flowering is the first step of sexual reproduction, thus the timing of the transition from vegetative growth to generative development is very important in agriculture, horticulture or plant breeding (Bernier et al., 1993; Nocker, 2001). During juvenile period plants are insensitive to any flowering inductive factor and are not able to form reproductive organs. The minimal developmental level to accept photo and thermo inductive factors for flowering induction for various plants differs (Duchovskis et al., 2003). Much of plant development occurs at the shoot apical meristem (SAM). The acquisition of reproductive competence is marked by changes in the morphology and physiology of vegetative structures, where SAM is the primordia for growth of vegetative organs (leaves) or is responsible for transition to reproductive development (Leavy and Dean, 1998).

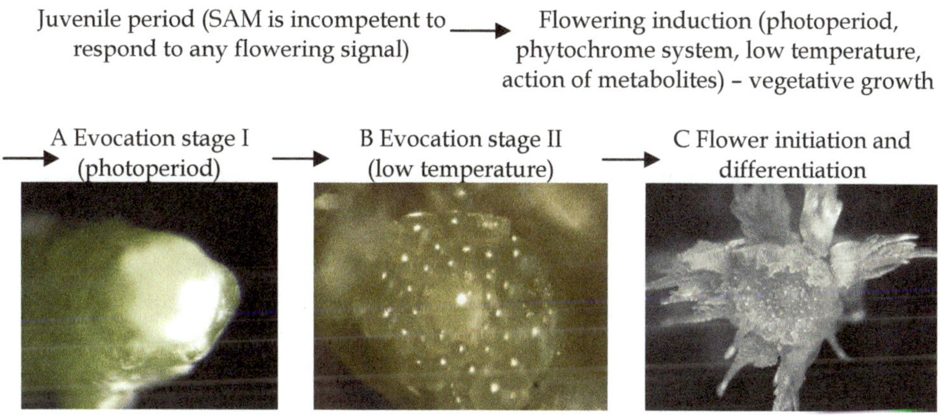

Fig. 1. Model of flowering initiation.

The flowering initiation of biennial plants is a multicomponent and multistep mechanism. It has several stages: flowering induction, evocation, flower initiation and differentiation, gamete initiation and destruction of flowering stimulus (Duchovskis, 2004). Furthermore, it is essential to know the limiting factors of flowering initiation and further development. Endogenous and exogenous factors affecting this process are the complex ones and they depend on solid action of photosynthesis and metabolism systems (Gibson, 2004). Plant flowering initiation processes are related to the duration of juvenile period. During this period plants are insensitive to any inductive factor and are not able to form reproductive organs (Bernier et al., 1993; Duchovskis et al., 2003; Leavy and Dean, 1998). The model of flowering initiation is presented in Fig. 1. According to this model, flowering can occur when certain limiting factors are present at the apex at the right time and in the appropriate concentrations. The multifactoral control model of flowering shows, that a number of promoters and inhibitors, including assimilates, are involved in controlling the developmental transition (Bernier et al., 1993; Gibson, 2004; Németh, 1998).

According to Németh (1998), for the majority of the most important vegetables of the *Apiacea* family temperature between 5–10°C proved to be the most effective for flowering, however both lower (5°C) and higher (15°C) temperatures might have an inductional effect. Duchovskis (2000) stated that the formation of inflorescence axis (5 formed leaves in rosette for carrots) means that photo induction ended and after that the processes of second evocation stage began. The thermo induction conditioned the formation of inflorescence axis elements (evocation stage II) when carrots had 8–9 leaves in rosette (Duchovskis et al., 2003). However, in opposite to high temperature, low positive temperature caused faster development rate of carrots independently from duration of photoperiod (Table 2). As for caraway, it seems that juvenile period is longer than in carrots. Németh (1998) noticed that optimal induction regime for caraway might lie between 5°C and 8°C, which is effective when lasting more than two weeks. In case of caraway, scientific data are very few. Putievsky (1983) examined the effect of day length and temperatures on the flowering of three *Apiacea* species: caraway, dill and coriander. The tree spices exhibited different reactions to the treatments. Caraway developed flowers under all experimental circumstances (18/12°C or 24/12°C day and night temperatures, with 10 h or 16 h photoperiods). Pursuant to other authors, a longer vegetative growth of caraway at low (4°C) temperature and short day (SD) (8 h) occurred, whereas earlier flowering was preceded by long day (LD) (16 h) and low temperature, and the duration of photoperiod did not affect flowering rate under treatment with high temperature (see Table 2). Thus, both a shorter period as well as high temperatures results in partial flowering of treated species. It might mean either that caraway does not need any short day induction for flower initiation at all, or that any photoperiodic response is effective only with interaction of low temperatures (Németh, 1998).
Confirming the conception of flowering induction and evocation of wintering plants, we assume that the mechanisms of photo and thermo induction in edible carrot and common caraway are independent and autonomous. That's why the minimal developmental level to accept these flowering inductive factors differs.

Species	Number of developed leaves	Treatment				
		+4°C-0h	+4°C-8h	+4°C-16h	+21/17°C - 8h	+21/17°C - 16h
Carrot	5	-	4	4	2	2
	9	-	5	5	3	3
Caraway	9	-	2	4	3	3

Table 2. The transition level from vegetative to generative stage in edible carrot and common caraway. Note. 1 point – the lowest development rate, 5 points – the most intensive.

According to our data there were no drastic changes in accumulation of photosynthetic pigments during transition from vegetative growth to generative development (Table 3). During flower initiation and differentiation the significant decrease of chlorophyll *a* and carotenoids was observed under low temperature and short photoperiod treatment, whereas normal temperature and SD photoperiod conditioned the significant increase of these pigments for carrots with 5 leaves in rosette. In opposite to younger plants, the accumulation of photosynthetic pigments was significantly effected by low temperature already in first evocation stage. Moreover during flower initiation and differentiation maturated plants accumulated significantly less photosynthetic pigments under LD photoperiod.

Development	Treatment	Eddible carrot						Common caraway		
		5 leaves			9 leaves			9 leaves		
		Chl *a*	Chl *b*	Carot.	Chl *a*	Chl *b*	Carot.	Chl *a*	Chl *b*	Carot.
Evocation stage I	4°C-8h	1.97ab	0.51ab	0.83ab	2.55b	0.56a	0.78b	1.62c	0.52c	0.55abc
	4°C-16h	1.92ab	0.51ab	0.90b	2.09a	0.65b	0.65a	1.42b	0.49abc	0.59c
	21°C-8h	2.08b	0.52ab	0.84ab	1.76a	0.53a	0.56a	1.22a	0.40a	0.48a
	21°C-16h	1.92ab	0.55b	0.73ab	1.98a	0.56a	0.61a	1.45b	0.50bc	0.53abc
Evocation stage II	4°C-8h	1.74ab	0.55ab	0.66ab	1.74ab	0.53a	0.65ab	1.40abc	0.38d	0.61b
	4°C-16h	1.97ab	0.63ab	0.72b	2.15b	0.67c	0.73ab	1.62c	0.45c	0.75c
	21°C-8h	2.03ab	0.63ab	0.61b	2.09ab	0.56abc	0.73b	1.22a	0.26a	0.46a
	21°C-16h	2.19b	0.68b	0.68ab	2.08ab	0.61abc	0.64ab	1.47bc	0.44c	0.64b
Flower init. and diff.	4°C-8h	1.70a	0.58ab	0.57a	2.09bc	0.69bc	0.67c	1.13b	0.25b	0.39bc
	4°C-16h	1.89abc	0.59ab	0.76bcd	2.16c	0.71c	0.68c	0.89a	0.18a	0.32a
	21°C-8h	2.47c	0.61ab	0.81d	2.00bc	0.69bc	0.60d	1.28c	0.31c	0.41c
	21°C-16h	1.87abc	0.78b	0.61ab	1.19a	0.37a	0.41a	1.06b	0.23b	0.40bc

Table 3. The distribution of photosynthetic pigment content during different flowering initiation stages in leaves of edible carrot and common caraway. The values with the same letters are not significantly different with $P \leq 0.05$

In relation to leaf developmental stages the repression of photosynthesis by non-structural carbohydrate accumulation was examined by Araya et al. (2006) in *Phaseolus vulgaris* L. They noticed that carbohydrate accumulation on photosynthesis repression is significant in the source leaves, but not in the young sink leaves. According to our data, in carrot with 5 leaves in rosette significantly the highest content of glucose was observed under low temperature and SD treatment on first evocation stage (Table 4). However the correlation between chlorophyll content and glucose was week. The accumulation of glucose in carrot leaves mostly was effected by low temperature and LD photoperiod. Meanwhile in caraway leaves significantly bigger contents of glucose were determined under LD and normal temperature during evocation. Paul and Stitt (1993) stated that at low temperature hexose does not induce sugar repression. It is likely that more than one trigger is involved in modulating the carbohydrate signal perception or transduction. Observing the transition from vegetative growth to generative development, linear correlation was noticed under LD treatment in carrot with 5 leaves in rosette. But for carrots with 9 leaves in rosette such linear correlation was affected by low temperature and SD photoperiod. This may occur due to higher contents of glucose in older plants leaves. Krapp and Stitt (1995) noticed that cold exposure prevents carbohydrate export from leaves.

Development	Treatment	Eddible carrot						Common caraway		
		5 leaves			9 leaves			9 leaves		
		Chl *a*+*b*	Glu	COREL	Chl *a*+*b*	Glu	COREL	Chl *a*+*b*	Glu	COREL
Evocation stage I	4°C-8h	2.48ab	3.18c	-0.3	2.67ab	5.75c	-0.6	2.14b	1.11c	-0.8
	4°C-16h	2.43ab	0.37ab	-0.4	2.74b	5.01c	-0.9	1.85a	0.40a	1.0
	21°C-8h	2.60b	0.72b	1.0	2.28ab	3.15a	-1.0	1.62a	0.87b	-1.0
	21°C-16h	2.47ab	0.02a	1.0	2.60ab	3.31a	-0.9	1.75a	1.58d	-0.5
Evocation stage II	4°C-8h	2.62abc	2.17c	0.2	2.46ab	1.00b	1.0	1.78a	0.82c	0.7
	4°C-16h	2.60abc	4.59d	1.0	2.82b	2.00c	0.4	2.07c	0.21a	0.9
	21°C-8h	2.43a	0.13a	0.8	2.69ab	0.49a	-0.6	1.58a	0.65b	-0.2
	21°C-16h	2.87c	1.39b	-0.1	2.69ab	1.06b	-0.3	1.81abc	1.10d	0.5
Flower nit and diff.	4°C-8h	2.28a	1.95a	-0.3	2.68bc	2.56b	-1.0	1.28abc	1.22d	0.9
	4°C-16h	2.51a	1.67a	1.0	2.87c	3.12c	0.2	1.13a	0.25a	0.8
	21°C-8h	3.26b	1.74a	-0.7	2.69bc	2.17b	-0.2	1.59c	0.80c	-1.0
	21°C-16h	2.45a	3.19b	0.9	1.49a	0.40a	-0.5	1.32abc	0.45b	-0.7

Table 4. The correlation between chlorophyll a and b sum and glucose during different flowering initiation stages in edible carrot and common caraway leaves. The values with the same letters are not significantly different with $P \leq 0.05$

Moreover, both chlorophyll and glucose contents were significantly higher in carrot with 9 leaves in rosette, but the correlation was week under low temperature and SD photoperiod in flower initiation and differentiation stage. This is in agreement that mature source leaves export primary metabolites to sink organs, while carbohydrates in young sink leaves are used for photosynthetic system construction. In case of caraway, low correlation in second evocation stage was observed under normal temperature and SD due to significantly low contents of chlorophylls and decrease of glucose concentration. In other developmental stages strong or linear correlation in caraway was observed. Araya et al. (2010) observed negative relationship between leaf carbohydrate content and photosynthetic rate. They showed that leaf photosynthesis is influenced by changing carbohydrate level rather than through modifying sensitivity of the leaf to the carbohydrate level.

Photosynthesis and synthesis of primary metabolites occurs in leaves. Plants of *Apiacea* family are interesting because they form crop-root. It is important to handle mechanisms which regulate photosynthates export from source leaves in response to the demand in the growing sink organs of plant. One of widely cited possibilities is that sugar repression is associated with Rubisco and its distribution to other parts of the plant (Lu et al., 2002; Paul and Stitt, 1993; Paul and Foyer, 2001). The changes in carbohydrate accumulation are frequently in response to changes in the balance between supply and demand for fixed carbon perceived from dark reactions (Farrar et al., 2000). Moreover sink regulation of photosynthesis is highly dependent on the physiology of the rest of the plant, which regulates photosynthesis through signal transduction pathways (Paul and Foyer, 2001).

According to our data (Table 5), under low temperature significantly higher concentrations of hexoses were determined in apical meristems of carrots with 5 and 9 leaves in rosette during transition from vegetative growth to generative development.

Meanwhile in caraway apical meristem zone low temperature conditioned significantly higher accumulation of fructose and under normal temperature exposure significantly higher contents of glucose were determined during transition from first to second evocation stage. This means that glucose can act as morphogenetic factor and regulate the mechanisms of plant development and flowering (Borisjuk et al., 2003; Gibson, 2005). This is in correlation with our data, because the best development of carrot was observed under low temperature, and high development rate of caraway was observed under normal temperature (Table 2). The best development rate of caraway was observed under low temperature and LD (Table 2), this can be explained by significantly high content of sucrose in apical meristem zone (Table 5) during evocation. Moreover there were no significant differences in sucrose content in caraway leaves and the concentrations were very low. Besides, high contents of sucrose were detected in root-crop of caraway. It is known that many plant developmental, physiological and metabolic processes are regulated by soluble sugars such as glucose and sucrose and by other signaling molecules (Gibson, 2004). A lot of scientists investigated the sucrose distribution in apex and in other plant tissues (Bodson and Outlaw, 1985; Lejeune et al., 1993; King and Ben-Tal, 2001). It is presumed that the supply of sucrose to apical meristemic tissues is important for flower induction. Still it may not be the specific flowering induction stimulus and independent from the response to the photoperiod duration. In agreement with other authors (Borisjuk et al., 2002), the highest sucrose concentrations were determined in cells which can actively divide (Tabe 5). Carrots with 9 leaves in a rosette can accept the stimulus of photo induction. Unlike caraway, during the evocation, high temperature disturbs sugar metabolism in carrot apical meristems. Such sugar metabolism and transport to apical meristems can determine the differences in plant development processes (Table 2). Also it may depend on the special plant requirement to photo and thermo induction for the acceptation of flowering stimulus. After transition to generative development during flower initiation and differentiation intensive accumulation of sucrose in apical meristems and transport from leaves to root-crop was observed (Table 5). It is known that sucrose may be converted to glucose and fructose by acid invertase (Mansoor et al., 2002) or it can be reduced from starch by sucrose synthase (Sturm et al., 1999). There is contrary data about feedback inhibition of photosynthesis.

Development	Treatment	Eddible carrot, 5 leaves								
		Root-crop			Apical meristem zone			Leaves		
		Fru	Glu	Suc	Fru	Glu	Suc	Fru	Glu	Suc
Evocation stage I	4°C-8h	7.40c	7.32b		6.31b	4.60a	0.33a	2.99c	3.18c	
	4°C-16h	1.62ab			12.90d	13.11d	0.53b	2.22c	0.37ab	
	21°C-8h	0.41a			4.56a	5.96b	1.93d	0.05a	0.72b	
	21°C-16h	2.62b	1.57a		10.07c	7.00c	0.95c	0.04a	0.02a	
Evocation stage II	4°C-8h	8.41c	5.23c		6.34a	3.63a	0.47a	4.26ab	2.17c	1.05a
	4°C-16h	3.82a	3.12a		12.91c	10.34c	0.75a	5.12b	4.59d	1.03a
	21°C-8h	3.88a	3.99b	3.38b	8.43ab	6.25b	1.68c	2.37ab	0.13a	1.52b
	21°C-16h	7.75c	6.28d	2.12a	10.11b	5.52b	1.34c	3.64ab	1.39b	3.04c
Flower init. and diff.	4°C-8h	4.91c	6.16d	8.41a	4.04c	4.43d	8.86b	1.60a	1.95a	0.16a
	4°C-16h	4.34d	4.47c	10.28a	2.91b	2.32b	10.59c	1.28a	1.67a	0.09a
	21°C-8h	0.60a	1.16a	25.37b	2.14a	0.24a	7.46a	2.71b	1.74a	1.06b
	21°C-16h	4.86c	3.99b	32.03c	4.66d	3.88c	16.36d	3.82c	3.19b	
Eddible carrot, 9 leaves										
Evocation stage I	4°C-8h	7.11c	8.81b	6.95a	7.10c	8.80b	6.92a		5.75c	
	4°C-16h	12.25d	7.85a	14.60b	12.22d	7.82a	14.53b		5.01c	
	21°C-8h	1.63a		16.49c	2.40a		16.43c		3.15a	0.77b
	21°C-16h	4.61b		18.50d	4.60b		18.43d		3.31a	0.08a
Evocation stage II	4°C-8h	4.19d		24.01d	6.48d	2.23c	11.12b	3.83d	1.00b	
	4°C-16h	3.49b		18.47c	4.07c		7.79a	2.76c	2.00c	
	21°C-	2.25a	2.07b	6.80a	1.83a	0.21a		2.34b	0.49a	

Development	Treatment	Eddible carrot, 5 leaves								
		Root-crop			Apical meristem zone			Leaves		
		Fru	Glu	Suc	Fru	Glu	Suc	Fru	Glu	Suc
	8h									
	21°C-16h	3.69bcd	1.39a	9.20b	3.57b	0.88b		1.11a	1.06b	
Flower nit. and diff.	4°C-8h	4.60c	3.42c	18.27a	5.82c	2.47c	5.10bcd	3.15d	2.56b	
	4°C-16h	4.32b	4.67d	17.41a	5.84c	5.32d	4.64b	2.95c	3.12c	
	21°C-8h	0.79a	1.71b	23.68b	3.59d	1.42b	1.95a	1.96b	2.17b	
	21°C-16h	1.03a	0.25a	26.33c	2.15a	0.42a	5.56d	1.35a	0.40a	
Common caraway, 9 leaves										
Evocation stage I	4°C-8h	0.67c	0.50b	3.58a	0.65c	0.63b	4.49a	2.63b	1.11c	0.34ab
	4°C-16h	1.16d	0.27a	5.43c	1.12d	0.33a	4.71b	2.19a	0.40a	0.60ab
	21°C-8h	0.40b	2.11d	7.85d	0.37b	2.63d	10.01d	3.28c	0.87b	0.87b
	21°C-16h	0.22a	1.02c	5.10b	0.23a	1.23c	6.51c	2.29a	1.58d	0.21ab
Evocation stage II	4°C-8h	0.40c	0.30b	2.03a	2.60d	0.50b	1.21a	2.20a	0.82c	0.34ab
	4°C-16h	0.70d	0.20a	2.97b	1.73c	0.20a	5.43c	2.00a	0.21a	0.60ab
	21°C-8h	0.23b	1.50d	6.00c	1.34b	1.12c	3.89b	2.97b	0.65b	0.87b
	21°C-16h	0.10a	0.60c	3.03b	0.71a	1.50d	6.82d	1.90a	1.10d	0.21ab
Flower init. and diff.	4°C-8h	0.26b	1.01c	5.43b	1.47c	0.24c		1.80d	1.22d	
	4°C-16h	0.31c	1.20d	7.40c	1.08b	0.97d	1.30b	1.05c	0.25a	0.38b
	21°C-8h	0.04a	0.50a	5.40b	0.77a	0.03a	0.08a	0.83a	0.80c	
	21°C-16h	0.04a	0.80b	2.43a	1.16b	0.10b	0.04a	0.93b	0.45b	0.37a

Table 5. The distribution of non-structural carbohydrates between edible carrot and common caraway organs. The values with the same letters are not significantly different with $P \leq 0.05$

It has been found that inhibition of photosynthesis occurs in both starch-storing and low accumulating species (Goldschmidt and Huber, 1992). Recent evidence suggests that night-time hexose content derived from starch may provide long term signals for gene expression responsible for feedback regulation. A lag in starch degradation lasts till there is a drop in leaf sucrose, thus the information on the carbohydrate status of the chloroplast may contribute information on the assimilate status of the leaf (Paul and Foyer, 2001).

The effects on Rubisco expression depends not only on distribution between plant organs (Lu et al., 2002), but also on leaf age (Aryae et al., 2006). Since carbohydrate accumulation in the leaves leads to nitrogen release from Rubisco, then sugar signaling is dependent on both nitrogen and carbon status of the plant (Paul and Foyer, 2001). Moreover glucose and sucrose is supposed to be signaling molecules for flowering initiation (Gibson, 2005; King and Ben-Tal, 2001) and their metabolism and supply to apical meristemic tissues determine the plant development. Thus sugars must be considered in the wider context of other important factors, not only as a part of signals that coordinate source-sink interaction.

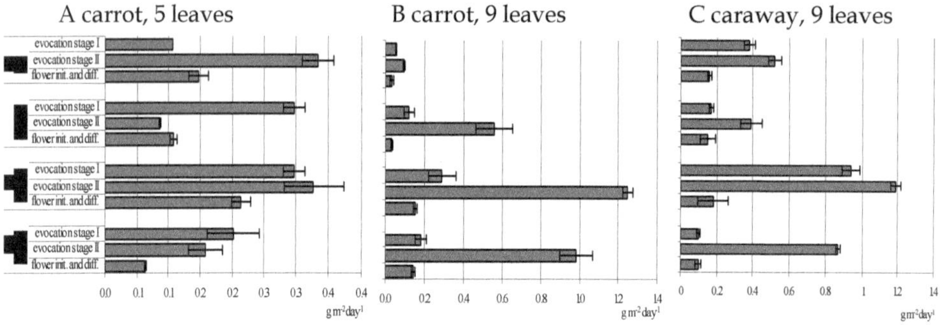

Fig. 2. The distribution of net assimilation rate during different flowering initiation stages in edible carrot and common caraway.

Net assimilation rate (NAR) is related to whole-plant net photosynthetic rate on an area basis and is primary determinant of relative growth rate, which can change daily and over the growing season (Shipley, 2006). Moreover NAR depends on biomass accumulation in a given area and on CO_2 assimilation during day, thus leaf area plays an important role in photosynthesis (Cho et al., 2007; Evans and Pooter, 2001). Therefore normalized photorespiration processes would improve net assimilation rate. Apart of light, temperature is one of indirect factors effecting photosynthesis as critically high temperature accelerates the los of water and stomata closing (Öpik, Rolfe, 2005). According to our data the common trend of NAR increase during second evocation stage was determined (Fig. 2). Such data confirms that plants can accept photo- and thermo- inductive stimulus, and the duration of juvenile period can be associated with maturity of photosynthetic pigments system. Besides normal temperature conditioned higher values of NAR in carrot and caraway with 9 leaves in rosette. The decrease of NAR in all cases was observed in flower initiation and differentiation stage (Fig. 2). This can be related with carbohydrate transport from leaves to root-crop and apical meristem (Table 5). Common trends allow presuming that photosynthetic apparatus of plant not only supply metabolites for morphogenetic processes but also proximately participates in these processes.

3.2 Photophysiological processes: Manipulation by light-emitting diodes

In agreement with other authors (Yanagi et al., 2006), our results show that the elimination of both red (669 nm) and far-red (731 nm) light, which is utilized in the reversible transformation of phytochrome, suppresses floral initiation (Table 6). Although plants were treated by low positive temperatures required for floral induction, after treatment with above-mentioned illumination (455, 638 nm or sole 638 nm) carrots remained in the vegetative stage (Table 6). Also, for these treatments the sugar concentration was low and distribution of hexoses and sucrose was the same in roots and in leaves (Table 8). The elimination of only far-red (445, 638 and 669 nm), red (455, 638 and 731 nm) or blue (638, 669 and 731 nm) light did not have such a dramatic effect on the suppression of plants flowering (Table 6). These results imply that at least far-red is required to invoke floral initiation, probably mediated by phytochrome response. This is in agreement with Ballaré and Casal (2000), where they states that both phytochromes and cryptochrome are involved in photoperiod sensing and accelerate flowering. The effect of the absence of blue light might be lower than that of the absence of far-red light (Table 2). Similarly to our results a study of the light wavelength range for floral induction is required, because floral initiation by red and far-red lighting seems to be mild in strawberry plants (Yanagi et al., 2006). In higher plants there are general pathways in the transduction of photoperiodic/photomorphogenetic signals. Effects of different environmental stimuli (e.g., flowering occurs in response to long photoperiod as well as to low red to far-red ratio) often result in the same developmental or morphogenetic pattern depending on the plant life strategy. According to our results, when red to far-red ratio was equal to zero (445, 638 and 669 nm) even 38% of plants formed the elements of inflorescence axis (flower initiation and differentiation started). In treatments where red to far-red ratio was equal to 50% (445, 638, 669 and 731 nm; 445, 438 and 731 nm; 638, 669 and 731 nm), the development of carrots was lower (Table 6). In other words, photoperiodic and photomorphogenetic light signals trigger similar stress-avoidance response. In long-day plants with competitor strategy the same conditions influence rapid flowering and bolting response (Tarakanov, 2006).

Treatment	Flowering initiation stages (%)			
	Vegetative	Evocation I	Evocation II	Flower initiation and differentiation
445, 638, 669, 731 nm		65	35	
445, 638, 669 nm		27	35	38
445, 638, 731 nm		40	60	
638, 669, 731 nm	22	11	67	
445, 638 nm	100			
638 nm	100			

Table 6. The development rate of edible carrot after one month LED treatment.

Photosynthetic system plays an important role in plant development. Moreover photosynthesis is the main process that interact between light and plant development (Carvalho et al., 2011) In photoperiodic plants, there is strong experimental evidence that leaves produce promoters and inhibitors of flowering when exposed to favorable and unfavorable conditions, respectively (Bernier and Prilleux, 2005). These signals are transported from the leaves to the apical meristem with metabolites of photosynthesis process. Plants are adapted to utilize a wide-spectrum of light to control photomorphogenetic responses (Björn, 1994). Various parts of light spectrum serve as signals providing organism with important information from their environment. Besides through photosynthesis light modulates several metabolic pathways which invoke photomorphogenetic response (Quail, 2007). Appropriate combinations of red and blue light have great potential for use as a light source to drive photosynthesis due to the ability to tailor irradiance output near the peak absorption regions of chlorophyll. There are close relations between plant photosynthesis and photoperiodic response based on source-sink relations (Tarakanov, 2006). Chlorophyll *b* and inactive phytochrome form (P_r) have absorption spectrum at 660 nm, the elimination of this red light (445, 638 and 731 nm) resulted in significantly low rate of photosynthetic pigments synthesis (Table 7) and due to transport from leaves conditioned significantly the highest sucrose content in carrot root-crop (Table 8). In the contrary, the elimination of far-red light (445, 638 and 669 nm) stimulated synthesis of chlorophylls and carotenoids (Table 7). The absorption spectrum of carotenoid and chlorophyll *a* is at blue light region. The elimination of this blue light did not influence the content of chlorophyll *a* but dramatically influenced sucrose transport from leaves and apical meristemic tissues: a high level was found in carrot root (Table 8).

Treatment	Chl *a*	Chl *b*	Carot.	Chl *a*+*b*	Glu	COREL
Before LED	1.48 ±0.13	0.50 ±0.03	0.58 ±0.05	1.98 ±0.17	7.73 ±0.06	-0.9
445, 638, 669, 731 nm	1.34ab	0.57bc	0.36a	1.54a	1.44b	-1.0
445, 638, 669 nm	1.67d	0.59bc	0.53d	2.27c	2.63c	0.1
445, 638, 731 nm	1.16a	0.42a	0.35a	1.42a	7.07e	-0.4
638, 669, 731 nm	1.49bcd	0.62c	0.46bcd	2.21bc	4.24d	-1.0
445, 638 nm	1.34ab	0.47a	0.42ab	1.85abc	0.58a	-0.8
638 nm	1.39abcd	0.56bc	0.38a	1.95abc	0.71a	0.8

Table 7. The content of photosynthetic pigments and correlation between chlorophyll *a* and *b* sum and glucose in edible carrot leaves before and after LED treatment. The values with the same letters are not significantly different with $P \leq 0.05$

According to King (2006), the transported sucrose is effective as a florigen even if its main action is the energy supply. Moreover, conversion of sucrose to glucose can control flowering. According to Bernier and Perilleux (2005), mutants that block photosynthetic carbon metabolism usually exhibit late flowering as could be expected for a plant that shows flowering due to a photosynthetic response. The stability of photosynthetic pigment contents in growth runs without blue (638, 669 and 731 nm) and only with 638 nm lighting (carotenoid absorption maximum) shows very strong participation of photosynthetic system antennal complex (Table 7). Thus antennas permit organisms to increase greatly the absorption cross section for light without having to build an entire reaction center and associated electron transfer system for each pigment, which would be very costly in terms of cellular resources. In photomorphogenesis the attention falls on red/far-red light absorbing phytochromes (Carvalho et al., 2011; Yanagi et al., 2006). These photomorphogenetic receptors operate through interactions with one another and with signaling systems, thus forming complex response networks (Spalding and Folta, 2005). Thus the elimination of far-red light (455, 638 and 669 nm) showed week correlation between chlorophyll content and glucose, and the elimination of red light (455, 638 and 731 nm) conditioned negative medial correlation (Table 7). Meanwhile the elimination of both active components, responsible for phytochrome reversion, showed strong correlation (455, 638 nm; sole 638 nm). Negative linear correlation was observed when both 669 and 731 nm components were combined in illumination treatment (Table 7).

Root-crop			Apical meristem zone			Leaves		
Fru	Glu	Suc	Fru	Glu	Suc	Fru	Glu	Suc
Before LED								
3.29 ±0.09	3.78 ±0.07	43.69 ±1.68	5.57 ±0.94	-	0.37 ±0.17	5.76 ±0.25	0.73 ±0.06	1.52 ±0.21
445, 638, 669, 731 nm (Basal)								
2.07b	20.86e	0.10a	3.20d	0.17a	2.25d	7.10e	1.44b	0.13a
445, 638, 669 nm								
2.34b	10.95d	0.86b	2.34c	10.95d	0.86b	9.85f	2.63c	0.07a
445, 638, 731 nm								
5.49d	6.63c	13.09d	1.98b	1.17b	0.37a	3.40c	7.07e	1.05c
638, 669, 731 nm								
3.08c	1.54b	25.78e	1.98b	1.67c	0.47a	4.44d	4.24d	0.90b
445, 638 nm								
1.21a	0.77a	1.93c	1.31a	0.87b	2.12d	1.02a	0.58a	5.37e
638 nm								
1.41a	0.75a	1.56c	1.21a	0.83b	1.60c	1.23b	0.71a	1.85d

Table 8. The distribution of non-structural carbohydrates among carrot organs before and after LED treatment. The values with the same letters are not significantly different with $P \leq 0.05$

During the reproductive phase of development a lot of new structures are formed, the photosynthetic apparatus is complete. Photosynthetic pigments can participate like structural material for carbohydrates biosynthesis. Moreover photomorphogenetic responses may involve changes in the partitioning of photoassimilates (Ballaré and Casal, 2000). Both, light quantity and quality are known to affect the contents and the ratio of individual proteins and pigment-protein complexes of the photosynthetic apparatus. It is well known that blue light promotes stomatal opening and influences the biochemical properties of photosynthesis. Table 8 shows that though the variation between hexoses and sucrose in different lighting treatments was different, but we found that when the total amount of monocarbohydrates is high, the levels of sucrose are low. Contrarily where the concentrations of monocarbohydrates were low, the amounts of sucrose were high. The significant decrease of sucrose content in leaves was observed under basal LED treatment and without far-red component. Significantly the lowest contents of hexoses and the highest contents of sucrose in leaves were detected under blue and red lighting (455 and 638 nm). Ballaré and Casal (2000) stated that increased export of photosynthates from leaves is due to low red and far-red ratio.

Menard et al. (2005) showed that plants grown under blue fluorescent lamps had higher chlorophyll *a*-to-*b* ratios, smaller amounts of light-harvesting chlorophyll *a*/*b*-binding protein of photosynthetic system II per unit chlorophyll content, and higher ribulose-1.5-

bisphosphate carboxygenase/oxygenase activities per unit leaf area compared to plants grown under red fluorescent lamps. Our data shows that elimination of blue light (638, 669 and 731 nm) influenced the highest net assimilation rate (Fig. 3). 2-3 times bigger values of NAR was observed in lighting treatments without red (669 nm) or without far-red (731 nm) components. Since NAR is closely related to biomass accumulation, CO_2 assimilation during day, thus the production of assimilates is one of the most important components (Cho et al., 2007; Evans and Pooter, 2001; Shipley, 2006). This correlates with our data, as the highest contents of total carbohydrates (44.1 mg g^{-1}, 40.25 mg g^{-1} and 40.85 mg g^{-1} in fresh weight respectively) was exactly under discussed lighting conditions (Table 8). Thus this indicates that net assimilation rate is correlated with changes in photoassimilates content but this increase was not involved by changes in photosynthetic pigment contents (Table 7).

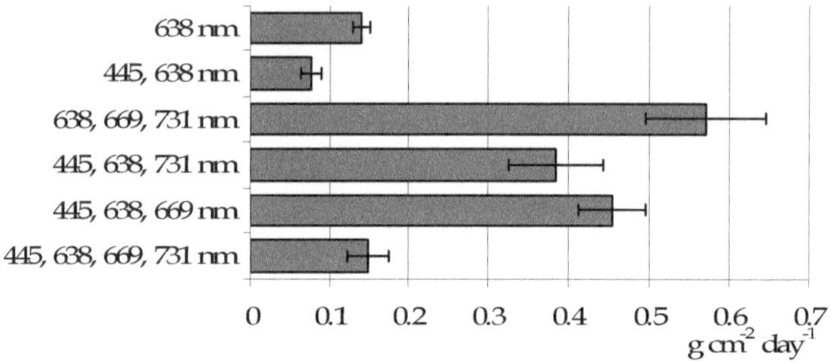

Fig. 3. The distribution of net assimilation rate in edible carrot after LED treatment.

4. Conclusion

To summarize, the following conclusions can be drawn.

The photosynthesis system and carbohydrate metabolism not only supplies morphogenetic processes with metabolites, but also tendentiously varies during ontogenesis, especially during transition from vegetative growth to generative development. Normal development of carrots and caraway was observed only when particular ratio of fructose, glucose and sucrose in different flowering initiation stages constituted. The active supply of glucose and sucrose to apical meristems is important not only for flower induction but as signalling molecule it also participates in source-sink interaction. Notwithstanding, it is not the specific induction stimulus for flowering and is independent from response to photoperiodic duration. Besides temperature make stronger effect on carrot flowering initiation, morphogenetic effects, and on non-structural carbohydrate metabolism neither photoperiod regimes during different ontogenesis stages. Long day and vernalization determines almost full flowering, high temperatures independently from photoperiod results in partial flowering and short day and vernalization is the limiting factor of caraway flowering.

Plant photosynthetic and photomorphogenetic processes, biosynthesis of primary and secondary metabolites and expression of genes can be controlled by light spectral composition. The elimination of both red (669 nm) and far-red (731 nm) or only blue (445 nm) light appeared to suppress floral initiation, under such conditions vegetative growth of carrots was observed. In contrary, the elimination of solely far-red light resulted in faster differentiation of influorescence axis. The elimination of solely red (669 nm) or blue (445 nm) light resulted in a low synthesis rate of photosynthetic pigments and conditioned carbohydrate transport from carrot leaves. Meanwhile, the elimination of solely far-red light resulted in the opposite effect. Dominating 638-nm light was found to considerably contribute to the excitation of the carotenoid antennal complex of the photosynthetic system. The ratio of blue and red light is less important for photosynthesis system than the ratio of different red lights. Blue (445 nm), red (638 and 669 nm) and far-red (730 nm) light and their ratio are very important for photomorphogenetic processes during different plant developmental stages.

Common trends allow presuming that light signals perceived by specific photoreceptor system control the morphology and development of plants and photosynthetic apparatus of plant not only supply metabolites for morphogenetic processes but also proximately participates in these processes. Studies with individual plants in pot of experiments are important steps, but the results of these studies, due to great species and varieties diversity, can not be directly scaled up to predict the impacts of photomorphogenetic manipulations at the biennual plant level.

5. References

Araya, T.; Noguchi, K. & Terashima, I. (2006). Effects of carbohydrate accumulation on photosynthesis differ between sink and source leaves of *Phaseolus vulgaris* L. *Plant and Cell Physiology*, Vol.47, pp. 644-652, ISSN 1471-9053

Araya, T.; Noguchi, K. & Terashima, I. (2010). Effect of nitrogen nutrition on the carbohydrate repression of photosynthesis in leaves of *Phaseolus vulgaris* L. *Journal of Plant Research*, Vol.123, pp. 371-379, ISSN 1618-0860

Ballaré, C.L. & Casal, J.J. (2000). Light signals perceived by crop and weed plants. *Field Crop Research*, Vol.67, pp. 149-160, ISSN 0378-4290

Bell, G.E.; Danneberger, T.K. & McMahon, M.J. (2000). Spectral irradiance available for turfgrass growth in sun and shade. *Crop Science*, Vol.40, pp. 189-195, ISSN 1435-0653

Bernier, G. & Prilleux, C. (2005). A physiological overview of the genetics of flowering time control. *Plant Biotechnology Journal*, Vol. 3, pp. 3-16, ISSN 1467-7644

Bernier, G.; Havelange, A., Housa, C., Petitjean, A. & Lejeune P. (1993). Physiological signals that induce flowering. *The Plant Cell*, Vol.5, pp. 1147–1155 ISSN 1531-298X

Björn, L.O. (1994). Modelling the light environment. In Photomorphogenesis in Plants (2nd ed., Kendrick, R.E. & Kronenberg, G.H.M., eds), pp. 537-551. Martinus Nijhoff Publishers. ISBN 0-7923-2550-8.

Blazquez, M.A. & Weigel, D. (2000). Interaction of floral inductive signals in *Arabidopsis*. *Nature*, Vol.404, pp. 889-892, ISSN 0028-0836

Bodson, M. & Outlaw, W. H. (1985). Elevation in the sucrose content of the shoot apical meristem of *Sinapis alba* at floral evocation. *Plant Physiology*, Vol.79, pp. 420–424, ISSN 1532-2548

Borisjuk, L., Walenta, S., Rolletschek, H., Mueller-Klieser, W., Wobus, U. & Weber, H. (2002). Spatial analysis of plant metabolism: sucrose imaging within *Vicia faba* cotyledons reveals specific developmental patterns. *The Plant Journal*, Vol.29, pp. 521–530, ISSN 1365-313X

Borisjuk, L.; Rolletschek, H., Wobus, U. & Weber H. (2003). Differentiation of legume cytoledons as related to metabolic gradients and assimilate transport in seeds. *Journal of Experimental Botany*, Vol.54, pp. 503–512, ISSN 1640-2431

Bradbulrne, J.A.; Kasperbauer, M.J. & Mathis, J.N. (1989). Reflected far-red light effects on chlorophyll and light-harvesting chlorophyll protein (LHC-II) contents under field conditions. *Plant Physiology*, Vol.91, pp. 800-803, ISSN 1532-2548

Carvalho, R.F.; Takakai, M. & Azevedo, R.A. (2011). Plant pigments: the many faces of light perception. *Acta Physiologiae Plantarum*, Vol.33, pp. 241-248, ISSN 1861-1664

Chen, S.; Hajirezaei, M., Peisker, M., Tschersch, H., Sonnewald, U. & Börnke, F. (2005). Decreased sucrose-6-phosphate phosphatase level in transgenic tobacco inhibits photosynthesis, alters carbohydrate partitioning and reduces growth. *Planta*, Vol.221, pp. 479-492, ISSN 1432-2048

Cho, Y.Y.; Oh, S., Oh, M.M. & Son, J.E. (2007). Estimation of individual leaf area, fresh weight, and dry weight of hydroponically grown cucumbers (*Cucumis sativus* L.) using leaf length, width, and SPAD value. *Scientia Horticulturae*, Vol.111, pp. 330-334, ISSN 0304-4238

Corbesier, L.; Lejeune, P. & Bernier G. (1998). The role of carbohydrates in the induction of flowering in *Arabidopsis thaliana*: comparision between the wild type and a atarchless mutant. *Planta*, Vol.206, pp. 131-137, ISSN 1432-2048

Dai, N.; Schaffer, A., Petrikov, M., Shahak, Y., Giller, Y., Ratner, K., Levine, A. & Granot, D. (1999). Over expression of *Arabidopsis* hexokinase in tomato plants, inhibits growth, reduces photosynthesis, and induces rapid senescence. *Plant Cell*, Vol.11, pp. 1253-1266, ISSN 1532-298X

Duchovskis, P. (2000) Conception of two-phase flowering induction and evocation in wintering plants. *Sodininkystė ir daržininkystė*, Vol.19, No.3, pp. 3–14, ISSN 0236-4212

Duchovskis, P. (2004). Flowering initiation of wintering plants. *Sodininkystė ir daržininkystė*, Vol.23, No.2, pp. 3-11, ISSN 0236-4212

Duchovskis, P.; Žukauskas, N., Šikšnianienė, J.B. & Samuolienė, G. (2003). Valgomųjų morkų (*Daucus sativus* Röhl.) juvenalinio periodo, žydėjimo indukcijos ir evokacijos procesų ypatumai. *Sodininkystė ir daržininkystė*, Vol.22, No.1, pp. 86–93, ISSN 0236-4212

Evans, J.R. & Poorter, H. (2001). Photosynthetic acclimation of plants to growth irradiance: the relative importance of specific leaf area and nitrogen partitioning in maximizing carbon gain. *Plant, Cell and Environment*, Vol.24, pp. 755-767, ISSN 1365-3040

Farrar, J.; Pollock, C. & Gallagher, J. (2000). Sucrose and the integration of metabolism in vascular plants. *Plant Science*, Vol.154, pp. 1-11, ISSN 0168-9452

Folta, K.M. & Childers, K.S. (2008). Light as a growth regulator: controlling plant biology with narrow-bandwidth solid-state lighting systems. *HortScience*, Vol.43, No.7, pp. 1957-1964, ISSN 0018-5345

Gavrilenko, V.Y. & Zigalova, T.V. (2003). Big practice of growing physiology. Moscow. (in Russian).

Gibson, S.I. (2004). Sugar and phytohormone response pathways: navigating a signalling network. *Journal of Experimental Botany*, Vol.55, pp. 253–264, ISSN 1460-2431

Gibson, S.I. (2005). Control of plant development and gene expression by sugar signaling. *Current Opinion in Plant Biology*, Vol.8, pp. 93–102, ISSN 1369-5266

Goins, G.D.; Yorio, N.C., Sanwo, M.M. & Brown, C.S. (1997). Photomorphogenesis, photosynthesis, and seed yield of wheat plants grown under red light-emitting diodes (LEDs) with and without supplemental blue lighting. *Journal of Experimental Botany*, Vol.48, pp. 1407-1413, ISSN 1460-2431

Goldschmidt, E.E. & Huber, S.C. (1992). Regulation of photosynthesis by end-product accumulation in leaves of plants storing starch, sucrose and hexose sugars. *Plant physiology*, Vol.99, pp. 1443-1448, ISSN 1532-2548

Hogewoning, S.W.; Trouwborst, G., Maljaars, H., Poorter, H., Ieperen, W. & Harbinson, J. (2010). Blue light dose-responses of leaf photosynthesis, morphology, and chemical composition of *Cucumis sativus* grown under different combinations of red and blue light. *Journal of Experimental Botany*, Vol.61, No.11, pp. 3107-3117, ISSN 1460-2431

ICH. (1996). Harmonized Tripartite Guideline, Validation of Analytical procedures - Methodology The International Conference on Harmonization of Technical Requirements for Registration of Pharmaceuticals for Human Use, 6 November.

King, R. (2006). Light-regulated plant growth and flowering; from photoreceptors to genes, hormones and signals. *Acta Horticulturae*, Vol.711, pp. 227-233, ISSN 0567-7572

King, R.W. & Ben-Tal, Y. (1998). A florigenic effect of sucrose in *Fuchsia hybrida* is blocked by gibberellin-induced assimilate competition. *Plant Physiology*, Vol.125, pp. 488–496, ISSN 1532-2548

Krapp, A. & Stitt, M. (1995). An evaluation of direct and indirect mechanisms fot the "sink-regulation" of photosynthesis in spinach: changes in gas exchange, carbohydrates, metabolites, enzyme activities and steady-state transcript levels after cold-girdling source leaves. *Planta*, Vol.195, pp. 313-323, ISSN 1432-2048

Leavy, Y. & Dean, C. (1998). The transition of flowering. *Plant Cell*, Vol.10, pp. 89-96, ISSN 1532-298X

Lejeune, P.; Bernier, G., Reguler, M. C. & Kinet, J. M. (1993). Sucrose increase during floral induction in the phloem sap collected at the apical part of the shoot of the long-day plant *Sinapis alba*. *Planta*, Vol.190, pp. 71–74, ISSN 1432-2048

Lu, C.; Koroleva, O.A., Farrar, J.F., Gallagher, J., Pollock, C.J & Tomos A.D. (2002). Rubisco small subunit, chlorophyll a/b-binding protein and sucrose:fructan-6-fructosyl transferase gene expression and sugar status in single barley leaf cells in situ. Cell type specificity and induction by light. *Plant Physiology*, Vol.130, pp. 1335-1348, ISSN 1532-2548

Mansoor, S.; Naqvi, F.N. & Salimullah, T. (2002). Soluble forms of invertases in mungbean (*Vigna radiata*) during germination and developmental stages of various tissues. *Pakistan Journal of Biological Sciences*, Vol.5, No.10, pp. 1063-1066, ISSN 1812-5735

Matsuda, R.; Ohashi-Kaneko, K., Fujiwara, K., Goto, E. & Kurata, K. (2004). Photosynthetic characteristics of rice leaves grown under red light with or without supplemental blue light. *Plant and Cell Physiology*, Vol.45, pp. 1870-1874, ISSN 1471-9053

Menard, C.; Dorais, M., Hovi, T. & Gosselin, A. (2006). Development and physiological responses of tomato and cucumber to aditional blue light. *Acta Horticulturae*, Vol.711, pp. 291-296, ISSN 0567-7572

Nakano, H.; Muramatsu, S., makino, A. & Mae, T. (2000). Relationship between the suppression of photosynthesis and starch accumulation in the pod-removed bean. Australian Journal of Plant Physiology, Vol.27, pp. 167-173, ISSN 0310-7841

Németh, E. (1998). Caraway. The genus *Carum*. ISBN 90-5702-395-4, UK, pp. 69–76

Nocker, S. (2001). The molecular biology of flowering. *Horticultural Reviews*, Vol.27, pp. 1-38, ISBN 0-471-38790-8, John Wiley & Sons, Inc.

Öpik, H. & Rolfe, S. (2005). The physiology of flowering plants. 4th Ed., pp. 392, ISBN 0521662516, Cambridge

Paul, M.J. & Driscoll, S.P. (1997). Sugar repression of photosynthesis: the role of carbohydrates in signaling nitrogen deficiency through source:sink imbalance. *Plant, Cell and Environment*, Vol.20, pp. 110-116, ISSN 1365-3040

Paul, M.J. & Foyer, C.H. (2001). Sink regulation of photosynthesis. *Journal of Experimental Botany*, Vol.52, No.360, pp. 1383-1400, ISSN 1460-2431

Paul, M.J. & Stitt, M. (1993). Effects of nitrogen and phosphorus deficiencies on levels of carbohydrates, resiratory enzymes and metabolites in seedlings of tobacco and their response to exogenous sucrose. *Plant, Cell and Environment*, Vol.16, pp. 1047-1057, ISSN 1365-3040

Putievsky, E. (1983). Effects of daylength and temperature on growth and yield components of three seed spices. *Journal of Horticulture Science*, Vol.58, pp. 271–275, ISSN 1611-4434

Quail, P.H. (2007). Phytochrome-regulated gene expression. *Journal of Integrative Plant Biology*, Vol.49, pp. 11-20, ISSN 1744-7909

Shipley, B. (2006). Net assimilation rate, specific leaf area and leaf mass ratio: which is most closely correlated with relative growth rate? A meta-analysis. *Functional Ecology*, Vol.20, pp. 565-574, ISSN 1365-2435

Spadling, E.P. & Folta, K.M. (2005). Illuminating topics in plant photobiology. *Plant Cell &Environment*, Vol.28, pp. 39-53, ISSN 1365-3040

Sturm, A.; Lienhard, S., Schatt, S. & Hardegger, M. (1999). Tissue-specific expression of two genes for sucrose synthase in carrot (*Daucus carota* L.). *Plant Molecular Biology*, Vol.39, pp. 349-360, ISSN 1572-9818

Suetsugu, N. & Wada, M. (2003). Cryptochrome B-light photoreceptors. *Current Opinion in Plant Biology*, Vol.6, pp. 91-96, ISSN 1369-5266

Tamulaitis, G.; Duchovskis, P., Bliznikas, Z., Breivė, K., Ulinskaitė, R., Brazaitytė, A., Novičkovas, A. & Žukauskas, A. (2005). High-power light-emitting diode based facility for plant cultivation. Journal of Physics D, Vol.38, pp. 3182-3187, ISSN 1361-6463

Tarakanov, I.G. (2006). Light control of growth and development in vegetable plants with various life strategies. *Acta Horticulturae*, Vol.711, pp. 315-321, ISSN 0567-7572

Yanagi, T.; Yachi, T., Okuda, N. & Okamoto, K. (2006). Lihgt quality of continuous illuminating at night to induce floral initiation of Fragaria chiloensis L. CHI-24-1. *Scientia Horticulturae*, Vol.109, pp. 309-314, ISSN 0304-4238

6

Remote Sensing of Photosynthetic Parameters

Natascha Oppelt
Kiel University
Germany

1. Introduction

Remote sensing cannot measure photosynthesis directly. However, remote sensing can provide information about parameters directly or indirectly connected to the photosynthetic activity of a plant or a vegetation canopy.

Since the launch of the first earth resource satellite in 1972 researchers focused on the relationship between vegetation and its radiometric response. Comparisons of ground measurements and data of the first generations of the Landsat series proved very soon the suitability of the red and near infrared bands of the sensors for vegetation analysis. In the following years, various mathematical combinations of the green, red and near infrared bands, the so called vegetation indices (Bannani et al., 1995) were developed to quantify properties of plant canopies such as biomass, productivity or vigour (Pearson & Miller, 1972; Kauth & Thomas, 1976; Misra et al., 1977; Huete, 1988). However, the low spectral resolution of those sensors was inadequate to derive biochemical properties of vegetation. The development of hyperspectral instruments with a high number of narrow bands first enabled the quantification of pigment concentration and indices related to the photosynthetic capacity of vegetation. While field spectrometers were used to derive pigment concentration of single leaves or small plots of vegetation, the advent of airborne imaging spectrometers such as the compact airborne spectral imager *casi*, the Airborne Visible/Infrared Imaging Spectrometer AVIRIS or the HyMAP™ Hyperspectral Scanner enabled the monitoring of vegetated canopies and small landscape sections from the 1980s. Since then a new generation of indices was developed considering limitations known from the first generation (Baret et al., 1989; Qi et al., 1994). However, due to the restriction to airborne instruments the acquisition of hyperspectral images was a cost-intensive task. Studies dealing with this kind of sensors assessing biochemical or biophysical plant properties were limited to a relatively small group of scientists and stakeholders.

With the advent of space-borne hyperspectral instruments (e.g. the Compact High Resolution Imaging Spectrometer CHRIS (since 2001), the Moderate Resolution Imaging Spectroradiometer MODIS (since 2002) and the Environmental Mapper EnMAP (launch in 2015)) monitoring of biophysical parameters related to photosynthesis becomes increasingly operational. The growing availability as well as the reduced costs for hyperspectral data had a great impact on the numbers of studies and literature dealing with pigment assessment of native and cultivated vegetation canopies on various spatial scales.

Vegetation type and species composition strongly affect the derivation of biochemical plant properties and components. Biochemical components (e.g. chlorophyll content) and related biophysical properties (e.g. light use efficiency (*LUE*)) are more species-specific than biome-specific (Ahl et al., 2004). Thus, biome-related parameter retrieval and canopy-related studies require different ambitions for parameter retrieval. Large scale and global approaches require the development of simple, generalized representations of the most important plant processes and can be used in different biomes with a minimum of modifications (Running & Hunt, 1993; Running et al., 2004). Contrary local studies focus on single crop canopies with specific approaches to estimate parameters affecting photosynthesis such as crop chlorophyll, nitrogen content or uptake, or stress factors for different species and phenological stages (e.g. Daughtry et al. 1992; Gamon et al., 1992; Gilabert et al., 1996; Rascher & Pieruschka, 2007; Yoder & Pettrigrew-Crosby, 1995). Photosynthetic parameter retrieval usually serves for applications related to precision farming, e.g. crop growth modelling, application of fertilizers, herbi- and fungicides and yield forecasting (e.g. Haboudane et al., 2002, 2004; Hank et al., 2007; Malenowski et al., 2009; Oppelt et al., 2007; Strachan et al., 2008).

This study focuses on the assessment of the chlorophyll content of wheat (*Triticum aestivum* L.) canopies using airborne hyperspectral data. Several types of indices were applied and resulting results are discussed. Then the estimation of crop growth, gross and net primary production using the indices is described exemplarily for two applications: the use of indices for operational remote sensing products as well as the integration of indices with physically based crop growth modelling

2. Vegetation radiometric properties

Plant leaves show typical characteristics in their reflectance in the visible (VIS), near-infrared (NIR) and shortwave infrared (SWIR) parts of the electromagnetic spectrum (Figure 1). In general, VIS is dominated by the absorption features of leaf pigments, mainly by the chlorophylls. In the near-infrared region, high reflectance is due to the internal structure of plant mesophyll. The internal structure of leaves with numerous refractive discontinuities and intercellular air spaces scatters incident radiation and results in a large proportion to be passed back as reflected radiation. Plant water absorption features affect the reflectance behaviour in the SWIR leading to a strongly decreased reflectance at high plant water contents.

Light reaction is commonly measured using the chlorophyll absorption features in the VIS, which are known to correspond well with the fraction of absorbed photosynthetically active radiation (*fAPAR*) (Schurr et al., 2006). At the canopy level, the efficiency of carbon fixation is denoted. *LUE* refers to the projected ground surface and describes the net canopy CO_2 fixation. The spatial variability of *LUE* results in enormous variations of net photosynthetic productivity (*NPP*), which ranges from 30 to 1000 g C m^{-2} in different ecosystems (Schurr et al., 2006). Thus knowledge of the spatial distribution of *LUE*, *fAPAR* or chlorophyll is essential for a realistic estimation of photosynthetic processes.

To extract pigment information, a range of other factors that influence vegetation reflectance must be taken into account. The leaf reflectance may vary independently of pigment concentrations due to differences in internal structure, surface characteristics and moisture content. Furthermore, the reflectance spectrum of a whole canopy is influenced by factors such as the effects of leaf area, the orientation of leaves, ground coverage, and presence of non-leaf elements, areas of shadow and soil surface reflectance. These factors obscure the relationship between spectral reflectance and chlorophyll concentration (Blackburn, 2006) and thus have to be considered.

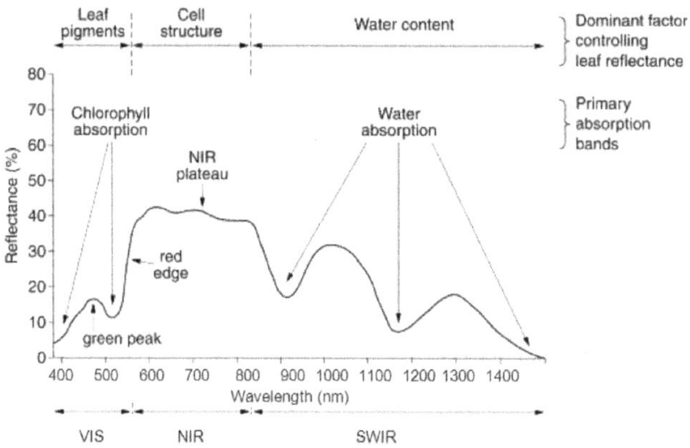

Fig. 1. Typical spectral reflectance characteristics of green leaves (Keyworth et al., 2009, modified)

3. Material and methods

To gather field and hyperspectral remote sensing data, intensive field campaigns were conducted during the growing seasons in 2004 and 2005 in a test site located in the Bavarian Alpine foothills. The study area, hyperspectral data assessment and processing, field measurements, data processing and analysis will be described in the following section.

3.1 Study sites

The study area is located in the county Gilching, 25 km south-west of Munich, Germany, (upper left corner 48°8′N, 11°17′E). The study area is characterized by an ever-moist and temperate climate with a mean temperature of 8.3° C (1961-1990), ranging from -5 °C and 2 °C in January to 12° C and 23° C in July. The mean annual sum of precipitation is 900 mm with 540 mm during the growing season of wheat (April to August). Two climatological stations of the Bavarian network of agro-meteorological stations enable access to local weather monitoring. The stations provide meteorological data such as precipitation, soil and air temperature, total irradiance and humidity (www.lfl.bayern.de/agm /start.php).

For each 2004 and 2005, one field of winter wheat grown with the unawned cultivar *Achat* was chosen as a test field. The plots were characterized by similar soil conditions (cambisols according to the FAO classifications scheme) and have been managed by the same farmer, who enabled information on management data such as fertilizer and herbicide application as well as the date of sowing, soil treatment and harvest. A detailed description of the test site and the field plots is provided by Oppelt (2010).

During 2004 and 2005, the growing conditions are similar regarding temperature and radiation which are within the standard deviation of the average values, but precipitation varied strongly. The area received 164 mm more precipitation in 2005 than in 2004. Frequent and heavy rainfalls during July and August of 2005 were exceptional and caused many crops to mould upon the field.

3.2 The Airborne Visible/Infrared Imaging Spectrometer AVIS

The Airborne Imaging/Infrared imaging Spectrometer AVIS was built at the Department for Geography of the Ludwig-Maximilians-University in Munich (Germany) (Oppelt & Mauser 2001, 2004, 2006, 2007). The second generation of the sensor, AVIS-2, was operated for this study. AVIS-2 is a CCD-based system operating in the VIS and NIR (400 nm to 900 nm) spectral range with a spectral resolution of 9 nm. The system is based on a spectrograph (SPECIM Imspector V9NIR), mounted to a black and white 2/3" CCD-video camera (Vosskühler 1300) and a filter-lens system. Table 1 summarizes the specifications of AVIS-2; the sensor, its specification and calibration are described in detail by Oppelt & Mauser (2007).

Nominal spectral range [nm]	400-900
Analyzable spectral range [nm]	420–880
Spectral resolution [nm]	9
Radiometric sampling [bits]	16
Number of bands	64
Signal to Noise Ratio (SNR) [dB]	65
Spatial resolution [pixels per line]	300
Spatial sampling [pixels per line]	640
Field of view (FOV) [rad]	1.0
Navigation systems	INS, GPS
Period of operation	since 2001

Table 1. AVIS-2 specifications

3.3 AVIS data

AVIS was designed to be operated on different platforms, such as Dornier DO-27 or DO-228 and microlight aircrafts, where the sensor is mounted on vibration dampening mounts. For this study AVIS-2 was operated on a Dornier DO-27 aircraft flown by pilots of the Bavarian armed forces.

Four AVIS-2 acquisitions are available for the growing periods in 2004 and 2005. In 2004, the sensor was flown on March 31, May 25 and June 8. One overflight was conducted in 2005, i.e. on July 6. To ensure comparable illumination and viewing geometries, the overflights were conducted in the principal plane with a flight azimuth of about 0°. The data were gathered in a time period between 10 and 14 hrs UTC resulting in sun azimuth angles between 31.5° and 37°. The ground sampling distance (GSD) for the overflights was 4 m.

The radiometrical pre-processing of the data included correction of sensor dark current, CCD homogeneity and smile effect (Oppelt & Mauser, 2007). Then the data were corrected atmospherically and reflectance calibrated using an approach based on MODTRAN-4 (Berk et al., 2000). The data for the parameterization of the atmosphere were estimated using climatological data of the Bavarian agro-meteorological network. The geometric processing of the AVIS data was carried out using data of a differential Global Positioning System (dGPS) and an Inertia Navigation System (INS), which were mounted on the sensor. INS and dGPS data were stored in the header of each image line and provide data of the aircraft location (latitude, longitude and altitude) and pointing information (roll, pitch and yaw). Combined with a digital elevation model, these data are used to compensate for the motion of the aircraft and rectify the data to a respective coordinate system. The geometric correction was carried out by means of the header information, aerial orthophotography and ground control points applying a second-order polynomial function in ESRI ARCGIS 9.1. Figure 2 presents an image stripe with exemplary reflectance spectra.

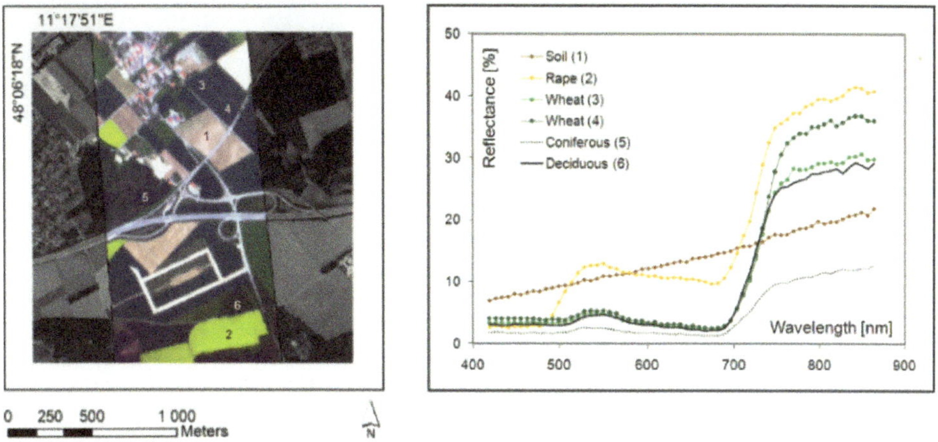

Fig. 2. AVIS real colour image acquired on May 25, 2004 (left hand side); the data were the processed geometrically and radiometrically as well as reflectance calibrated; the numbers in the AVIS image indicate the location of the reflectance spectra presented on the right hand side.

The reflectance spectra of the different wheat pixels unveil the potential of hyperspectral remote sensing for vegetation monitoring. In the VIS, the pixels reflect the solar radiation nearly identical. Thus, for the human eye, which is represented by the real colour composite, these wheat patches look nearly identical. The differences become obvious when looking at the NIR part of the reflectance spectra. The higher reflectance of wheat (4) indicates that the vegetation is more densely developed than the wheat (3) patch.

3.4 Field measurements

For every field, five sampling points were selected diagonally in the field and fixed via handheld GPS (GARMIN VISTA). The locations of the sampling points were selected in close cooperation with the farmers in order to gather ground truth data from areas in the field that exhibited differences in plant development and yield over recent years.

At the sampling points weekly measurements were conducted between April and harvest (beginning of August) including phenological stage, plant height and DM (separated into stem, leave and fruit fraction). Measurements of leaf area were conducted using a LI-COR LAI 2000 plant canopy analyzer. Leaf chlorophyll was measured from April to the beginning of ripening. After sampling the leaves were frozen immediately in liquid nitrogen and taken to the laboratory, where the chlorophyll analysis was conducted according to the procedure described by Porra et al. (1989). The resulting chlorophyll concentration per weighted portion was multiplied with the leaf dry matter resulting in the total chlorophyll which is stored in the leaves within a square metre on ground [mg m^{-2}]. Field spectrometer measurements (Ocean Optics SD-2000 combined with SD-2000 NIR) were conducted concurrent to the AVIS overpasses to validate AVIS processing results (Oppelt, 2010).

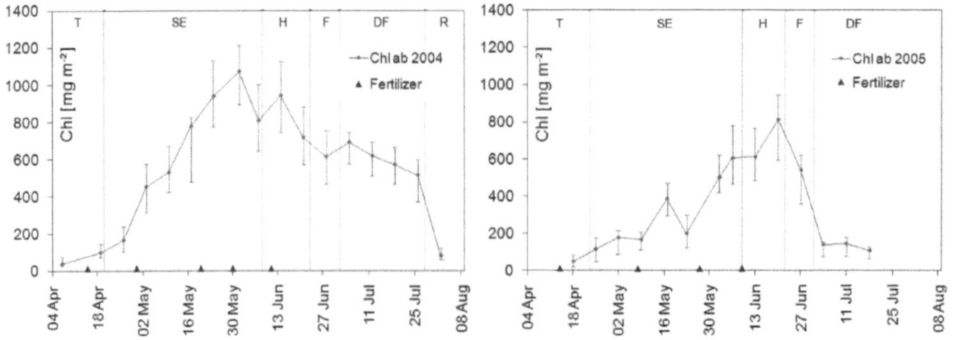

Fig. 3. Results of total chlorophyll measurements during the growing seasons 2004 and 2005; the rhombuses correspond to the mean field values during the phenological stages (T = tillering; SE = stem elongation; H = heading; F = flowering; DF = development of fruit; R= ripening); the error bars mark the minimum and maximum values measured at the sampling points indicating the scattering of the chlorophyll measurements

Figure 3 presents the results of the in situ chlorophyll measurements. Despite the nearly parallel phenological development, the chlorophyll content varies considerably in both amount and chronology. In 2004, canopy leaf chlorophyll develops with a general increase during the vegetative growth to 1100 mg m^{-2} chlorophyll and a decrease during the generative growing phase. This pattern is obviously superimposed by the development of leaf biomass, which increases during spring and decreases after flowering. Figure 3 also indicates that fertilization affects the general course of canopy chlorophyll; chlorophyll content increases after the application of fertilizer. The error bars indicate the variability of chlorophyll within a vegetation stand. As anticipated, the heterogeneity within the field generally is high during vegetative growth while during ripening the canopy reveals a more homogeneous chlorophyll distribution.

In 2005, the canopy chlorophyll increases with a relatively soft increase until May, where a notable decrease can be observed. Late frost events led to a reduced metabolic activity resulting in decreased chlorophyll contents (Oppelt, 2010). After the frost events the wheat canopy exhibits a strong increase in canopy chlorophyll to 682 mg m^{-2}, which was promoted by high temperatures and a high amount of incoming radiation in June. Due to the wet conditions in July, the plants began to mould, which led to a considerable reduced yield in 2004 (average of 7.01 t ha^{-1} in comparison to 8.3 t ha^{-1} in 2005 (Oppelt, 2010)). The decay of the plants is accompanied by a rapid decomposition of chlorophyll during anthesis. These results underline that a general assumption of the canopy chlorophyll can be deceptive for the assessment of vegetation photosynthesis. Variations in crop chronology due to different weather conditions as well as existing spatial heterogeneities may distort an expected universal course of the metabolic activity even at a single crop stand.

3.5 Indices for the derivation of canopy chlorophyll

A large variety of approaches has been developed for estimating chlorophyll content. Modelling studies provided good evidence that reflectance is more sensitive to high pigment concentrations at wavelengths where pigment absorption is low. Contrary, spectral regions with high absorption are more sensitive to low pigment concentrations (Blackburn, 2006;

Jacquemoud & Baret, 1990). Empirical studies have confirmed this statement and demonstrated that reflectance at wavelengths corresponding to the lower and upper flanks of the major chlorophyll absorption feature in the Red is most sensitive to the normal range pigment concentrations in most leaves (Carter & Knapp, 2001) and canopies (Filella et al., 1995; Yoder & Pettrigrew-Crosby, 1995). In young and senescent leaves and canopies bands at the centre of absorption features are most sensitive to low pigment concentrations (Sari et al., 2005).

To deal with the difficulties in relating reflectance to individual bands due to variations in the controlling factors on canopy reflectance, many approaches use reflectance in multiple bands. Most of them have employed ratios of narrow bands in spectral regions that are sensitive to pigments and those areas that are not sensitive. They were proposed as a means of solving the problems of overlapping absorptions of different pigments (Chappelle et al., 1992) and the effects of leaf and canopy structure (Peñuelas et al., 1995). Many indices have been derived for chlorophyll quantification and are based on ratios of bands in the VIS and NIR (Filella et al., 1995; Sims & Gamon, 2002), in the Red (Vogelman et al., 1993), or in the NIR and red edge region (Gitelson & Merzlyak, 1997).

As mentioned previously, canopy reflectance results from a complex interaction between pigment concentrations, canopy structure, background signal and illumination conditions (sun-sensor-target geometry). Moreover, vegetation indices that are insensitive to soil optical properties seem to be relatively insensitive to chlorophyll variations. Conversely, most indices sensitive to chlorophyll content variability are strongly affected by the differences in the canopy vegetation cover (Haboudane et al., 2002). Various indices have been developed to be both sensitive to a broad range of chlorophyll content and robust towards different types of noise. Four approaches, which represent different types of indices, are discussed in this paper, i.e. a hyperspectral derivate of the Normalized Vegetation Difference Index, the Optimized Soil Adjusted Vegetation Index, the Photochemical Reflectance Index and the Chlorophyll Absorption Integral.

3.5.1 Normalized Difference Vegetation Index

Rouse and colleagues (1974) published the probably best common well-known index, the Normalized Difference Vegetation Index (*NDVI*) (Equation 1). The *NDVI* belongs to the first generation of indices which were based on empirical methods designs for a specific sensor, i.e. Landsat MSS.

$$NDVI = \frac{(R_{NIR} - R_{Red})}{(R_{NIR} + R_{Red})} \quad (1)$$

This index is still used for studying the state of vegetation with various sensors in regional to global applications (Prince & Tucker, 1996; Hame et al., 1997). However, several studies note that its usefulness depends strongly on noise associated with view angle differences, soil background influences, clouds and cloud shadow, atmospheric influences and topographic effects (Carlson & Ripley, 1997; Huete et al., 1997; Kim et al., 2010). In addition, saturation of the vegetation index in high biomass conditions or pigment concentrations limits quantitative vegetation assessments (Kim et al., 2010; Oppelt & Mauser, 2004). Nevertheless, the *NDVI* is still one of the most common indices. A hyperspectral variant the *NDVI* was used in this study:

$$hNDVI = \frac{(R_{827} - R_{668})}{(R_{827} + R_{668})} \quad (2)$$

where R_{827} and R_{668} correspond to the centre wavelength of the respective AVIS bands.
The *hNDVI* showed high correlations for wheat canopies with medium chlorophyll content, but it becomes insensitive to chlorophyll contents at canopies with low LAI and dense vegetation (Oppelt & Mauser, 2004).

3.5.2 Optimized Soil Adjusted Vegetation Index OSAVI

OSAVI is a derivative of the *NDVI* and, as indicated by the name, includes a soil adjustment factor. To compensate for the effects of background and soil reflectance, particularly for sparse vegetation cover Rondeaux et al. (1996) introduced the *OSAVI*:

$$OSAVI = \frac{(R_{800} - R_{668})}{(R_{800} + R_{668} + 0.16)} \quad (3)$$

where R_{800} and R_{668} correspond to the centre wavelength of the respective AVIS bands.
OSAVI was proposed to reduce the background reflectance contributions and to enhance sensitivity to leaf chlorophyll variability. Its determination requires no knowledge of soil properties resulting in an easy application in the context of operational observations. However, some studies note that *OSAVI* also becomes insensitive to high chlorophyll contents (Oppelt & Mauser, 2004; Wu et al., 2008).

3.5.3 Photochemical Reflectance Index

The Photochemical Reflectance Index (*PRI*) was developed by Gamon et al. (1992) to minimize the effects of xanthophylls signal overlapping the chlorophyll spectral features due to sun angle variation. As many other indices the *PRI* was based on measurements in the laboratory and was then successfully applied and tested on field, air and spaceborne imaging spectrometers (e.g. Gamon & Qiu, 1999; Penuelas et al., 1997; Sims & Gamon, 2002; Stylinski et al., 2002; Thenot et al., 2002; Trotter et al., 2002; Weng et al., 2006).
The *PRI* is calculated as follows (Gamon et al., 1997)

$$PRI = \frac{(R_{531} - R_{570})}{(R_{531} + R_{531})} \quad (4)$$

where R corresponds to the reflectance at the wavelength considered. The 531nm waveband is sensitiv to pigment concentration while the 570nm waveband is used as a reference.
The PRI provides an easy measurement of chlorophyll content or *LUE* (Gamon et al., 1992). Moreover, it can be used for a wide range of species (Gamon et al., 1997). One problematic feature is its high sensitivity to soil reflectance, which has to be taken into account in areas or times with low vegetation cover (Mänd et al., 2010).

3.5.4 Chlorophyll Absorption Integral

The Chlorophyll Absorption Integral (*CAI*) is an approach based on the measurement of the chlorophyll absorption feature depth obtained by fitting a continuum to vegetation reflectance. Kokaly & Clark (1999) first described this method to assess nitrogen, lignin and cellulose for leaves of different tree and crop species. They used linear segments to approximate the continuum. Once the continuum is established, the continuum-removed spectra are calculated by dividing the original reflectance values by the corresponding values of the continuum line. From the continuum-removed reflectance R′ [%], the depth D [%] in the absorption feature is computed with a uniform interval of 0.1 nm:

$$D = 1 - R' \tag{5}$$

The small interval for calculating the continuum removal was used to overcome difficulties with varying band settings of different sensors which affect *CAI* values for the same target on the ground (Oppelt, 2008). To minimize the influence of extraneous factors such as atmosphere, soil or topography, the absorption depths are normalized (D_n in Equation 6) (Curran et al., 2001; Kokaly & Clark, 1999). This is calculated by dividing the absorption depth of each band by the absorption depth at the centre of the absorption D_c.

$$D_n = \frac{D}{D_c} \tag{6}$$

Kokaly & Clark (1999) demonstrated that the normalized index exhibits a low sensitivity to background effects due to atmosphere, soil and topography. These results were confirmed at the leaf level by Curran et al. (2001) as well as by Oppelt & Mauser (2004) for canopy chlorophyll.

The start and end point of the continuum can be chosen according to the band setting of the instrument. The AVIS *CAI* extends from the Red (600 nm) to the NIR (740 nm), whereby the former includes the chlorophyll a absorptions and the latter is an area insensitive to chlorophyll (Gitelsen & Merzlyak, 1997). Another advantage of *CAI* is that it includes both the lower and upper flanks of the chlorophyll absorption in the Red as well as the central absorption. Thus it includes wavelengths sensitive to a wide range of chlorophyll contents (Oppelt, 2002, 2010).

4. Chlorophyll assessment

The results of the chlorophyll assessment are summarized in Figure 5. As mentioned previously, indices become saturated at high chlorophyll contents. While indices such as the *hNDVI* and *OSAVI* saturate at chlorophyll a contents at about 1.0 g m^{-2}, the *CAI* is known to saturate at chlorophyll contents higher than 1.5 g m^{-2} (Oppelt 2002). With increasing chlorophyll content its absorption feature at 680 nm flattens and narrows. *OSAVI* and *hNDVI* are directly affected by reflectance in the Red and tend to saturate by an increase in this spectral region. The high correlations between the *CAI* and chlorophyll can be ascribed to the fact that the *CAI* is based on an integrated measurement and therefore is less affected by an increase of reflectance in single wavelengths. The *CAI* becomes insensitive when the narrowing of the absorption feature leads to a shift of the red edge position towards the Blue (Oppelt, 2002).

The effect of saturation indicates a non-linear relationship, thus an exponential relationship should be expected. However, the chlorophyll contents monitored are below the saturation levels and thus the results can be approximated assuming linear relationships. The regression equations indicate that they do not cross the ordinate at zero, but show an offset, which is caused by the range of chlorophyll contents measured. Hence, the valid range of the chlorophyll estimation using the equations given in Figure 5 strictly is limited for chlorophyll contents between 200 mg m^{-2} and 800 mg m^{-2}.

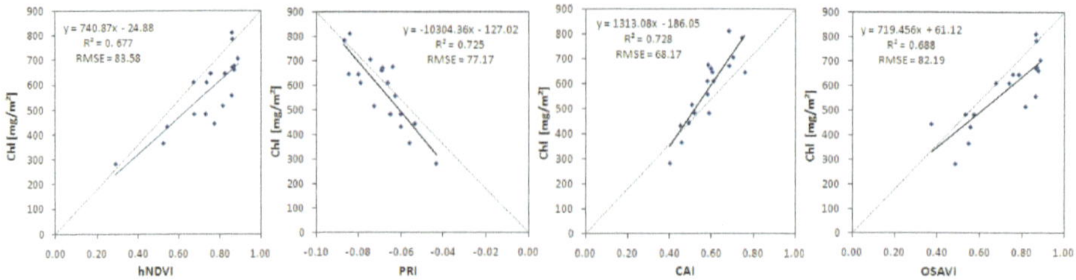

Fig. 5. Linear relationships between vegetation indices and measured canopy chlorophyll; the regression equations are given as well as the coefficient of determination (R^2) and the root mean square error (rmse).

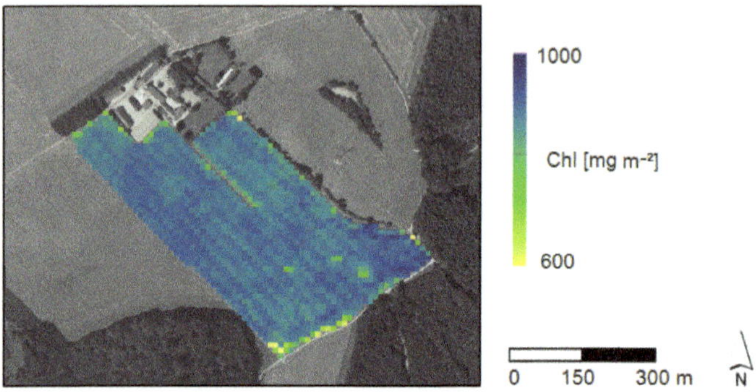

Fig. 6. CAI derived distribution of total canopy chlorophyll as monitored with the AVIS sensor on May 25 2004

However, the results presented in Figure 5 indicate that *hNDVI* and *OSAVI* already saturate at chlorophyll contents above 600 mg m^{-2}, while *PRI* and *CAI* are not affected by saturation. *PRI* shows a significant negative correlation with the canopy chlorophyll while *CAI* is highly positively correlated. The coefficient of determination (R^2) is relatively high for both *CAI* and *PRI* with *CAI* reveals best results with the lowest root mean square error. Moreover, PRI seems to be not affected by the low leaf area during the AVIS March data. The low saturation level of *hNDVI* and *OSAVI* apparently prevents higher coefficients of determination for the crop stands investigated. Figure 6 presents the spatial distribution of canopy chlorophyll content using the regression equation of CAI given in Figure 5 exemplarily for one AVIS acquisition. The Figure shows the existing heterogeneity in the wheat stand at May 25 at the end of stem elongation. The dominant structural pattern is given by the tractor lanes. However, specific areas with similar chlorophyll contents also become visible. Zones of high chlorophyll in the western part as well as in the very east of the field are apparent. The northern and southern field margins are characterized by low chlorophyll contents. A fertilization window is visible in the south-eastern part of the field. This area is not fertilized and therefore is characterized by low chlorophyll contents.

5. Estimation of photosynthesis and primary productivity

Canopy photosynthesis is defined to equal the integrated sum of photosynthesis by leaves in a canopy (Amthor et al., 2001). Jarvis (1993) defined three classes of canopy photosynthesis models; two of them are so-called "big-leave" models which define the canopy as a single layer of vegetation covering the soil. The third model class divides the canopy into multiple layers, which underlie different microclimates and simulate the impact of spatial gradients within the canopy (Baldocchi & Amthor, 2001).

The terms Gross Primary Production (GPP) and Net Primary Production (NPP) are the most important parameters related to the photosynthetic activity from plant to canopy level. Most crop-growth as well as biome-related models therefore deal with these parameters. GPP is the total amount of carbon fixed by plants through photosynthesis. NPP is the net amount of carbon fixed by plants after the costs of respiration are included (McGuire et al., 1993).

5.1 Big-leave gross primary productivity

The simplest big-leaf model assesses canopy GPP based the assumption that photosynthetic assimilation or NPP is proportional to the amount of solar radiation intercepted by vegetation (Monteith, 1972). Thus, canopy photosynthesis can be calculated as a linear function of the photosynthetically active radiation absorbed by the canopy (*APAR*). The slope of the equation is *LUE* (Monteith & Moss, 1977; Ruimy et al., 1995).

$$GPP = LUE * APAR \tag{7}$$

APAR is the product of incoming *PAR* and the fraction absorbed by the canopy (*fAPAR*). The measures of *APAR* integrate the geographic and seasonal variability of day length and potential incident radiation with daily cloud cover and aerosol attenuation. In addition, *APAR* implicitly quantifies the amount of leave canopy that is displayed to absorb radiation (Running et al., 2004). This model approach is very simple and enables the estimation of GPP with a very limited number of parameters.

Time and space variability of *LUE* and *APAR* can directly be derived using remote sensing and meteorological data (Hilker et al., 2008; McCallum et al., 2009). *LUE* is influenced by many factors and thus varies in space and time. It is known to vary among crops (Gosse et al.; 1986, Prince, 1991) and nutrient status (Balakrishnan et al. 2001; Oppelt, 2002; Penuelas & Filella, 1995); however, *LUE* often is assumed to be constant when growth is not limited by water or nutrient shortage or climate conditions (Ruimy et al., 1995). Some authors propagate the use of *PRI* as a proxy for *LUE* (Gamon et al., 1992; Nichol et al. 2000), but Gitelson et al. (2006) stated that *PRI* is most sensitive to *LAI* and therefore is difficult to apply on a canopy or even on global scales. Thus, the assessment of LUE differs between authors and application; advantages and disadvantages of the different approaches have to be considered when using vegetation indices as proxies for *LUE*.

Monteith's logic is fundamental on a suite of operational remote sensing products, e.g. the MODIS GPP, NPP and photosynthesis (PSN) products. Running et al. (2004) described the MODIS algorithms in detail; a simple model based on look-up tables for different biomes is combined with meteorological data. *APAR* is estimated using the *NDVI* through Equation 8 (Myneni et al., 1999).

$$fAPAR = \frac{APAR}{PAR} \alpha \, NDVI \tag{8}$$

This expression is based on results of several studies which found that under specified canopy reflectance properties *APAR* can be estimated using the *NDVI* (Asrar et al., 1992; Sellers et al. 1987; Sellers et al., 1992). Myneni et al. (1999) demonstrated that *fAPAR* is proportional to *NDVI* if soil background is ideally black and therefore introduced a factor of proportionality which accounts for soil contribution. The linear relationship has been discussed in literature; Ruimy et al. (1995) demonstrated that linearity between *APAR* and *NDVI* is only valid during vegetative growth. Moreover, comparison with modelled *APAR* unveiled that the *NDVI*-related *APAR* is significantly lower than independently modelled *APAR* (Ruimy et al., 1999). A likely explanation is the saturation of the *NDVI* to high chlorophyll or *fAPAR*, which also can be observed in Figure 5. Therefore, one of the main problems of GPP assessment using remote sensing data is caused by the uncertainty of a linear *NDVI*/*fAPAR* relationship (Gitelson et al., 2006). Besides the controversial discussion, *NDVI* still is fundamental to MODIS products.

The GPP product is used to calculate net photosynthesis PSN, which is computed as

$$PSN_{net} = GPP - R \tag{8}$$

where R is an estimate of daily respiration of leaves and roots (Running et al., 2004). MODIS *LAI* is used to estimate the biomass for the purpose of estimating R (Myneni et al., 1997, 1999). The PSN and GPP products have an 8-day temporal resolution while NPP is an annual value. Figure 6 presents a series of PSN_{net} products where the test area is located. Each PSN image covers an area of 10° degrees in latitude and longitude and enables the monitoring of large scale photosynthesis and carbon uptake.

However, with a spatial resolution of 1km the MODIS products are suitable rather for large and global scale issues. To gather primary productivity on a smaller scale, crop growth models can be applied.

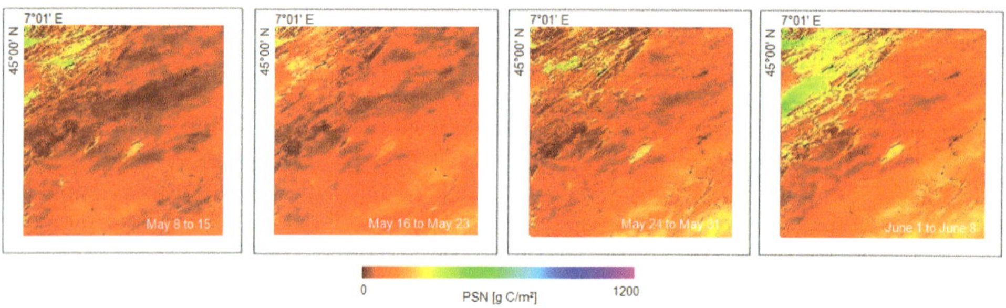

Fig. 6. Series of MODIS 8-day PSN (GSD = 1 km) composites from Mai to June 2004 (data source: US Geological Survey, Earth Observation and Science Center)

5.2 Derivation of NPP using a Vegetation Growth Model Approach

Multi-layer models are able to consider the impacts of nonlinear physiological and physical processes on canopy photosynthesis (Baldocchi et al., 2001; Wang & Jarvis, 1990). Yin and Struik (2009) provided an overview of photosynthesis models available for C_3 and C_4 crop modelling. The approach used for this study is the advanced biological sub-model of the process of radiation mass and energy transfer model PROMET (Mauser & Bach, 2009).

The core model is based on eight components (meteorology; land surface energy and mass balance; vegetation; snow and ice; soil hydraulics and soil temperature; ground water; channel flow; man-made hydraulic structures (Figure 7) to simulate the water and energy fluxes for variable time steps.

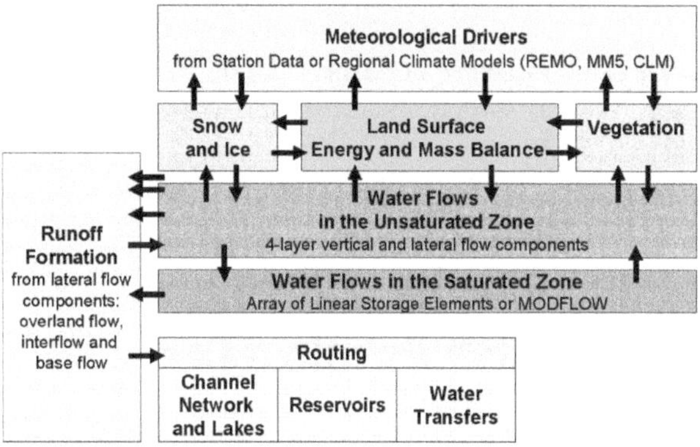

Fig. 7. PROMET core components (Mauser & Bach, 2009)

5.2.1 The biological sub-model of PROMET

The modular structure of PROMET facilitates improvements in particular sub-models. The biological sub-model calculates photosynthesis according to a biochemical approach introduced by Farquhar, von Caemmerer and Berry (1980). NPP is modelled on the basis of the temperature-dependent rate of Ribulose-1.5-Biphosphate (RuBP) reproduction and the availability of Ribulose-1.5-Biphosphate-Carboxylase-Oxigenase (Rubisco), i.e. by simulating the Calvin cycle. The availability and transport of CO_2 is regulated by the stomatal resistance of the leaf described by Ball and colleagues (1987). The rate of leaf photosynthesis is modelled in dependence of *APAR* and the temperature dependent rate of RuBP regeneration. All processes are modelled in two vegetation layers, i.e. a sunlit and a shade layer. PROMET calculates an optimum photosynthesis under given environmental conditions (Hank et al., 2007). The model is able to reproduce effects on plant development that are caused mainly by variations in radiation regime. Effects due to relief, exposition and differences in the soil type or texture can also be modelled well (Hank, 2008; Oppelt, 2010).

The Farquhar and von Caemmerer approach does not require chlorophyll contents but leaf absorptance (*abs*), which is directly related to *CAI* (Oppelt, 2010). *Abs* is dimensionless and refers to the mean relative quantum yield in the range of *PAR*. The quantum yield is the CO_2 assimilation in the absence of photorespiration and represents the maximum efficiency with which light can be converted to chemical energy by photosynthesis (Farquhar & von Caemmerer, 1982). In the model, *abs* usually is used as a constant value, which now can be dynamised using the *CAI*-*abs* remote sensing product. Then, DM is modelled using the constant value (abs_{const} = 0.83, Oppelt (2010)) until a remotely sensed *abs* (abs_{RS}) map is available. Then on, the abs_{RS} is used to calculate DM.

Lack of remote sensing data or large time gaps between two acquisition dates lead to *abs* values being inadequate for the specific growing period. These problems can be avoided if the abs_{RS} values are traced back to abs_{const} after a certain period of time. Assuming average growing conditions, the nitrogen in the fertilizer is metabolized within 21 to 30 days

(Döhler, 2007). Thus the abs_{RS} is used for a time period of three weeks before abs_{const} is set again, but is replaced again if an additional abs_{RS} map is available for a later day.

To fit the grid size of the other input data of the model (i.e. the limiting resolution of the digital elevation model) the data were resampled to a resolution of 10 m using a nearest neighbour approach. The resulting root mean square errors (rmse) from the ground control points were less than 0.5 pixels along track and 0.6 pixels across track.

To provide appropriate soil moisture conditions at the beginning of the measurement periods, the simulation run was started about four months prior to the sowing date of the crops (i.e. August 2003 and 2004 respectively).

5.2.2 Model results and discussion

NPP is defined as the rate at which vegetation fixes carbon from the atmosphere minus the plant respiration (McGuire et al., 1993); therefore NPP demonstrates its link to DM development during plant growth (Gitelson et al., 2006; Wu et al., 2010). Box et al. (1989) described the relationship between NPP and biomass as follows

$$NPP = LF + \Delta DM + H \tag{9}$$

where LF represents the biomass discarded periodically (e.g. litter or dead leaves), ΔDM is the increment of dry matter and H represents the dry matter lost to herbivores or harvest. For a precision farming managed crop canopy, the loss of biomass due to herbivores and decay of plant material are assumed to be negligible. Therefore, in this study aboveground dry biomass is used as proxy for NPP, but it has to be mentioned that, due to the lack of root biomass measures, it is restricted to above ground biomass and above ground NPP.

Figure 8 presents the average field values of modelled and measured DM in 2004 and 2005. In 2004, PROMET reproduces the field average plant development well, but tends to overestimate DM when used with the constant *abs* value. The general course of DM is based on a standardized development of canopy leaf area. The LAI measurements conducted during the growing season were used to adapt the standardized canopy leaf area. They are mainly responsible for the excellent results even when PROMET was run without remote sensing data. The modelling of optimum photosynthesis results in increased DM at the early stages of plant growth. The integration of CAI_{abs} results in a decrease of the modelled average DM at the beginning of tillering (due to March 31 AVIS data) and a slower increase during anthesis and ripening (due to the May 25 and June 8 data). However, only a slight increase in model performance could be observed when looking at the field average.

The results for 2005 clearly demonstrate the limitation of crop modelling. Results shown in Figure 6 demonstrate that PROMET cannot correctly reproduce average DM when factors, which are not driven by radiation regime, influence plant growth and health. PROMET was able to trace the reduced development of leaf area, but could not reproduce the moldering and decay. Unfortunately, the AVIS acquisition in July was far too late to adjust plant development to a more realistic level. The modelled DM is on a lower level compared to 2004, which is due to the influence of the *LAI* based modelling of the canopy. However, the resulting model performance was poor with high deviations up to 0.46 kg m^{-2} (Figure 9).

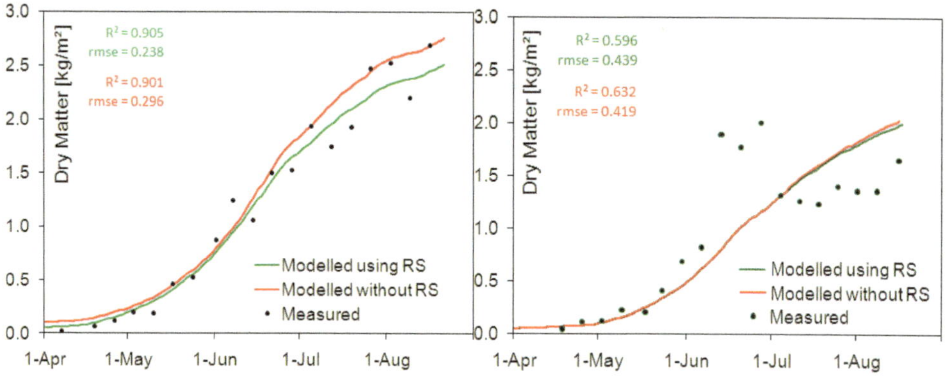

Fig. 8. Comparison of modelled and measured mean field values of DM with or without the integration of remote sensing (RS) data for the growing season in 2004 (left hand side) and 2005 (right hand side)

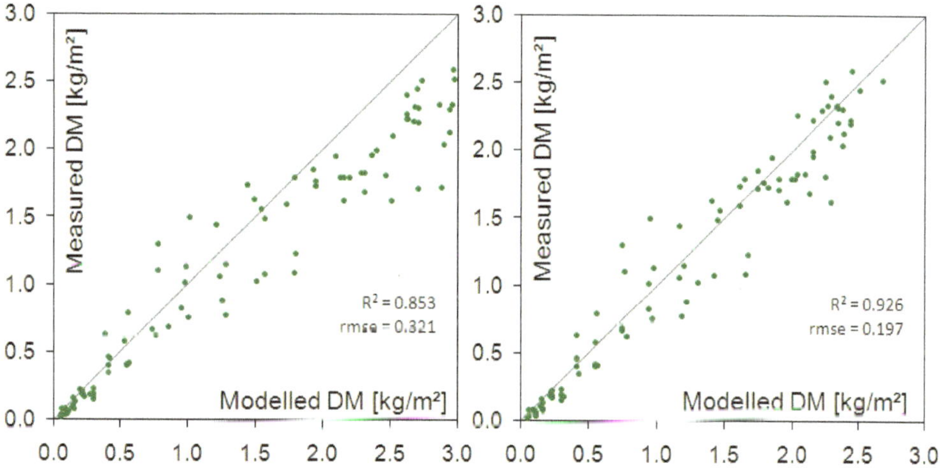

Fig. 9. Results of modelled and measured above-ground DM in 2004 when PROMET is used stand alone (left hand side) and with integration of AVIS data regulating the amount and spatial distribution of chlorophyll (right hand side)

The potential of integrating remote sensing data becomes obvious when looking at the results for the field sampling points (Figure 8). Plant development at the different sampling points can be reproduced more realistically when abs_{RS} is used instead of abs_{const}. The implementation of AVIS data results in a slightly higher coefficient of determination, but rmse was reduced by approximately 30%.

The use of a dynamic *abs* enables a more realistic modelling of dry matter, i.e. DM production is lowered. However, the results depend strongly - as could have been expected - on the time (or developmental stage of the plants) when remote sensing data are available. The time when a remotely sensed *abs* distribution can be integrated is crucial for two reasons: firstly, if spatially distributed *abs* is available at the beginning of the vegetation period, this period can be modelled more realistically, because in the early growth stages the *abs* values turned out to be much lower than the constant value. With progressive plant development the chlorophyll content and therefore the absorptance increase, resulting in a low modelled DM.

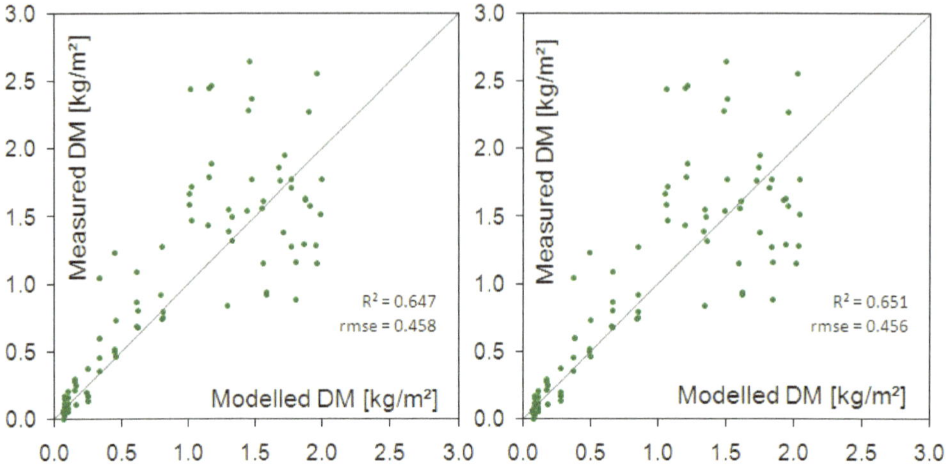

Fig. 9. Results of modelled and measured above-ground DM in 2005 when PROMET is used stand alone (left hand side) and with AVIS data acquired on 5 July (right hand side)

The importance of the phenological stage at which abs_{RS} is introduced is obvious. The advantage of implementing additional remote sensing data is the adjustment of the canopy absorptance properties, which would generally result in a decrease of DM. Thus model results improve at most when remote sensing data during the early growing stages are available. The AVIS acquisition in 2005 was too late for a proper characterization of the stand, because nearly the whole plant development was modelled with abs_{const}.

6. Conclusions

Remote sensing can provide information about parameters directly or indirectly connected to the photosynthetic activity of a plant or a vegetation canopy. Different types of vegetation indices were applied to estimate total chlorophyll of wheat canopies using airborne hyperspectral data. Validation with field measurements showed that *OSAVI* and *hNDVI* tend to saturate at chlorophyll contents above 600 mg m^{-2} while *PRI* and *CAI* were not affected by saturation. *PRI* showed the highest degree of correlation ($R^2 = 0.725$), but CAI proved the most precise estimation (rmse = 81.1 mg m^{-2}).

Vegetation indices can be used as input parameter for calculating photosynthesis from small to global scale. MODIS PSN, GPP and NPP products are based on *NDVI* measurements and provide information with a spatial resolution of 1 km. Examples of MODIS PSN provide valuable information of photosynthesis at regional to large scales.

To provide NPP on a field scale, the Farquhar - based biological sub-model of PROMET was used as vegetation growth model. PROMET integrates *CAI* derived leaf absorptance values as input parameter to calculate canopy photosynthesis. To validate model results, canopy dry matter was used as a proxy for NPP. Under standard growing conditions, PROMET reproduced average field biomass development well, even without integration of remote sensing data ($R^2 = 0.91$). The model calculates optimum photosynthesis under given meteorological conditions and therefore tends to overestimate DM. The integration of remote sensing data adapts varying chlorophyll condition occurring in the field to the model. The results show a general decrease of modelled average DM. However, the heterogeneities in the wheat canopies could be reproduced better when a *CAI* based absorptance was integrated in PROMET; the resulting degree of correlation increased ($R^2 = 0.93$ compared to $R^2 = 0.85$) while the prediction error decreased by 30%.

The advantage of implementing additional remote sensing data lies in the adjustment of the canopy absorptance properties, e.g. on a deficit in nutrient supply, mechanical inflictions or plant diseases or moulding. Still, the acquisition time is a crucial task for the enhancement of crop growth modelling. If remote sensing data were not available directly after a mechanical inflict or the appearance of diseases, the model is not able to reproduce the changing plant metabolism ($R^2 = 0.65$).

This paper demonstrates that the use of remote sensing data to adapt "real conditions" to models of photosynthesis is very promising, both at field and coarse scale. The success and progress of photosynthetic related MODIS products and the model results emphasize the need for space-borne instruments to enable an operational monitoring with regular acquisitions on a regional and local scale. The advent of the EnMAP instrument in 2015 will hopefully close this gap. In addition, sun induced chlorophyll fluorescence becomes increasingly important in the monitoring of photosynthetic processes. Instruments that measure sun induced fluorescence such as FLEX (candidate for the Earth Explorer mission of the European Space Agency) will contribute significantly to the remote sensing based research in the field of photosynthesis.

7. Acknowledgments

The author would like to thank the German Research Foundation (DFG) for funding the project "Coupled analysis of vegetation chlorophyll and water content using hyperspectral, bidirectional remote sensing". Thanks are also due for Mr. Stürzer for continuing access to the test fields and for supporting the study with detailed management data.

8. References

Balakrishnan, K., Rajendran, C. & Kulandaivelu, G. (2001). Differential Responses of Iron, Magnesium, and Zinc Deficiency on Pigment Composition, Nutrient Content, and Photosynthetic Activity in Tropical Fruit Crops. *Photosynthetica*, 38. 3, pp.477–479.

Baldocchi, D.D. & Amthor, J. (2001). Canopy Photosynthesis: History, Measurements, and Models. In J. Roy, B. Saugier, & H. Mooney (Eds.), *Terrestrial Global Productivity:* Academic Press. San Diego, California, USA.

Ball, J., Woodrow, I. & Berry, J. (1987). A model predicting stomatal conductance and its contribution to the control of photosynthesis under different environmental conditions. *Progress in Photosynthesis Research*, 4, pp.221-224.

Bannari, A., Morin, D., Bonn, F. & Huete, A.R. (1995). A review of vegetation indices. *Remote Sensing Reviews*, 13/1, pp.95–120.

Baret, F., Guyot, G. & Major, D. (1989). TSAVI - A vegetation index which minimizes soil brightness effects on LAI and APAR estimation. In *Proceedings of IGARSS '89, Vancouver (Canada)*, pp. 1355–1358.

Barton, C.V.M. & North, P.R.J. (2001). Remote sensing of canopy light use efficiency using the photochemical reflectance index: Model and sensitivity analysis. *Remote Sensing of Environment*, 78. 3, pp.264–273.

Berk, A., Anderson, G., Acharya, P., Bernstein, L., Chetwynd, J., Hoke, M., Matthew, M., Shettle, E. & Adler-Golden, S. (2000). MODTRAN 4 Version 2 User Manual. Air Force Research Laboratory, Space Vehicle Directorate.

Blackburn, G. (1998). Quantifying chlorophylls and carotenoids at leaf and canopy scales: an evaluation of some hyperspectral approaches. *Remote Sensing of Environment*, 66, pp.273–285.

Blackburn, G. (2006). Hyperspectral remote sensing of plant pigments. *Journal of Experimental Botany*, 10.1093, pp.1–13.

Box, E., Holben, B. & Kalb, V. (1989). Accuracy of the AVHRR vegetation index as a predictor of biomass, primary productivity and net CO_2 flux. *Vegetation*, 80, pp.71–89.

Carlson, T.N. & Ripley, D.A. (1997). On the relation between NDVI, fractional vegetation cover, and leaf area index. *Remote Sensing of Environment*, 62. 3, pp.241–252.

Carter, G. & Knapp, A. (2001). Leaf optical properties in higher plants: linking spectral characteristics to stress and chlorophyll concentrations. *American Journal of Botany*, 88, pp.677–684.

Chen, J.M., Liu, J., Cihlar, J. & Goulden, M.L. (1999). Daily canopy photosynthesis model through temporal and spatial scaling for remote sensing applications. *Ecological Modelling*, 124/2-3, pp.99–119.

Curran, P., Dungan, J. & Peterson, D. (2001). Estimating the foliar biochemical chemistry of leave reflectance spectrometry: Testing the Kokaly and Clark methodology. *Remote Sensing of Environment*, 76, pp.349–359.

Daughtry, C.S.T., Gallo, K.P., Goward, S.N., Prince, S.D. & Kustas, W.P. (1992). Spectral estimates of absorbed radiation and phytomass production in corn and soybean canopies. *Remote Sensing of Environment*, 39, pp.141–152.

Davidson, M., Berger, M., Moya, I., Moreno, J., Laurila, T., Stoll, M. & Miller, J. (2003). Mapping Photosynthesis from Space. a new vegetation-fluorescence technique. *ESA bulletin*, pp.34–37.

DeRose, R.J., Long, J.N. & Ramsey, R.D. (2011). Combining dendrochronological data and the disturbance index to assess Engelmann spruce mortality caused by a spruce beetle outbreak in southern Utah, USA. *Remote Sensing of Environment*, 115. 9, pp.2342–2349.

Döhler, H. (2007). *Energiepflanzen (Energy Plants)*. Darmstadt: KTBL Verlag.

Farquhar, G.D. (2001). Models of Photosynthesis. *Plant Physiology*, 125, pp.42–45.

Farquhar, G.D., Caemmerer, S. von & Berry, J. (1980). A biochemical model of photosynthetic CO_2 assimilation in leaves of C3 species. *Planta*, 149, pp.78–90.

Féret, J.B. & Asner, G.P. (2011). Spectroscopic classification of tropical forest species using radiative transfer modeling. *Remote Sensing of Environment*, 115. 9, pp.2415–2422.

Filella, I., Serrano, L., Serra, J. & Penuelas, J. (1995). Evaluating wheat nitrogen status with canopy reflectance indices and discriminant analysis. *Crop Science*, 35, pp.1400–1405.

Galvão, L.S., dos Santos, J.R., Roberts, D.A., Breunig, F.M., Toomey, M. & Moura, Y.M. (2011). On intra-annual EVI variability in the dry season of tropical forest: A case study with MODIS and hyperspectral data. *Remote Sensing of Environment*, 115. 9, pp.2350–2359.

Gamon, J., Serrano, L. & Surfus, J. (1997). The Photochemical Reflectance Index: An Optical Indicator of Photosynthetic Radiation Use Efficiency across Species, Functional Types, and Nutrient Levels. *Oecologia*, 112, pp.492–501.

Gamon, J.A., Peñuelas J. &. Field C. B. (1992). A narrow-wave band spectral index that track diurnal changes in photosynthetic efficiency. *Remote Sensing of Environment*, 41, pp.35–44.

Gamon, J.A. & SURFUS, J.S. (1999). Assessing leaf pigment content and activity with a reflectometer. *New Phytologist*, 143, pp.105–117.

Gilabert, M.A., Soledad, G. & Melia, J. (1996). Analyses of spectral-biophysical relationships for a corn canopy. *Remote Sensing of Environment*, 55. 1, pp.11–20.

Giles, A.B., Massom, R.A., Heil, P. & Hyland, G. (2011). Semi-automated feature-tracking of East Antarctic sea ice from Envisat ASAR imagery. *Remote Sensing of Environment*, 115. 9, pp.2267–2276.

Gitelson, A.A. & Merzlyak, M.N. (1997). Remote estimation of chlorophyll content in higher plant leaves. *International Journal of Remote Sensing*, 18, pp.2691–2697.

Glenn, E., Huete, A., Nagler, P. & Nelson, S. (2008). Relationship Between Remotely-sensed Vegetation Indices, Canopy Attributes and Plant Physiological Processes. *Sensors*, pp.2136–2160.

Goerner, A., Reichstein, M., Tomelleri, E., Hanan, N., Rambal, S., Papale, D., Dragoni, D. & Schmullius, C. (2011). Remote sensing of ecosystem light use efficiency with MODIS-based PRI. *Biogeosciences*, 8, pp.189–202.

Haboudane, D., Miller, J., Tremblay, N., Zarco-Tejada, P. & Dextraze, L. (2002). Integrated narrow-band vegetation indices for prediction of crop chlorophyll content for application to precision agriculture. *Remote Sensing of Environment*, 81, pp.416–426.

Haboudane, D., Miller, J.R., Pattey, E., Zarco-Tejada, P.J. & Strachan, I.B. (2004). Hyperspectral vegetation indices and novel algorithms for predicting green LAI of crop canopies: Modeling and validation in the context of precision agriculture. *Remote Sensing of Environment*, 90, pp.337–352.

Hank, T. (2008). A Biophysically based Coupled Model Approach for the Assessment of Canopy Processes under Climate Change Conditions. PhD thesis. Department of Geography. Munich.

Hank, T., Oppelt, N. & Mauser, W. (2007).Physically based modelling of photosynthetic processes. In. Stafford (Ed.). *Precision Ariculture 07*. Wageningen Acad. Publishers.

Hernández-Clemente, R., Navarro-Cerrillo, R.M., Suárez, L., Morales, F. & Zarco-Tejada, P.J. (2011). Assessing structural effects on PRI for stress detection in conifer forests. *Remote Sensing of Environment*, 115, pp.2360–2375.

Hilker, T., Coops, N., Black, T., Wulder, M.A. & Guy, R.D. (2008). The use of remote sensing in light use efficiency based models of gross primary production: A review of current status and future requirements. *Science of The Total Environment*, 404, pp.411–423.

Hilker, T., Coops, N.C., Wulder, M.A., Black, T.A. & Guy, R.D. (2008). The use of remote sensing in light use efficiency based models of gross primary production: A review of current status and future requirements. Biochemistry of Forested Ecosystems. *Science of The Total Environment*, 404. 2-3, pp.411–423.

Holben, B.N., Kaufman, Y.J. & Kendall, J.D. (1990). NOAA-11 AVHRR visible and near-IR inflight calibration. *International Journal of Remote Sensing*, 11, pp.1511–1519.

Huete, A.R. (1988). A soil-adjusted vegetation index (SAVI). *Remote Sensing of Environment*, 25, pp.295–309.

Huete, A.R., Liu, H.Q., Batchily, K. & van Leeuwen, W. (1997). A comparison of vegetation indices over a global set of TM images for EOS-MODIS. *Remote Sensing of Environment*, 59, pp.440–451.

Jaquemoud, S. & Baret, F. (1990). PROSPECT: a model of leaf optical properties spectra. *Remote Sensing of Environment*, 34, pp.75–91.

Jin, Y., Randerson, J.T. & Goulden, M.L. (2011). Continental-scale net radiation and evapotranspiration estimated using MODIS satellite observations. *Remote Sensing of Environment*, 115. 9, pp.2302–2319.

Johnson, I.R. & Thornley, J.H. (1984). A model of instantaneous and daily canopy photosynthesis. *Journal of Theoretical Biology*, 107. 4, pp.531–545.

Keyworth, S., Jarmann, M. & Medcalf, K. (2009). Assessing the Extent and Severity of Erosion on the Upland Organic Soils of Scotland using Earth Observation. Environment Systems Limited. Aberystwyth.

Kim, Y., Huete, A.R., Miura, T., & Jiang, Z. (2010). Spectral compatibility of vegetation indices across sensors: band decomposition analysis with Hyperion data. *Journal of Applied Remote Sensing*, 4, pp.43520-43531.

Kim, S.-H. (2003). A Coupled Model of Photosynthesis, Stomatal Conductance and Transpiration for a Rose Leaf (Rosa hybrida L.). *Annals of Botany*, 91, pp.771–781.

M. Begon and A.H. Fitter (Ed.) (1995). Advances in Ecological Research. ISBN 0065-2504: Academic Press.

Malenovsky, Z., Mishra, K.B., Zemek, F., Rascher, U. & Nedbal, L. (2009). Scientific and technical challenges in remote sensing of plant canopy reflectance and fluorescence. *Journal of Experimental Botany*, 60, pp.2987–3004.

Mänd, P., Hallik, L., Peñuelas, J., Nilson, T., Duce, P., Emmett, B.A., Beier, C., Estiarte, M., Garadnai, J. & Kull, O. (2010). Responses of the reflectance indices PRI and NDVI to experimental warming and drought in European shrublands along a north-south climatic gradient. *Remote Sensing of Environment*, 114. 3, pp.626–636.

Mauser, W. & Bach, H. (2009). PROMET – Large scale distributed hydrological modelling to study the impact of climate change on the water flows of mountain watersheds. *Journal of Hydrology*, 376, pp.362–377.

McCallum, I., Wagner, W., Schmullius, C., Shvidenko, A., Obersteiner, M., Fritz, S. & Nilsson, S. (2010). Comparison of four global FAPAR datasets over Northern Eurasia for the year 2000. *Remote Sensing of Environment*, 114, pp.941–949.

McGuire, A.D., Joyce, L.A., Kicklighter, D.W., Melillo, J.M., Esser, G. & Vorosmarty, C.J. (1993). Productivity response of climax temperate forests to elevated temperature and carbon dioxide: a north american comparison between two global models. *Climatic Change*, 24, pp.287–310.

Melillo, J.M., McGuire, A.D., Kicklighter, D.W., Moore, B., Vorosmarty, C. & Schloss, A. (1993). Global climate change and terrestrial net primary production. *Nature*, 363. 6426, pp.234–240.

Meroni, M., Rossini, M., Guanter, L., Alonso, L., Rascher, U., Colombo, R. & Moreno, J. (2009). Remote sensing of solar-induced chlorophyll fluorescence: Review of methods and applications. *Remote Sensing of Environment*, 113, pp.2037–2051.

Monteith, J. (1972). Solar Radiation and Productivity in Tropical Ecosystems. *Journal of Applied Ecology*, pp.747–766.

Montheith, J. & Moss, C. (1977). Climate and the Efficiency of Crop Production in Britain. *Phil. Trans. R. Soc. Lond. B*, 281, pp.277–294.

Myneni, R., Knyazikhin, Y., Zhang, Y., Tian, Y., Wang, Y., Lotsch, A., Privette, J., Morisette, J., Running, R., Nemani, R. & Glassy, J. (1999). MODIS Leaf Area Index and Fraction of Photosynthetically Active Radiation Absorbed by Vegetation Product. Algorithm Theoretical Basis Document. http://eospso.gsfc.nasa.gov/atbd/modistables.html.

Myneni, R., Ramakrishna, R., Nemani, R. & Running, S. (1997). Estimation of global leaf area index and absorbed par using radiative transfer models. *IEEE Transactions on Geoscience and Remote Sensing*, 35, pp.1380–1393.

Newton, P.C.D., Clark, H., Bell, C.C., Glasgow, E.M., Ross, D.J., Yeates, G.W. & Saggar, S. (1995). Plant Growth and Soil Processes in Temperate Grassland Communities at Elevated CO_2. *Journal of Biogeography*, 22. 2/3, pp.235–240.

Nichol, C.J., Lloyd, J., Shibistova, O., Arneth, A., Roser, C., Knohl, A., Matsubara, S. & Grace, J. (2002). Remote sensing of photosynthetic light use efficiency of a Siberian boreal forest. *Tellus B*, 54, pp.677–687.

Oppelt, N. (2002).Monitoring of Plant Chlorophyll and Nitrogen Status Using the Airborne Imaging Spectrometer AVIS. Dissertation. Department of Geography, University of Munich.

Oppelt, N., Hank, T. & Mauser, W. (2007). GVIS - ground based hyperspectral remote sensing. In *Proc of the 2nd International Symposium on Optics, Informatics and Cyber-Technology (OIC 2007)*. 8-11 July 2007, Orlando/Florida (USA). CD-Rom publication.

Oppelt, N. & Mauser, W. (2001). The Chlorophyll Content of Maize (*Zea Mays*) derived with the Airborne Imaging Spectrometer AVIS. In *ISPRS 8th International Symposium "Physical Measurements and Signatures in Remote Sensing"*, 8-12 January 2001; Aussoir, France), pp.407-412.

Oppelt, N. & Mauser, W. (2006). Three Generations of the Airborne Imaging Spectrometer AVIS. Expectations – Applications – Results. In *Proceedings of SPIE Europhotonics*

Europe. Strasbourg (France), 02.04.-07.04.2006, Vol. 6189, SPIE, Bellingham, Wash. ISBN 97-80819462459

Oppelt, N. & Mauser, W. (2004). Hyperspectral monitoring of physiological parameters of wheat during a vegetation period using AVIS data. *International Journal of Remote Sensing*, 25, pp.145–159.

Oppelt, N. & Mauser, W. (2007). The Airborne Visible/Infrared imaging Spectrometer AVIS: design, characterization and calibration. *Sensors*, 7, pp.1934–1953.

Pearson, R. & Miller, L. (1972). Remote Mapping of Standing Crop Biomass for Estimation of the Productivity of the Shortgrass Prairie. In Willow Run Laboratories (Ed.), *Proceedings of the Eighth International Symposium on Remote Sensing of Environment:* Willow Run Laboratories. Ann Arbor. pp. 1355

Peng, C. & Apps, M. (1999). Modelling the response of net primary productivity (NPP) of boreal forest ecosystems to changes in climate and fire disturbance regimes. *Ecological Modelling*, 122. 3, pp.175–193.

Penuelas, J. & Filella, I. (1998). Visible and near-infrared reflectance techniques for diagnosing plant physiological status. *Trends in Plant Science*, 3. 4, pp.151–156.

Penuelas, J., Filella, I. & Gamon, J. (1995). Assessment of Photosynthetic Radiation-Use Efficiency with Spectral Reflectance. *New Phytologist*, 131, pp.291–296.

Penuelas, J., Llusia, J., Pinol, J. & Filella, I. (1997). Photochemical reflectance index and leaf photosynthetic radiation-use-efficiency assessment in Mediterranean trees. *International Journal of Remote Sensing*, 18. 13, pp.2863–2868.

Porra, E.J., Kriedman, P.E. & Thomson, W. (1989). Determination of accurate extinction coefficients and simultaneous Eq.s for assaying chlorophyll a and b extracted with four different solvents. *Biochemical and Biophysical Acta*, 975, pp.384-394.

Pugnaire, F.I. & Valladares, F. (2007). *Functional plant ecology*. (2nd ed.). Boca Raton, FL. 084937488X: CRC Press.

Qi, J., Chehbouni, A., Huete, A.R., Kerr, Y.H. & Sorooshian, S. (1994). A modified soil adjusted vegetation index. *Remote Sensing of Environment*, 48. 2, pp.119–126.

Qiu, H.-L., Sanchez-Azofeifa, A. & Gamon, J. Ecological Applications of Remote Sensing at Multiple Scales. In. Valladares, F. & Pugnaire, I. (Eds). *Functional Plant Ecology*, 2nd edition. pp. 655-678.

Rahman, A.F., GAMON, J.A., Fuentes, D.A. & Roberts, D.A. (2001). Modeling spatially distributed ecosystemflux of boreal forest using hyperspectral indices from AVIRIS imagery. *Journal of Geophysical Research*, pp.33579–33591.

Rascher, U., Nichol, C.J., Small, C. & Hendricks, L. (2007). Monitoring Spatio-temporal Dynamics of Photosynthesis with a Portable Hyperspectral Imaging System. *Photogram-metric Engineering & Remote Sensing*, 73, pp.45–65.

Rascher, U., & Pieruschka, R. (2008). Spatio-temporal variations of photosynthesis: the potential of optical remote sensing to better understand and scale light use efficiency and stresses of plant ecosystems. *Precision Agriculture*, 9, pp.355–366.

Román, M.O., Gatebe, C.K., Schaaf, C.B. & King, M.D. (2011). Variability in surface BRDF at different spatial scales over a mixed agricultural landscape as retrieved from airborne and satellite spectral measurements. *Remote Sensing of Environment*, 115. 9, pp.2184–2203.

Rondeaux, G., Steven, M. & Baret, F. (1996). Optimization of soil-adjusted vegetation indices. *Remote Sensing of Environment*, 55. 2, pp.95–107.

Rouse, J., Haas, R., Schell, J., Deering, D. & Harlan, J. (1974). Monitoring the vernal advancements and retrogradation of natural vegetation. NASA/Gsfc final report.

Ruimy, A., Jarvis, P.G., Baldocchi, D.D. & Saugier, B. (1995). CO_2 Fluxes over Plant Canopies and Solar Radiation: A Review. In M. Begon & A.H. Fitter (Eds.). *Advances in Ecological Research*, 26. Pp.1-68.

Running, S.W., Nemani, R., Glassy, J. & Thornton, P. (1999). MODIS Daily Photosynthesis (PSN) and Annual Net Primary Production (NPP) Product. Algorithm Theoretical Basis Document. http://modis.gsfc.nasa.gov/data/atbd/land_atbd.php.

Running, S.W., Nemani, R., Heinsch, F.A., Zhao, M., Reeves, M.A. & Hashimoto, H. (2004). A Continuous Satellite-Derived Measure of Global Terrestrial Primary Production. *BioScience*, 54, p.547-560.

Sari, M., Somnez, N. & Kurklu, A. (2005). Determination of seasonal variations in solar energy utilization by the leaves of Washington navel orange trees (*Citrus sinensis* L. Osbeck). *International Journal of Remote Sensing*, 26, pp.3295–3307.

Schurr, U., Walter A. & Rascher, U. (2006). Functional dynamics of plant growth and photosynthesis - from steady-state to dynamics - from homogeneity to heterogeneity. *Plant, Cell and Environment*, 29, pp.340–352.

Steven, M.D., Malthus, T.J., Baret, F., Xu, H. & Chopping, M.J. (2003). Intercalibration of vegetation indices from different sensor systems. *Remote Sensing of Environment*, 88. 4, pp.412–422.

Strachan, I., Pattey, E., Salustro, C. & Miller, J. (2008). Use of hyperspectral remote sensing to estimate the gross photosynthesis of agricultural fields. Canadian Journal of Remote Sensing. *Canadian Journal of Remote Sensing*, 34. 3, pp.333–341.

Suárez, L., Zarco-Tejada, P.J., González-Dugo, V., Berni, J.A.J., Sagardoy, R., Morales, F. & Fereres, E. (2010). Detecting water stress effects on fruit quality in orchards with time-series PRI airborne imagery. *Remote Sensing of Environment*, 114. 2, pp.286–298.

Thenot, F., Methy, M. & Winkel, T. (2002).The Photochemical Reflectance Index (PRI) as water-stress index. *International Journal of Remote Sensing*, 23, pp.5135–5139.

Tilstone, G.H., Angel-Benavides, I.M., Groom, S. & Sathyendranath, S. (2011). An assessment of chlorophyll-a algorithms available for SeaWiFS in coastal and open areas of the Bay of Bengal and Arabian Sea. *Remote Sensing of Environment*, 115. 9, pp.2277–2291.

Trotter, G.M., Whitehead, D. & Pinkney, E.J. (2002).The photochemical reflectance index as a measure of photosynthetic light use efficiency for plants with varying foliar nitrogen contents. *International Journal of Remote Sensing*, 23. 6, pp.1207–1212.

Weng, J., Liao, T.H.M., Chung, C., Lin, C. & Chu, C. (2006). Seasonal variation in photosystem II efficiency and photochemical reflectance index of evergreen trees and perennial grasses growing at low and high elevations in subtropical Taiwan. *Tree Physiology*, 26, pp.1097–1104.

Wu, C.; Niu, Z.; Tang, Q. & Huan, W. (2008). Estimating chlorophyll content from hyperspectral vegetation indices: Modeling and validation. *Remote Sensing of Environment*, 148, pp. 1230-1241.

Yin, X. & Struik, P.C. (2009). C3 and C4 photosynthesis models: An overview from the perspective of crop modelling. Recent Advances in Crop Growth Modelling. *NJAS - Wageningen Journal of Life Sciences*, 57. 1, pp.27–38.

Yoder, B., & Pettrigrew-Crosby, R. (1995). Predicting nitrogen and chlorophyll content and concentrations from reflectance spectra (400-2500nm) at leaf and canopy scales. *Remote Sensing of Environment*, 53, pp.191–211.

Zhao, M., Heinsch, F., Nemani, R. & Running, S.W. (2005). Improvements of the MODIS terrestrial gross and net primary production global data set. *Remote Sensing of Environment*, 95. 2, pp.164–176.

7

Nitrate Assimilation: The Role of *In Vitro* Nitrate Reductase Assay as Nutritional Predictor

Fungyi Chow
University of Sao Paulo, Department of Botany, Institute of Bioscience
São Paulo, SP
Brazil

1. Introduction

Macroalgae or macrophytes are a heterogeneous assemblage of macroscopic eukaryotes belonging to various evolutionary lineages, which live predominantly in aquatic habitats. They have undifferentiated vegetative bodies organized in pseudoparanchymatous and parenchymatous bodies. As with higher plants, marine macroalgae or seaweeds are photosynthetic species that, by harvesting sunlight energy, convert carbon dioxide in oxygen to produce organic compounds, especially carbohydrates. In addition, they require mineral nutrients, essential for growth, development and reproduction, which are incorporated into carbon skeletons.

In natural aquatic ecosystems, 95% of the nitrogen which occurs as dissolved dinitrogen gas (N_2), is not directly accessible to most photosynthetic-oxygen organisms. Dissolved inorganic nitrogen (DIN) includes the ions, ammonium (NH_4^+), nitrite (NO_2^-), and nitrate (NO_3^-). In seawater, and under natural conditions, about 3,5% is NO_3^- (*ca.* 0.35 mg $NO_3^-.L^{-1}$), which, near to coastal zones, appears in abundance as a product of upwelling or pollution, whence, their importance as the predominant cause of local eutrophication.

Thus, nitrate constitutes the prevailing available nitrogen source for macroalgae in the marine environment. The available DIN may be supplemented by dissolved organic nitrogen (DON), this including urea and amino acids. For all eukaryotic photoautotrophs, NH_4^+, NO_2^-, and NO_3^- are the only directly assimilated sources.

Nitrogen, which is rapidly taken up, is a key element in several compounds present in the cells. It is used to build up amino acids, proteins, nucleoside phosphates, nucleic acids, and other organic N-containing macromolecules. The availability of nutrients, especially nitrogen, in marine habitats is one of the main regulating factors that limit growth, morphology, development, reproduction, distribution, and biochemical composition in seaweeds. The importance of nitrogen for biological life is evident, in that only oxygen, carbon, and hydrogen are more abundant in the cells of photosynthetic organisms.

Macroalgae and photoautotrophic organisms have considerable intracellular capacity for storing nitrogen as soluble nitrogen and organic molecules, whereby growth and development can be regulated and limited according to nitrogen uptake. This characteristic for storing and assimilating nutrients, when available and at high concentrations, besides

facilitating their use under external, restrictive conditions, provides certain species with ecological advantages for persistence and prolife during limiting stress periods.
Seasonal nutrient limitation in macroalgal growth is well known (Lobban & Harrison, 1994). The importance of nitrogen for the growth-cycle is related to life-strategy. In perennial species, nutrient availability is a determinant factor that plays an important role in the seasonal reproductive life-cycle (Kain, 1989). In annual species, nitrogen, when available, is uptake and stored inside the cells, whence its rapid conversion into new biomass.
During the past decades, substantial efforts have been made to understand the biochemistry, molecular biology and regulation of nitrate reductase (NR) in higher plants, to so further ecophysiological information and applied botany. The basic-action mechanisms and importance of NR in seaweeds are no different from higher plants. Notwithstanding, the amount of knowledge is still superficial and phycological studies scarce.
Based on higher plants and microalgae, the purpose is to briefly point out the role of NR in nitrogen metabolism, with a focus on macroalga research, and highlight the importance of *in vitro* NR assay optimization and its value as a physiological tool.

2. Overview of nitrate assimilation

Inorganic nitrogen availability plays a critical role in the physiology of marine macroalgae and the productivity of complete ecosystems (Lapointe & Duke, 1984). Nitrogen depletion has been shown to increase photoinhibitory responses in the photosynthesis of marine organisms, including macroalgae (Korbee-Peinado et al., 2004; Huovinen et al., 2006). On the other hand, photosynthetic pigments, through generally being positively correlated with nitrogen availability, rapidly respond to varying nitrogen levels (Davison et al., 2007). For example, phycobiliproteins, reported as nitrogen-storage compounds in N-rich conditions, act as nitrogen sources under N-limiting conditions (Lobban et al., 1985).
Much of what is known on nitrogen metabolism is based on studies with microalgae; the number with macroalgae is still few in comparison.
As nitrogen uptake by macroalgae are usually studied by monitoring the disappearance of the nutrient from the culture medium and is influenced by irradiance, temperature, water motion, desiccation, and age.
Macroalgae have either the plasticity or preference to uptake several forms of nitrogen. Hanisak (1983) noted that the uptake rate of ammonium generally exceeds that of nitrate. Chow et al. (2001) registered the same trend for *Gracilaria chilensis,* with total uptake of only ammonium, when present together with nitrate and nitrite. However, very high concentrations of ammonium (> 30 -50 μM) can saturate nitrate transport, thereby inducing toxification by ammonium. As ammonium can be used directly to synthesize amino acids, and nitrate stored inside vacuoles, the energetic cost of ammonium assimilation is lower than that of nitrate.
Nitrate uptake normally involves saturating kinetics (DeBoer, 1981). As greater concentrations of intracellular nitrate than that of the surrounding seawater constitute a negative gradient for transport, active transport can be considered as a primary process. Some authors have proposed that the NR plasma membrane acts as protein transporter to within the cell.
Considering the relative abundance of environmental nitrate in seawater, the nitrogen metabolism is usually commanded by nitrate assimilation fitness. Nitrate (NO_3^-) is taken up

by the cells and translocated across the plasmalemma by energy-dependent processes. Once inside the cells, any excess can be stored within vacuoles, while a fraction is being metabolized in the cytoplasm by reduction to nitrite, via the enzyme nitrate reductase (**NR**), and using NAD(P)H as electron donor. In turn nitrite (NO_2^-) is transported to chloroplasts and reduced to ammonium, prior to assimilation into organic compounds by enzyme nitrite reductase (**NiR**), by means of reduced ferredoxin (Fd_{red}) as electron source. Thus, the nitrogen assimilation pathway is a two-step process, first with NO_3^- reduction to NO_2^- (1) and then to ammonium (NH_4^+) (2) as described below:

Nitrate reductase (NR)

$$NO_3^- + NAD(P)H + 2e^- + H^+ \rightarrow NO_2^- + NAD(P)^+ + H_2O \quad (1)$$

Nitrite reductase (NiR)

$$NO_2^- + 6Fd_{red} + 6e^- + 8H^+ \rightarrow NH_4^+ + 6Fd_{ox} + 2\,H_2O \quad (2)$$

Nitrite and ammonium ions can not be accumulating inside cells, as they are cytotoxic through producing pH change and inducing reactive nitrogen species (RNS) and oxidative damage. Consequently, their incorporation into organic compounds must be relatively fast, in order to prevent accumulation and toxicity. In the case of photosynthetic organisms, fungi, and bacteria present a variety of mechanisms to regulate and control the expression of those enzymatic activities involved in nitrogen assimilatory pathways.

The assimilation of ammonia-N into carbon compounds (amino acids) primarily takes place through the sequential actions of glutamine synthetase (GS) and glutamine 2-oxoglutarate aminotransferase (GOGAT), inside chloroplasts, where both are localized, although isozymes of both may also be found in cytosol. Ammonium assimilation by GS requires glutamate (Glu) as substrate and ATP input to form glutamine (Gln) (3).

Glutamine synthetase (GS)

$$NH_4^+ + \text{glutamate (Glu)} + ATP \rightarrow \text{glutamine (Gln)} + ADP + Pi \quad (3)$$

The amino-N of glutamine, subsequently transferred to 2-oxoglutarate (2-OXG), is reduced by GOGAT to form two molecules of glutamate (4).

Glutamine 2-oxoglutarate aminotransferase (GOGAT)

$$\text{2-oxoglutarate} + \text{glutamine (Gln)} + NADPH \rightarrow 2\,[\text{glutamate}]\,(\text{Glu}) + NADP^+ \quad (4)$$

The production of glutamate can be through two pathways. The first involves the reductive amination of α-ketoglutarate catalyzed by the enzyme glutamate dehydrogenase (GDH), which is found in chloroplasts and mitochondria. In the latter, α-ketoglutarate is normally continually produced by the Krebs cycle.

Independent of the location of GS and GOGAT, glutamate is exported from chloroplasts to cytosols, where transamination reactions can proceed, thereby facilitating the synthesis of other amino acids.

The control of the nitrate assimilatory rate is attributed to NR action, as this is the first enzyme in the specific pathway. Consequently, it is of increasing interest to study its molecular and catalytic properties, as well as the physiological responses to environmental stressing conditions and intracellular factors. Nitrate reductase activity has been proposed as an index of the rate of nitrate incorporation, with the additional inference that nitrate reduction is a rate limiting process for nitrogen assimilation, since any reduction in enzymatic activity results in a relative drop in nitrogen assimilation.

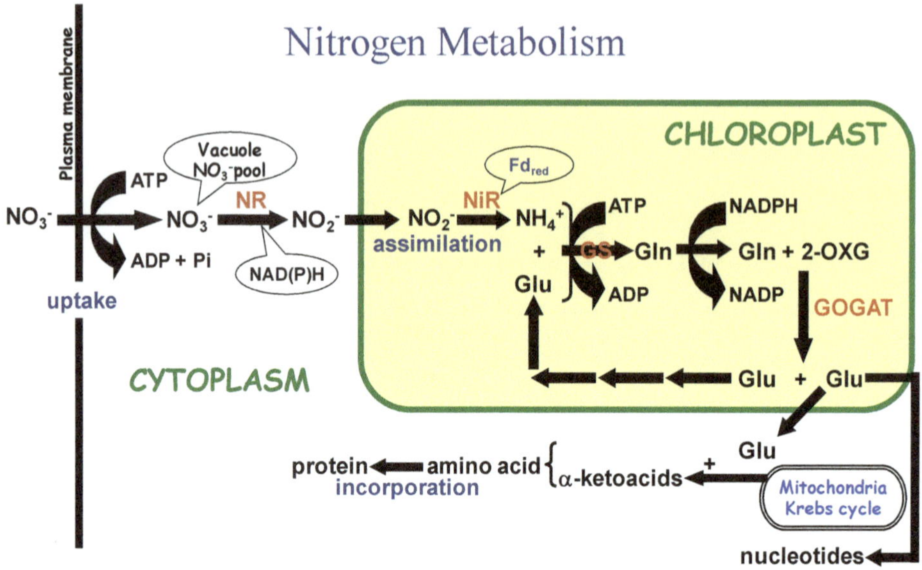

Fig. 1. General brief of nitrate assimilation pathway. Nitrate (NO_3^-) is actively transported from the external medium across the plasma membrane into the cytoplasm. It can be stored into vacuoles or reduced to incorporating in carbon skeletons. Nitrate is reduced to nitrite (NO_2^-) in the cytoplasm by nitrate reductase (NR) that uses NAD(P)H. Nitrite is transported inside the chloroplast and reduced to ammonium (NH_4^+) by nitrite reductase (NiR) that uses reduced ferredoxin (Fd_{red}). Ammonium is incorporated into glutamate (Glu) to form glutamine (Gln) via the action of glutamine synthetase (GS). The amino-N of Gln is then transferred to 2-oxoglutarate (2-OXG) via the action of glutamine 2-oxoglutarate aminotransferase (GOGAT). This reaction produces two molecules of Glu, one of them reenters to the assimilation pathway as substrate for GS and the second molecule of Glu is exported to the cytoplasm and will participate of transamination reactions with α-ketoacids to produces other amino acids and proteins (modified from Falkowski & Raven, 1997).

3. Interaction between nitrate assimilation and carbon metabolism

Nitrate assimilation is intrinsically dependent on the organic carbon substrates, reductants, and ATPs that are supplied for both processes photosynthesis and respiratory processes (Turpin, 1991). When nitrogen is limited, or photosynthesis and respiration are negatively affected, this dependency becomes multifactorial, whereupon compensatory mechanisms or regulation must be activated.

It is clearly evident that nitrogen metabolism in macroalgae is closely linked to photosynthetic carbon metabolism (Vergara et al., 1998). Nitrate reductase activity presents maximal enzymatic rates during the diurnal phase, and minimal during dark phase, with a narrow relationship of maximal and minimal photosynthetic rates. These responses indicate regulatory activation of both processes by light. The intracellular toxic conditions of nitrite and ammonium make the urgent incorporation of both into the carbon skeleton essential, to so avoid toxicity.

Carbon molecules and reductant sources proceeding from photosynthesis, is an indication of the existence of related regulatory mechanisms between both processes. Carbon and nitrogen metabolic pathways consume large amounts of photosynthetic carbon and energy sources. Both metabolic forms are connected, either by the organic carbon and energy that are directly supplied for photosynthetic electron transport and fixated CO_2, or by the respiration of fixed carbon, via glycolysis by the Krebs cycle, and the electron transport chain of mitochondria. Therefore, the integration of these important metabolic processes must have integrated regulatory mechanisms.

Nitrogen limitation affects photosynthetic processes. Under N-limiting conditions, there is a concomitant decrease in PSII photochemical efficiency, as a consequence of the dissipation of absorbed excitation energy in the center of pigments. Depleted photosynthetic efficiency appears to occur through a drop in the number of functional PSII reaction centers relative to the antennae system (Falkowski, 1992). On the other hand, the reduction in photosynthetic energy conversion under N-limiting conditions appears to affect amino acid biosynthetic processes. Nitrogen limitation also affects the respiration rate. The molecular basis of nitrogen limitation and respiratory rate alteration is unclear, but it appears to be related to the demand for carbon skeletons and ATP, two of the major products of respiratory pathways. Thus, the depletion of nitrogen creates a chain reaction that decompensates energy metabolism and amino acid biosynthesis, thereby affecting photosynthesis, respiration and growth.

4. Nitrate reductase in macroalgae

Nitrate reductase, a relatively large molecule, is usually composed of two or four subunits, each of which with approximately 100 kDa. Nakamura & Ikawa (1993) reported these subunits in *Porphyra yezoensis* at close to 100 kDa. The four in *Gracilaria tenuistipitata* var. *liui* (Lopes et al., 2002), were also of the same size, as were the possibly two in *Kappaphycus alvarezii* (Granbom et al., 2007).

Three assimilatory NR forms are recognized in eukaryotes: (a) EC 1.6.6.1 NADH-specific and (b) EC 1.6.6.2 NADP/NADPH, both occurring in eukaryotic algae and higher plants, and (c) EC 1.6.6.3 NADPH-specific.

The enzyme is preferentially present in the cytoplasm, although there is growing evidence of NR associated to chloroplast membranes (Solomonson & Barber, 1990) and plasmalemma (Tischner et al., 1989; Fernandez-Lopez et al., 1996).

Nitrate reductase becomes interesting through its usefulness as a model for prospecting multi-component interaction mechanisms related to redox enzymes. Nitrate reductase is one of the few inducible/repressible enzymatic systems reasonably well-characterized in photoautotrophs, especially in higher plants and microalgae, thus making it a fantastic biological model for physiological studies. On the other hand, it is of concern to use NR as an ecophysiological parameter for predicting nutritional rates of nitrate assimilation and growth. Furthermore, as NR-protein characteristics differ among algal groups, this diversity may be of relevance in revealing evolutionary adaptation patterns (Zhou & Kleinhofs, 1996; Howartha & Baumb, 2002; Stolz & Basu, 2002).

Marine macrophytes inhabiting intertidal coastal regions are exposed to extreme fluctuations in physicochemical parameters, midday increased irradiance levels, UV radiation (UVR), nitrogen depletion, desiccation, high-temperature stress, etc. (Lobban et al., 1985). Individually or together these stressing conditions result in drastic physiological responses and acclimation. Thus, comprehension of macroalgal responses to daily and seasonal fluctuations in these abiotic factors is critical for a better understanding of the regulation of nitrogen metabolism.

Nitrate reductase-activity assaying has been used for indicating algal capacity in using nitrate and the internal nutritional index. A newer approach is to determine the protein at the molecular level by studies of gene NR-mRNA expression.

The use of NR in an ecological context is particularly relevant for marine environments where nitrogen is often limiting, thereby providing relevant information regarding the physiological nitrogen status of organisms (Hernández et al., 1993).

Nitrate reductase is considered as a key enzyme in nitrogen metabolism, through being, not only the rate-limiting enzyme in inorganic nitrogen assimilation, but also the major regulatory step in nitrogen metabolism (Crawford, 1995; Berges, 1997; Davison & Stewart, 1984; Lartigue & Sherman, 2005; Young et al., 2009). Changes in NR activity, both in the field or in laboratory, have been examined in very few macroalgae.

Nitrate reductase expression is a complex process regulated by various factors, such as levels of nitrate, CO2, light, carbon skeletons and nitrogen metabolites (Crawford, 1995, Lopes et al., 2002). Furthermore, it is highly regulated in multiple steps, transcriptionally, post-transcriptionally, translationally and post-translationally. These regulatory mechanisms can act individually or synergically, and are correlated to short and long-term NR response. Thus, NR activity can be modified rapidly in response to nitrate availability and other controlling factors.

This intricate control can be shown experimentally by adding nitrate to the medium. In *Gracilaria chilensis*, NR activity was thus rapidly stimulated within a few minutes (Chow et al., 2007; Chow & Oliveira, 2008), probably by post-translational NR-protein regulation. Lartigue & Sherman (2005) observed the same trend in *Enteromorpha* sp. A like inducing response, under the same conditions, has also been observed in other macroalgae (Gao, Smith & Alberte 1995; Lartigue & Sherman, 2005; Young et al., 2007; Martins et al., 2009; Cabello-Pasini et al., 2011).

Nitrate reductase activity in Arctic species appears to be directly enhanced by nitrate addition, with relatively little feedback from the N-status of the cell (Gordillo et al., 2006). Communities of Laminariales species apparently possess a high degree of resilience to disruption in natural nutrient-availability patterns.

Inactivation of NR activity and degradation of NR-protein have been observed in microalgae, macroalgae and higher plants undergoing nitrate deficiency or other forms of reduced nitrogen (*e.g.* ammonium and urea) (Weidner & Kiefer, 1981; Vergara et al., 1998; Solomonson & Barber, 1990; Balandin & Aparicio, 1992; Crawford &Arst, 1993; Berges et al., 1995; Vergara et al., 1998; Campbell, 1999; Gao et al., 2000; Chow & Oliveira, 2007; Nicodemus et al., 2008). The NR-inhibiting action mechanism is unknown, although it is thought to occur via feedback-regulation of ammonium assimilation by metabolites, *e.g.* glutamine (Flynn 1991, Vergara et al., 1998), or indirectly by inhibition of nitrate-uptake (Collos, 1989).

Synthesis of the NR protein is also regulated by its substrate with a half-life of a few hours. The induction of NR activity is preceded by an increase in NR mRNA, which is activated in a question of hours (Granbom et al., 2007).

Light is a regulatory factor of nitrogen metabolism. Light provides the energy to produce/reduce the power and ATP utilized in nitrate transport, nitrate and nitrite reduction and ammonium fixation into amino acids. Furthermore, light increases the production of carbon skeletons essential for nitrogen assimilation. Additionally, light may play a signaling role in controlling the activity-level and NR-protein synthesis.

As NR-protein degradation normally takes a few hours, near to the interphase light:dark cycle, NR is rapidly regulated by phosphorylation and dephosphorylation mechanisms. Transient induction of NR activity in *G. chilensis* by light pulse, during the dark phase, was inhibited by the addition of calyculin A (Chow et al., 2008), thereby inferring this rapid NR regulation, as already reported in higher plants (Kaiser & Spill, 1991; Huber et al., 1992a, b; Kaiser & Huber, 1994; Campbell, 1996).

Probably the quick-NR regulatory mechanisms by phosphorylation and dephosphorylation constitute an essential system for tolerating changes in environmental stress, especially for macroalgae, which undergo tidal variation that affects nutrient availability and irradiance intensity. Under permanent stressing-conditions on a scale of hours to weeks, the synthesis and degradation of NR proteins and mRNA would be the most functional mechanism in preventing unnecessary energy expenditure.

Nitrate reductase is also influenced by irradiance (Davison & Stewart, 1984; Gao et al., 1995; Lopes et al., 1997; Vergara et al., 1998; Lartigue & Sherman, 2002; Chow & Oliveira, 2008), the rapid suppression in darkness probably arising from the availability of carbon skeletons, ATP, and NAD(P)H from photosynthesis and respiration.

Ultraviolet radiation also affects NR activity in macroalgae (Figueroa & Viñegla, 2001). Nevertheless, studies are scarce, and the mechanisms of action unknown.

NR and NR-protein activities manifest a daily rhythm with circadian influence (Lillo, 1983; Deng et al., 1991; Lopes et al., 1997, 2002; Granbom et al., 2004, 2007; Chow et al., 2004, 2007; Granbom et al., 2007). Under the light:dark cycle, NR activity reaches a plateau around the middle of the photoperiod, as well as a nocturnal minimum (Weidner and Kiefer 1981, Gao et al. 1992, Ramalho et al. 1995, Lopes et al. 1997, Chow et al. 2004, 2007; Granbom et al. 2004), presumably the normal behavior in photosynthetic species.. The circadian diel cycle of NR seems to be primarily regulated transcriptionally and correlated to the rate of mRNA protein-synthesis (Smith et al., 1992; Ramalho et al., 1995; Granbom et al., 2007). The peaks of enzymatic activity appear to be in concert with maximal photosynthetic flow, when intracellular carbohydrates and end products of photosynthesis begin to accumulate. At this point, the importance of light in promoting NR activity and synthesis is indirectly linked to carbon metabolic requirements.

Low NR activities during darkness have been suppressed by the artificial apply of light-mimicry carbohydrate sources. Furthermore, a light-pulse of 15 minutes during the dark phase, also induced NR activity to levels similar to those of the light phase in *G. chilensis* (Chow & Oliveira, 2008), possibly indicating NR inducible behavior by post-translational mechanisms. When intracellular carbohydrates begin to accumulate in excess, NR mRNA transcription is suspended. Inversely, when carbohydrates become depleted, NR expression is enhanced.

Higher NR activities are also associated with parts with active metabolic rates, and probably, to high levels of photosynthetic activity. In macroalgae with apical growth, NR activity in the tips is higher than in the basal parts (Granbom et al., 2004, 2007; Chow, 2004), thereby implying post-translational regulation. Nevertheless, NR protein content in the basal parts is the highest (Granbom et al., 2007), possibly indicating that a large part of NR is in an active form, compared to the basal part of the thallus.

Post-translational regulatory mechanisms are common in NR enzymes, especially during short-term response. This mechanism includes phosphorylation (Huber et al., 1992) involving specific protein kinases, protein phosphatases and a protein inativator (MacKintosh et al., 1995; Glaab & Kaiser, 1996).

In most studies of macroalgae NR activity, it is possible to establish the same trend of physiological response, i.e., the pronounced dependence on the external and internal pool of available nitrate, light stimulation and low or constitutive dark activity, temperature range of action according to the natural habitat of the seaweed, and correlations with carbon and ATP availability from photosynthesis and respiration. The slight differences in NR behavior can be attributed to species-specific response depending on particular environmental conditions, and may reflect special turnover of NR activity, as a product of acclimation and adaptation response to the extremely changeable intertidal environment.

5. *In vitro* nitrate reductase assay (optimization)

For several years, we have been studying the physiology of the red macroalgae, gracilariods (Rhodophyta, Gracilariales), from different view-points. In most cases, these have been cultured in PES (Provasoli Enrichment Medium) and VSES (von Stosch Enrichment Solution), at different concentrations and biomass densities.

Nitrogen repletion and starvation causes, not only alterations in pigment content, nitrogen assimilation and photosynthesis, but also morphological changes in growth and ultrastructure. Consequently, certain studies in our laboratory were directed towards characterizing and understanding nitrogen assimilation regulation, especially as regards NR behavior. Undoubtedly, physiological changes in nitrate assimilation and NR activity are involved during alga growth and development. Knowledge of these responses would contribute towards a better optimization of laboratory efforts and in-field cultures for physiological studies and biomass yield, for possible economical usage.

Methods for estimating *in vitro* NR activity have been developed by Weidner & Kiefer (1981), Chapman & Harrison (1988), Thomas & Harrison (1988), Brinkhuis et al. (1989) and Chow et al. (2001). Assaying procedures of NR activity are based on quantifying the reduction rate of nitrate to nitrite during the reaction catalyzed by intracellular NR enzymes.

Enzymatic NR assays have been used without considering adequate optimization of the method. The low NR activity observed during assaying could be due to the loss of enzymatic cofactors during extraction, inhibition of activity by phenolic compounds or other inhibitors, and the presence of endogenous proteases. The key points in NR assaying depend on adequate enzyme extraction, preservation, and stability during the whole procedure. There are two important considerations in enzymatic assays: (a) the enzyme must to be completely or nearly completely extracted, and (b) assaying conditions, such as pH, temperature, substrate and electron donor concentrations, and protectant, must be optimal for maximal activity, in order to preserve enzymatic activity during the process. In most cases, NR activity is measured by saturating the enzymatic system with substrate. Under saturation conditions, NR activity is used as a nutritional estimator of nitrogen metabolism capacity. Some researchers prefer to use *in vitro* NR data as "potential activity", this representing a theoretical maximum of real NR activity.

Furthermore, the amount of detectable NR activity depends on protocol optimization, since enzymatic activities can vary between and within species, according to environmental conditions, circadian fluctuation and endogenous nutritional status, as well as thallus portion, age and size. Thus, for reliable comparison and evaluation, it is important to optimize the enzymatic assay, and clearly establish the conditions of the biological material to be studied.

Nitrate redutase assay is based on defining, spectrophotometrically, nitrite concentrations of the product from nitrate substrate reduction by the NR enzyme at constant temperature and time (Eppley et al., 1969). Spectrophotometric *in vitro* NR assay, while relatively easy and fast, and conferring the advantage of kinetic dependence between activity and time, confirmable by quantification of nitrate reduction, is not sensitive to any enzymatically crude extract. Nevertheless, it is recommended for very active and induced extracts. The comparison between deficient N samples cannot be sensitive enough. Therefore, even for studies of limited N, it is recommended to supply nitrate before sampling, to thus guarantee NR induction during the assay.

Appropriate extraction of the complete NR enzyme for *in vitro* assaying was achieved by grinding the biological material in liquid nitrogen. Sample grinding under liquid nitrogen increases cell disruption, thus facilitating the extraction of larger amounts of NR enzymes. The addition of bovine serum albumin (BSA) is advisable for protection against proteolytic enzymes or phenolic compounds, although additional protectants must be used in the case of complex species, so as to avoid enzyme denaturation by proteases, or phenolic and other compounds (*e.g.* high phenolic-containing brown algae). Another precaution for preserving enzymatic activity is to maintain the crude extract at a low temperature (4°C or on ice), and protected from light, to so prevent activity-degradation until assaying.

In vitro assays require fixed conditions, as regards pH, temperature, and the concentration of substrate and reductant source, as well as strict testing of each parameter of all the biological material to be studied. Macroalgal pH assaying varies slightly between species, maximal activity having been observed close to pH 8.0 (Lopes et al., 1997; Chow et al., 2004, 2007; Granbom et al., 2004).

Nitrate reductase activity is temperature sensitive over a narrow range, with several optimum temperatures for the various species, depending on the habitat. For temperate macroalgae, this ranges from 10 to 25 °C (Gao et al., 2000; Berges et al., 2002; Chow et al., 2004). Activities at low temperatures may require a larger amount of NR protein or higher catalytic rates to so maintain the same catalytic activity, as during low winter temperatures enzymes function below the optimum. Species with high NR activity during the winter may present a cold acclimation component. This high activity has been reported in *Laminaria saccharina* (Davison & Davison 1987), *Fucus vesiculosus* (Collén & Davison 2001), and *L. digitata*, *Fucus serratus*, *Fucus vesiculosus* and *Fucus spiralis* (Young et al. (2007). Macroalgae in tropical and sub-tropical environments presented high temperature tolerance during assaying (Lopes et al., 1997; Chow et al., 2007; Granbom et al., 2004; Martins et al., 2009), the optimum usually being higher than normal.

Nitrate reductase activity can also vary considerably among and within species, depending on natural or laboratory growing conditions. Therefore, optimal concentrations of electron donor (NAD(P)H) and substrate must be verified for each of the species studied, to so guarantee saturating conditions for assaying.

As nitrate reductase activity is daily cyclical, this requires care with the period of collecting samples, since differences among treatments can arise from natural circadian behavior, and not from treatment effects. Furthermore, NR is more active in areas with highly active metabolic rates, usually the meristematic region. For example, Granbom et al. (2004, 2007) and Chow (2004) detected higher NR activity in apical than basal parts in *Kappaphycus alvarezii* and *G. chilensis*, respectively. Thus, the appropriate choice of biological material for enzymatic assaying is also extremely important, as the amount of extractable enzyme varies drastically among and within species, with thallus part and age, internal nutritional stock, environmental nitrogen availability, culture conditions, etc.

Optimized *in vitro* NR activity in some macroalgae was studied and important highlights described (Lopes et al., 1997; Lartigue & Sherman, 2002; Chow et al., 2004, 2007; Granbom et al., 2004).

There is a growing interest in applying macroalga NR activity to evaluating nutritional physiology in the laboratory and field, as a useful ecophysiological index. On the other hand, NR is regarded as a focal point in regulating the nitrogen assimilation pathway and for integrating the control of carbon and nitrogen metabolism. However, the potential applicability of *in vitro* NR assaying, as an important parameter of N and C metabolism, must to be carefully considered, in which case optimal assaying procedures are required.

6. Important remarks

In general, the importance of nitrogen metabolism in the marine environment, particularly nitrate assimilation, is based on the frequent identification of nitrogen as limiting nutrients for macroalgal growth. Various species under the same environmental conditions have developed special strategies for remaining in the habitat and benefit from the adversity of nitrogen limitation, either by taking advantage of nitrogen pulses or learning to live with low nutrient levels.

On the other hand, the increasing eutrophication of coastal aquatic environments, associated with anthropogenic nitrogen inputs, is a global reality. Opportunistic green macroalgae, other bloom species and species susceptible to growing nitrogen concentration, can be rapidly affected by ammonium and nitrate availability altering ecological dynamics of both populations and communities.

Nitrate reductase, through being the first enzyme in the nitrogen assimilatory pathway, assumes the responsibility for controlling the nitrate assimilatory rate in all algal cells. Thus, due to its importance in the general metabolism connected to N and C pathways, there is a constantly growing interest in studying the molecular and catalytic properties of NR enzymes and physiological responses to environmental stressing conditions and intracellular factors.

Previous studies on NR activity in micro and macroalgae, encountered the contradiction of using assay protocols without optimization, thus making comparison difficult, with little emphasis being placed on appropriate NR extraction and assaying. Nitrate reductase, through being a sensitive enzyme, rapidly inducible and repressive at various molecular levels and with diverse internal and external factors, is important for establishing minimum optimal conditions, both for comparison and acquiring an understanding of nitrogen metabolism.

The constant accumulation of knowledge on NR activities, together with studies of nutrient uptake, will facilitate the collection of tools for: (1) identifying limiting, defective and saturating levels of growth-nutrients; (2) regulating pathways of nitrogen assimilation and incorporation; (3) providing environmental indicators for monitoring; (4) identifying macroalgae with high nitrate-reduction potential for biofilter application in eutrophized environments and increasing the standing crop in polycultures; (5) optimizing culture systems by regulating the reduction rate of nitrate with optimal nitrogen uptake and reduction; (6) developing management strategies for the culture of economically important algae; and (7) understanding evolutive patterns that support the adaptation of macroalgae to their environment.

Further studies of NR activity, both in the field and laboratory, are necessary as a contribution, both to understanding seaweed physiology, as well as to clarify the importance and role of these algae in near-shore biogeochemical cycling. Moreover, the constant changes in the coastal environment, brought about by anthropic action (artificial eutrophication), and variations arising from global climate change, will undoubtedly influence the ranges of tolerance and acclimation of algae, whereby the necessity for monitoring changes in coastal environments and communities.

7. Acknowledgment

The author would thank the research financial support of São Paulo Research Foundation (FAPESP).

8. References

Balandin, T. & Aparicio, P. J. (1992). Regulation of nitrate reductase in *Acetabularia mediterranea*. J. Exp. Bot. 43:625–631.

Berges, J.A. (1997). Algal nitrate reductases. Eur. J. Phycol. 32: 3-8.

Berges, J. A. & Harrison, P. J. (1995). Nitrate reductase activity quantitatively predicts the rate of nitrate incorporation under steady state light limitation: a revised assay and characterization of the enzyme in three species of marine phytoplankton. Limnol. Oceanogr. 40: 82-93.

Berges, J. A., Cochlan, W. P. & Harrison, P. J. (1995). Laboratory and field responses of algal nitrate reductase to diel periodicity in irradiance, nitrate exhaustion, and the presence of ammonium. Mar. Ecol. Prog. Ser. 124: 259-269.

Berges J.A.,Varela D.E. & Harrison P.J. (2002) Effects of temperature on growth rate, cell composition and nitrogen metabolism in the marine diatom *Thalassiosira pseudonana* (Bacillariophyceae). Marine Ecology Progress Series 225, 139–146.

Brinkhuis, B. H.; Renzhi, L.; Chaoyuan, W. & Xun-sen, J. (1989). Nitrite reductase transients and consequences for *in vivo* algal nitrate reductase assays. J. Phycol. 25: 539–45.

Cabello-Pasini, A.; Macías-Carranza, V.; Abdala, R.; Korbee, N. & Figueroa, F. L. (2011). Effect of nitrate concentration and UVR on photosynthesis, respiration, nitrate reductase activity, and phenolic compounds in *Ulva rigida* (Chlorophyta). J. Appl. Phycol. 23: 363–369.

Campbell, W. H. (1996). Nitrate reductase biochemistry comes of age. Plant Physiol. 111: 355–361.

Campbell, W. H. (1999). Nitrate reductase structure, function, and regulation: bridging the gap between biochemistry and physiology. Annu. Rev. Plant Physiol. Plant Mol. Biol. 50: 277-303.

Chapman, D.J. & P.J. Harrison. 1988. Experiment 22. Nitrogen metabolism and measurement of nitrate reductase. In: Experimental Phycology. A Laboratory Manual. Lobban, C.S., D.J. Chapman & B.P. Kremer (eds). Cambridge University Press, Cambridge, USA. Pp. 196-202.

Chow, F. & Oliveira, M. (2008). Rapid and slow modulation of nitrate reductase activity in the red macroalga *Gracilaria chilensis* (Gracilariales, Rhodophyta): influence of different nitrogen sources. J. Appl. Phycol. 20: 775–782.

Chow, F.; Oliveira, M. C. & Pedersén, M. (2004). *In vitro* assay and light regulation of nitrate reductase in the red alga *Gracilaria chilensis*. J. Plant Physiol. 161: 769–776.

Chow, F.; Capociama, F. V.; Faria, R. & Oliveira, M. C. (2007). Characterization of nitrate reductase activity in vitro in *Gracilaria caudata* J. Agardh (Rhodophyta, Gracilariales). Rev. Brasil. Bot. 30: 123-129.

Chow, F.; Macchiavello, J.; Santa-Cruz, S.; Fonck, E. & Olivares, J. (2001). Utilization of *Gracilaria chilensis* (Rhodophyta, Gracilariaceae) as a biofilter in the depuration of effluents from tank cultures of fish, oysters, and sea urchins. J. World Aquacul. Soc. 32: 215–20.

Collén, J. & Davison, I. R. (2001). Seasonality and thermal acclimation of reactive oxygen metabolism in *Fucus vesiculosus* (Phaeophyceae). J. Phycol. 37: 474–481.

Collos, Y. (1989). A linear model of external interactions during uptake of different forms of inorganic nitrogen by microalgae. J. Plankton. Res. 11: 521–523.

Corzo, A. & Niell, F. X. (1991). Determination of nitrate reductase activity in *Ulva rigida* C. Agardh by the in situ method. J. Exp. Mar. Biol. Ecol. 146: 181-191.

Crawford, N. M. (1995). Nitrate—nutrient and signal for plant growth. Plant Cell 7: 859–868.

Crawford, N. M. & Arst, H. N., Jr. (1993). The molecular genetics of nitrate assimilation in fungi and plants. Ann. Rev. Genet. 27: 115-46.

Davison I.R. & Davison J.O. (1987) The effect of growth temperature on enzyme activities in the brown alga *Laminaria saccharina*. British Phycol. J. 22: 77–87.

Davison I.R. & Stewart W.D.P. (1984) Studies on nitrate reductase activity in *Laminaria digitata* (Huds.) Lamour. II. The role of nitrate availability in the regulation of enzyme activity. J. Exp. Mar. Biol. Ecol. 79: 65–78.

Davison I.R., Andrews M. & Stewart W.D.P. (1984) Regulation of growth in *Laminaria digitata*: use of in vivo nitrate reductase activities as an indicator of nitrogen limitation in field populations of *Laminaria* spp. Mar. Biol. 84: 207–217.

Davison, I.; Jordan, T.; Fegley, J. & Grobe, C. (2007). Response of *Laminaria saccharina* (Phaeophyta) growth and photosynthesis to simultaneous ultraviolet radiation and nitrogen limitation. J. Phycol. 43: 636–646.

DeBoer, J.A. (1981). Nutrients. In: Lobban, C. S. & Wynne, M. J. (eds.). The Biology of Seaweeds, Bot. Monogr., pp. 356–392.

Deng, M.-D.; Moureaux, T.; Cherel, I.; Boutin, J.-P. & Caboche, M. (1991). Effects of nitrogen metabolites on the regulation and circadian expression of tobacco nitrate reductase. Plant Physiol. Biochem. 29: 239–247.

Eppley, J.; Coastworth, L. & Solorzado, L. (1969). Studies of nitrate reductase in marine phytoplankton. Limnol. Oceanogr. 14: 194-205.

Falkowski, P. G. (1992). Molecular ecology of phytoplankton photosynthesis. In: Falkowski, P. G. & Woodhead, A. (eds.). Primary Productivity and Biogeochemical Cycles in the Sea, pp. 47-67. Plenum Press, New York.

Falkowski, P. G. & Raven, J. A. (1997). Aquatic Photosynthesis. Blackwell Science, UK. 375 pp.

Fernández-López, M.; Olivares, J. & Bedmar, E. J. (1996). Purification and characterization of the membrane-bound nitrate reductase isoenzymes of *Bradyrhizobium japonicum*. FEBS Lett. 392: 1-5.

Figueroa, F.L. & Viñegla, B. (2001). Effects of solar radiation on photosynthesis and enzyme activities (carbonic anhydrase and nitrate reductase) in marine macroalgae from southern Spain. Rev. Chil. Hist. Nat. 74: 237–249.

Flynn, K. J. 1991. Algal carbon-nitrogen metabolism: a biochemical basis for modeling the interactions between nitrate and ammonium uptake. J. Plankton Res. 13: 373-387.

Gao, Y.; Smith, G. J. & Alberte, R. S. (1993). Nitrate reductase from the marine diatom *Skeletonema costatum* biochemical and immunological characterization. Plant Physiol. 103: 1437-1445.

Gao Y., Smith G.J. & Alberte R.S. (1995) Induction of nitrate reductase activity by light and NO_3^- in the marine diatom *Skeletonema costatum*. Plant Physiol. 108: 71.

Gao Y., Smith G.J. & Alberte R.S. (2000) Temperature dependence of nitrate reductase activity in marine phytoplankton: biochemical analysis and ecological, implications. J. Phycol. 36: 304–313.

Gao, Y.; Smith, G. J. & Alberte, R. S. (1992). Light regulation of nitrate reductase in *Ulva fenestrata* (Chlorophyceae). I. Influence of light regimes on nitrate reductase activity. Mar. Biol. 112: 691–696.

Glaab, J. & Kaiser, W. M. (1996). The protein kinase, protein phosphatase and inhibitor protein of nitrate redutase are ubiquitous in higher plants and independent of nitrate reductase turnover. Planta 199: 57-63.

Gordillo, F. J. L.; Aguilera, J. & Jiménez, C (2006). The response of nutrient assimilation and biochemical composition of Arctic seaweeds to a nutrient input in summer. J. Exp. Bot. 57: 2661–2671.

Granbom, M.; Pedersén, M.; Kadel, P. & Lüning, K. (2001). Circadian rhythm of photosynthesis in the red macroalga *Kappaphycus alvarezii*: dependence on light quantity and quality. J. Phycol. 37: 1020–1025.

Granbom, M.; Chow, F.; Lopes, P. F.; Oliveira, M. C.; Colepicolo, P.; Paula, J. P. & Pedersén, M. (2004). Characterisation of nitrate reductase in the marine macroalga *Kappaphycus alvarezii* (Rhodophyta). Aquat. Bot. 78: 295–305.

Granbom, M.; Lopes, P. F.; Pedersén, M. & Colepicolo, P. (2007). Nitrate reductase in the marine macroalga *Kappaphycus alvarezii* (Rhodophyta): oscillation due to the protein level. Bot. Mar. 50: 106-112.

Hanisak, M. D. (1983). The nitrogen relationships of marine macroalgae. In: Carpenter, E. J. & Capone, D. G. (eds.). Nitrogen in the Marine Environment. Academic Press, New York, pp. 699–730.

Hernández I., Corzo A., Gordillo F.J., Robles M.D., Saez E., Fernández J.A. & Niell F.X. (1993). Seasonal cycle of the gametophytic form of *Porphyra umbilicalis* - nitrogen and carbon. Mar. Ecol. Progress Ser. 99: 301-311.

Howartha, D. G. & Baumb, D. A. (2002). Phylogenetic utility of a nuclear intron from nitrate reductase for the study of closely related plant species. Molec. Phylog. Evol. 23: 525–528.

Huber, J. L.; Huber, S. C.; Campbell, W. H. & Redinbaugh, M. G. (1992a). Reversible light/dark modulation of spinach leaf nitrate reductase activity involves protein phosphorylation. Arch. Biochem. Biophys. 296: 58–65.

Huber, J. L.; Huber, S. C.; Campbell, W. H. & Redinbaugh, M. G. (1992b). Reversible light dark modulation of spinach leaf nitrate reductase-activity involves protein-phosphorylation. Arch. Biochem. Biophys. 296: 58–65.

Huovinen, P.; Matos, J.; Sousa-Pinto, I. & Figueroa, F. (2006). The role of nitrogen in photoprotection against high irradiance in the Mediterranean red alga *Grateloupia lanceola*. Aquat. Bot. 84: 208–316.

Hurd, C. L.; Berges, J. A.; Osborne, J. & Harrison, P. J. (1995). An in vitro nitrate reductase assay for marine macroalgae: optimization and characterization of the enzyme for *Fucus gardneri* (Phaeophyta). J. Phycol. 31: 835–43.

Kain, J. M. (1989). The seasons in the subtidal. Br. Phycol. J. 24: 203-215.

Kaiser, W. & Huber, S. C. (1994). Posttranslational regulation of nitrate reductase in higher plants. Plant Physiol. 106: 817-821.

Kaiser, W. M. & Spill, D. (1991). Rapid modulation of spinach leaf nitrate reductase by photosynthesis. II. *In vitro* modulation by ATP and AMP. Plant Physiol. 96: 368-375.

Korbee-Peinado, N.; Abdala-Díaz, R.; Figueroa, F. & Helbling, W. (2004). Ammonium and UV radiation stimulate the accumulation of mycosporine-like amino acids in Porphyra columbina (Rhodophyta) from Patagonia, Argentina. J. Phycol. 40: 248–259.

Lapointe, B. E. & Duke, C. S. (1984). Biochemical strategies for growth of *Gracilaria tikvahiae* (Rhodophyta) in relation to light intensity and nitrogen availability. J. Phycol. 20: 488–495.

Lartigue J. & Sherman T.D. (2002) Field assays for measuring nitrate reductase activity in *Entermorpha* sp. (Chlorophyceae), *Ulva* sp. (Chlorophyceae) and *Gelidium* sp. (Rhodophyceae). J. Phycol. 38: 971–982.

Lartigue J. & Sherman T.D. (2005) Response of *Enteromorpha* sp. (Chlorophyceae) to a nitrate pulse: nitrate uptake, inorganic nitrogen storage, and nitrate reductase activity. Mar. Ecol. Progress Ser. 292: 147–157.

Lillo, C. (1983). Studies of diurnal variations of nitrate reductase activity in barley leaves using various assay methods. Physiol. Plant. 57: 357-362.

Lillo, C. (1994). Light regulation of nitrate reductase in green leaves of higher plants. Physiol. Plant. 90: 616-620.

Lobban, C. S. & Harrison, P. J. (1994). Seaweed Ecology and Physiology. Cambridge University Press, Cambridge. 366 pp.

Lobban, C. S.; Harrison, P. J. & Duncan, M. J. (1985). The physiological ecology of seaweeds. Cambridge University Press, Cambridge.

Lopes P.F., Oliveira M.C. & Colepicolo P. (1997). Diurnal fluctuation of nitrate reductase activity in the marine red alga. J. Phycol. 33: 225–231.

Lopes, P. F.; Oliveira, M. C. & Colepicolo P. (2002). Characterization and daily variation of nitrate reductase in *Gracilaria tenuistipitata* (Rhodophyta). Biochem. Biophys. Res. Commun. 295: 50–54

MacKintosh, C.; Douglas, P. & Lillo, C. (1995). Identification of a protein that inhibits the phosphorylated form of nitrate reductase from spinach (*Spinacia oleracea*) leaves. Plant Physiol. 107: 451–457.

Martins, A. P., Chow, F. & Yokoya, N. S. (2009). Ensaio *in vitro* da enzima nitrato redutase e efeito da disponibilidade de nitrato e fosfato em variantes pigmentares de *Hypnea musciformis* (Wulfen) J.V. Lamour. (Gigartinales, Rhodophyta). Rev. Brasil. Bot. 32: 635-645.

Nakamura, Y. & Ikawa, T. (1993). Purification and properties of NADH: nitrate reductase from red alga Pophyra yezoensis. Plant Cell Physiol. 34: 1239-49.

Nakamura, Y., Saji, H., Kondo, N. & Ikawa, T. (1994). Preparation of monoclonal antibodies against NADH-nitrate reductase from the red alga *Porphyra yezoensis*. Plant Cell Physiol. 35: 1185-1198.

Nicodemus, M. A.; Salifu, K. F. & Jacobs, D. F. (2008). Nitrate reductase activity and nitrogen compounds in xylem exudate of *Juglans nigra* seedlings: relation to nitrogen source and supply. Trees Struct. Funct. 22: 685–695.

Ramalho, C. B., Hastings, J. W. & Colepicolo. P. (1995). Circadian oscillation of nitrate reductase activity in *Gonyaulax polyedra*. Plant Physiol. 107: 225-231.

Smith, G. J.; Zimmerman, R. C. & Alberte, R. S. (1992). Molecular and physiological responses of diatoms to variable levels of irradiance and nitrogen availability: growth of *Skeletonema costatum* in simulated upwelling conditions. Limnol. Oceanogr. 37: 989-1007.

Solomonson, L. P. & Barber, M. J. (1990). Assimilatory nitrate reductase: functional properties and regulation. Annu. Rev. Plant Physiol. Plant Mol. Biol. 41: 225–253.

Stolz, J. F. & Basu, P. (2002). Evolution of nitrate reductase: molecular and structural variations on a common function. Chem. BioChem. 3: 198–206.

Thomas, T. E. & Harrison, P. J. (1988). A comparison of in vitro and in vivo nitrate reductase assays in three intertidal seaweeds. Bot. Mar. 31: 101-107.

Tischner, R.; Ward, M. R. & Huffaker, R. C. (1989). Evidence for a plasma-membrane-bound nitrate reductase involved in nitrate uptake of *Chlorella sorokiniana*. Planta 178: 19-24.

Vergara, J. J.; Berges, J .A. & Falkowski, P. G. (1998). Diel periodicity of nitrate reductase activity and protein levels in the marine diatom *Thalassiosira weissflogii* (Bacillariophyceae). J. Phycol. 34: 952–961.

Weidner, M.; & Kiefer, H. (1981). Nitrate reduction in the marine brown algae *Giffordia mitchellae* (Harv.) Ham. Z. Pflazenphysiol. Bd. 104: 341–351.

Young, E. B., Dring, M. J., Savidge, G., Birkett, D. A & Berges, J. A. (2007). Seasonal variations in nitrate reductase activity and internal N pools in intertidal brown algae are correlated with ambient nitrate concentrations. Plant, Cell and Environ. 30: 764–774.

Young, E. B., Berges, J. A. & Dring, M. J. (2009). Physiological responses of intertidal marine brown algae to nitrogen deprivation and resupply of nitrate and ammonium. Physiol. Plant. 135: 400–411.

Zhou, J. & Kleinhofs, A. (1996). Molecular evolution of nitrate reductase genes. J. Mol. Evol. 42: 432-442.

8

Effects of Bioactive Natural and Synthetic Compounds with Different Alkyl Chain Length on Photosynthetic Apparatus

Katarína Kráľová and František Šeršeň
Faculty of Natural Sciences, Comenius University in Bratislava
Slovak Republic

1. Introduction

Many compounds containing alkyl substituent(s) in their molecule exhibit a wide spectrum of biological activities. Such compounds, mainly water soluble amphiphilic compounds are frequently used in the industry and households as detergents or as disinfectants due to their solubilizing and antimicrobial properties. Consequently they can enter in the environment via waste waters and so affect photosynthetic processes in algae and plants. In general, the biological activity of compounds with alkyl substituent(s) was found to depend on the alkyl chain length. Good correlations between photosynthesis-inhibiting and antimicrobial activity were estimated. For amphiphilic compounds such as surfactants quasi-parabolic course of the dependence of biological activity on the length of alkyl substituent is characteristic. The decrease of biological activity observed from certain chain elongation is called "cut-off" effect. The main mechanism of biological action of the discussed compounds is closely connected with their membrane-damaging effects (Devinsky et al., 1990; Balgavý & Devínsky, 1996).

In photosynthetic electron transfer from water to $NADP^+$ three integral membrane protein complexes operating in series: the photosystem 2 (PS 2) reaction centre, the cytochrome *bf* complex and the photosystem 1 (PS 1) reaction centre are involved. In two reaction centres primary charge separation occurs, in which light energy or excitation energy is transformed into redox free energy (Whitmarsh, 1998). Using an artificial electron acceptor, e.g. 2,6-dichlorophenol-indophenol (DCPIP) with the known site of action in plastoquinone on the acceptor side of PS 2 (Izawa, 1980), inhibition of photosynthetic electron transport (PET) in PS 2 by PET inhibitors can be monitored. On the other hand, application of artificial electron donors with known site of action to chloroplasts activity of which was inhibited by PET inhibitors enables more nearly to specify the site of action of tested inhibitors.

EPR is very useful method for investigation of the effects of organic and metal inhibitors in the photosynthetic apparatus because intact chloroplasts of algae and vascular plants exhibit EPR signals in the region of free radicals (g = 2.00), which are stable during several hours and could be registered at laboratory temperature by conventional continual wave EPR apparatus. EPR spectrum of intact chloroplasts is composed of two components, so-called signal I and signal II, belonging to PS 1 and 2, respectively (Hoff, 1979). Signal II consists of two parts, namely from (i) EPR signal II_{slow} (g = 2.0046, ΔB_{pp} = 1.9 mT) which is clearly visible in Fig. 1a (full line) and belongs to the intermediate D•, i.e. to the tyrosine radical in the 161th position of D_2 protein on the donor side of PS 2 (Debus et al., 1988a) and (ii) EPR signal $II_{very\ fast}$ (g = 2.0046, ΔB_{pp} = 1.9 mT) which is observable as an increase of signal II in light (Fig. 1a, difference between the dashed and full lines). This signal belongs to the intermediate Z•, i.e. to the tyrosine radical in the 161th position of D_1 protein (Debus et al., 1988b) which is also situated on the donor side of PS 2. Further EPR signal is associated with cation-radical of chlorophyll (Chl) *a* dimer situated in the core of PS 1 (Hoff, 1979).

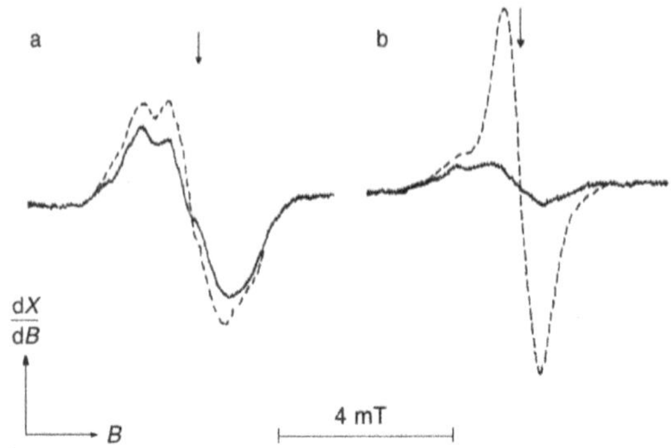

Fig. 1. EPR spectra of untreated spinach chloroplasts (a) and of chloroplasts treated with 0.05 mol dm^{-3} of 2-*n*-butylsulfanyl-4-pyridinecarbothioamide (b). The full lines correspond to chloroplasts kept in the dark, the dotted lines to the illuminated chloroplasts (Source: Kráľová et al., 1997).

Due to treatment of plant chloroplasts and algae with PET inhibitors the intensity and the shape of the above mentioned EPR signals can be changed (Fig. 1b). From these changes the site of action of studied inhibitors can be determined. Due to interaction of inhibitors with the oxygen evolving complex (OEC) also release of manganese Mn^{2+} ions from OEC into interior of thylakoid membranes can occur, which can be registered by EPR spectroscopy.
This contribution will be focused on comprehensive review related to inhibition of photosynthetic electron transport by inhibitors of natural origin as well as synthetic inhibitors containing alkyl chain(s) of different length in their molecules. The mechanism and the site of their action in the photosynthetic apparatus will be discussed as well.

2. Inhibitors of photosynthetic electron transport of natural origin

Natural products represent a vast repository of materials and compounds with evolved biological activity, including phytotoxicity. The two fundamental approaches to the use of natural products for weed management are their application as herbicides or leads for synthetic herbicides and their use in allelopathic crops or cover crops (Duke et al., 2002a, 2002b). Structures of some further discussed PET inhibitors of natural origin with alkyl substituent are presented in Fig. 2.

2.1 Fatty acids

Natural fatty acids are important components of biological membranes. They commonly have a chain of 4 to 28 carbons (usually unbranched and even numbered), which may be

saturated or unsaturated. In natural unsaturated fatty acids the double bonds are all *cis* and are usually not conjugated (Bhalla et al., 2009).
The saturated fatty acids (palmitic acid [16:0] as well as stearic acid [18:0]) applied at concentration 20 and 50 μmol dm^{-3} did not inhibit electron transport activity, whereas the unsaturated fatty acids (oleic acid [18:1], linoleic acid [18:2] and α-linolenic acid [18:3]) inhibited electron transport activity by ~50%. The monounsaturated fatty acid completely inhibited electron transport at 50 μmol dm^{-3}. It could be stressed that the extent of PET-inhibiting activity was not dependent on the degree of unsaturation since all the unsaturated fatty acids inhibited PET to the same magnitude (Peters & Chin, 2003). Krogmann & Jagendorf (1959) observed PET inhibition by unsaturated C_{18} fatty acids. A few years later McCarty & Jagendorf (1965) and Molotkovsky & Zheskova (1965) showed that linolenic acid (LA) can induce damage in freshly isolated chloroplasts resembling that in chloroplasts after inactivation by gentle heating. Katoh & San Pietro (1968) suggested that high concentrations of LA inhibit PS 1 as well as PS 2. Brody (1970) observed a decline in the population of PS 2 reaction centres due to treatment with LA. Golbeck et al. (1980) localized the site of LA inhibitory action on the donor side of PS 1 and at two functionally distinct sites in PS 2. A reversible site and an irreversible site of inhibition have been located in PS 2. At the irreversible site a time-dependent loss of the loosely bound pool of Mn in the oxygen evolving complex occurred whereas at the reversible site, the photochemical charge separation was rapidly inhibited (<20 s) but after washing of LA-treated chloroplasts a resumption of artificial donor activity from diphenylcarbazide (DPC) to DCPIP was observed. The fact that inhibition of the Hill reaction by linolenic acid may be partially reversible was described also by Okamoto & Katoh (1977) and Okamoto et al. (1977).
According to Golbeck et al. (1980) the mechanism of inhibition of the photoactivity may ultimately lie in the ability of LA to penetrate its hydrophobic tail into the lipid membrane and change the orientation of electron donor and acceptor complexes relative to one another. Alternatively, the hydrophobic tail might interact with the antenna chlorophylls and inhibit transfer of energy to the photoactive trap. Due to direct interaction of free fatty acids with membrane proteins conformational changes in these proteins occur. Since PS 2 and its accessory pigments are most likely bound in a membrane protein complex, it could be supposed that alteration of the membrane structure would induce organizational changes in the associated peptides resulting in inhibition of charge separation between electron donor and acceptor complexes.
Linolenic acid was found to exhibit several effects on thylakoid membrane resulting in: i) modification of the membrane surface-charge density; ii) uncoupling of photophosphorylation (McCarty & Jagendorf, 1965); iii) release of manganese ions from water-oxidizing complex of PS 2 (Golbeck et al., 1980); iv) inhibition of artificial donor-assisted electron transport in PS 1 and PS 2 (Golbeck et al., 1980; Siegenthaler, 1974). Whereas Golbeck & Warden (1984) situated the site of LA action in Q_A on the donor side of PS 2, Vernotte et al. (1983) stated that its site of action is similar to that of 3-(3,4-dichlorophenyl)-1,1-dimethylurea (DCMU), namely Q_B. Warden and Csatorday (1987) indicated that interactions of the unsaturated fatty acid, linoleic acid, with PS 2 have two principal regions of inhibition: one associated with the donor complex (signal II_{fast} or D_1) to the reaction centre, and the other located on the reducing side between pheophytin and Q_A, whereby linoleic acid inhibits secondary electron transport in PS 2 via displacement of endogenous quinone Q_A from quinone binding peptides.

Peters & Chin (2003) found that palmitoleic acid (PA), a monounsaturated fatty acid [16:1], caused rapid PET inhibition (within 30 s). The oxidizing side of PS 2 was up to 90% inactivated, whereas no inhibition occurred on the reducing side of the PS 2 complex and PS 1 activity was ~65% inhibited. These researchers did not observed correlation between PET inhibition and lipid peroxidation. On the other hand, PA caused the loss of proteins from the thylakoid membrane which was exacerbated by the light. The proteins were found to be lost in the following order: plastocyanin (PC) < 1 min, manganese stabilizing protein (MSP) ~5 min, cytochrome *f* (Cyt*f*) ~10 min, D_1 protein ~60 min and D_2 protein ~60 min. The timing of the loss of a PS 1 associated protein, PC, overlapped with that of the inhibition of PS 1. According to Peters & Chin (2003) the loss of PC is the cause of PS 1 inhibition. Because MSP loss from the OEC occurs later than the inhibition, it could not be considered as the cause of PET inhibition by PA on the oxidizing side of PS 2 and inhibition occurs due to the loss of Mn^{2+} ions.

Unsaturated fatty acids have been found to remove the Mn^{2+} ions from OEC (Kaniuga et al., 1986) and depletion of Mn^{2+} is correlated to inactivation of oxygen evolution (Garstka & Kaniuga, 1988) and to an inhibition of PS 2 activity (Krieger et al., 1998). Because addition of Mn^{2+} ions can effectively reverse PET inhibition at the donor side of PS 2 caused by PA, it can be assumed that loss of Mn^{2+} was the cause of this inhibition. It can be also supposed that at removing of Mn^{2+} ions form OEC the negatively charged unsaturated fatty acids may act as chaotropic agents or as chelating agents (Peters & Chin, 2003).

Nakai et al. (2005) found that *Myriophyllum spicatum* released fatty acids, specifically nonanoic, tetradecanoic, hexadecanoic, octadecanoic and octadecenoic acids. Anti-cyanobacterial effects of these fatty acids considerably depended upon their chain length: nonanoic, *cis*-6-octadecenoic, and *cis*-9-octadecenoic acids significantly inhibited growth of *Microcystis aeruginosa*, whereas tetradecanoic, hexadecanoic and octadecanoic acids did not show any effect.

2.2 Fischerellin, sorgoleone and resorcinolic compounds isolated from sorghum

Fischerellin A, a secondary metabolite of the benthic cyanobacterium *Fischerella muscicola* (Thur.), was found to inhibit the photosynthetic but not the respiratory electron transport of cyanobacteria and chlorophytes and its site of action is located in PS 2. It is the most active allelochemical compound of *F. muscicola,* which is toxic to other cyanobaceria and photoautotrophic organisms (Bagchi & Marwah, 1994; Hagmann & Jüttner, 1996). Structural elements of fischerellin A ((3*E*)-1,5-dimethyl-3-{(3*R*,5*S*)-3-methyl-5-[(4*E*)-2-methylpentadec-4-ene-6,8-diyn-1-yl]pyrrolidin-2-ylidene}pyrrolidine-2,4-dione) are an enediyne moiety and two heterocyclic ring systems. This compound exhibits a unique structure composed of two cyclic amines and a C_{15} substituent that contains a double bond in the (*Z*)-configuration and two triple bonds. Srivastava et al. (1998) found that fischerellin A affects the fluorescence transients, as well as oxygen evolution by the cyanobacterium *Anabaena P9*. The green alga, *Chlamydomonas reinhardtii,* and higher plants were also affected by fischerellin A in a concentration- and time-dependent fashion. It acts at several sites which appear with increasing half-time of interaction in the following sequence: (i) effect on the rate constant of Q_A^- reoxidation; (ii) primary photochemistry trapping; (iii) inactivation of PS 2 reaction centre; (iv) segregation of individual units from grouped units. However, fischerellin A does not affect the photosynthetic activity of purple bacteria, *Rhodospirillum rubrum*. Fischerellin B was determined to be (3*R*,5*S*)-3-methyl-5-((5*E*)-pentadec-5-ene-7,9-diynyl)-pyrrolidin-2-one.

Sorgoleone (2-hydroxy-5-methoxy-3-[(8´Z,11´Z)-8´,11´,14´-pentadecatriene]benzo-1,4-quinone) is one of the major components of the oily substance exuding from the roots of sorghum (*Sorghum bicolor* (L.) Moench) and it is one of the most studied allelochemicals (Dayan et al., 2010). This natural herbicide (bioherbicide) repressing the growth of other plants present in its surroundings (Dayan, 2006) was found to be a potent inhibitor of PS 2 in isolated chloroplasts, being as effective as diuron at inhibiting photosynthetic electron transport (IC_{50} ~ 100 nmol dm^{-3}) (Gonzalez et al.,1997). Its efficacy is not reduced in triazin-resistant pigweed (Dayan et al., 2009) because the common mutation of Ser264 to Gly or Ala in PS 2 causes resistance to triazines, but not to the quinone inhibitors (Oettmeier et al., 1982). Czarnota et al. (2001) performed three-dimensional structure analysis to characterize sorgoleone's mode of action and the results of their studies indicated that sorgoleone required about half the amount of free energy (493.8 kcal mol^{-1}) compared to plastoquinone (895.3 kcal mol^{-1}) to dock into the Q_B-binding site of the PS 2 complex of higher plants.

Rimando et al. (2003) observed PS 2 inhibition by resorcinolic compounds having the characteristic three double bonds in terminal methylene lipid side chain as sorgoleone, i.e. 4,6-dimethoxy-2-[(8´Z,11´Z)-8´,11´,14´-pentadecatriene]resorcinol (IC_{50} = 0.09 μmol dm^{-3}) and 4-methoxy-6-ethoxy-2-[(8´Z,11´Z)-8´,11´,14´-pentadecatriene]resorcinol (IC_{50} = 0.20 μmol dm^{-3}). A new benzoquinone derivative, 2-hydroxy-5-ethoxy-3-[(8'Z,11'Z)-8´,11´,14'-pentadecatriene]-rho-benzoquinone, which was isolated from the root exudates of sorghum was found to be less effective PS 2 inhibitor than sorgoleone (Rimando et al., 1998).

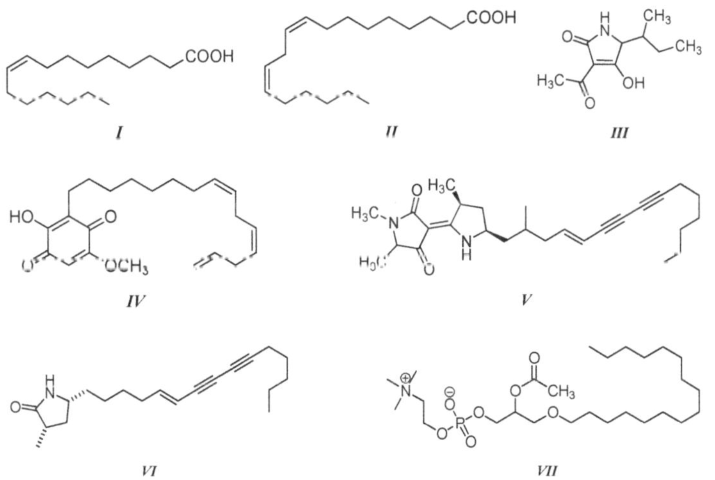

Fig. 2. Structures of some PET inhibiting compounds of natural origin: palmitoleic acid (*I*), linoleic acid (*II*), tenuazonic acid (*III*), sorgoleone (*IV*), fischerellin A (*V*), fischerellin B (*VI*) and platelet activating factor (1-*O*-alkyl-2-acetyl-sn-glycero-3-phosphocholine) (*VII*).

2.3 Hydroxydietrichequinone, tenuazonic acid and platelet activating factor

The natural quinone, hydroxydietrichequinone ((8Z)-3-heptadec-8-enyl-2-hydroxy-5-methoxybenzo-1,4-quinone) is a secondary metabolite of *Cyperus javanicus*. This natural quinone has a long aliphatic chain (C_{17}) including an unsaturated bond at its midpoint. Morimoto et al. (2001) found that this quinone inhibited both mitochondrial respiration and photosynthesis in their electron transport systems. In chloroplasts prepared from spinach leaves this natural quinone inhibited PET in PS 2 in a similar way to that of the triazin herbicide, atrazine, which belongs to PS 2 herbicides.

Tenuazonic acid (TeA), a nonhost-specific phytotoxin produced by *Alternaria alternata*, is the first toxin from a phytopathogen which was reported as a natural PS 2 inhibitor with several action sites (Chen et al., 2007, 2008). This bioherbicide with relatively short alkyl chain (*sec*-butyl) mainly interrupts PS 2 electron transport beyond Q_A (primary quinone acceptor) by competing with Q_B (secondary quinone acceptor) for Q_B-niche of the D_1 protein. Competition experiments between non-labeled TeA and [^{14}C] atrazine showed that TeA has a similar site of action as atrazine, which binds to the Q_B-site since atrazine binding to Q_B-site could be prevented by TeA (Chen et al., 2007). After TeA treatment an increase of non-Q_A reducing centres was observed. Non-Q_A reducing centres, also so-called heat sink centres or silent centres, are radiators and often are used to protect the system from over excitation and over reduction which would create dangerous reactive oxygen species (ROS) (Chen et al., 2010). TeA also had a visible effect on electron flow at PS 1 acceptor. Because TeA interrupts PS 2 electron flow and ATP synthesis, it is regarded as an inhibitor of redox energy conservation and therefore also is expected to increase the energization levels in thylakoid, which can result in a large generation of ROS (Chen et al., 2010).

Barr et al. (1988) observed very efficient PET inhibition in PS 2 of spinach chloroplasts by the platelet activating factor 1-*O*-alkyl-2-acetyl-sn-glycero-3-phosphocholine, a phospholipid, which is an ether analogue of phosphatidylcholine. Application of 2.8 to 3.5 μg cm^{-3} of this compound resulted in > 90% PET inhibition. The inhibition site for platelet activating factor was localized close to the reaction centre of PS 2, based on the inhibition of the donor reaction, DPC → DCPIP, in Tris-treated chloroplasts. On the other hand, treatment by phorbol myristate acetate, 1,2-dipalmitin or fatty acid esters gave up to 17–32% PET inhibition, which was observed only at higher concentrations of these compounds.

2.4 Rhamnolipid biosurfactants produced by *Pseudomonas aeruginosa*

Wang et al (2005) investigated the algicidal activity of the rhamnolipid biosurfactants (the mixture of mono and dirhamnolipids Rha-Rha-C_{10}-C_{10} and Rha-C_{10}-C_{10}) produced by *Pseudomonas aeruginosa*. These biosurfactants were found to have potential algicidal effects on *Heterosigma akashiwo*. The growth of *H. akashiwo* was strongly inhibited in medium containing rhamnolipids (0.4–3.0 mg dm^{-3}) and at higher concentrations (≥ 4.0 mg dm^{-3}) the rhamnolipids showed strong lytic activity toward *H. akashiwo*. The extent of ultrastructural damage of the alga was severe at high concentrations of rhamnolipids and during extended periods of contact, whereby the plasma membrane was partly disintegrated. The lack of membrane facilitated entry of the rhamnolipid biosurfactants into the cells and allowed damage to other organelles, which resulted in the injury of chloroplasts.

3. Synthetic inhibitors of photosynthetic electron transport

Many synthetic compounds disrupt photosynthesis by blocking electron transfer in photosynthetic apparatus. Herbicides which inhibit Hill reaction belong to the group of PS 2 herbicides. Experimental studies have established that many PS 2 herbicides bind noncovalently to a 32 kDa protein in the PS 2 complex and inhibit electron transfer between primary electron acceptor - quinone Q_A and the secondary electron acceptor - quinone Q_B on the reducing side of PS 2 (Shipman, 1981). Many PS 2 herbicides contain hydrophobic components as well as a flat polar component. The function of the hydrophobic components is to increase the lipid solubility of the entire herbicide molecule and to fit the hydrophobic surface of the herbicide binding site. It is assumed that the flat polar component binds electrostatically at a highly polar protein site (Shipman, 1981).

3.1 Alkyl substituted 2,4,6-trihydroxybenzamides and thiobenzamides

Highly potent PET inhibitors, alkyl substituted derivatives of 3-nitro-2,4,6-trihydroxybenzamide (NTHBA) and thiobenzamide (NTHTBA) (alkyl = ethyl - pentadecyl, benzyl, phenyl) were synthesized and tested by Honda et al. (1990a, 1990b). Comparison of PET-inhibiting activity expressed by pI_{50} value (negative logarithm of IC_{50}, i.e. compound concentration causing 50 % inhibition with respect to the control) showed that dodecyl (pI_{50} = 8.4) and tridecyl (pI_{50} = 8.1) NTHBA derivatives as well as octyl (pI_{50} = 8.7) and decyl (pI_{50} = 8.3) NTHTBA derivatives were ten times more active than DCMU (pI_{50} = 7.3) (Honda et al., 1990a). In general, thiobenzamide derivatives were found to be more active than benzamide derivatives, presumably due to greater variance in the electron withdrawing potency of the two groups on the nucleus. For both series the PET-inhibiting activity largely depended on the overall lipophilicity of the molecules, however optimal chain length in NTHBA and NTHTBA was different. High activity of NTHBA and their thioamide analogues (NTHTBA) for PET inhibition indicated that they should interact with D_1 protein of the PET system in a specific manner (Honda et al., 1990b). Free amino hydrogen atom in NTHBA and NTHTBA may be needed for binding to the receptor site, possibly by forming a hydrogen bond. *N*-octyl-3-nitro-2,4,6-trihydroxybenzamide (PNO8), classified as a phenol-type PS 2 inhibitor, caused degradation of the D_1 protein of PS 2 reaction centre into two fragments of 23 and 9 kDa in complete darkness while the D_2 protein was not affected at all by incubation with this inhibitor. Occupation by another PS 2 inhibitor, DCMU, of the binding site of the secondary quinone acceptor, Q_B, prevented the D_1 protein from PNO8-induced degradation. These results indicate a selective and specific cleavage of the D_1 protein triggered by binding of PNO8 to the Q_B site (Nakajima et al., 1995).

Yoneyama et al. (1989a, 1989b) synthesized and tested a set of alkyl substituted 3-acyl-2,4,6-trihydroxybenzamides and thiobenzamides for their PET-inhibiting activity. The thioamide derivatives were found to be much more active than the corresponding amide derivatives and some of the compounds were as active as DCMU. The activity of the 3-propionyl-2,4,6-trihydroxybenzamide derivatives with varying *N*-alkyl group was enhanced by increasing the length of the *N*-alkyl group reaching maximum with the *N*-heptyl group (pI_{50} = 7.4) and then remaining at constant level (pI_{50} = 7.3) until the chain reached C_{10}. The activity of the *N*-phenylalkyl derivatives was also enhanced by increasing number of methylene groups indicating that the PET-inhibiting activity of these compounds depended largely on the overall lipophilicity of the amide derivatives (Yoneyama et al., 1989a). Slightly lower activity of thiobenzamide compound which had long side chains in both functional groups (R^1 = R^2 = hexyl) compared with that which had the same total number of carbon atoms in the side chains (R^1 = ethyl, R^2 = decyl), suggested that an asymmetric distribution of lipophilic groups about the phloroglucinol nuclei might be preferable for high activity (Yoneyama et al., 1989b).

3.2 2-Hydroxy-3-alkyl-1,4-naphthoquinones and *n*-alkyl-substituted ubiquinones

Jewess et al. (2002) found that the main mode of herbicidal activity of 2-hydroxy-3-alkyl-1,4-naphthoquinones is the inhibition of PS 2. The length of the 3-*n*-alkyl substituent for optimal activity differed between herbicidal and *in vitro* activity. The maximum *in vitro* activity was given by the nonyl to dodecyl homologues (log K_{ow} between 6.54 and 8.12), whereas herbicidal activity peaked with the *n*-hexyl compound (log K_{ow} = 4.95). The compounds did not show any activity on PS 1 and did not generate toxic oxygen radicals by redox cycling reactions.

Warncke et al. (1994) studied the influence of hydrocarbon tail structure on quinone binding and electron transfer performance at the Q_A and Q_B sites of the photosynthetic reaction-center protein isolated from *Rhodobacter sphaeroides* and solubilized in aqueous and in hexane solutions. It was found that contributions of the same tail structures to the binding free energies of quinones at the Q_A and Q_B sites are comparable, suggesting that the binding domains share common features. Comparison of the affinities of a homologous series of 10 *n*-alkyl substituted ubiquinones resolves the binding forces along the length of the tail binding domain and shows that strong steric constraints oppose accommodation of the tail in its extended conformation. One- and two isoprene-substituted quinones bind more tightly than analogues substituted with saturated alkyl tail substituents. Thus, the sites exhibit binding specificity for the native isoprene tail structure. Calculations indicated that the binding specificity aroused primarily from a lower integrated torsion potential energy in the bound isoprene tails.

3.3 Triazine and phenylurea derivatives with tail-like substituents

Reifler et al (2001) investigated effects of tail-like substituents on the binding of competitive inhibitors to the Q_B site of PS 2. They synthesized triazine and phenylurea derivatives with tail-like substituents and tested the effects of charge, hydrophobicity and size of the tail on binding properties. If the tail was attached to one of the alkylamino groups of triazine-type herbicides or to the *para* position of phenylurea-type herbicides, loss of binding was not observed. Consequently, the herbicides must be oriented in the Q_B site such that these positions point toward the natural isoprenyl tail-binding pocket that extends out of the Q_B site. The requirement that the tail must extend out of the Q_B site, constrains the size of the other herbicide substituents in the pocket. When longer hydrophobic tails are used, the binding penalty that occurs upon adding a charged substituent at the distal end is reduced. This allows the use of a series of tail substituents possessing a distal charge as an approximate molecular ruler to measure the distance from the Q_B site to the aqueous phase.

3.4 Alkyl-*N*-phenylcarbamates and thiocarbamates and amphiphilic alkoxyphenyl cabamates

Derivatives of phenylcarbamic acids, which contain biologically active–NH-CO- group are biologically active compounds applied mainly as herbicides (Moreland, 1993). Phenylcarbamates were found to be mitotic poisons that killed roots by inhibiting cell division (e.g. Nurit et al., 1989).

The action of alkyl-*N*-phenylcarbamates on the photolytic activity of isolated chloroplasts was already studied in the late 50s of last century by Moreland & Hill (1959). Later it was found that alkyl-*N*-phenylcarbamates (alkyl = methyl - octyl) and alkyl-*N*-

phenylthiocarbamates (alkyl = methyl - butyl) interact with the intermediate D$^{\bullet}$ situated in D_1 protein on the donor side of PS 2, however OEC, the intermediate Z$^{\bullet}$ and PS 1 were not injured (Šeršeň et al., 2000).

The most active PET inhibitor was methylthio derivative (IC_{50} = 8.5 μmol dm^{-3}), whereby for compounds with linear alkyl substituent PET inhibiting activity showed a linear decrease with increasing lipophilicity of the studied compounds. The inhibitory activity of compounds with branched alkyl substituents (R = isopropyl, *tert*- butyl, isobutyl) was lower than that of their linear isomers. The lower effectiveness can be connected with the fact that for achievement of the site of action in the photosynthetic apparatus, the branched substituents represent a higher hindrance than their linear isomers (Šeršeň et al., 2000). On the other hand, Hansch & Deutsch (1966) found that the inhibition of the Hill reaction in chloroplasts produced by ethyl and isopropyl derivatives of *N*-phenylcarbamates with different substituents in positions 3 and 4 on the benzene ring showed an increase with increasing lipophilicity of the compounds. This indicates that for PET-inhibiting activity of these compounds not only the lipophilicity of the compounds but also electronic properties of the substituents were determinant.

Esters of 2-, 3- and 4-alkoxy substituted phenylcarbamic acids (alkyl = methyl - decyl) were found to inhibit photosynthetic electron transport in spinach chloroplasts and to reduce chlorophyll content in alga *Chlorella vulgaris*. The inhibitory effectiveness strongly depended on the alkyl chain length of the alkoxy substituent showing a typical quasi-parabolic dependence with maximum effect at 6-8 carbon atoms in the alkyl chain of piperidinoethylesters (Kráľová et al., 1992a), 7-9 carbon atoms in the alkyl chain of dimethylaminoethylesters (Kráľová et al., 1992b) and 8-9 carbon atoms in the alkyl chain of piperidinopropylesters of alkoxyphenylcarbamic acids (Kráľová et al., 1995a; Šeršen & Kráľová, 1996). Similar results were obtained with morpholinoethylesters of 2-, 3- and 4-alkoxy substituted phenylcarbamic acids (Kráľová et al., 1994a) for which bilinear dependence of photosynthesis-inhibiting activity upon lipophilicity of compounds was confirmed. The alkoxy substitution in the position 2 decreased the inhibitory activity of the compounds when compared with their 3- and 4-substituted analogues. Similar dependences of photosynthesis-inhibiting activity on the length of alkyl substituent were obtained also for 22 alkyl substituted aryloxyaminopropanols (Mitterhauszerová et al., 1991a).

The dependence of photosynthesis-inhibiting activity of 1,3-diamino-2-propylesters of 2- and 3-substituted alkoxyphenylcarbamic acids on the dibasic part of the molecule was weak and the activity was more strongly affected by the position of the alkoxy substituent on the benzene ring of molecule as well as by the length of the alkoxy chain (Mitterhauszerová et al., 1991b).

The PET-inhibiting and algicidal activity of *N*-alkyl-4-piperidinoethylesters (alkyl = ethyl - butyl) and *N*-ethylpyrrolidinylmethylesters of 2- and 3-substituted alkoxyphenylcarbamic acids (alkoxy = butyloxy - heptyloxy) strongly depended on the lipophilicity of the whole molecule, whereby lower inhibitory activity was determined for 2-alkoxy substituted derivatives (Kráľová et al., 1992c). Strong effect of the chain length of alkyl substituent on photosynthesis-inhibiting activity was found also for quaternary ammonium salts of heptacaine, i.e. for *N*-[2-(2-heptyloxyphenylcarbamoyloxy)-ethyl]-*N*-alkylpiperidinium bromides (Kráľová et al., 1994b). Moreover, it was confirmed that piperidinopropylesters of 2-, 3- and 4-alkoxy substituted phenylcarbamic acids stimulated oxygen evolution rate

(OER) in spinach chloroplasts at relatively low effector concentration, causing photophosphorylation uncoupling due to protonophore properties of these amphiphilic compounds (Šeršeň & Kráľová, 1996).

Using EPR spectroscopy it was confirmed that the above mentioned alkoxyphenylcarbamates interacted with Z•/D• intermediates situated on the donor side of PS 2 and with OEC, causing release of Mn^{2+} ions into interior of thylakoid membranes (Mitterhauszerová et al., 1991b; Kráľová et al., 1992a, 1992c; Šeršeň & Kráľová, 1996). The quasi-parabolic course of the dependence of **P** parameter evaluated from EPR spectra of effector-treated chloroplasts (which could be considered as a measure of photosynthesis inhibition in plant chloroplasts) on the length of alkoxy substituents of tested alkoxyphenyl carbamates correlated with dependences obtained for inhibitory activity of 1,3-diamino-2-propylesters of 2- and 3-alkoxyphenylcarbamic acids (Mitterhauszerová et al., 1991b). **P** parameter was evaluated from the ratio of EPR signal I intensity determined for effector-treated chloroplasts in the light and in the dark related to such ratio obtained for untreated chloroplasts. From the above results it can be concluded that lower inhibitory activity of more lipophilic compounds with long alkyl substituents can be connected with the fact that these compounds predominantly remain incorporated in the lipid part of the membrane and only limited number of effector molecules will reach the membrane proteins. On the other hand, low lipophilicity of the compounds with short alkyl chains will result in their limited transition through membrane and in their more difficult access to interaction with PS 2 proteins situated on the inner side of thylakoid membranes.

3.5 2-Alkylsulfanyl-4-pyridinecarbothioamides and anilides of 2-alkylsulfanylpyridin-4-carboxylic acids

The IC_{50} values related to PET inhibition in spinach chloroplasts by 2-alkylsulfanyl-4-pyridinecarbothioamides (APCT) varied for the investigated set (alkyl = methyl – dodecyl) in the range from 72 μmol dm^{-3} (for octyl) to 6.24 mmol dm^{-3} (for methyl derivative) and it was found that the Hansch´s parabolic model is suitable for description of the correlation between photosynthesis-inhibiting activity and lipophilicity of APCT. For more precise determination of the site of APCT action in the photosynthetic apparatus of spinach chloroplasts EPR spectroscopy was used (Kráľová et al., 1997).

Fig. 1 presents EPR spectra of untreated spinach chloroplasts (Fig. 1a) and of chloroplasts treated with 0.05 mol dm^{-3} of 2-*n*-butylsulfanyl derivative (Fig. 1b). From Fig. 1b it is evident that the intensity of EPR signal II, mainly of its constituent signal II_{slow}, has been decreased by the studied compound indicating that APCT interact with D• intermediate. Due to the interaction of APCT with this part of PS 2, PET between PS 2 and PS 1 is impaired and consequently a pronounced increase of signal I intensity in the light (Fig. 1b, dashed line; g = 2.0026, ΔB = 0.7 mT) belonging to chlorophyll *a* dimer in the core of PS 1 can be observed. Upon addition of DPC, an artificial electron donor with the known site of action in the intermediate Z•/D• on the donor side of PS 2 (Izawa, 1980) to chloroplasts activity of which was inhibited by APCT, PET was practically completely restored. Consequently, it can be assumed that in the presence of APCT the own core of PS 2 (P 680) and a part of the electron transport chain – at least up to plastoquinone – remain intact.

Anilides of 2-alkylsulfanylpyridin-4-carboxylic acids were found to inhibit oxygen evolution rate in *C. vulgaris*, whereby the lipophilicity of the compound was determining for OER-inhibiting activity (Kráľová et al., 2001). They inhibited also PET in spinach chloroplasts and the corresponding IC_{50} values varied in the range from 4.8 μmol dm^{-3} to 69.1 μmol dm^{-3} and the lipophilicity of the most active compounds was about log P = 5.0-5.5 (Miletín et al., 2001). EPR spectroscopy confirmed that these anilides interacted with the intermediate $D^{\bullet}$ (Tyr_D) and in a pronouncedly lesser extent also with the intermediate $Z^{\bullet}$ (Tyr_Z). The intensive interaction of these compounds with Tyr_D which is situated on the donor side of PS 2 in less polar environment of the thylakoid membranes can be connected with the presence of hydrophobic alkylsulfanyl substituent in their molecules (Miletín et al., 2001).

3.6 Alkyl substituted benzothiazole derivatives

Several series of alkyl substituted benzothiazole derivatives were investigated also for their photosynthesis-inhibiting activity (Kráľová et al., 1992d, 1993; Sidóová et al., 1998, 1999; Šeršeň et al., 1993). The dependence of the negative logarithm of IC_{50} values on the alkyl chain length of 2-alkylthio-6-R-benzothiazoles determined in the system of plant chloroplasts with partially damaged membranes showed a significant role of the substituent in position 6 with respect to the studied inhibitory activity. The increasing of its lipophilicity in comparable series leads to higher activity of compounds having shorter alkyl chains, with subsequent strong decrease of the activity with the further prolongation of the alkyl chain. The drop of the activity at derivatives with longer alkyl chains was the most pronounced in series having the most lipophilic substituent in position 6 (see Fig. 3A) (Kráľová et al., 1992d).

On the other hand, the inhibition of chlorophyll production in algae *C. vulgaris* was in the case of 6-formamido-, 6-acetamido- and 6-benzoylamino derivatives more strongly affected by the presence of compounds with lower lipophilicity of the substituent in position 6 (the inhibitory efficiency decreased in the order 6-formamido-, 6-acetamido- and 6-benzoylamino derivatives) (Fig. 3B). Thus, it can be assumed that higher lipophilicity of the substituent in the position 6 at equal alkyl chain length of the alkyl substituent diminishes the possibility of the compounds to penetrate through the intact outer algal cell membrane resulting in decreased inhibitory activity of the compounds. All bicyclo[2.2.1]hept-5-ene-2,3-dicarboximidomethylamino derivatives showed higher photosynthesis inhibition in plant chloroplasts as well as in *C. vulgaris* than the corresponding bicyclo[2.2.1]hept-5-ene-2,3-dicarboximido derivatives (Fig. 3) (Kráľová et al., 1992d).

Inhibition of PET in spinach chloroplasts and chlorophyll synthesis in *C. vulgaris* was observed also with 2-alkylthio-6-aminobenzothiazoles and their 6-*N*-substituted derivatives 3-(2-alkylthio-6-benzothiazolylaminomethyl)-2-benzothiazolinethiones and 3-(2-alkylthio-6-benzothiazolinone)-6-bromo-2-benzothiazolinones (Kráľová et al., 1993).

The dependence of inhibitory activity of these compounds on the alkyl chain length of the thioalkyl substituent showed quasi-parabolic course indicating decrease of biological activity for compounds with higher lipophilicity. It was found that in the presence of these compounds no reduction of PS 1 occurred, however interaction with the intermediates $Z^{\bullet}/D^{\bullet}$ was not confirmed. Since DPC practically completely restored PET through PS 2 in spinach chloroplasts activity of which was inhibited by these inhibitors it can be assumed that these benzothiazole derivatives did not damage PET between photosynthetic centres PS 2 and PS 1 and their site of action is situated at the donor side of PS 2 (Kráľová et al., 1993).

PET inhibition in spinach chloroplasts was observed also by 2-(6-acetamido-benzothiazolethio)acetic acid esters (Sidóová et al., 1998). The dependence of PET inhibiting activity on the lipophilicity of the derivatives with R = *n*-alkyl and allyl showed a quasi-parabolic course, the most active compound was the hexyl derivative (IC_{50} = 47 μmol dm^{-3}). The effect of 2-(alkoxycarbonylmethylthio)-6-aminobenzothiazoles on photosynthetic apparatus was similar to that of above discussed 2-alkylthio-6-R-benzothiazoles and as probable site of their action the oxygen evolving complex was suggested (Šeršeň et al., 1993).

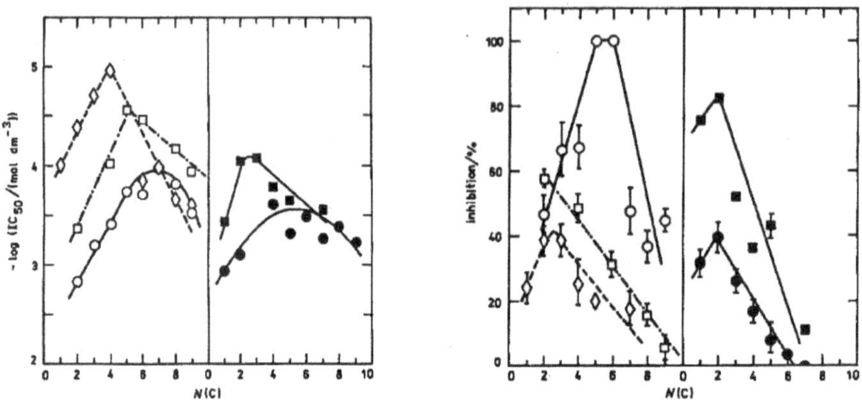

A B

Fig. 3. Dependence of negative logarithm of IC_{50} values related to PET inhibition in spinach chloroplasts on the number of carbons of the alkyl chain of 2-alkylthio-6-R-benzothiazoles (A) and dependence of inhibition of chlorophyll production in *C. vulgaris* in the presence of 0.1 mmol dm^{-3} of 2-alkylthio-6-R-benzothiazoles (B). (R = formamido (o), acetamido (□), benzoylamino (◊), bicyclo[2.2.1]hept-5-ene-2,3-dicarboximino (•) and bicyclo[2.2.1]hept-5-ene-2,3-dicarboximidomethylamino (▪). (Source: Kráľová et al., 1992 d).

3.7 Alkyl substituted *N*-oxides

Amine oxides, also known as amine-*N*-oxides or *N*-oxides are chemical compounds which contain the functional group R_3N^+-O^-, an N-O bond with three additional hydrogens and/or hydrocarbon side chains attached to nitrogen atom. The structure of alkyl substituted *N*-oxides predestine them to have membrane damaging as well as PET-inhibiting properties. Avron (1961) found that the Hill reaction and associated phosphorylation in Swiss chard chloroplasts were sensitive to heptyl- and nonylhydroxyquinoline-*N*-oxides. Among the photosynthetic reactions studied, light-induced cyclic phosphorylation with phenazine methosulfate as cofactor was least sensitive to the inhibitors, whereas the most sensitive was the Hill reaction and coupled phosphorylation in the presence of ferricyanide. Avron suggested that these inhibitors blocked electron transport at a site similar to that of DCMU. Bamberger et al. (1963) also showed that $NADP^+$ photoreduction with reduced DCPIP as electron donor was highly resistant to these inhibitors, whereas photoreduction through the normal Hill reaction system was not. Izawa et al. (1966) suggested that the inhibition site of 2-heptyl-4-hydroxyquinoline-*N*-oxide (HOQNO) in non-cyclic electron flow in chloroplasts is somewhat different from that of DCMU, as inferred from the difference in dependency of the actions of these two inhibitors on light intensity. Later, Gromet-Elhanan (1969) described HOQNO as acting at two different sites in the electron transport chain. At a low concentration it inhibited near PS 2, and at higher concentrations, somewhere near PS 1, what was effectively by-passed by phenazine methosulfate. The latter inhibition site of HOQNO has also been suggested by Hind & Olson (1966) who showed that HOQNO increased the magnitude of reversible changes of cytochrome *b6* and proposed that the inhibitor blocked electron flow between cytochrome *b6* and PS 1.

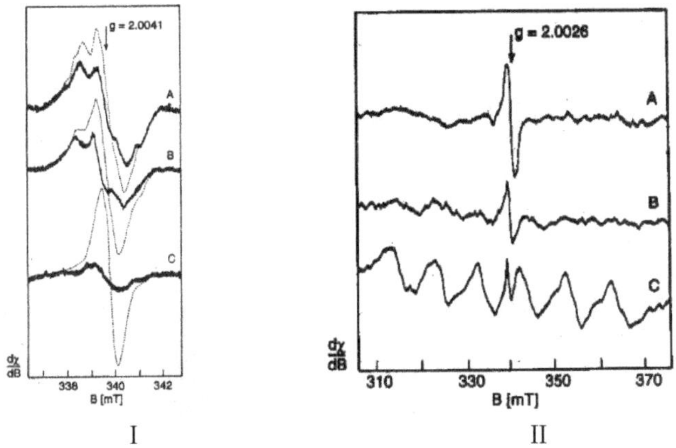

Fig. 4. EPR spectra of spinach chloroplasts (I) registered in the dark (full line) and under irradiation (dotted line) for control sample (A) and for chloroplasts treated with 0.01 mol dm^{-3} 1-octylpiperidine-*N*-oxide (B) or 1-dodecylpiperidine-*N*-oxide (C); EPR spectra of Mn^{2+} ions (II) in untreated spinach chloroplasts (A) and in chloroplasts treated with 0.05 mol dm^{-3} 1-hexyl-1-ethylpiperidium bromide (B) or with 1 tetradecyl-1-ethylpiperidinium bromide (C). (Source: Kráľová et al. 1992e).

Surfactants of homologous series of 1-alkylpiperidine-*N*-oxides (APNO) with alkyl = hexyl – octadecyl were found to inhibit PET in spinach chloroplasts and chlorophyll synthesis in algal suspensions of *C.vulgaris* (Kráľová et al., 1992e). The dependence of algicide effects of APNO (expressed by minimum inhibitory concentration, MIC) on the surfactant alkyl chain length showed quasi-parabolic course and the highest algicide activity (MIC about 10 μmol dm^{-3}) exhibited derivatives in which alkyl chain varied from tridecyl to hexadecyl. The algicide effect of surfactants was not connected with their associative properties since all MIC values were far below the critical micelle concentration of these compounds. Similar results were obtained also for PET inhibiting activity of APNO in spinach chloroplasts whereby the more pronounced decrease of activity was associated with the prolongation of the alkyl chain (C_{15}-C_{18}).

Decreased intensity of both components of signal II as well as the rise of signal I in the light in EPR spectra of APNO-treated chloroplasts indicated that these compounds caused damage of PS 2 and electron flow to PS 1 was interrupted (Fig. 4(I), C). From this Figure it is evident that the changes in EPR spectra caused by dodecyl derivative were considerably higher than those caused by octyl derivative, which is in accordance with the results related to algicide and PET-inhibiting activity of APNO. Moreover, in EPR spectra of APNO-treated chloroplasts occurrence of six lines of fine structure belonging to free Mn^{2+} ions was observed (similarly as it is documented in Fig. 4(II), C), indicating injury of OEC, which is situated on the donor side of PS 2 (Kráľová et al., 1992e). The amount of released Mn^{2+} ions from the above mentioned OEC - at constant chlorophyll and surfactant concentration - is proportional to the inhibitory activity of surfactant.

N-alkyl-*N,N*-dimethylamine oxides (ADAO) (alkyl = hexyl – octadecyl) were also found to inhibit PET in plant chloroplasts and chlorophyll synthesis in green alga *C. vulgaris* (Šeršeň et al., 1992). The dependence of the biological activity (expressed by log IC_{50}) on the length of alkyl substituent showed quasi-parabolic course with maximum activity for tetradecyl derivative. From EPR spectra it was evident that the site of ADAO action is PS 2, mainly $Z^{\bullet}/D^{\bullet}$ intermediates or its neighbouring surroundings. The release of Mn^{2+} ions into thylakoid membranes due to ADAO treatment was also confirmed. In order to find what effects are exhibited by ADAO on chloroplast membranes, an EPR study using spin labels CAT 16 (*N*-hexadecyl-*N*-tempoyl-*N,N*-dimethylammonium bromide) and 16 DSA (16-doxylstearic acid) was performed. The motion of spin labels after their incorporation into membranes will be limited and consequently changes in their EPR spectra occur.

Incorporation of ADAO into thylakoid membranes causes also perturbation in membrane structure depending on ADAO concentration and it is evident that the arrangement of thylakoid membrane expressed by order parameter **S** with increasing ADAO concentration decreased (Fig. 5A). The order parameter **S** evaluated from the above mentioned changes in EPR spectrum (in detail see in Šeršeň et al. (1989)) reflects relative membrane perturbation. From the dependence of order parameter **S** of hexyl-, dodecyl- and hexadecyl derivatives (obtained with the spin label CAT 16) on the ADAO concentration it is evident that the order parameter **S** decreased as follows: hexyl, hexadecyl and dodecyl.

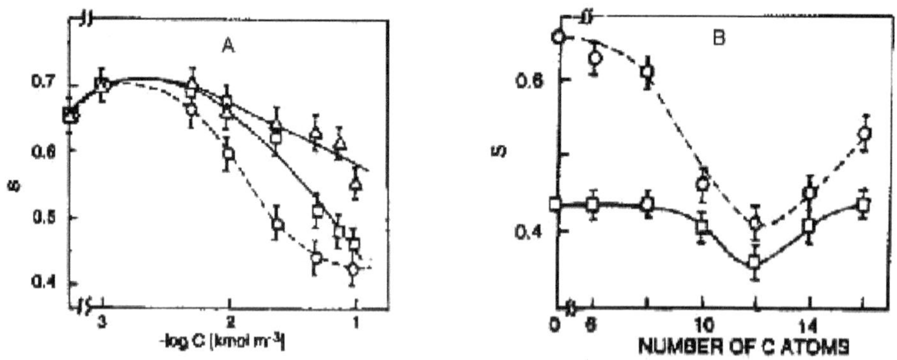

Fig. 5. Dependences of the order parameter of thylakoid membranes **S** (determined from EPR spectra of CAT 16) on the concentration of ADAO for hexadecyl- (□), dodecyl- (o) and hexyl- (Δ) derivatives (A) and on the alkyl chain length of ADAO at the constant concentration of ADAO 50 μmol dm^{-3} (B); **S** was evaluated from EPR spectra of CAT 16 (o) and 16 DSA (□) spin labels. (Source: Šeršeň et al., 1992).

Fig. 5B presents dependence of order parameter **S** on the alkyl chain length of ADAO at constant compound concentration (50 mmol dm^{-3}). The experiments with both spin labels confirmed that the most effective derivative related to perturbation of membrane arrangement was dodecyl, which is in accordance with the above mentioned results obtained for PET inhibition in spinach chloroplasts and chlorophyll content reduction in alga. After adding of ADAO to chloroplasts containing spin labels, an increase in the rate of molecular reorientation of spin label was observed. This was manifested by a decrease in the rotational correlation time values (τ_c) with increasing ADAO concentration (Fig. 6A). The rotational correlation time is linearly proportional to the microviscosity of the environment in which the spin label is located., i.e. ADAO decrease the microviscosity of thylakoid membranes. The course of τ_c of 16 DSA spin label located in the thylakoid membrane on the alkyl chain length of ADAO at constant compound concentration 50 mmol dm^{-3} (Fig. 6B) was similar to that obtained for order parameter **S** (Fig. 5B), i.e. the lowest τ_c exhibited dodecyl derivate. The dependence of IC_{50}, **S** and τ_c (characterizing the effect of ADAO on photosynthetic apparatus) on the alkyl chain length showed a typical "cut-off" dependence with maximal effects for dodecyl or tetradecyl derivatives. This can be explained by the incorporation of ADAO in the lipid phase of membranes whereby the greatest perturbation in the membrane is caused by ADAO with middle chain length (approximately dodecyl), which create sufficiently great free volume in the membrane and their partition coefficients between chloroplast organelles and aqueous phase have also sufficiently high values.

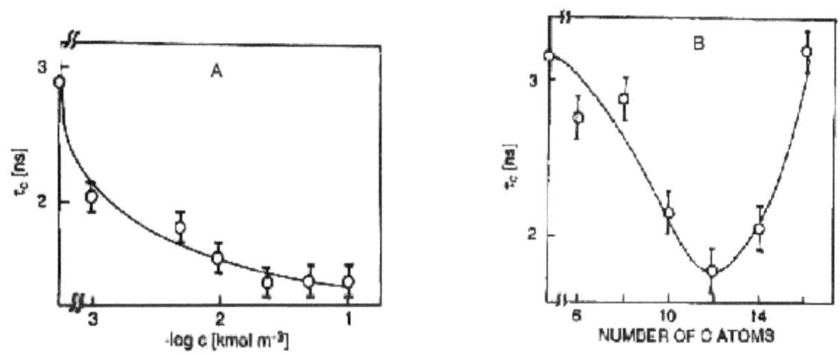

Fig. 6. Dependences of rotational time τ_c of 16 DSA spin label located in the thylakoid membranes on the concentration of *N*-dodecyl-*N,N*-dimethylamine oxide (A) and on the alkyl chain length of ADAO at the constant concentration of ADAO 50 μmol dm^{-3} (B). (Source: Šeršeň et al., 1992).

3.8 Cationic surfactants of the type of alkyl substituted quaternary monoammonium and diammonium salts

Lower concentrations of 1-alkyl-1-ethylpiperidinium bromides (AEPBr; alkyl = hexyl - octadecyl) stimulated OER in spinach chloroplasts (Šeršeň & Devínsky, 1994; Šeršeň & Lacko, 1995) indicating their activity with respect to thylakoid membranes. Previously it was found that anionic surfactant sodium dodecyl sulphate (SDS), if applied at low concentrations stimulated photoreduction of potassium ferricyanide, but application of higher surfactant concentrations exhibited inhibitory effects. Stimulation was caused by increased permeability of chloroplast envelope membrane and inhibitory effects were connected with changes in the chloroplast membrane organization, induced by treatment with surfactants. Application of SDS led to an inhibition of the light-induced proton uptake (ΔpH) due to deterioration of ATP-ase (Apostolova, 1988). The possibility that AEPBr enhance OER by a damage of ATP-ase and so prevent photophosphorylation is unlikely, because all AEPBr derivatives stimulated Hill reaction irrespective of their alkyl chain length (Šeršeň & Devínsky, 1994). It is also unlikely that AEPBrs can act as protonophores because their structure excludes such effect. Thus, it is probable that AEPBr after their incorporation into thylakoid membranes cause changes in their organization leading to an increase of OER. Similar effect was observed with other surfactants (Apostolova, 1988) and linolenic acid (Golbeck et al., 1980) but changes in the arrangement of the membrane caused by structurally different surfactants need not be the same. Siegenthaler & Packer (1965) found that higher concentrations of decenyl and dodecenyl succinic acids inhibited ferredoxin-NADP photoreduction and photophosphorylation but the application of low concentrations increased photophosphorylation and NADP photoreducion as well.

The arrangement of the thylakoid membranes in spinach chloroplasts was investigated by the spin label method, using CAT 16 as spin label. For characterization of the arrangement of thylakoid membranes the order parameter **S** was calculated from EPR spectrum of spin label incorporated into thylakoid membranes according to Šeršeň et al. (1989). The dependences of order parameter **S** and OER on the concentration of 1-dodecyl-1-ethylpiperidinium bromide (DEPBr) is shown in Fig. 7. At certain DEPBr concentrations enhancement of both parameters (**S** and OER) was observed related to control samples. This stimulating effect was observed at different DEPBr concentrations in OER and in EPR experiment, which was connected with different chlorophyll content in chloroplast suspensions used for individual experiments. Using DEPBr partition coefficient between chloroplast organelles and aqueous environment, DEPBr concentration within chloroplast organelles was calculated. In such terms, OER stimulation and increase of order parameter **S** occurred in the same concentration range (Fig. 7B). Based on these findings it could be assumed that OER stimulation is caused by changes in the arrangement of thylakoid membranes.

As mentioned above, the interaction of amphiphilic molecules, including alkyl substituted quaternary ammonium salts, with the hydrophobic parts of cell membranes leads to damage

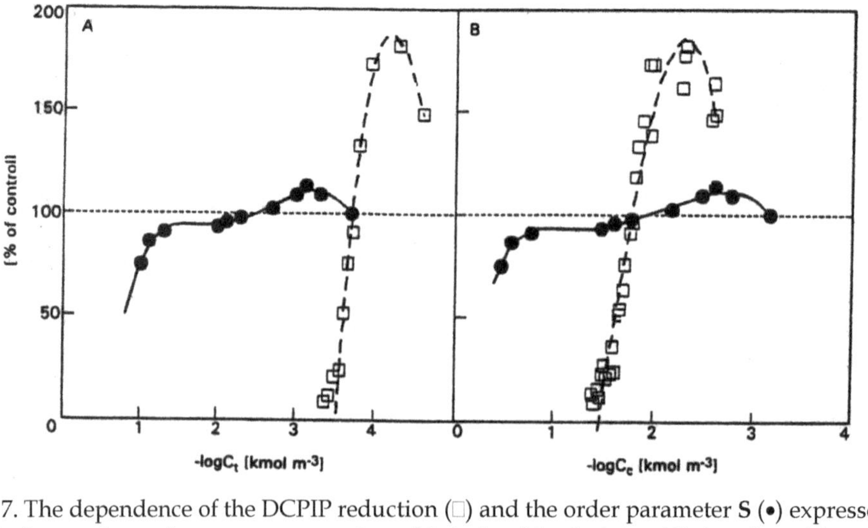

Fig. 7. The dependence of the DCPIP reduction (□) and the order parameter **S** (•) expressed as % of control sample upon concentration of 1-dodecyl-1-ethylpiperidinium bromide (DEPBr) in chloroplast suspension (A) and in chloroplast organelles (B). (Source: Šeršeň & Lacko, 1995).

of membrane structure (Apostolova, 1988; Devínsky et al., 1990; Balgavý & Devínsky, 1996). Detergent properties of surfactants enable to solubilize functional components bonded with the organized structures (Loach, 1980) and to form mixed micelles with them (Hermann et al., 1988). This can affect the key biochemical and energy yielding processes and subsequently cause biostatic and biocidal effects in lower organisms.

AEPBr (alkyl = hexyl - octadecyl) applied at higher concentrations inhibited PET in spinach chloroplasts and reduced chlorophyll content in algal suspensions of *C.vulgaris*. The photosynthesis-inhibiting activity of these surfactants showed quasi-parabolic dependence upon the length of alkyl substituent (Kráľová et al., 1992e). From changes in EPR spectra of spinach chloroplasts treated with AEPBr **P** parameter was evaluated which is available as a measure of photosynthesis inhibition in plant chloroplasts for homologous series of inhibitors having the same site and mechanism of action. The value of **P** parameter showed very strong dependence on the alkyl chain length of AEPBr. While the values of **P** parameter for AEPBr with shorter alkyl chain (pentyl, hexyl, heptyl) were < 1 and for octadecyl derivative $\mathbf{P} = 6$ was determined, the highest value of this parameter was obtained for tetradecyl derivative ($\mathbf{P} = 27.7$). These results are in accordance with the finding related to PET inhibition in spinach chloroplasts determined by the use of artificial electron acceptor DCPIP.

The decrease of a biological activity observed for amphiphilic compounds upon elongation of their hydrophobic (hydrocarbon) part is called "cut-off" effect (Devínsky et al., 1990; Balgavý & Devínsky, 1996). These hydrophobic parts of surfactants interact with lipid parts of biological (including thylakoid) membranes. However, penetration of surfactants with longer alkyl through hydrophilic (aqueous) regions of biological membranes is restricted due to their low aqueous solubility causing lower concentration of such surfactants in the membrane in comparison with the concentration of surfactants with shorter alkyl chain. Consequently, the biological activity of long-chain surfactants decreases. It is suggested that the lateral expansion of the phospholipid bilayer of biological membranes caused by the intercalation of long-chain amphiphilic molecules between the phospholipid molecules and the mismatch between their hydrocarbon chain lengths results in the creation of free volume in the bilayer hydrophobic region. According to the free volume theory the extent of membrane disturbance due to surfactant incorporation depends on the size of free volume created under its alkyl chain which can be then filled up with chains of neighbouring lipids as well as on the partition coefficient of the surfactants, i.e. on the number of surfactant molecules in the membrane (Devínsky et al., 1990; Balgavý & Devínsky, 1996).

From EPR spectra of AEPBr-treated chloroplasts it was evident that these compounds interacted with $Z^{\bullet}/D^{\bullet}$ intermediate on the donor side of PS 2 and with OEC as well (Kráľová et al., 1992e). Due to interaction of AEPBr with OEC, Mn^{2+} ions were released into interior of thylakoid membranes, which was manifested in EPR spectra of AEPBr-treated chloroplasts by the presence of six lines of fine structure (Fig. 4(II), C). In OEC, which is situated on the donor side of PS 2, four ions of manganese are bound in the 33kDa protein (Blankenship & Sauer, 1974; Cheniae, 1980), however due to intense spin-spin interaction of protein-bound manganese in untreated chloroplasts is the EPR signal of Mn^{2+} ions very low (Fig. 4(II), A). The amount of released manganese ions (at constant chlorophyll and surfactant concentration) was proportional to the inhibitory activity of surfactant (compare corresponding signals for hexyl (Fig. 4(II), B) and tetradecyl (Fig. 4(II), C) derivatives).

Several homologous series of cationic gemini surfactants, namely *N,N´*-bis(alkyldimethyl)-1,6-hexanediammonium dibromides (HDDBr) (Kráľová & Šeršeň, 1994), isosteric *N,N´*-(alkyldimethyl)-3X-1,5-pentanediammonium dibromides (PDDBr) (Kráľová et al., 1995b) and 3,8-diaza-4,7-dioxodekane-1,10-diylbis(alkyldimethylammonium) bromides (DDDBr) (Kráľová et al., 2010) were tested for their PET-inhibiting activity and also for inhibition of chlorophyll synthesis in *C. vulgaris.*

For isosteric *N,N´*-(alkyldimethyl)-3X-1,5-pentanediammonium dibromides (PDDBr) with X = CH_2, NCH_3, O or S it was found that their critical micelle concentration did not reflect small differences in lipophilicity of the compounds with a very similar structure sensitively enough. The differences in biological activities (IC_{50} values related to PET inhibition in spinach chloroplasts and MIC values related to reduction of chlorophyll content in *C. vulgaris*) of PDDBr isosters with the same alkyl chain were very small as well, indicating that modification of spacer did not affect the mode of action of these gemini surfactants (Kráľová et al., 1995b).

Similarly to AEPBr, the above mentioned gemini surfactants (PDDBr, HDDBr and DDDBr) interacted with $Z^{\bullet}/D^{\bullet}$ intermediates situated on the donor side of PS 2 and with OEC as well. For HDDBr surfactants the highest **P** values (about 16.7) reached undecyl and tridecyl derivatives which were found to be the most effective PET inhibitors (Kráľová & Šeršeň, 1994; Kráľová et al., 2010). The **P** parameter determined for dodecyl derivative of DDDBr was 12.4 indicating that insertion of two NHCO groups into spacer resulted in partial decrease of this parameter as well as in the decrease of PET inhibition (Kráľová et al., 2010).

For ascertaining whether two CONH groups in the spacer of DDDBr could affect the resulting inhibitory activity of these surfactants, DCPIP photoreduction by the base *N,N´*-bis(2-dimethylaminoethyl)ethanediamide was estimated (Kráľová et al., 2010). This compound did not contain long alkyl chain and so its PET-inhibiting activity is connected only with two CONH groups in its molecule. While the determined IC_{50} value for this diamide was found to be 4.0 mmol dm^{-3} and the corresponding IC_{50} value estimated for nonyl derivative was only 1.74 mmol dm^{-3}, it can be supposed that CONH groups in the spacer participate on the resulting inhibitory effects. Moreover, the extent of the contribution of amide groups to the total inhibitory effect of DDDBr surfactants with longer alkyl chains was much lower, which was reflected in the corresponding IC_{50} values of these surfactants (e.g. 69.7 μmol dm^{-3} for dodecyl derivate). Using EPR spectroscopy it was found that the interaction of tested base with the $Z^{\bullet}$ and $D^{\bullet}$ intermediates was much weaker than this of DDDBr and not even release of Mn^{2+} ions into thylakoid membrane was observed after diamide treatment. Weaker interaction was reflected also by very low value of **P** parameter (**P** = 2.9) obtained with tested base. These results indicate that for amphiphilic compounds from the group of cationic gemini surfactants it is more easy to reach the sites of their action in the photosynthetic apparatus (which can be situated in the regions of thylakoid membranes with different polarity) than for their non-polar bases.

4. Conclusion

Herbicidal effects of compounds having alkyl chain(s) in their molecule are caused either by interaction of the compound with membrane (destruction, re-arrangement, change of the viscosity, etc.) or by its interaction with proteins occurring in plant cells. Interaction with membranes is characteristic mainly for amphiphilic compounds (amine oxides, quaternary ammonium salts and fatty acids) in which alkyl chains could be incorporated into membrane, what results in structure modification of photosynthetic proteins, which bind individual components of the PET chain. Due to such changes the photosynthetic electron transport through photosynthetic centres will be interrupted what results in the decrease of OER. The most effective disturbance of the membrane and thus the highest inhibitory effect will be exhibited by compounds with middle alkyl chain length ensuring not only sufficiently high free volume under alkyl chain but also high concentration of the surfactant in the membrane due to suitable value of surfactant partition coefficient. Moreover, it was found that due to interaction of amphiphilic membrane-active compounds with manganese cluster release of manganese ions from the oxygen evolving complex occurs and dysfunction of intermediates $Z^\bullet/D^\bullet$ occurring at 161th position of D_1 and D_2 proteins situated on the donor side of PS 2 is manifested.

However, herbicides can act also by direct interaction with cell proteins. Such herbicides have in their structure certain functional groups which can interact with some amino acid residues of proteins. Due to such interaction interruption of PET through photosynthetic centres occurs and photosynthesis is inhibited.

Many bioactive natural and synthetic compounds with alkyl chain(s) in their molecules act as PS 2 herbicides. The site of their action is usually situated in Q_B on the reducing side of PS 2 (e.g. sorgoleone, tenuazonic acid, *N*-octyl-3-nitro-2,4,6-trihydroxybenzamide) and/or in $Z^\bullet/D^\bullet$ intermediates on the donor side of PS 2 (e.g. fatty acids, alkyl-*N*-phenylcarbamates, amphiphilic alkoxyphenylcarbamates, 2-alkylsulfanyl-4-pyridinecarbothioamides, alkyl substituted *N*-oxides and quaternary monoammonium and diammonium salts).

5. Acknowledgments

This chapter is dedicated to the memory of Prof. Ľudovít Krasnec (1913–1990), former leader of our research team. The work was supported by the grant VEGA No. 1/0612/11 and by grant APVV-0416-10. The authors wish to thank the journals *Chemical Papers*, *Collection of Czechoslovak Chemical Communications*, *General Physiology and Biophysics* and *Photosynthetica* for providing us with the original data and figures published in these journals.

6. References

Apostolova, E. L. (1988). Effect of detergent treatment and glutaraldehyde modification on the photochemical activity of chloroplasts. *Comptes Rendus of Bulgarian Academy of Sciences*, Vol. 41, No. 7, pp. 117-120, ISSN 0366-8681

Avron, M. (1961). The inhibition of photoreactions of chloroplasts by 2-alkyl-4-hydroxyquinoline *N*-oxides. *Biochemical Journal*, Vol. 78, No. 4, pp. 735-739, ISSN 0264-6021

Balgavý, P. & Devínsky, F. (1996). Cut-off effects in biological activities of surfactants. *Advances in Colloid and Interface Science*, Vol. 66, (August 1996), pp. 23-63, ISSN 0001-8686

Bamberger, E. S.; Black C. C.; Fewson, C. A. & Gibbs, M. (1963). Inhibitor studies on carbon dioxide fixation, adenosine triphosphate formation, and triphosphopyridine nucleotide reduction by spinach chloroplasts. *Plant Physiology*, Vol. 38, No. 4, pp. 483-487, ISSN 0032-0889

Barr, R.; Floreani, M.; Weiss, J. & Crane, F. L. (1988). The effects of platelet activating factor on electron transport of spinach chloroplasts. *Biochemical and Biophysical Research Communications,* Vol. 155, No. 2, (September 1988), pp. 576-582, ISSN 0006-291X

Bagchi, S. N. & Marwah, J. B. (1994). Production of an algicide from cyanobacterium *Fischerella* species which inhibits photosynthetic electron transport. *Microbios,* Vol. 79, No. 320, pp. 187-193, ISSN 0026-2633

Bhalla, R.; Singh, A. K.; Pradhan, S. & Unnikumar, K. R. (2009). Lipids: Structure, Function and Biotechnology Aspects. In: *A Textbook of Molecular Biotechnology,* A. K. Chauhan & A. Varma, (Ed.), 173-209, I.K. International Publishing House Pvt. Ltd., ISBN 978-93-80026-37-4, New Delhi, India

Blankenship, R. E. & Sauer, K. (1974). Manganese in photosynthetic oxygen evolution. I. Electron paramagnetic resonance study of the environment of manganese in Tris washed chloroplasts. *Biochimica et Biophysica Acta-Bioenergetics,* Vol. 357, No. 2, pp. 252-266, ISSN 0005-2728

Brody, S. S.; Brody, M. & Dőring, G. (1970). Effects of linolenic acid on system II and system I associated light induced changes in absorption of chloroplasts. *Zeitschrift für Naturforschung,* Vol. 25b, No. 4, pp. 367-372, ISSN 0932-0776

Chen, S. G.; Xu, X. M.; Dai X. B.; Yang, C. L. & Qiang, S. (2007). Identification of tenuazonic acid as a novel type of natural photosystem II inhibitor binding in Q_B-site of *Chlamydomonas reinhardtii. Biochimica et Biophysica Acta-Bioenergetics,* Vol. 1767, No. 4, pp. 306–318, (April 2007), ISSN 0005-2728

Chen, S. G.; Yin, C. Y.; Dai, X. B.; Qiang, S. & Xu, X. M. (2008). Action of tenuazonic acid, a natural phytotoxin, on photosystem II of spinach. *Environmental and Experimental Botany,* Vol. 62, No. 3, pp. 279–289, (April 2008), ISSN 0098-8472

Chen, S.; Yin, C.; Qiang, S.; Zhou, F. & Dai, X. (2010). Chloroplastic oxidative burst induced by tenuazonic acid, a natural photosynthesis inhibitor, triggers cell necrosis in *Eupatorium adenophorum* Spreng. *Biochimica et Biophysica Acta-Bioenergetics,* Vol. 1797, No. 3, (March 2010), pp. 391-405, ISSN 0005-2728

Cheniae, G. M. (1980). Manganese binding sites and presumed manganese proteins in chloroplasts. In: *Methods in Enzymology,* S. P. Colowick & N. O. Kaplan, (Ed.), Vol. 69, Part C, 349-363, Academic Press, ISBN 0121819698, New York - London

Czarnota, M. A.; Paul, R. N.; Dayan, F. E.; Nimbal, C. I. & Weston, L. A. (2001). Mode of action, localization of production, chemical nature, and activity of sorgoleone: A potent PS II inhibitor in *Sorghum* spp. root exudates. *Weed Technology,* Vol. 15, No. 4, (October-December 2001), pp. 813 -825, ISSN 0890-037X

Dayan, F. E. (2006). Factors modulating the levels of the allelochemical sorgoleone in *Sorghum bicolor. Planta,* Vol. 224, No. 2, (July 2006), pp. 339-346, ISSN 0032-0935

Dayan, F. E.; Howell, J. L. & Weidenhamer, J. D. (2009). Dynamic root exudation of sorgoleone and its in planta mechanism of action. *Journal of Experimental Botany,* Vol. 60, No. 7, (May 2009), pp. 2107–2117, ISSN 0022-0957

Dayan, F. E.; Rimando, A. M.; Pan, Z. Q.; Baerson, S. R.; Gimsing, A. L. & Duke, S. O. (2010). Sorgoleone. *Phytochemistry,* Vol. 71, No. 10, (July 2010), pp. 1032-1039, ISSN 0031-9422

Debus, R. J.; Barry, D. A.; Babcock, G. T. & McIntosh, L. (1988a). Site-directed mutagenesis identifies a tyrosine radical involved in the photosynthetic oxygen-evolving system. *Proceedings of the National Academy of Sciences of the United States of America*, Vol. 85, No. 2, (January 1988), pp. 427–430, ISSN 0027-8424

Debus, R. J.; Barry, D. A.; Sithole, I.; Babcock, G. T. & McIntosh, L. (1988b). Directed mutagenesis indicates that donor to P680+ in photosystem II is tyrosine-161 of the D1 polypeptide. *Biochemistry*, Vol. 27, No. 26, (December 1988), pp. 9071–9074, ISSN 0006-2960

Devínsky, F.; Kopecká-Leitmanová, A.; Šeršeň, F. & Balgavý, P. (1990). Interaction of surfactants with model and biological membranes. 8. Amine oxides. 24. Cut off effect in antimicrobial activity and in membrane perturbation efficiency of the homologous series of *N,N*-dimethylalkylamine oxides. *Journal of Pharmacy and. Pharmacology*, Vol. 42, No. 11, (November 1990), pp. 790-794, ISSN 0022-3573

Duke, S. O.; Rimando, A. M.; Baerson, S. R.; Scheffler, B. E.; Ota, E. & Belz, R. G. (2002a). Strategies for the use of natural products for weed management. *Journal of Pesticide Science*, Vol. 27, No. 3, pp. 298-306, ISSN 0385-1559

Duke, S. O.; Dayan, F. E.; Rimando, A. M.; Schrader, K. K.; Aliotta, G.; Oliva, A. & Romagni, J. G. (2002b). Chemicals from nature for weed management. *Weed Science*, Vol. 50, No. 2, pp. 138-151, (March-April 2002), ISSN 0043-1745

Garstka, M. & Kanuiga, Z. (1988). Linolenic acid-induced release of Mn, polypeptides and inactivation of oxygen evolution in photosystem II particles. *FEBS Letters*, Vol. 232, No. 2, (May 1988), pp. 372-376, ISSN 0014-5793

Golbeck, J. H.; Martin, I. F. & Fowler, C. F. (1980). Mechanism of linolenic acid induced inhibition of photosynthetic electron transport. *Plant Physiology*, Vol. 65, No. 4, pp. 707-713, ISSN 0032-0889

Golbeck, J. H. & Warden, J. T. (1984). Interaction of linolenic acid with bound quinone molecules in photosystem II. Time-resolved optical and electron spin resonance studies. *Biochimica et Biophysica Acta -Bioenergetics*, Vol. 767, No. 2, pp. 263-271, ISSN 0005-2728

Gonzalez, V. M.; Kazimir, J.; Nimbal, C.; Weston, L. A. & Cheniae, G. M. (1997). Inhibition of a photosystem II electron transfer reaction by the natural product sorgoleone. *Journal of Agricultural and Food Chemistry*, Vol. 45, No. 4, (April 1997), pp. 1415-1421, ISSN 0021-8561

Gromet-Elhanan, Z. (1969). Two types of cyclic electron transport in isolated chloroplasts. In: *Progress in Photosynthesis Research 3*, H. Metzner, (Ed.), 1197-1202, Publ. sponsored by International Union of Biological Sciences, Tübingen, Germany

Hagmann, L. & Jüttner, F. (1996). Fischerellin A, a novel photosystem-II-inhibiting allelochemical of the cyanobacterium *Fischerella muscicola* with antifungal and herbicidal activity. *Tetrahedron Letters*, Vol. 37, No. 36, (September 1996), pp. 6539-6542, ISSN 0040-4039

Hansch, C. & Deutsch, E. W. (1966). The structure-activity relationship in amides inhibiting photosynthesis. *Biochimica et Biophysica Acta -Biophysics including Photosynthesis*, Vol. 112, No. 3, pp. 381-391, ISSN 0926-6585

Hermann, P.; Hladík, J. & Sofrová, D. (1988). Effect of detergents in thylakoid membranes of chloroplasts. *Photosynthetica*, Vol. 22, No. 3, pp. 411-422, ISSN 0300-3604

Hind, G. & Olson, J. M. (1966). Light-induced changes in cytochrome b6 in spinach chloroplasts. *Brookhaven Symposia in Biology*, No. 19, pp. 188-194, ISSN 0068-2799

Hoff, A. J. (1979). Application of ESR in photosynthesis. *Physics Reports*, Vol. 54, No. 2, pp. 75-200, ISSN 0370-1573

Honda, I.; Yoneyama, K.; Iwamura, H.; Konnai, M.; Takahashi, N. & Yoshida, S. (1990a). Structure-activity relationship of 3-nitro-2,4,6-trihydroxybenzamide derivatives in photosynthetic electron transport inhibition. *Agricultural Biology and Chemistry*, Vol. 54, No. 5, (May 1990), pp. 1227-1233, ISSN 0002-1369

Honda, I.; Yoneyama, K.; Konnai, M.; Takahashi, N. & Yoshida, S. (1990b). Synthesis of 3-nitro-2,4,6,-trihydroxybenzamide and thiobenzamide derivatives as highly potent inhibitors of photosynthetic electron transport. *Agricultural Biology and Chemistry*, Vol. 54, No. 5, (April 1990), pp. 1071-1072, ISSN 0002-1369

Izawa, S.; Connolly, T. N.; Winget, G. D. & Good, N. E. (1966). Inhibition and uncoupling of photophosphorylation in chloroplasts. *Brookhaven Symposia in Biology*, No.19, pp. 169-187, ISSN 0068-2799

Izawa, S. (1980). Acceptors and donors for chloroplast electron transport. In: *Methods in Enzymology*, P. Colowick & N. O. Kaplan, (Ed.), Vol. 69, Part C, 413-434, Academic Press, ISBN 0121819698, New York - London

Jewess, P. J.; Higgins, J.; Berry, K. J.; Moss, S. R.; Boogaard, A. B. & Khambay, B. P. S. (2002). Herbicidal action of 2-hydroxy-3-alkyl-1,4-naphthoquinones. *Pest Management Science*, Vol. 58, No. 3, (May 2002), pp. 234-242, ISSN 1526-498X

Kaniuga, Z.; Gemel, J. & Zablocka, B. (1986). Fatty acid induced release of manganese from chloroplasts. *Biochimica et Biophysica Acta-Bioenergetics*, Vol. 850, No. 3, (July 1986), pp. 473–482, ISSN 0005-2728

Katoh, S. & San Pietro, A. (1968). Comparative study of the inhibitory action of the oxygen-evolving system of various chemical and physical treatments of *Euglena* chloroplasts. *Archives of Biochemistry and Biophysics*, Vol. 128, No. 2, pp. 378-386, ISSN 0003-9861

Kráľová, K; Šeršeň, F. & Čižmárik, J. (1992a). Inhibitory effect of piperidinoethylesters of alkoxyphenylcarbamic acids on photosynthesis. *General Physiology and Biophysics*, Vol. 11, No. 3, (June 1992), pp. 261-267, ISSN 0231-5882

Kráľová, K.; Šeršeň, F. & Čižmárik, J. (1992b). Dimethylaminoethyl alkoxyphenylcarbamates as photosynthesis inhibitors. *Chemical Papers*, Vol. 46, No. 4, pp. 266-268, ISSN 0366-6352

Kráľová, K.; Šeršeň, F. & Csöllei, J. (1992c). Inhibitory effects of some esters of 2- and 3-substituted alkoxyphenylcarbamic acids on photosynthetic characteristics. *Biologia Plantarum*, Vol. 34, No. 3-4, pp. 253-258, ISSN 0006-3134

Kráľová, K.; Šeršeň, F. & Sidóová, E. (1992d). Photosynthesis inhibition produced by 2-alkylthio-6-R-benzothiazoles. *Chemical Papers*, Vol. 46, No. 5, pp. 348-350, ISSN 0366-6352

Kráľová, K.; Šeršeň, F.; Mitterhauszerová, Ľ.; Krempaská, E. & Devínsky, F. (1992e). Effect of surfactants on growth, chlorophyll content and Hill reaction activity. *Photosynthetica*, Vol. 26, No. 2, pp. 181-187, ISSN 0300-3604

Kráľová, K.; Šeršeň, F. & Sidóová, E. (1993). Effect of 2-alkylthio-6-aminobenzothiazoles and their 6-*N*-substituted derivatives on photosynthesis inhibition in spinach chloroplasts. *General Physiology and Biophysics*, Vol. 12, No. 5, (October 1993), pp. 421-427, ISSN 0231-5882

Kráľová, K. & Šeršeň, F. (1994). Long chain bisquaternary ammonium salts - effective inhibitors of photosynthesis. *Tenside Surfactants Detergents*, Vol. 31, No. 3, pp. 192-194, ISSN 0932-3414

Kráľová, K.; Loos, D. & Čižmárik, J. (1994a). Correlation between some lipophilicity characteristics of morpholinoethylesters of 2-, 3- and 4-alkoxysubstituted phenylcarbamic acids, their inhibitory activity in photosynthesizing organisms. *Photosynthetica*, Vol. 30, No. 1, pp. 155-159, ISSN 0300-3604

Kráľová, K.; Loos, D.; Šeršeň, F. & Čižmárik, J. (1994b). Correlation between physico-chemical properties of quaternary ammonium salts of heptacaine and their inhibitory activity against photosynthesizing organisms. *Biologia Plantarum*, Vol. 36, No. 2, pp. 313-316, ISSN 0006-3134

Kráľová, K.; Bujdáková, H. & Čižmárik, J. (1995a). Antifungal and antialgal activity of piperidinopropyl esters of alkoxy substituted phenylcarbamic acids. *Pharmazie*, Vol. 50, No. 6, (June 1995), pp. 440-441, ISSN 0031-7144

Kráľová, K.; Loos, D.; Devínsky, F. & Lacko, I. (1995b). Correlation between structure and biological activity of bis-quaternary isosters of 1,5-pentanediammonium dibromides. Algicidal activity and inhibition of photochemical activity of chloroplasts. *Chemical Papers*, Vol. 49, No. 1, pp. 39-42, ISSN 0366-6352

Kráľová, K.; Šeršeň, F.; Klimešová, V. & Waisser, K. (1997). Effect of 2-alkylthio-4-pyridinecarbothiamides on photosynthetic electron transport in spinach chloroplasts. *Collection of Czechoslovak Chemical Communications*, Vol. 62, No.3, (March 1997), pp. 516-520, ISSN 0010-0765

Kráľová, K.; Miletín, M. & Doležal, M. (2001). Inhibition of oxygen evolution rate in freshwater algae *Chlorella vulgaris* by some anilides of substituted pyridine-4-carboxylic acids. *Chemical Papers*, Vol. 55, No. 4, pp. 251-253, ISSN 0366-6352

Kráľová, K.; Šeršeň, F.; Devínsky, F. & Lacko, I. (2010). Photosynthesis-inhibiting effects of cationic biodegradable gemini surfactants. *Tenside Surfactants Detergents*, Vol 47, No. 5, (September-October 2010), pp. 288-293, ISSN 0932-3414

Krieger, A.; Rutherford, W. A.; Vass, I. & Hideg, E. (1998). Relationship between activity, D1 loss, and Mn binding in photoinhibition of photosystem II. *Biochemistry*, Vol. 37, No. 46, (November 1998), pp. 16262-16269, ISSN 0006-2960

Krogmann, D. W. & Jagendorf, A. T. (1959). Inhibition of the Hill reaction by fatty acids and metal chelating agents. *Archives of Biochemistry and Biophysics*, Vol. 80, No. 2, pp. 424-430, ISSN 0003-9861

Loach, P. A. (1980). Bacterial reaction center (RC) preparations. In: *Methods in Enymology*, S. P. Colowick & N. O. Kaplan, (Ed.), Vol. 69, Part C, 151-172, Academic Press, ISBN 0121819698, New York - London

McCarty, R. E. & Jagendorf, A. J. (1965). Chloroplast damage due to enzymatic hydrolysis of endogenous lipids. *Plant Physiology*, Vol. 40, No. 4, pp. 725-735, ISSN 0032-0889

Miletín, M.; Doležal, M.; Kráľová, K. & Šeršeň, F. (2001). Inhibition of oxygen evolution rate in spinach chloroplasts by some anilides of substituted pyridine-4-carboxylic acids. *Folia Pharmaceutica Universitatis Carolinae*, Vol. 26, pp. 27–32, ISSN 1210-9495

Mitterhauszerová, Ľ.; Kráľová, K.; Šeršeň, F.; Blanáriková, V. & Csöllei, J. (1991a). Effects of substituted aryloxyaminopropanols on photosynthesis and photosynthesizing organisms. *General Physiology and Biophysics*, Vol. 10, No. 3, (June 1991), pp. 309-319, ISSN 0231-5882

Mitterhauszerová, Ľ.; Kráľová, K.; Šeršeň, F.; Csöllei, J. & Krempaská, E. (1991b). Influence of dibasic local anaesthetics on photosynthesis, green algae and plants. [In Slovak]. *Ceskoslovenska. Farmacie,* Vol. 40, No. 6-7, pp. 209-213, ISSN 0009-0530

Molotkovsky, Y. G. & Zheskova, I. M. (1965). The influence of heating on the morphology and photochemical activity of isolated chloroplasts. *Biochemical and Biophysical Research Communications,* Vol. 20, No. 4, pp. 411-415, ISSN 0006-291X

Moreland, D. E. & Hill, K. L. (1959). Herbicide structure and activity, the action of alkyl *N*-phenylcarbamates on the photolytic activity of isolated chloroplasts. *Journal of Agricultural and Food Chemistry,* Vol. 7, No. 12, pp. 832–837, ISSN 0021-8561

Moreland, D. E. (1993). Research on biochemistry of herbicides - an historical overview. *Zeitschrift für Naturforschung,* Vol. 48c, No. 3, (March-April 1993), pp. 121-131, ISSN 0939-5075

Morimoto, M.; Shimomura, Y.; Mizuno, R. & Komai, K. (2001). Electron transport inhibitor in *Cyperus javanicus. Bioscience, Biotechmology and Biochemistry,* Vol. 65, No. 8, (August 2001), pp. 1849-1851, ISSN 0916-8451

Nakai, S.; Yamada, S. & Hosomi, M. (2005). Anti-cyanobacterial fatty acids released from *Myriophyllum spicatum. Hydrobiologia,* Vol. 543, No. 1, (July 2005), pp. 71-78, ISSN 0018-8158

Nakajima, Y.; Yoshida, S.; Inoue, Y.; Yoneyama, K. & Ono, T. A. (1995). Selective and specific degradation of the D1 protein induced by binding of a novel photosystem-II inhibitor to the Q_B binding site. *Biochimica et Biophysica Acta-Bioenergetics,* Vol. 1230, No. 1-2, (June 1995), pp. 38-44, ISSN 0005-2728

Nurit, F.; Gomes de Melo, E.; Ravanel, P. & Tissut, M. (1989). Specific inhibition of mitosis in cell suspension cultures by a *N*-phenylcarbamate series. *Pesticide Biochemistry and Physiology,* Vol. 35, No. 3, (November 1989), pp. 203-210, ISSN 0048-3575

Oettmeier, W.; Masson, K.; Fedtke, C.; Konze, J. & Schmidt, R. R. (1982). Effect of different photosystem II inhibitors on chloroplasts isolated from species either susceptible or resistant toward s-triazine herbicides. *Pesticide Biochemistry and Physiology,* Vol.18, No. 3, pp. 357-367, ISSN 0048-3575

Okamoto, T. & Katoh, S. (1977). Linolenic acid binding by chloroplast. *Plant and Cell Physiology,* Vol. 18, No. 3, pp. 539-550, ISSN 0032-0781

Okamoto, T.; Katoh, S. & Murakami, S. (1977). Effects of linolenic acid on spinach chloroplast structure. *Plant and Cell Physiology,* Vol. 18, No. 3, pp. 551-560, ISSN 0032-0781

Peters, J. S, & Chin, C. K. (2003). Inhibition of photosynthetic electron transport by palmitoleic acid is partially correlated to loss of thylakoid membrane proteins. *Plant Physiology and Biochemistry,* Vol. 41, No. 2, (February 2003), pp. 117–124, ISSN 0981-9428

Reifler, M. J.; Szalai, V. A.; Peterson, C. N. & Brudvig, G. W. (2001). Effects of tail-like substituents on the binding of competitive inhibitors to the Q_B site of photosystem II. *Journal of Molecular Recognition,* Vol. 14, No. 3, (May-June 2001), pp. 157-165, ISSN 0952-3499

Rimando, A. M.; Dayan, F. E.; Czarnota, M. A.; Weston, L. A. & Duke, S. O. (1998). A new photosystem II electron transfer inhibitor from *Sorghum bicolor. Journal of Natural Products,* Vol. 61, No. 7, (November 1998), pp. 927-930, ISSN 0163-3864

Rimando, A. M.; Dayan, F. E. & Streibig, J. C. (2003). PS II inhibitory activity of resorcinolic lipids from *Sorghum bicolor*. *Journal of Natural Products*, Vol. 66, No. 1, (January 2003), pp. 42-45, ISSN 0163-3864

Šeršeň, F.; Leitmanová, A.; Devínsky, F.; Lacko I. & Balgavý, P. (1989). A spin label study of perturbation effects of *N*-(1-methyldodecyl)-*N,N,N*-trimethylammonium bromide and *N*-(1-methyldodecyl)-*N,N*- dimethylamine oxide on model membranes prepared from *Escherichia coli*-isolated lipids. *General Physiology and Biophysics*, Vol. 8, No. 2, (April 1989), pp. 133-156, ISSN 0231-5882

Šeršeň, F.; Gabunia, G.; Krejčíová, E. & Kráľová, K. (1992). The relationship between lipophilicity of *N*-alkyl-*N,N*-dimethylamine oxides and their effects on the thylakoid membranes of chloroplasts. *Photosynthetica*, Vol. 26, No. 2, pp. 205-212, ISSN 0300-3604

Šeršeň, F.; Kráľová, K. & Sidóová, E. (1993). Inhibition of photosynthetic electron transport in spinach chloroplasts by 2-(alkoxycarbonylmethylthio)-6-aminobenzothiazoles. *Photosynthetica*, Vol. 29, No. 1, pp. 147-150, ISSN 0300-3604

Šeršeň, F. & Devínsky, F. (1994). Stimulating effect of quaternary ammonium salts upon the photosynthetic electron transport. *Photosynthetica*, Vol. 30, No. 1, pp. 151-154, ISSN 0300-3604

Šeršeň, F. & Lacko, I. (1995). Stimulation of photosynthetic electron transport by quaternary ammonium salts. *Photosynthetica*, Vol. 31, No. 1, pp. 153-156, ISSN 0300-3604

Šeršeň, F. & Kráľová, K. (1996). Concentration-dependent inhibitory and stimulating effects of amphiphilic ammonium salts upon photosynthetic activity of spinach chloroplasts. *General Physiology and Biophysics*, Vol. 15, No. 1, (February 1996), pp. 27-36, ISSN 0231-5882

Šeršeň, F.; Kráľová, K. & Macho, V. (2000). New findings about the inhibitory action of phenylcarbamates and phenylthiocarbamates on photosynthetic apparatus. *Pesticide Biochemistry and Physiology*, Vol. 68, No. 2, (October 2000), pp. 113-118, ISSN 0048-3575

Shipman, L. L. (1981). Theoretical study of the binding site and mode of action for photosystem II herbicides. *Journal of Theoretical Biology*, Vol. 90, No. 1, pp. 123–148, ISSN 0022-5193

Siegenthaler, P. A. (1974). Inhibition of photosystem II electron transport by fatty acids and restoration of its activity by Mn^{2+}. *FEBS Letters*, Vol. 39, No. 3, pp. 337-340, ISSN 0014-5793

Siegenthaler, P. A. & Packer, L. (1965). Light-dependent volume changes and reactions in chloroplasts. I. Action of alkenylsuccinic acids and phenylmercuric acetate and possible relation to mechanisms of stomatal control. *Plant Physiology*, Vol. 40, No. 5, pp. 785-791, ISSN 0032-0889

Sidóová, E.; Kráľová, K. & Loos, D. (1998). Synthesis of 2-(6-acetamidobenzothiazolethio) acetic acid esters as photosynthesis inhibitors. *Molecules*, Vol. 3, No. 4, (April 1998), pp. 135-140, ISSN 1420-3049

Sidóová, E.; Kráľová, K. & Loos, D. (1999). 3-(2-Alkylsulfanyl-6-benzothiazolylamino-methyl)-2-benzothiazolethiones. *Molecules*, Vol. 4, No. 3, pp. 73-80, ISSN 1420-3049

Srivastava, A; Jüttner, F. & Strasser, R. J. (1998). Action of the allelochemical, fischerellin A, on photosystem II. *Biochimica et Biophysica Acta -Bioenergetics*, Vol. 1364, No. 3, (May 1998), pp. 326-336, ISSN 0005-2728

Vernotte, C.; Solis, C.; Moya, I.; Maison, B.; Briantais, J. M.; Arrio, B. & Johannin, G. (1983). Multiple effects of linolenic acid addition to pea thylakoids. *Biochimica et Biophysica Acta-Bioenergetics*, Vol. 725, No. 2, pp. 376-383, ISSN 0005-2728

Wang, X. L.; Gong, L. G.; Liang, S. K.; Han, X. R.; Zhu, C. J. & Li, Y. B. (2005). Algicidal activity of rhamnolipid biosurfactants produced by *Pseudomonas aeruginosa*. *Harmful Algae*, Vol. 4, No. 2, pp. 433-443, ISSN 1568-9883

Warncke, K.; Gunner, M. R.; Braun, B. S.; Gu, L. Q.; Yu, C. A.; Bruce, J. M. & Dutton, P. L. (1994). Influence of hydrocarbon tail structure on quinone binding and electron-transfer performance at the Q_A and Q_B sites of the photosynthetic reaction-center protein. *Biochemistry*, Vol. 33, No. 25, (June 1994), pp. 7830-7841, ISSN 0006-2960

Warden, J. T. & Csatorday, K. (1987). On mechanism of linoleic acid inhibition in photosystem II. *Biochimica et Biophysica Acta-Bioenergetics*, Vol. 890, No. 2, (February 1987), pp. 215-223, ISSN 0005-2728

Whitmarsh, J. (1998). Electron transport and energy transduction. In: *Photosynthesis: A Comprehensive Treatiseed*, A. S. Raghavendra, (Ed.), 87-110, Cambridge University Press, ISBN 0-521-57000-X, Cambridge

Yoneyama, K.; Konai, M.; Takematsu, T.; Iwamura H.; Asami T.; Takahashi, N. & Yoshida, S. (1989a). Photosynthetic electron transport inhibition by 3-acyl-2,4,6-trihydroxybenzamide derivatives. *Agricultural and Biological Chemistry*, Vol. 53, No. 7, (July 1989), pp. 1953-1959, ISSN 0002-1369

Yoneyama, K.; Konai, M.; Takematsu, T.; Asami, T.; Takahashi, N. & Yoshida, S. (1989b). Inhibition of photosynthesis by 3-acyl-2,4,6-trihydroxythiobenzamide derivatives. *Agricultural and Biological Chemistry*, Vol. 53, No. 8, (August 1989), pp. 2281-2282, ISSN 0002-1369

9

Leaf Photosynthetic Responses to Thinning

Qingmin Han
Forestry and Forest Products Research Institute
Japan

1. Introduction

Plants, as sessile organisms living in a changing light environment, exhibit a remarkable capacity to adjust their morphology and physiology to a particular set of light conditions by acclimation or, more broadly, phenotypic plasticity (Oguchi et al., 2005). Many studies have documented remarkable light-driven structural and functional modifications, with responses occurring from the level of the chloroplast to the whole plant. For instance, plants are able to adjust leaf area per unit biomass invested in leaves (i.e., specific leaf area) by altering leaf thickness (Niinemets, 1999). A linear relationship has been observed between growth irradiance and the reciprocal of specific leaf area between species, within species grown in different habitats and within individual tree crowns (Bond et al., 1999; Han et al., 2003). Photosynthetic capacity, measured in the form of either light saturated photosynthetic rate or the maximum rate of carboxylation, also varies according to growth irradiance.

Nitrogen, another key resource involved in the process of carbon fixation, plays an important role in the dynamics of the leaf canopy (Hikosaka, 2004; 2005). A linear relationship between photosynthetic capacity and leaf nitrogen concentration has been found in various species and canopies (Bond et al., 1999; Brooks et al., 1996; Han et al., 2004; Wilson et al., 2000); this is because of the large amount of nitrogen in the photosynthetic machinery (Evans & Seemann, 1989). In addition, fractional nitrogen within the leaf photosynthetic apparatus responds to changes in light and nitrogen availability. For example, the proportion of total leaf nitrogen partitioned into light harvesting proteins has been found to be higher in the lower crowns of deciduous broadleaf trees (Koike et al., 2001). Moreover, the proportion of total leaf nitrogen partitioned into ribulose-1,5-bisphosphate carboxylase/oxygenase (a key enzyme for carbon fixation) has been found to be higher at increased light levels in *Picea abies* (L.) Karst. (Grassi & Bagnaresi, 2001). These structural and functional responses enhance light capture, resource use and the photosynthetic efficiency of the whole tree crown, thus increasing plant performance and productivity.

Thinning, a common forest management practice used to reduce stand density, has been employed to enhance wood quality, for successive harvesting and to increase stand productivity (Cutini, 2001; Fujimori, 2001; Muñoz et al., 2008; Skovsgaard, 2009). Although the basic theory of the relationship between stand density and plant growth was developed in the early 1960s, the physiological mechanisms behind this traditional forest management practice are not well understood. At present, only 14 references with the words "thinning" and "photosynthesis" or "physiology" in the title can be found in the ISI Web of Science.

Species	Site location	DBH (cm)	Thinning treatments	Gas exchange measured after thinning	Source
Chamaecyparis obtusa	36°3′N, 140°7′E	6.5	From 3000 to 1500 trees ha^{-1} at 10 years old	1st growing season	Han et al 2006
				2nd to 4th growing season	Han and Chiba 2009
Pinus taeda	31°11′N, 92°41′E	na	From 2990 to 731 trees ha^{-1} at 8 years old	5th growing season	Gravatt et al 1997
				6th growing season	Tang et at 1999
				8th growing season	Tang et al 2003
	36°78′N, 80°08′W	14.2	From basal area 16.8 to 9.4 m^2 ha^{-1} at 8 years old	1st to 2nd growing season	Ginn et al 1991
				3rd to 7th growing season	Peterson et al 1997
Pinus ponderosa	35°15′N, 111°42′W	na	Basal area thinned from 33.0, 38.0 to 18.0, 22.0 m^2 ha^{-1} at 250 years old	1st to 2nd growing season	Skov et al 2004
	46°5′N, 114°15′W	na	From 420 to 280 trees ha^{-1} at 70 years old	8th to 9th growing season	Sala et al 2005
	44°25′-30′N, 121°37′-40′W	58.4 - 79.2	Basal area thinned from 31.4-60.0 to 8.4-12.2 m^2 ha^{-1} at 250 years old	7th to 15th growing season	McDowell et al 2003
	35°16′N, 111°44′W	40.9, 30.5, 26.7, 10.7	Basel area maintained to 6.9, 18.4, 27.6, 78.2 m^2 ha^{-1} by thinning for 32 years commencing at 43 years old	32nd growing season	Kolb et al 1998
Eucalyptus nitens	43°21′S, 146°54′E	na	From 1430 to 250 trees ha^{-1} at 8 years old	2nd to 3rd growing season	Medhurst and Beadle 2005
Betula papyrifera	na	na	From 11000-31000 to 400, 1000 and 3000 stems ha^{-1} at 9-13 years old	2nd to 3rd growing season	Wang et al 1995
Juglans nigra	40°19′N, 86°42′W	23.1	From 612 to 74 trees ha^{-1} at 19 years old	1st to 2nd growing season	Gauthier and Jacobs 2009

Table 1. List of studies on the effect of thinning on photosynthesis

However, global warming has heightened awareness of the importance of estimating the amount of carbon that is fixed by forests and, in addition, separating the effects on carbon fixation caused by silvicultural manipulation from those due to climate change. The accurate estimation of carbon gain requires an understanding of the processes and the allocation of photosynthates, as well as the variation in canopy physiology in response to environmental changes and forest management practices (Chiba, 1998; Johnsen et al., 2000; Simioni et al., 2008). In this chapter, changes in resource availability and leaf photosynthetic response after thinning are summarized, based on information from the limited number of reports available (Table 1). The focus is on the effects of thinning on (1) the availability of light, nitrogen and water, and (2) photosynthetic acclimation to changes in resource availability.

2. Changes in resource availability

Stand productivity is closely related to resource availability, especially when the latter is limited. Total leaf area and its distribution within a tree crown, the unit of photosynthesis at the individual level, are important determinants of canopy photosynthesis (Monsi & Saeki, 1953). The optimum leaf area index (LAI, total leaf area per ground area), when canopy photosynthesis is maximized, is closely related to light attenuation within the canopy and nitrogen distribution between leaves. When the stomata open to allow carbon dioxide (CO_2) to diffuse into the leaf, water is lost through transpiration. Water supply and demand interactions regulate conductance of CO_2 via the stomata and thus affect photosynthesis. Therefore, soil water availability is as important a resource as light and nutrients in relation to stand productivity. Thinning improves the availability of resources such as light, mineral nutrients and water available to the retained trees.

2.1 Light

Thinning obviously increases light intensity. The magnitude of light increase varies between canopy positions and depends on the stand density before thinning, thinning intensity, tree size/age and species. For example, in an eight-year-old *Pinus taeda* L. stand (see Table 1 for detailed site information), photosynthetic photon flux density (PPFD) increased by 28-52% as a result of thinning (Tang et al., 1999). In most studies, PPFDs are measured at various points down through the canopy or individual tree crown. As trees grow taller, a wide range of horizontal variation in leaf PPFD develops from the outer to the inner crown at the same vertical height (Brooks et al., 1996; Han et al., 2003). Therefore, light improvement after thinning also varies with respect to horizontal crown position, especially in shade tolerant species with a large proportion of leaves along first-order branches. For example, relative PPFD in the lower crown in a completely closed 10-year-old *Chamaecyparis obtusa* (Sieb. Et Zucc.) Endl. stand (Table 1, Figure 1), varied from 0.05-0.24 with average of 0.11 in the thinned stand, while the proportion was 0.03 in the unthinned control stand (Han et al., 2006). As shown in Figure 1, the improved rPPFD in all crown positions decreased progressively with height growth and the process of crown re-closure in the thinned stand, although values were still higher than those at the same heights in the control stand three years after the thinning operation.

2.2 Nutrients

In most regions, nutrients limit forest productivity. Thinning influences the nutrient cycle and thus has a direct impact on soil fertility. Therefore, thinning is an important management practice for both the production of high-quality timber and the conservation of soil fertility. For example, Blanco et al (2008) found that thinning seems to affect nutrient returns mainly by reducing aboveground biomass and litterfall in *Pinus sylvestris* L. forests in northern Spain. Inagaki et al (2008) reported that net nitrogen mineralization in the surface soil and fresh-leaf nitrogen concentration in thinned plots was significantly greater than in control plots in two *Chamaecyparis obtusa* plantations in southern Japan. The increase in nitrogen mineralization rate after thinning is considered to be related to soil microbial activity, caused by changes in soil temperature and soil water content (see next section), and increases in the pool of soil inorganic nitrogen and other nutrients due to reduced plant uptake. In addition, Inagaki et al considered increases in the input of fresh leaves after thinning because of the high frequency of severe typhoons in the study region. Moreover, both studies were conducted one to two years after thinning. Such short-term effects are usually temporary, and the ammonium or nitrate pool size, net mineralization, microbial activity and potentially mineralizable nitrogen usually return to pretreatment values within only a few years of thinning (Sala et al., 2005).

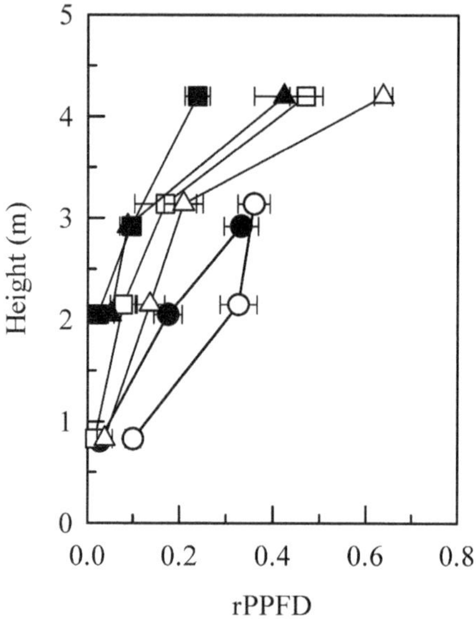

Fig. 1. Comparison of the relative photosynthetic photon flux density (rPPFD, monthly averaged value in October) in the crowns of thinned (open symbols) and non-thinned control (filled symbols) stands of *Chamaecyparis obtusa* in 2004 (circles), 2005 (triangles), and 2006 (squares). Values are means ± SE (n = 12 to 24 photodiodes). With kind permission from Springer Science+Business Media: <*Journal of Forest Research*, Leaf photosynthetic responses and related nitrogen changes associated with crown reclosure after thinning in a young *Chamaecyparis obtusa* stand, 14, 2009, 349-357, Q. Han & Y. Chiba, Fig.1 >.

Thinning also improves the nutrient supply indirectly due to reduced competition between the retained trees. However, thinning may result in an increase in understory biomass, which also competes for nutrients. Consequently, nutrient resources for retained trees may be more limited than expected. For example, in a study of *Pinus ponderosa* var. *scopulorum* Engelm. stands (Table 1), basal area was maintained at 6.9, 18.4, 27.6 and 78.2 m^2 ha^{-1} by frequent thinnings for 32 years (Kolb et al., 1998). Leaf nitrogen concentration was greatest in trees in the intermediate basal area plots, probably because the total competition for nitrogen between a tree and its neighbors, herbaceous plants and shrubs was less than the total competition for nitrogen at either higher or lower basal areas.

2.3 Water

Thinning leads to an increase in transpiration by the retained trees because a larger portion of the crown is exposed to the sun, leading to increased air temperature and a greater vapor pressure deficit (Jiménez et al., 2008; Lagergren et al., 2008; Morikawa et al., 1986). In contrast, stand transpiration decreases because of the reduction in number of trees per ground area (Jiménez et al., 2008; Morikawa et al., 1986). Less water consumption at the stand level results in increased soil water availability (Bréda et al., 1995; Inagaki et al., 2008; Jiménez et al., 2008; Lagergren et al., 2008). The improvement in tree water status is more significant at dry sites, in dry climates, and for species that are sensitive to soil water availability (Moreno & Cubera, 2008).

Soil water availability is affected by thinning not only through changes in transpiration, but also through canopy interception, soil evaporation and stem throughfall. A reduction in leaf area reduces the capacity of the canopy to store intercepted water and, therefore, the evaporation of intercepted water is reduced (Bréda et al., 1995; Stogsdili et al., 1992). The effect on throughfall could have a greater impact, since an increase in throughfall will increase the amount of water that infiltrates into the soil. For example, soil water content increased by 2.9-4.1% in a 50-year-old *Pinus sylverstris* forest in central Sweden after thinning, despite an increase in transpiration; this was interpreted as being the result of reduced interception and increased throughfall (Lagergren et al., 2008).

It is worth noting that soil water behaves differently at different depths and the distribution of the rooting zone is species-specific. Ignoring the link between these two issues would lead to different conclusions when exploring the effect of thinning on transpiration and soil water availability. For example, Sala et al (2005) reported increased predawn leaf water potential in thinned *Pinus ponderosa* Dougl. Ex. Laws stands, but no significant difference in soil water content was found at a depth of 15 or 45 cm, indicating that soil water measurements may not accurately reflect soil water potentials in rooting zones.

3. Photosynthesis acclimation

Photosynthetic responses and acclimation to changes in resource availability after thinning occur at the single leaf as well as the whole crown level. The occurrence, duration and magnitude of photosynthetic acclimation are apparently related to stand density and thinning intensity, site quality, and tree species, age, and vigor. In this section, the photosynthetic response to thinning is summarized with respect to three types of tree: coniferous species, evergreen broadleaf species, and deciduous broadleaf species. The site location, species and thinning information for the research examined are summarized in Table 1.

3.1 Coniferous species

Photosynthetic rates and stomatal conductance were found to increase in the lower crown in 8-year-old *Pinus taeda* trees from the first to the third growing season after thinning (Ginn et al., 1991; Peterson et al., 1997). However, this thinning effect on needle physiological parameters was not observed from the fourth growing season onwards (Peterson et al.,

1997). Thinning was not found to change leaf nitrogen concentration. The enhanced photosynthetic rates were attributed to increased light in the lower crown because the improved light availability disappeared in the fourth growing season. In contrast, at another site with the same species and with trees of the same age, this enhancement in photosynthesis persisted into the sixth and eighth growing seasons (Tang et al., 1999; Tang et al., 2003). The different durations of thinning effect between the two studies were due to different thinning intensities: the former site was thinned by 50% whereas the latter was subjected to 75% thinning. Thus, thinning intensity appears to be a key factor in determining the length of time that the impact of thinning on crown physiology persists. Like *Pinus taeda*, experimental maintenance of basal area by frequent thinnings in 43-year-old *Pinus ponderosa* stands over 32 years demonstrated that photosynthetic rate and predawn leaf water potential increased as basal area decreased from 78.2 to 6.9 m^2 ha^{-1} (Kolb et al., 1998). Similar results were observed 8 years after thinning in a 70-year old *Pinus ponderosa* forest (Sala et al., 2005), and 7-15 years after thinning in stands containing 250-year-old trees of the same species (Mcdowell et al., 2003). At these two sites, photosynthetic increase was accompanied by higher predawn leaf water potential and stomatal conductance but more limited changes in leaf nitrogen concentration, indicating that changes in water availability were the dominant control over gas exchange in these old-growth forests. Furthermore, predawn leaf water potential, canopy gas exchange, and subsequent growth remained higher as long as leaf area index at stand level remained lower than the pretreatment value, indicating that forest managers can manipulate old-growth stands effectively on an infrequent basis. Moreover, ontogeny alters a plethora of plant physiological and structural traits in coping with limited light, including allometry and crown architecture. Foliage photosynthetic capacities and specific leaf areas invariably decrease with increasing tree size and age (Bond, 2000). Tree height-related enhancement in water use efficiency occurs at the expense of nitrogen due to greater leaf hydraulic limitation (Han, 2011). These ontogenic adjustments suggest that tree size and age modify the effect of thinning on photosynthetic response (Skov et al., 2004).

Although photosynthetic rate was enhanced after thinning in all the species mentioned above, none of them were found to exhibit a significant increase in leaf nitrogen concentration or a decrease in specific leaf area. The enhancements in photosynthetic rate after thinning were mainly the result of changes in light and water availabilities. It is not clear where photosynthetic capacity changes after thinning. Only recently, in a 10-year-old *Chamaecyparis obtusa* stand (Table 1), the effect of thinning on photosynthetic capacity was intensively studied over a period of four years in combination with variations in nitrogen concentration at different locations within the crown (Han et al., 2006; Han & Chiba, 2009).

The maximum rate of carboxylation exhibited significant increases in the middle and lower crown three months after thinning (Figure 2a). Although thinning did not affect leaf nitrogen concentration per unit area at any of the crown positions (Figure 2b), the thinned stand exhibited an increase in nitrogen partitioned to ribulose-1,5-bisphosphate carboxylase/oxygenase in the lower and middle crowns relative to the control stands three and five months after thinning (Figure 2c). These results indicate that nitrogen reallocation occurs within the leaves' photosynthetic apparatus: from light harvesting to carboxylation in response to improved light availability. Specific leaf area increased significantly in all crown positions five months after thinning (Figure 2d). This was a consequence of a decrease in leaf dry mass due to rapid shoot growth, which may relate to the species' unique scaly leaves that can quickly switch between quiescence and growth depending on the sink-source relationship. These results indicate that the increase in photosynthesis in the first year after thinning mainly occurred as a result of redistribution of nitrogen within but not between leaves.

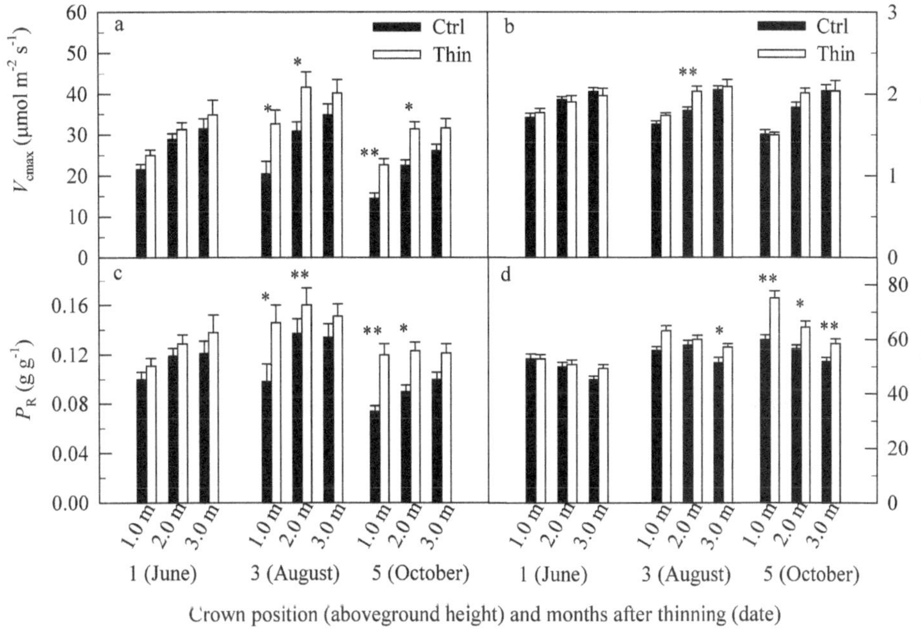

Fig. 2. Comparison of (a) the maximum rate of carboxylation (V_{cmax}), (b) leaf nitrogen concentration (N_a), (c) the proportion of leaf nitrogen in ribulose-1,5-bisphosphate carboxylase/oxygenase per total leaf nitrogen (P_R) and (d) specific leaf area (SLA) in the lower (aboveground height, AGH: 1.0m), middle (AGH: 2.0 m) and upper (AGH: 3.0 m) crown of *Chamaecyparis obtusa* between the thinned (open bars) and unthinned control (filled bars) stands one, three and five months after thinning. Significant differences between the respective values of the two groups: * $p < 0.05$; ** $p < 0.01$. Each value is the mean ± SE (n = 12-24 leaves). With kind permission from Springer Science+Business Media: <*Photosynthetica*, Acclimation to irradiance of leaf photosynthesis and associated nitrogen reallocation in photosynthetic apparatus in the year following thinning of a young stand of *Chamaecyparis obtusa*, 44, 2006, 523-529, Q. Han, M. Araki & Y. Chiba, Figs.2-4 >.

In the second year after thinning, nitrogen concentration in newly developed leaves was found to be generally higher in a thinned stand than in an un-thinned control (Han & Chiba, 2009). Thus, the redistribution of nitrogen between leaves enhanced photosynthetic capacity even in the upper crown in the thinned stand of *Chamaecyparis obtusa*. Furthermore, thinning affected the slope of the linear relationship between nitrogen concentration and photosynthetic capacity (Figure 3, Table 2). In the first year, the slope of the linear regression line representing the relationship between photosynthetic capacity and nitrogen concentration in the thinned stand was shallower than in the control stand, reflecting enhanced photosynthetic capacity in the lower crowns (Figs. 2a and 3a, Table 2). In the second year, however, the slope in the thinned stand was higher than in the control stand, reflecting the enhanced photosynthetic capacity in the upper crown (Fig. 3b, Table 2). As is the case with the response to light conditions, redistribution of nitrogen between leaves appears to act over a longer time scale than reallocation within leaves (Brooks et al. 1996; Han et al. 2006). These results suggest that photosynthetic acclimation after thinning involves different leaf nitrogen reallocation/redistribution mechanisms driven by light changes during the process of crown re-closure. These results indicate the importance of considering the time since thinning when evaluating acclimation of photosynthesis and associated nitrogen relations with respect to silvicultural manipulation.

In the third year, the effect of thinning on the slope of the linear regression line representing the relationship between photosynthetic capacity and leaf nitrogen concentration became negligible again (Fig. 3c and Table 2), although both leaf nitrogen concentration and photosynthetic capacity were still higher in the thinned stand than in the control stand. The higher nitrogen concentration was due to improved nitrogen availability for the retained trees. Based on the leaf nitrogen difference and total leaf biomass difference in the two stands, the authors concluded that nitrogen redistribution driven by light changes, along with the nitrogen supply and demand balance in conjunction with the process of crown re-closure, resulted in the relations between leaf nitrogen and photosynthetic capacity in the thinned stand converging with those in the control stand, three years after thinning.

3.2 Evergreen broadleaf species

Eucalyptus species, in fast-growing commercial plantations, play an important role worldwide in satisfying both an increasing demand for wood and the provision of environmental services. Practices to manage growth have been intensively studied, but little research has examined the associated physiological responses. The only study to date was conducted in the Creekton plantation in southern Tasmania (43° 21′S, 146° 54′E), which was thinned when it was eight years old from a density of 1254 to 250 trees ha^{-1} (Medhurst & Beadle, 2005). Thinning enhanced light-saturated photosynthetic rates throughout the crown and the greatest increases were recorded in the lower and middle crowns. Thinning increased leaf nitrogen and phosphorus concentrations because of a significant decrease in specific leaf area after thinning. Photosynthetic rate was positively related to leaf nitrogen and phosphorus concentrations per unit area. All these changes were mainly driven by increased light levels after thinning, as found for the coniferous species described above.

3.3 Deciduous broadleaf species

Enhanced photosynthesis has been detected in the second growing season after thinning in two deciduous species (Table 1) (Gauthier & Jacobs, 2009; Wang et al., 1995). Like *Eucalyptus*, photosynthetic acclimation in deciduous species has also been attributed to increases in light availability and nitrogen concentration as well as a decrease in specific leaf area. In comparison with the second growing season after thinning, decreases in photosynthetic rates in *Betula papyrifera* were found in the third growing season, indicating a gradual reduction of thinning effect due to expansion of the trees' crowns. In contrast to *Chamaecyparis obtusa*, no photosynthetic response was found immediately after thinning in *Juglans nigra* in which thinning was carried out approximately 80 days after bud burst. There are two types of leaf development in deciduous species: indeterminant and determinant (Koike, 1990). *Juglans nigra* exhibits the later type. In mature leaves, photosynthetic

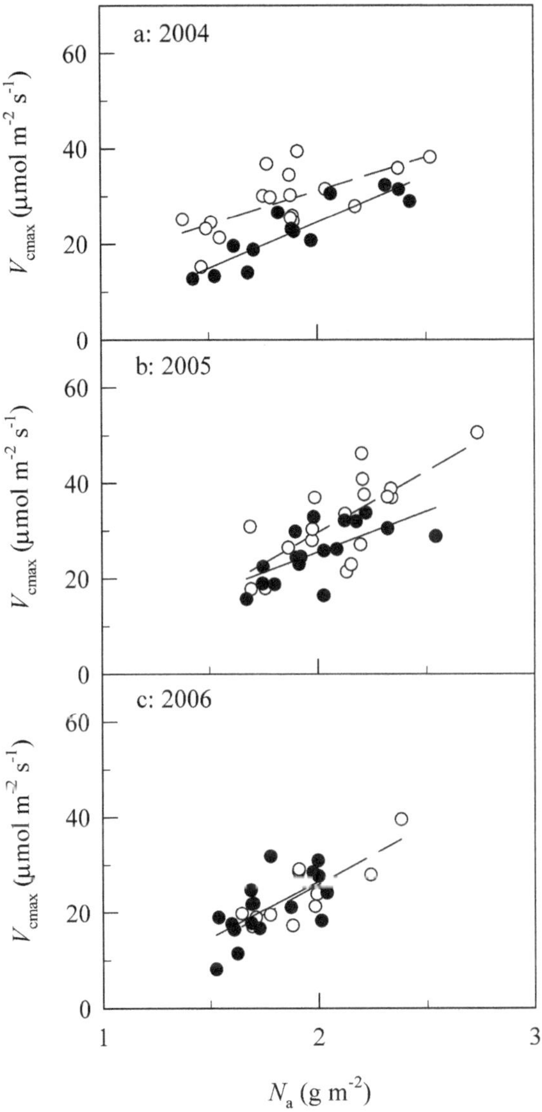

Fig. 3. The relationship between leaf nitrogen concentration (N_a) and the maximum rate of carboxylation (V_{cmax}) within a *Chamaecyparis obtusa* crown in thinned (open circles and dashed line) and unthinned control (closed circles and solid line) stands after thinning in May 2004. Regression results are presented in Table 2. Each value is the mean ± SE (n = 3-6 leaves) measured in October. With kind permission from Springer Science+Business Media: <*Journal of Forest Research*, Leaf photosynthetic responses and related nitrogen changes associated with crown reclosure after thinning in a young *Chamaecyparis obtusa* stand, 14, 2009, 349-357, Q. Han & Y. Chiba, Fig.7 >.

acclimation to sudden increases in light intensity differs between species because of differences in leaf anatomy (Oguchi et al., 2005). In species that have empty spaces along the mesophyll cell surface that are not occupied by chloroplasts or other organelles (i.e. *Betula ermanii* Cham.), the photosynthetic capacity of mature leaves developed in low light acclimates rapidly to high light levels. On the other hand, in species where little of the mesophyll cell surface is unoccupied by chloroplasts (i.e. *Fagus crenata*), the photosynthetic capacity of mature leaves is unable to respond to light increases. These species-specific physiological traits are important when selecting appropriate times to conduct thinning in commercial forestry operations.

Year	Thinned stand			Control stand		
	slope	intercept	r^2	slope	intercept	r^2
2004	14.11[a]	2.92	0.45	19.34[b]	-14.06	0.80
2005	25.74[a]	-21.91	0.53	17.02[b]	-8.50	0.43
2006	24.91[a]	-23.91	0.68	23.78[a]	-21.02	0.43

Table 2. Linear regression results for the data shown in Figure 3 describing inter-annual variations in the slope of the leaf nitrogen (N_a; g m^{-2}) to the maximum rate of carboxylation (V_{cmax}; µmol m^{-2} s^{-1}) in both the thinned and control stands after thinning in 2004. All slopes were significant ($P < 0.01$), but none of the non-zero intercepts were significant ($P > 0.05$). Different letters adjacent to each slope indicate a significant difference between the two stands in the respective year ($P < 0.05$, ANCOVA test).

4. Conclusion

Thinning of closed-canopy stands dramatically alters the light regime experienced by the retained trees. Enhanced leaf photosynthesis after thinning can be attributed to an increase in light availability or improved water and/or nutrient supply. The magnitude of the increase in leaf photosynthesis is species-specific and site-specific, and is also dependent on the thinning method employed. Species-specific physiological traits are important when selecting appropriate timings and intensities of thinning. Enhancements in canopy gas exchange and subsequent growth continue as long as the leaf area index at the stand level remains lower than the pretreatment value. Enhanced photosynthesis as a result of thinning virtually disappears after the canopy re-closes, providing further evidence that light is crucial for driving thinning responses. All these factors should be taken into account by land owners and managers when considering stand management and thinning regimes.

5. Acknowledgements

The author wishes to thank Dr. Daisuke Kabeya for his valuable comments on an earlier version of the manuscript. This work was supported, in part, by a research grant from the Forestry and Forest Products Research Institute (no. A1P01).

6. References

Bond, B. J.; Farnsworth, B. T.; Coulombe, R. A. & Winner, W. E. (1999). Foliage physiology and biochemistry in response to light gradients in conifers with varying shade tolerance. *Oecologia*, Vol.120, No.2, pp. 183-192, ISSN 0029-8549

Bond, B. J. (2000). Age-related changes in photosynthesis of woody plants. *Trends in Plant Science*, Vol.5, No.8, pp. 349-353, ISSN 1360-1385

Bréda, N.; Granier, A. & Aussenac, G. (1995). Effects of thinning on soil and tree water relations, transpiration and growth in an oak forest (*Quercus petraea* (Matt.) Liebl.). *Tree Physiology*, Vol.15, pp. 295-306, ISSN 0829-318X

Brooks, J. R.; Sprugel, D. G. & Hinckley, T. M. (1996). The effects of light acclimation during and after foliage expansion on photosynthesis of *Abies amabilis* foliage within the canopy. *Oecologia*, Vol.107, pp. 21-32, ISSN 0029-8549

Chiba, Y. (1998). Simulation of CO_2 budget and ecological implications of sugi (*Cryptomeria japonica*) man-made forests in Japan. *Ecological Modelling*, Vol.111, pp. 269-281, ISSN 0304-3800

Cutini, A. (2001). New management options in chestnut coppices: an evaluation on ecological bases. *Forest Ecology and Management*, Vol.141, No.3, pp. 165-174, ISSN 0378-1127

Evans, J. R. & Seemann, J. R. (1989). The allocation of protein nitrogen in the photosynthetic apparatus: costs, consequences, and control, In: *Photosynthesis*, W.R. Briggs, (Ed.), 183-205, Wiley-Liss, ISBN 0845118072, Stanford

Fujimori, T. (2001). *Ecological and silvicultural strategies for sustainable forest management*, Elsevier, ISBN 0-444-50534-2, Amsterdam, London, New York, Oxford, Paris, Shannon, Tokyo

Gauthier, M.-M. & Jacobs, D. F. (2009). Short-term physiological responses of black walnut (*Juglans nigra* L.) to plantation thinning. *Forest Science*, Vol.55, No.3, pp. 221-229, ISSN 0015-749X

Ginn, S. E.; Seiler, J. R.; Cazell, B. H. & Kreh, R. E. (1991). Physiological and growth responses of eight-year-old loblolly pine stands to thinning. *Forest Science*, Vol.37, No.4, pp. 1030-1040, ISSN 0015-749X

Grassi, G. & Bagnaresi, U. (2001). Foliar morphological and physiological plasticity in *Picea abies* and *Abies alba* saplings along a natural light gradient. *Tree Physiology*, Vol.21, No.12-13, pp. 959-967, ISSN 0829-318X

Han, Q.; Kawasaki, T.; Katahata, S.; Mukai, Y. & Chiba, Y. (2003). Horizontal and vertical variations in photosynthetic capacity in a *Pinus densiflora* crown in relation to leaf nitrogen allocation and acclimation to irradiance. *Tree Physiology*, Vol.23, No.12, pp. 851-857, ISSN 0829-318X

Han, Q.; Kawasaki, T.; Nakano, T. & Chiba, Y. (2004). Spatial and seasonal variability of temperature responses of biochemical photosynthesis parameters and leaf nitrogen content within a *Pinus densiflora* crown. *Tree Physiology*, Vol.24, No.7, pp. 737-744, ISSN 0829-318X

Han, Q.; Araki, M. & Chiba, Y. (2006). Acclimation to irradiance of leaf photosynthesis and associated nitrogen reallocation in photosynthetic apparatus in the year following thinning of a young stand of *Chamaecyparis obtusa*. *Photosynthetica*, Vol.44, No.4, pp. 523-529, ISSN 0300-3604

Han, Q. & Chiba, Y. (2009). Leaf photosynthetic responses and related nitrogen changes associated with crown reclosure after thinning in a young *Chamaecyparis obtusa* stand. *Journal of Forest Research*, Vol.14, No.6, pp. 349-357, ISSN 1341-6979

Han, Q. (2011). Height-related decreases in mesophyll conductance, leaf photosynthesis and compensating adjustments associated with leaf nitrogen concentrations in *Pinus densiflora*. *Tree Physiology*, Vol.31, No.9, pp. 976-984, ISSN 0829-318X

Hikosaka, K. (2004). Interspecific difference in the photosynthesis-nitrogen relationship: patterns, physiological causes, and ecological importance. *Journal of Plant Research*, Vol.117, No.6, pp. 481-494, ISSN 0918-9440

Hikosaka, K. (2005). Leaf canopy as a dynamic system: ecophysiology and optimality in leaf turnover. *Annals of Botany*, Vol.95, No.3, pp. 521-533, ISSN 0305-7364

Inagaki, Y.; Kuramoto, S.; Torii, A.; Shinomiya, Y. & Fukata, H. (2008). Effects of thinning on leaf-fall and leaf-litter nitrogen concentration in hinoki cypress (*Chamaecyparis obtusa* Endlicher) plantation stands in Japan. *Forest Ecology and Management*, Vol.255, No.5-6, pp. 1859-1867, ISSN 0378-1127

Jiménez, E.; Vega, J. A.; Pérez-Gorostiaga, P.; Cuiñas, P.; Fonturbel, T.; Fernández, C.; Madrigal, J.; Hernando, C. & Guijarro, M. (2008). Effects of pre-commercial thinning on transpiration in young post-fire maritime pine stands. *Forestry*, Vol.81, No.4, pp. 543-557, ISSN 0015-752X

Johnsen, K.; Samuelson, L.; Teskey, R.; McNulty, S. & Fox, T. (2000). Process models as tools in forestry research and management. *Forest Science*, Vol.47, No.1, pp. 2-8, ISSN 0015-749X

Koike, T. (1990). Autumn coloring, photosynthetic performance and leaf development of deciduous broad-leaved trees in relation to forest succession. *Tree Physiology*, Vol.7, pp. 21-32, ISSN 0829-318X

Koike, T.; Kitao, M.; Maruyama, Y.; Mori, S. & Lei, T. T. (2001). Leaf morphology and photosynthetic adjustments among deciduous broad-leaved trees within the vertical canopy profile. *Tree Physiology*, Vol.21, No.12-13, pp. 951-958, ISSN 0829-318X

Kolb, T. E.; Holmberg, K. M.; Wagner, M. R. & Stone, J. E. (1998). Regulation of ponderosa pine foliar physiology and insect resistance mechanisms by basal area treatments. *Tree Physiology*, Vol.18, No.6, pp. 375-381, ISSN 0829-318X

Lagergren, F.; Lankreijer, H.; Kučera, J.; Cienciala, E.; Mölder, M. & Lindroth, A. (2008). Thinning effects on pine-spruce forest transpiration in central Sweden. *Forest Ecology and Management*, Vol.255, No.7, pp. 2312-2323, ISSN 0378-1127

McDowell, N.; Brooks, J. R.; Fitzgerald, S. A. & Bond, B. J. (2003). Carbon isotope discrimination and growth response of old Pinus ponderosa trees to stand density reductions. *Plant, Cell and Environment*, Vol.26, No.4, pp. 631-644, ISSN 0140-7791

Medhurst, J. L. & Beadle, C. L. (2005). Photosynthetic capacity and foliar nitrogen distribution in *Eucalyptus nitens* is altered by high-intensity thinning. *Tree Physiology*, Vol.25, No.8, pp. 981-991, ISSN 0829-318X

Monsi, M. & Saeki, T. (1953). ber den Lichtfaktor in den Pflanzengesellschaften und seine Bedeutung fur die Stoffproduktion. *Japanese Journal of Botany*, Vol.14, pp. 22-52, ISSN 0021-5007

Moreno, G. & Cubera, E. (2008). Impact of stand density on water status and leaf gas exchange in *Quercus ilex*. *Forest Ecology and Management*, Vol.254, No.1, pp. 74-84, ISSN 0378-1127

Morikawa, Y.; Hattori, S. & Kiyono, Y. (1986). Transpiration of a 31-year-old *Chamaecyparis obtusa* Endl. stand before and after thinning. *Tree Physiology*, Vol.2, pp. 105-114, ISSN 0829-318X

Muñoz, F.; Rubilar, R.; Espinosa, M.; Cancino, J.; Toro, J. & Herrera, M. (2008). The effect of pruning and thinning on above ground aerial biomass of *Eucalyptus nitens* (Deane & Maiden) Maiden. *Forest Ecology and Management*, Vol.255, No.3-4, pp. 365-373, ISSN 0378-1127

Niinemets, Ü. (1999). Components of leaf dry mass per area- thickness and density- alter leaf photosynthetic capacity in reverse directions in woody plants. *New Phytologist*, Vol.144, pp. 35-47, ISSN 0028-646X

Oguchi, R.; Hikosaka, K. & Hirose, T. (2005). Leaf anatomy as a constraint for photosynthetic acclimation: differential responses in leaf anatomy to increasing growth irradiance among three deciduous trees. *Plant, Cell and Environment*, Vol.28, No.7, pp. 916-927, ISSN 0140-7791

Peterson, J. A.; Seiler, J. R.; Nowak, J.; Ginn, S. E. & Kreh, R. E. (1997). Growth and physiological responses of young loblolly pine stands to thinning. *Forest Science*, Vol.43, No.4, pp. 529-534, ISSN 0015-749X

Sala, A.; Peters, G. D.; McIntyre, L. R. & Harrington, M. G. (2005). Physiological responses of ponderosa pine in western Montana to thinning, prescribed fire and burning season. *Tree Physiology*, Vol.25, No.3, pp. 339-348, ISSN 0829-318X

Simioni, G.; Ritson, P.; McGrath, J.; Kirschbaum, M. U. F.; Copeland, B. & Dumbrell, I. (2008). Predicting wood production and net ecosystem carbon exchange of *Pinus radiata* plantations in south-western Australia: Application of a process-based model. *Forest Ecology and Management*, Vol.255, No.3-4, pp. 901-912, ISSN 0378-1127

Skov, K. R.; Kolb, T. E. & Wallin, K. F. (2004). Tree size and drought affect Ponderosa pine physiological response to thinning and burning treatments. *Forest Science*, Vol.50, No.1, pp. 81-91, ISSN 0015-749X

Skovsgaard, J. P. (2009). Analysing effects of thinning on stand volume growth in relation to site conditions: a case study for even-aged Sitka spruce (*Picea sitchensis* (Bong.) Carr.). *Forestry*, Vol.82, No.1, pp. 88-104, ISSN 0015-752X

Stogsdili, J. W. R.; Wittwer, R. F.; Hennessey, T. C. & Dougherty, P. M. (1992). Water use in thinned loblolly pine plantations. *Forest Ecology and Management*, Vol.50, No.3-4, pp. 233-245, ISSN 0378-1127

Tang, Z.; Chambers, J. L.; Guddanti, S. & Barnett, J. P. (1999). Thinning, fertilization, and crown position interact to control physiological responses of loblolly pine. *Tree Physiology*, Vol.19, No.2, pp. 87-94, ISSN 0015-752X

Tang, Z.; Chambers, J. L.; Sword, M. A. & Barnett, J. P. (2003). Seasonal photosynthesis and water relations of juvenile loblolly pine relative to stand density and canopy position. *Trees*, Vol.17, No.5, pp. 424-430, ISSN 0931-1890

Wang, J. R.; Simard, S. W. & Kimmins, J. P. H. (1995). Physiological responses of paper birch to thinning in British Columbia. *Forest Ecology and Management*, Vol.73, No.1-3, pp. 177-184, ISSN 0378-1127

Wilson, K. B.; Baldocchi, D. D. & Hanson, P. J. (2000). Spatial and seasonal variability of photosynthetic parameters and their relationship to leaf nitrogen in a deciduous forest. *Tree Physiology*, Vol.20, No.9, pp. 565-578, ISSN 0015-752X

10

Measuring Photosynthesis in Symbiotic Invertebrates: A Review of Methodologies, Rates and Processes

Ronald Osinga[1], Roberto Iglesias-Prieto[2] and Susana Enríquez[2]
[1]*Wageningen University, Aquaculture & Fisheries, Wageningen,*
[2]*Unidad Académica Puerto Morelos, Instituto de Ciencias del Mar y Limnología, Universidad Nacional Autónoma de México, Apdo,*
[1]*The Netherlands*
[2]*México*

1. Introduction

Several marine invertebrate species live in symbiosis with phototrophic organisms, mostly cyanobacteria and dinoflagellate algae. Such symbioses occur among different animal phyla, such as Porifera, Cnidaria, Mollusca and Plathyhelminthes (Venn et al. 2008). Animal host and phototrophic symbiont together are usually referred to as holobiont. Many of these holobionts act as net primary producers when growing in shallow waters (Wilkinson 1983; Falkowski et al. 1984) and thus have an important role in element cycling in marine ecosystems (Muscatine 1990). In addition, symbiont photosynthesis is often very important for the energy budget of the host animal (Davies 1984; Falkowski et al. 1984; Edmunds & Davies 1986; Anthony & Fabricius 2000). Hence, both from an ecological and a physiological point of view, it is important to have accurate, quantitative estimations of photosynthesis in symbiotic animals. In this chapter, we will provide an overview of methods to characterize photosynthesis in these animals, highlight important data obtained with these methods and present conceptual frameworks that describe how photosynthesis is controlled in marine symbiotic invertebrates. Hereby, we will be particularly focusing on zooxanthellate Scleractinia (stony corals, Fig. 1), a symbiosis that will be described in the next section.

2. The phototrophic symbiosis in stony corals

The fact that light affects the growth of many stony corals has been described already in the first half of the previous century (Vaughan 1919). The discovery that unicellular algae reside in the living tissue of corals (Fig. 2) even dates back further, to the late nineteenth century. The algae were termed zooxanthellae and the symbiosis was extensively studied in the first half of the twentieth century (e.g. Boschma 1925; Yonge & Nicholls 1931a,b). More recent works have established that all zooxanthellae found in scleractinian corals are dinoflagellates belonging to the genus *Symbiodinium* (see reviews by Baker 2003 and Stat et al. 2006). *Symbiodinium* is subdivided in several phylotypes (clades), termed A, B, C, D, E, F, G and H, which all have different properties in terms of pigmentation and heat tolerance,

thus providing their coral hosts with flexibility with regards to differences in light and temperature regimes, an aspect that will be further outlined in Section 4. *Symbiodinium* has also been found in other taxa (e.g. other cnidarian groups, Molluscs). In sponges, the photosynthetic microsymbionts are usually cyanobacteria (Venn et al. 2008).
The photosynthetic processes taking place in zooxanthellae inside coral cells in principle do not differ largely from photosynthetic processes in other plants. The main difference is the constraints impeded by the animal cell environment on the zooxanthellate cell. All supplies that are needed to perform photosynthesis have to cross several additional cell membranes (cell membranes of the different coral cell layers and the symbiosome). Therefore, the host cell can modify the surrounding environment and control the activity of the symbiont.

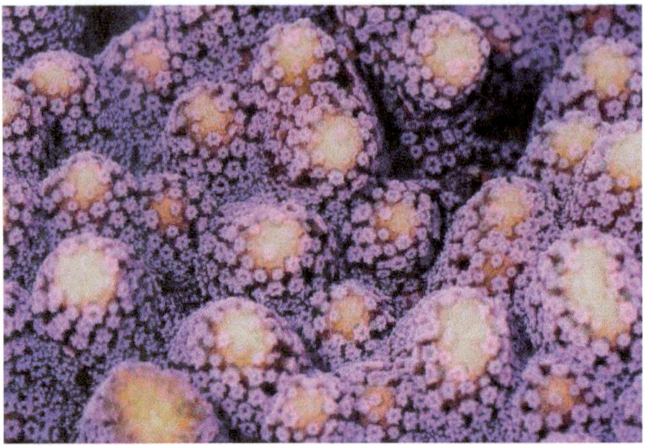

Fig. 1. Branch tips of the branching zooxanthellate stony coral *Stylophora pistillata*. The white tips of the branches, where the most active accretion of new calcium carbonate skeleton takes place, do not contain zooxanthellae. Photography by Tim Wijgerde.

Fig. 2. Zooxanthellae *in hospite* in *Porites asteroides*. Photography by M. en C. Alejandro Martínez Mena, Laboratorio de Microcine, Facultad de Ciencias, UNAM.

Coral-zooxanthellae associations are regarded as mutualistic symbioses, implying that there are benefits for both components. The coral host provides a sheltered environment for the symbiont and provides the algae with essential nutrients such as nitrogen and phosphorus, whereas the coral receives carbohydrate fuel and amino acids for protein synthesis (Dubinsky & Jokiel 1994). Initially, it was believed that the coral host would mainly acquire nutrients from the algae by digesting them (Boschma 1925; Goreau & Goreau 1960). Muscatine & Cernichiari (1969) were the first to demonstrate *in hospite* that photosynthesis products were actively translocated from the zooxanthellae to the host cells. Later, it became apparent that this process of translocation represents the main carbon flux between symbiont and host, and that zooxanthellae digestion is quantitatively of minor importance (Muscatine et al. 1984). The translocation of photoassimilates between symbiont and host is a host-controlled process. Muscatine (1967) discovered that host homogenates were able to release organic molecules from zooxanthellae suspensions, glycerol being the main constituent of the excreted materials. Later, Ritchie et al. (1993) added ^{14}C labelled glycerol to isolated zooxanthellae in suspension and found that glycerol was metabolized rapidly by zooxanthellae. It was concluded from this study that the hitherto unidentified host factor induces changes in the metabolism of the zooxanthellae rather than altering membrane permeability, as had previously been suggested. Studies on isolated zooxanthellae may not always reflect their behaviour *in hospite*, as was demonstrated by Ishikura et al. (1999). The composition of the translocated carbon in the intact host-symbiont association may vary among species and comprises sugars, glycerol, amino acids, fatty acids and other organic acids (see overview by Venn et al. 2008).

In zooxanthellate stony corals, photosynthesis is closely coupled to calcification (Gattuso et al 1999; Moya et al. 2006). The mechanism responsible for this coupling, which is also termed Light Enhanced Calcification (LEC) has been debated since its early discovery by Kawaguti & Sakamoto (1948). Most likely, calcification is stimulated in the light due to the simultaneous supply of oxygen and metabolic energy through photosynthesis (Rinkevich & Loya 1983; Colombo-Palotta et al. 2010), herein aided by the concurrent increase in pH in the calicoblastic fluid layer (Al Horani et al. 2003), which facilitates the deposition of calcium carbonate.

3. Measurement of photosynthetic processes in corals

The following sections provide descriptions of methods that are commonly used to characterize photosynthesis-related parameters in symbiotic marine invertebrates, including critical reflections on the use of these methods.

3.1 Relating photosynthesis to light

Measurements on photosynthetic processes are usually related to a quantification of the light field under which the photosynthesis takes place. Albert Einstein introduced the concept of photons, universal minimal quantities of light, comparable to molecules of mass. As such, light can be quantified in mole photons or mole quanta, also termed Einsteins (E). To describe the light field on a projected surface area, it is recommended to use the parameter Quantum Irradiance (E), to be expressed in μmole quanta m^{-2} s^{-1} (often written as μE m^{-2} s^{-1}). Quantum Irradiance is often referred to as Photon Flux Density (PFD) or light intensity.

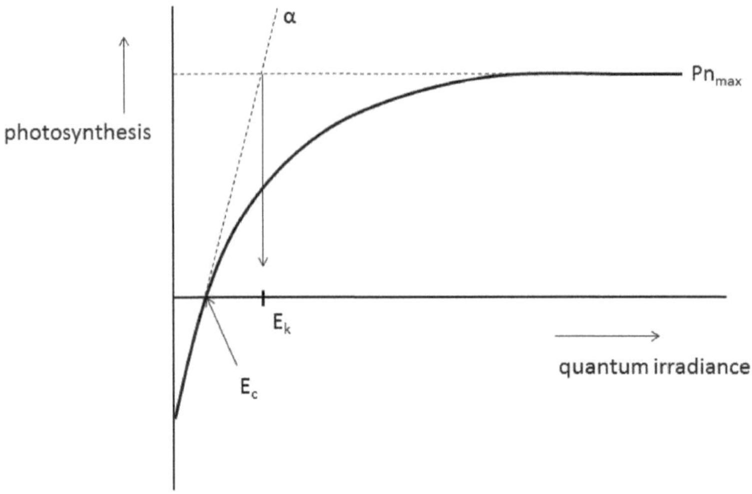

Fig. 3. Example of a curve that relates net photosynthesis to irradiance, showing the compensation point E_c, the Talling Index E_k, α and Pn_{max}.

Only photons with a wavelength between approximately 400 to 700 nm can be used in photosynthesis. This part of the light spectrum is termed photosynthetic active radiation (PAR) or photosynthetic usable radiation (PUR) if the available light spectrum is corrected for the spectrally variable capacity of the organism to absorb light (PAR x Absorptance). Throughout this text, the term quantum irradiance will be used to denominate the flux of photons within the PAR spectrum per m^2 per s, a parameter that is often referred to as Photosynthetic Photon Flux density (PPFD). Each photon carries a level of energy that is determined by its wavelength. Despite the energetic differences among photons, each photon within the PAR range is able to promote one photochemical excitation event. The energy that is in excess of the energy required for excitation transfer is dissipated as heat. For a more detailed list of quantities and units relating to light, we refer to a mini-review by Osinga et al. (2008).

Many studies on coral photosynthesis show photosynthesis rates under different irradiances, resulting in curves such as presented in Fig 3. Curves of this type are generally referred to as photosynthesis/irradiance curve, or shortly: P/E curve. It should be noted here that P/E curves obtained in the field (see for example the data obtained by Anthony & Hoegh-Guldberg 2003) reflect the actual photosynthetic response of corals to a natural light field (PAR) and daily light cycle, whereas P/E curves measured on aquarium corals (e.g. Goiran et al. 1996; Houlbrèque et al. 2004; Schutter et al. 2008) under laboratory conditions usually reflect the photosynthetic potential of the coral, since most aquarium corals are grown under a fixed quantum irradiance. Several descriptors can be derived from the P/E curves: 1) the compensation point (E_c), which is the irradiance at which photosynthesis equals respiration; 2) the maximal photosynthesis (P_{max}); 3) the photosynthetic efficiency (α), which is the slope of the linear increase in the photosynthetic rates as irradiance increases, and 4) the onset of saturation or Talling index (E_k), which is the point on the X-axis of the curve where the initial, linear slope of the curve intersects with the horizontal asymptote resembling P_{max}.

The Talling index is often used as a measure to characterize the acclimation of a photosynthetic organism to its light regime (Steeman-Nielsen 1975). The most commonly used model to describe P/E curves and to estimate P_{max} and E_k is the hyperbolic tangent function (Chalker & Taylor 1978):

$$P = P_{max} \times \tanh(E/E_k) \quad (1)$$

where P is the actual rate of photosynthesis measured at a given irradiance, P_{max} is the maximal photosynthetic capacity and E_k is the Talling Index.

Although this model provides accurate estimations of P_{max} and E_k, its assumption that photosynthesis increases with irradiance up till a horizontal asymptote is false: at high quantum irradiance levels, photosynthesis will decrease due to damage of the photosystems by excess light. An alternative (and probably better) approach is to use independent estimations of these parameters to allow better comparison between determinations obtained from the use of different equations (i.e., quadratic, exponential, etc.). A linear regression is required to determine α, paying attention that the incubations at low light levels need to be in the linear phase of the photosynthetic increment with irradiance. A minimum of four determinations are needed for a confident determination of α. Assessment of Pmax requires at least three consecutive measurement points at saturating irradiance. E_k and E_c can then be derived from those determinations as follows:

$$E_k = P_{max}/\alpha \quad (2)$$

and

$$E_c = R_d/\alpha \quad (3)$$

where R_d is the dark respiration.

3.2 Relating photosynthesis to size

When determining photosynthetic rates, it is necessary to relate the data to a size parameter. Several methods have been developed to assess the size of sedentary organisms such as corals. Selection of an appropriate method depends on the species (sensitive or robust, small or large, plate, boulder or branched), the circumstances (aquarium, *in situ*, is the colony fixed or can it be taken out of its environment) and the desired precision.

Two approaches to quantify coral size can be distinguished: estimating biomass (i.e. volume and weight-related parameters) and estimating surface area (see overview of related techniques in Table 1). The major difference between both approaches is that biomass measurements document mostly added coral skeleton, which does not contribute to biological activity, whereas surface-area measurements mainly reflect the amount of live coral tissues. As such, the surface area parameters serve as prime descriptors for standardization and quantification of physiological and biochemical parameters allowing comparative work of different experimental conditions, in particular work that relates to photosynthesis. When determining P/E curves, rates should always be related to surface parameters. Nevertheless, measurement of corals' surface area can be restricted by morphological variation and highly complex architectural structures that reduce accuracy. In addition, in complex structures such as branching corals, the light field within the colony varies largely. Hence, overall colony photosynthesis is an average rate reflecting

Size parameter	method	description	applications	advantages	limitations
Surface area	2D photometry	Top view pictures of corals are taken and analysed either by hand or by software to estimate the projected 2D surface area. A reference object of defined length (usually used is a ruler) must be included on the pictures to normalize the surface to values in square meters.	Encrusting and plate-shaped species with a regular shape (such as hemispherical boulders); assessment of % coverage; crude assessment of coral biomass on natural reefs.	Accurate for species with a flat surface or a simple, regular shape; applicable everywhere; non-destructive; no impact on organisms.	Not easily applicable to branching species and species with complex morphologies; the 2D projection of 3D organisms may lead to underestimations of the real surface area.
Volume	3D photometry[1]	Stereo-photography (3D) with two underwater cameras mounted on a fixed frame. The pictures are analysed using specific software.	Measurements of *in situ* growth rates of irregularly shaped species.	Very accurate, applicable everywhere, non-destructive, no impact on organisms.	Expensive, time-consuming technique.
Volume	replacement	The coral is positioned in a beaker glass (or similar) after filling this beaker up to an indicated level. Due to the incoming coral, the water level in the beaker will rise. The volume of the coral is now determined by siphoning off the excess water above the indicator level with a syringe (or similar) and by measuring the volume of the water in the syringe, which is equal to the coral volume.	Estimation of the volume of corals for incubation studies; a biomass estimate for branching and plate-shaped corals that can be related to a biological activity.	A quick, low-cost, non-destructive method with a reasonable precision.	The method can only be applied to corals that can be removed from their environment; the corals have to be taken out of the water to remove water that is attached to the coral surface - this step causes variability in the results and may inflict stress to the corals.
Volume	ecological volume[2]	A coral colony is converted to a cylindrical shape using 2D photography of coral colonies (one picture taken from the side, another from the top of the colony). A reference object of defined length must be included on the pictures to assess the height and the width of the colony. Image analysis (assessment of height and width) is done using simple software (note: height and width can also be measured directly by hand using a ruler). Biological volume is defined as $V = \pi r^2 h$, in which V is the ecological volume, r is the radius (calculated by dividing the average width of the colony by 2) and h is the height of the colony.	"Quick and dirty" measurements of (*in situ* and *ex situ*) growth rates and/or standing stocks. May be used as an efficient and comparative estimator for the total volume in the aquarium that is occupied by resident corals.	Relatively simple method using commonly available non-specialist materials; applicable everywhere, non-destructive, no impact on organisms.	Not a true size estimate: colonies having the same ecological volume may differ considerably in volume, weight, surface area and polyps numbers; slightly interpretation sensitive (comparative measurements should preferably done by the same person).
Weight	(drip dry) wet weight	Direct weighing of corals that are taken out of the water and have been shaken until no more drops fall off.	Measurements of *in situ* growth rates of irregularly shaped species.	Quick and easy, non-destructive method.	The method can only be applied to corals that can be removed from their environment; the corals have to be taken out of the water to remove water that is attached to the coral surface - this step causes variability in the results and stresses the corals.

Size parameter	method	description	applications	advantages	limitations
Weight	Buoyant weight3	Corals are weighed underwater, using a balance with an under-weighing device; a ring which is positioned on the bottom side of the balance to which a hook can be attached. The balance is positioned above an aquarium, bucket or anything that can hold a volume of water. The coral is hung onto the under-weighing ring-device using fishing line and two metal hooks.	Measurements of growth rates and/or standing stocks in aquaria.	Reasonably precise, easy, non-destructive method with low impact on organisms (that can be kept underwater; hence, water attached to the surface of the coral is also no issue here).	The method can only be applied to corals that can be removed from their environment; buoyant weight is a relative measure (although it can be converted into dry weight if the specific density of the coral material is known).
Weight	Dry weight	Corals are oven-dried (130 °C; incubation time varies depending on the size of the sample and the type of coral, but is usually around 24 h) and weighed regularly until they do not decrease in weight anymore.	Determination of biomass without water; first step to determine the separated weights of organic, living tissue and skeleton.	Very accurate.	Destructive, the coral has to be sacrificed.
Weight	Ash-free dry weight	Second step to determine separated weights of organic, living tissue and skeleton. Dried corals are ashed in a muffle furnace at 550 °C. By subtracting the weight of the remaining ashes (skeletal weight) from the previously obtained dry weight (skeleton + organic tissue), the dry weight of the organic fraction is obtained.	Comparative studies on growth, reference value for biological processes.	Very accurate; discriminates between organic tissue and skeletal weight	Destructive, the coral has to be sacrificed.

Table 1. Overview of methods to determine coral size.

photosynthetic activity under a wide range of quantum irradiance levels. P/E curves obtained for larger branching colonies as reported in the literature should be considered as the relation between colony photosynthesis and the ambient light field and should be termed differently.

3.3 Oxygen evolution techniques

Oxygen evolution techniques estimate net photosynthesis and dark respiration from changes in the oxygen concentration in an enclosure holding the targeted primary producer (Fig 4). When incubated in darkness, respiration rates can be assessed from the measured decrease in oxygen concentration. Incubation in light provides estimates on net photosynthesis (Pn): the observed change in the concentration of oxygen is the sum of production of oxygen due to photosynthesis and concurrent consumption of oxygen due to respiration by the algal population and the host. Determinations of respiration require distinction between respiration rates of dark-adapted samples (dark respiration, R_{dark}) and respiration rates of previously illuminated samples (light respiration R_{light}).

Incubations for oxygen evolution measurements should always be run concurrent with a blank control to correct for background activity in the water surrounding the targeted organism. In the case of corals, we found that the actual background in the water surrounding the coral is prone to change after one hour of incubation (Fig. 5). Production of mucus by the corals may stimulate bacterial activity in the surrounding water, thus increasing the actual background activity when compared to the blank control. Therefore, we recommend a regular (hourly) exchange of incubation water when performing oxygen evolution measurements on corals.

When related to an appropriate size measure (see Section 3.2), oxygen evolution measurements are currently the best method available to obtain quantitative data on net primary production in corals. Oxygen evolution can also be used to estimate gross primary production (Pg). To calculate Pg, it is necessary to add values for respiration losses during illumination (R_{light}) to values for Pn. This is due to the fact that light respiration is usually higher than dark respiration as photosynthesis stimulates both algal and host respiration. Light respiration is highly variable in corals and can be six times higher than dark respiration (Kuhl et al. 1995). Hence, in order to asses Pg from oxygen evolution measurements, an adequate measurement of light respiration is required. A suitable approach is to measure respiration immediately upon darkening at the end of the incubation under the highest irradiance level, which is termed Post-illumination respiration (R_{pI}). The average between R_d and R_{pI} is used to calculate gross-photosynthesis from net photosynthesis determinations. Section 3.4 deals with alternative techniques to assess Pg.

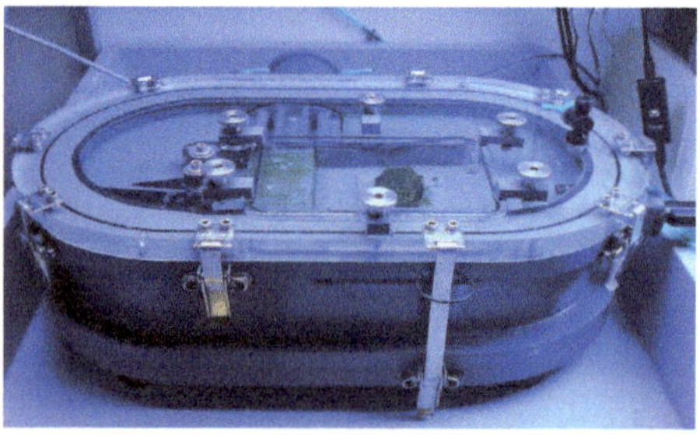

Fig. 4. Measurement of oxygen evolution. A coral colony is held in an enclosure equipped with an oxygen sensor, a paddle wheel, and a water jacket for temperature control. Light is supplied from the top.

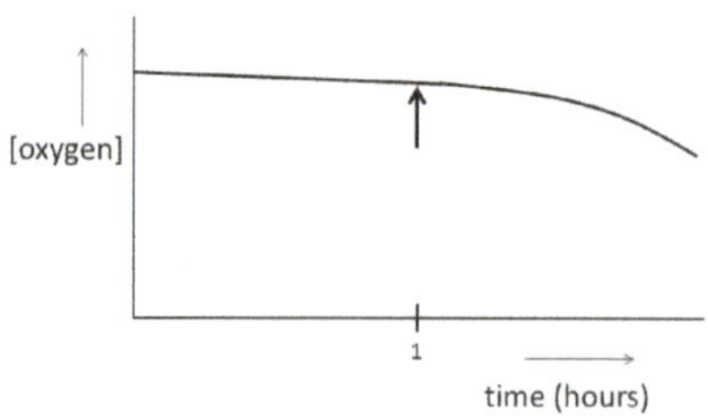

Fig. 5. Real-life example of increasing background respiration: the arrow indicates when the background sample was replaced with water in which a coral had been incubated for 1 hour.

When oxygen evolution is applied to generate P/E curves, the curve will show the relation between net photosynthesis and irradiance. The following modification of the hyperbolic tangent function (Equation 1) can be used to describe the data (Barnes & Chalker 1990):

$$Pn = R_{dark} + Pg_{max} \times \tanh(E/E_k) \tag{4}$$

where Pn is the net rate of photosynthesis, R_{dark} is the rate of respiration measured in darkness and Pg_{max} is the maximum gross photosynthetic rate (defined as Pn_{max} - R_{dark}, i.e. maximum net photosynthetic rate minus dark respiration). This equation also allows for calculation of the compensation point, at which Pn equals zero.

3.4 Measuring light respiration and gross photosynthesis

The rate of light respiration can be assessed either through the use of oxygen microsensors or by applying methods based on stable isotopes of oxygen. Oxygen microsensors can be applied to characterize the oxygen profile within biologically active layers that either produce or consume oxygen, such as sediments, microbial mats and living tissue (Revsbech & Jorgensen 1983). Oxygen production and consumption are deduced from the oxygen profiles using Fick's first law of diffusion:

$$J_{(x)} = -D_0 \, (\delta C_{(x)} / \, \delta_x) \tag{5}$$

where $J_{(x)}$ is the diffusive flux of oxygen at depth x, D_0 is the temperature- and salinity-dependent molecular diffusion coefficient for oxygen in water, and $\delta C_{(x)} / \, \delta_x$ is the slope of the oxygen profile at depth x.

Light respiration can be assessed using the so-called light/dark shift method (Revsbech & Jørgensen 1983), hereby measuring the depletion of oxygen immediately after an abrupt switch from ambient light to full darkness. During the first few seconds to minutes (the time depending on the thickness of the tissue layer involved) after the onset of darkness, the respiration rate will shift from a stable light respiration value to a stable dark respiration value. Hence, the initial rate of oxygen depletion will closely resemble the preceding rate of light respiration (Fig 6). Oxygen microsensors were used by Kuhl et al. (1995) to assess

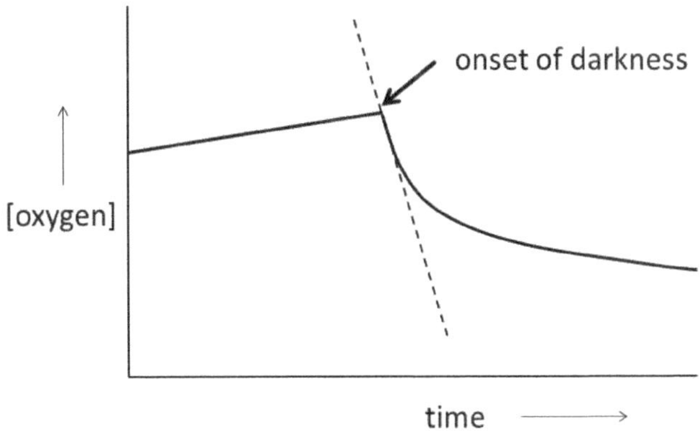

Fig. 6. Measuring light respiration using the light/dark shift method. The dashed line indicates the rate of light respiration.

photosynthetic parameters in *Acropora* sp. and *Favia* sp., hereby including measurements of the light respiration using the light/dark shift method. Their results show that the often used approach to take the dark respiration value for calculation of Pg may lead to a considerable underestimation of Pg and hence, provide an incorrect view on coral energetics, a consideration that had already been made several decades earlier in plant sciences.

Kana (1990) established an alternative technique where O^{18} labelled oxygen is used to assess respiration independently from oxygen production through photosynthesis. The technique is based upon two principles: 1) photosynthetic oxygen is produced from water through the Hill reaction (not from CO_2), and 2) during respiration, oxygen is used to produce CO_2 (and not water). Therefore, the O^{18} label is not likely to re-appear rapidly as photosynthetic oxygen. Using these principles, Mass et al. (2010) measured light respiration in colonies of *Favia veroni* by online membrane inlet mass spectrometry as the decline in O^{18} after a small spike of O^{18} labelled oxygen to the incubation chamber.

3.5 Measuring light respiration and gross photosynthesis

Pulsed Amplitude modulated (PAM) fluorometry has become one of the standards for the research on photosynthesis. The technique is based upon measuring the chlorophyll *a* fluorescence emission by Photosystem II of short, saturating pulses of light emitted onto photosynthetic active surfaces in relation to the fluorescence signal of continuously emitted light (Schreiber et al. 1986; Van Kooten & Snel 1990). PAM fluorometry is nowadays routinely applied to estimate a series of photosynthesis-related parameters, such as the maximum and effective photochemical efficiency (Fv/Fm and ΔF/Fm′) of photosystem II (PSII), non-photochemical quenching (the amount of excess excitation energy dissipated as heat), sustained quenching of fluorescence (qI) and the proportion of PSII that remain temporally or permanently closed and fail to reduce Q_B, the second quinone electron acceptor. PAM fluorometry has also become increasingly popular among coral scientists. Since the first application of this technique to corals in the late nineties of the previous century, a plethora of papers has been published on this topic.

The principles of PAM fluorometry and its suitability for application on marine organisms were reviewed recently in the book by Sugget et al. (2010). In this book, Enríquez & Borowitzka (2010), provide a thorough analysis of PAM fluorometry, which is briefly summarized below.

The principle of PAM measurements is depicted in Fig. 7. First, background fluorescence (F_0) is measured by supplying a moderated quantity of background light that is insufficient to induce photochemistry (all photosystems are open). Then, a saturated pulse is given, leading to a peak that represents maximal fluorescence (Fm; all photosystems are closed). The ratio between the observed increase in fluorescence over the background level (Variable fluorescence, $Fv = Fm - F_0$) and the maximal fluorescence (Fm) is a proxy for the probability for a photochemical event to occur (photochemical efficiency, Fv/Fm). If this parameter is determined after the sample was maintained under dark conditions and all the non-photochemical quenching activity has been relaxed, Fv/Fm represents the maximum photochemical efficiency of PSII. If this descriptor is determined under steady-state illuminated conditions, it represents the effective photochemical efficiency of PSII for this specific irradiance (ΔF/Fm′). A decrease in the maximum Fv/Fm over time implies that the rate of photodamage is exceeding the rate of repair of the damaged PSII. Photosynthetic organisms exposed to light usually show a decrease in ΔF/Fm′ associated with the increment in light exposure and a recovery after the peak of irradiance at noon. The initial maximum of the day Fv/Fm can be reached at the end of the light period or incomplete recovery or higher values can be observed depending on the amount of light in excess absorbed during the day: higher values lead to incomplete recovery and the accumulation of photodamage, but lower values allow better recovery and lead to higher initial Fv/Fm values. A recent study by Schutter et al. (2011b) shows that corals are exposed to continuous light (i.e. 24 hours per day) bleached and died after 7 weeks.

Under a range of assumptions, PAM can also be applied to estimate rates of electron transport (ETR), which is a proxy for gross photosynthesis under subsaturating light conditions (Genty et al. 1989). The apparent quantum yield is converted to ETR by multiplying it by the quantum irradiance, the absorptance (i.e. the fraction of the quantum irradiance absorbed by the photosynthetic apparatus) and the fraction of the corresponding energy delivered to PSII or the absorbed light that is utilized by PSII. Absorptance can be assessed through reflection measurements (Shibata 1969; Enríquez et al. 2005; see next subsection). The fraction of light energy delivered to by PSII is generally assumed to be 50%, based on an assumed balanced condition between PSII and PSI. ETR can simply be converted into oxygen evolution rates by assuming that four electrons are needed to produce one molecule of oxygen from water. Whereas this approach might work under non-saturating quantum irradiance levels, ETR values may overestimate oxygen evolution rates under saturating light due to an increasing proportion of non-photochemical quenching as a sink for excitation energy. In addition, since a PAM measurement represents only a small fraction of the total surface of an organism, many measurements will be needed to accurately quantify gross photosynthesis at the level of the whole organism if the colony has a large variation in tissue condition. Such an analysis should also take into account that the light field and tissue photoacclimatory condition will be highly variable in colonies with a complex architecture.

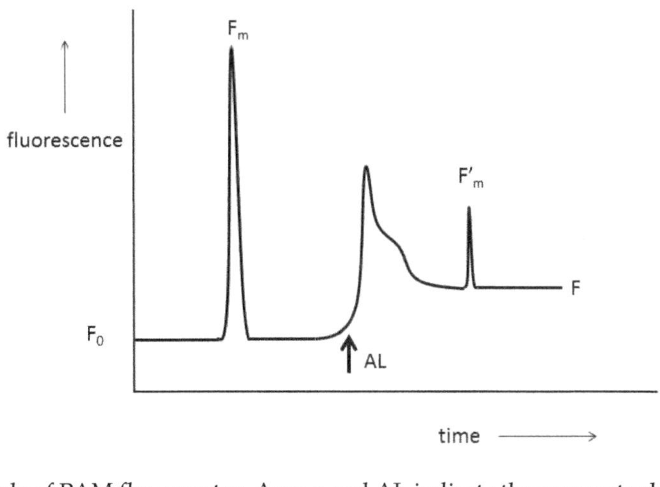

Fig. 7. Principle of PAM fluorometry. Arrow and AL indicate the moment when the actinic light is switched on. See text for further explanation.

As a more qualitative measure, indicative for changes in photosynthetic activity (i.e. to generate P/E curves), Beer et al. (2001) introduced the **relative ETR** (rETR). The use of this approach is limited and not recommended if the absorption cross-section of coral tissue changes among organisms, treatments or during the experimental approach. As the amount of light absorbed by a photosynthetic tissue is highly variable in most marine organisms the lack of control of this source of variation over $\Delta F/Fm'$ changes does not allow the comparison of relative changes in ETR. The use of this approach is limited and not recommended, because adsorptance is highly variable in most marine organisms. Acquiring rETR values through the making of rapid light curves (RLC, Ralph & Gademann 2005) is not recommended in particular, because the time needed for a photosynthetic organism to reach photosynthetic steady-state is much longer than the time intervals applied in the RLC method (Enríquez & Borowitzka 2009).

3.6 Measuring light respiration and gross photosynthesis

Reflection of light falling onto photosynthetic active biological surfaces can be quantified using a spectrometer and a small waveguide detector attached to it to collect the reflected light (Enríquez 2005, Enríquez et al. 2005, Terán et al 2010). It is hereby important to relate the measurement to a standard representing 100% reflection, for example white reference materials such as Teflon or a high reflecting materials supplied by manufacturers of light meters and spectrometers. Reflection measurements can provide information to quantify the light dose absorbed by a coral surface from absorptance determinations. Absorptance (A) is defined as the fraction of incident light absorbed by a surface and can be quantified from reflectance (R) measurements assuming that coral skeleton has minimal transmission (Enríquez et al. 2005) as $A = 1 - R$. In addition measurement of the percentage of light reflected can serve to quantify coral acclimation, coral paling and adaptive coral bleaching, and potentially, to indicate the onset of non-adaptive coral bleaching.

4. Rates and mechanisms: What controls coral photosynthesis?

Many researchers have carried out experiments to unravel how photosynthesis is controlled in stony corals. Limiting factors include the photon flux density and the availability of inorganic nutrients such as dissolved inorganic carbon (DIC), nitrogen (DIN), phosphorous (DIP) and iron. Additional factors that have been reported to influence coral photosynthesis are temperature, water flow, pH and oxygen. Table 2 summarizes photosynthetic rates that have been measured in stony corals *in situ* and *ex situ* in aquaria under a large variety of conditions. Using the data and experiments listed in Table 2, we will outline the mechanisms by which the different factors influence photosynthesis rates in stony corals.

4.1 Light

Light is the primary factor distinguishing photosynthesis from other assimilatory metabolic processes. Light is harvested by antenna molecules that are part of large light harvesting complexes termed Photosystem I (PSI) and Photosystem II (PSII), which are present on the thylakoid membranes of the chloroplasts residing in photosynthetically active cells. In the antenna molecules, an electron is excited by a photon to a higher energetic state. Through a cascade of events that take place in the thylakoid membrane and will not be discussed in

Species	conditions	Rdark	P and/or Pmax*	I_k ($\mu E\ m^{-2}\ s^{-1}$)	reference
Goniastrea retiformis	Shaded, incubated with filtered seawater	25.9 ($\mu g\ O_2\ cm^{-2}\ h^{-1}$)	127.6* ($\mu g\ O_2\ cm^{-2}\ h^{-1}$)	263.4	Anthony & Fabricius 2000
	Shaded, incubated with high loads of suspended matter	29.4	137.6* ($\mu g\ O_2\ cm^{-2}\ h^{-1}$)	261.4	
	Unshaded, incubated with filtered seawater	25.1	105.7*	310.6	
	Unshaded, incubated with high loads of suspended matter	22.3	104.1*	321.2	
Porites cylindrica	Shaded, incubated with filtered seawater	22.3	141.4*	350.3	
	Shaded, incubated with high loads of suspended matter	21.1	130.6*	306.8	
	Unshaded, incubated with filtered seawater	21.4	130.0*	390.7	
	Unshaded, incubated with high loads of suspended matter	23.0	92.2*	334.0	
Montipora monasteriata	Open water corals, simulated natural light cycle	1.33 ($\mu mol\ O_2\ cm^{-2}\ h^{-1}$)	3.92* ($\mu mol\ O_2\ cm^{-2}\ h^{-1}$)	211	Anthony & Hoegh-Guldberg 2003
	Corals growing under overhang, simulated natural light cycle	0.70	3.24*	127	
	Corals from cave, simulated natural light cycle	0.43	2.94*	80.8	
Porites porites	Branch tips obtained from 10 m depth	11.91 ($\mu l\ O_2\ cm^{-2}\ h^{-1}$)	82.34 ($\mu l\ O_2\ cm^{-2}\ h^{-1}$)	456	Edmunds & Davies 1986
	Constant PPFD of 500 $\mu E\ m^{-2}\ s^{-1}$, 0.5 mM HCO_3^-		12 (nmol O_2 mg chl *a*$^{-1}$ h^{-1})		Herfort et al. 2008
	Constant PPFD of 500 $\mu E\ m^{-2}\ s^{-1}$, 1.0 mM HCO_3^-		24		
	Constant PPFD of 500 $\mu E\ m^{-2}\ s^{-1}$, 2.0 mM HCO_3^-		43.5		
	Constant PPFD of 500 $\mu E\ m^{-2}\ s^{-1}$, 4.0 mM HCO_3^-		52.5		
	Constant PPFD of 500 $\mu E\ m^{-2}\ s^{-1}$, 6.0 mM HCO_3^-		61.5		
	Constant PPFD of 500 $\mu E\ m^{-2}\ s^{-1}$, 8.0 mM HCO_3^-		56		
Acropora sp.	Constant PPFD of 500 $\mu E\ m^{-2}\ s^{-1}$, 1.0 mM HCO_3^-		5		
	Constant PPFD of 500 $\mu E\ m^{-2}\ s^{-1}$, 2.0 mM HCO_3^-		8		
	Constant PPFD of 500 $\mu E\ m^{-2}\ s^{-1}$, 4.0 mM HCO_3^-		12		
	Constant PPFD of 500 $\mu E\ m^{-2}\ s^{-1}$, 6.0 mM HCO_3^-		15		
	Constant PPFD of 500 $\mu E\ m^{-2}\ s^{-1}$, 8.0 mM HCO_3^-		12		
Stylophora pistillata	Constant PPFD of 350 $\mu E\ m^{-2}\ s^{-1}$, low feeding, low flow (0.6-1 cm s^{-1}), measured after 3 weeks	0.44 ($\mu mol\ O_2\ cm^{-2}\ h^{-1}$)	0.57* ($\mu mol\ O_2\ cm^{-2}\ h^{-1}$)	403	Houlbrèque et al. 2004
	Identical as above, high feeding	0.43	0.20*	203	
	Constant PPFD of 300 $\mu E\ m^{-2}\ s^{-1}$, low feeding, low flow (0.6-1 cm s^{-1}), measured after 9 weeks	0.229	0.30		Houlbrèque et al. 2003
	Constant PPFD of 300 $\mu E\ m^{-2}\ s^{-1}$, high feeding	0.495	1.20		
	Constant PPFD of 200 $\mu E\ m^{-2}\ s^{-1}$, low feeding	0.134	0.22		
	Constant PPFD of 200 $\mu E\ m^{-2}\ s^{-1}$, high feeding	0.449	0.70		
	Constant PPFD of 80 $\mu E\ m^{-2}\ s^{-1}$, low feeding	0.186	0.15		
	Constant PPFD of 80 $\mu E\ m^{-2}\ s^{-1}$, high feeding	0.438	0.20		

Species	conditions	Rdark	P and/or Pmax*	I_k (μE m^{-2} s^{-1})	reference
Coral assemblage	Mesocosm experiment, high flow (20 cm s^{-1}), natural solar cycle, peak PPFD ~ 1200, ambient CO_2		37 (mmol O_2 m^{-2} h^{-1})	586	Langdon & Atkinson 2005
	Mesocosm experiment, high flow (20 cm s^{-1}), natural solar cycle, peak PPFD ~ 1200, 1.7 x ambient CO_2		36		
	Mesocosm experiment, high flow (20 cm s^{-1}), natural solar cycle, peak PPFD ~ 700, ambient CO_2		23		
	Mesocosm experiment, high flow (20 cm s^{-1}), natural solar cycle, peak PPFD ~ 700, 1.3 x ambient CO_2		31		
	Mesocosm experiment, high flow (20 cm s^{-1}), natural solar cycle, peak PPFD ~ 700, 2 x ambient CO_2		21		
	Mesocosm experiment, high flow (20 cm s^{-1}), natural solar cycle, peak PPFD ~ 700, enriched with DIN & DIP, 0.6-1.9 x ambient CO_2		29-34		
Montastrea annularis	Constant PPFD of 250 μE m^{-2} s^{-1}, natural oligotrophic seawater	14.8 (μl O_2 cm^{-2} h^{-1})	39.5* (μl O_2 cm^{-2} h^{-1})	119	Marubini & Davies 1996
	Constant PPFD of 250 μE m^{-2} s^{-1}, seawater + 1 μM NO_3	13.6	37.9*	111	
	Constant PPFD of 250 μE m^{-2} s^{-1}, seawater + 5 μM NO_3	14.4	46.4*	88	
	Constant PPFD of 250 μE m^{-2} s^{-1}, seawater + 20 μM NO_3	15.0	49.5*	104	
Porites porites	Constant PPFD of 250 μE m^{-2} s^{-1}, natural oligotrophic seawater	10.9	44.2*	215	
	Constant PPFD of 250 μE m^{-2} s^{-1}, seawater + 1 μM NO_3	10.2	45.8*	232	
	Constant PPFD of 250 μE m^{-2} s^{-1}, seawater + 5 μM NO_3	9.6	58.6*	304	
	Constant PPFD of 250 μE m^{-2} s^{-1}, seawater + 20 μM NO_3	8.5	61.8*	382	
Stylophora pistillata	Constant PPFD of 300 μE m^{-2} s^{-1}, pH = 7.6, 2 mM HCO_3^-		38 (μmol O_2 g^{-1} buoyant weight d^{-1})		Marubini et al. 2008
	Constant PPFD of 300 μE m^{-2} s^{-1}, pH = 8.0, 2 mM HCO_3^-		47		
	Constant PPFD of 300 μE m^{-2} s^{-1}, pH = 8.2, 2 mM HCO_3^-		38		
	Constant PPFD of 300 μE m^{-2} s^{-1}, pH = 7.6, 4 mM HCO_3^-		66		
	Constant PPFD of 300 μE m^{-2} s^{-1}, pH = 8.0, 4 mM HCO_3^-		66		
	Constant PPFD of 300 μE m^{-2} s^{-1}, pH = 8.2, 4 mM HCO_3^-		80		
	In situ, 5 m depth, winter	0.25 (μmol O_2 cm^{-2} h^{-1})	0.87* (μmol O_2 cm^{-2} h^{-1})	659.5	Mass et al. 2007
	In situ, 65 m depth, winter	0.04	0.15*	5.8	
	In situ, 5 m depth, summer	0.41	1.19*	1084.5	
	In situ, 65 m depth, summer	0.08	0.42*	108.9	
	Constant PPFD of 380 μE m^{-2} s^{-1}, 450 μatm CO_2, T = 25.3 °C	0.34 (μmol O_2 mg $protein^{-1}$ h^{-1})	0.24 (μmol O_2 mg $protein^{-1}$ h^{-1})		Reynaud et al. 2003
	Constant PPFD of 380 μE m^{-2} s^{-1}, 470 μatm CO_2, T = 28.2 °C	0.39	0.41		
	Constant PPFD of 380 μE m^{-2} s^{-1}, 734 μatm CO_2, T = 25.1 °C	0.39	0.20		
	Constant PPFD of 380 μE m^{-2} s^{-1}, 798 μatm CO_2, T = 28.3 °C	0.44	0.27		

Species	conditions	Rdark	P and/or Pmax*	I_k (µE m^{-2} s^{-1})	reference
Galaxea fascicularis	Constant PPFD of 90 µE m^{-2} s^{-1}, flow = 10 cm s^{-1}	9.0 (nmol O_2 cm^{-2} min^{-1})	11.7 (nmol O_2 cm^{-2} min^{-1})		Schutter et al. 2010
	Constant PPFD of 90 µE m^{-2} s^{-1}, flow = 20 cm s^{-1}	10.0	10.4		
	Constant PPFD of 90 µE m^{-2} s^{-1}, flow = 25 cm s^{-1}	10.4	8.2		
	Constant PPFD of 280 µE m^{-2} s^{-1}, flow = 5 cm s^{-1}		49		Schutter et al. 2011
	Constant PPFD of 560 µE m^{-2} s^{-1}, flow = 20 cm s^{-1}		42		
	Constant PPFD of 280 µE m^{-2} s^{-1}, flow = 5 cm s^{-1}		30		
	Constant PPFD of 560 µE m^{-2} s^{-1}, flow = 20 cm s^{-1}		57		

Table 2. Overview of photosynthetic rates measured in zooxanthellate stony corals.

detail here, the excitation energy is converted to metabolic energy (ATP) and reducing power (NADPH), which enables the conversion of CO_2 to organic carbon and the release of oxygen. Eight photochemical events, four per reaction centre in each photosystem, are needed to release one molecule of oxygen (i.e. eight photons), hence the maximal theoretical yield (also termed maximal quantum yield) of photosynthetic oxygen production from light energy is 12.5% (the evolution of 1 oxygen molecule requires 8 photons).

In the sea, the availability of light for photosynthesis depends largely on depth. In clear tropical waters, at a depth of 100 m, the PFD is only 2% of the PFD at the surface (Lesser et al. 2009). Zooxanthellate corals have been found up till depths exceeding 150 m. According to Lesser et al (2009), the deepest living photosynthetic coral specimen ever found so far being a colony of *Leptoseris hawaiiensis* growing at a depth of 165 m at Johnston Atoll (Maragos & Jokiel 1986). In order to cope with these highly variable light conditions, corals and their symbionts have developed a myriad of adaptation mechanisms (generally referred to as **photoadaptation** mechanisms), such as variation in the level of pigmentation, the number of zooxanthellae, pigmentation per cell, antenna size, coral morphology (Tissue and skeleton), polyp size and polyp behaviour (Dustan 1982, Iglesias-Prieto & Trench 1994; 1997a,b; Titlyanov et al. 2001; Levy et al. 2003; 2006; Hennige et al. 2008). When a coral is transferred to another location, the symbiotic population will respond to this change by modifying its photosynthetic apparatus to the new light conditions (Titlyanov et al. 2001). Changes in photosynthetic activity and the underlying photophysiology induced by diverse adjusting mechanisms upon changes in the light regime are generally referred to as **photoacclimation** processes. For example, in specimen of *Stylophora pistillata*, both the number of zooxanthellae per cm^2 coral surface and the amount of chlorophyll *a* per zooxanthellate cell doubled within 40 days upon translocation from an area exposed to 95% of the ambient surface irradiance to an area exposed to 30% of surface irradiance. The same response was observed upon translocation of corals from 30% surface irradiance to 0.8% surface irradiance (Titlyanov et al. 2001).

Under high light, down-regulation mechanisms are needed to prevent that the excess light causes large levels of damage to the photosynthetic apparatus. Several mechanisms for photoprotection have evolved in photosynthetic organisms (see mini-review by Niyogi 1999). Among these are pathways to safely dissipate the excess of energy absorbed as heat (non-photochemical quenching) or that allow to maintain the flow of electrons through both photosystems under Ferredoxin sinks limitation such as the cyclic electron flow and the water-water cycle (Asada 2000). The water-water cycle acts as a sink for electrons through a cyclic series of reactions involving superoxide dismutase and ascorbate. Photorespiration, which is basically the oxigenase instead of the carboxilase activity of the enzyme Rubisco, can also be considered as an alternative photoprotective mechanism that enables the maintenance of Rubisco activity under carbon limitation conditions (Niyogi 1999). These general photoprotective mechanisms also exist in zooxanthellate corals (e.g. Jones et al. 2001; Gorbunov et al. 2001), and will not be explained in detail here. Other adaptive mechanisms to high light include the synthesis of antioxidant molecules and host pigments that protect the symbiont photosynthetic membranes (Dove et al. 2008), efficient systems to repair photodamage, and the controlled expelling of zooxanthellae, which is also termed **adaptive bleaching** (Fautin & Buddemaier 2004).

Anthony & Hoegh-Guldberg (2003) studied photoacclimation by measuring photosynthesis and respiration in specimen of *Montipora monasteriata* from different habitats, hereby comparing corals growing in an open area (peak irradiance > 600 µmole quanta m^{-2} s^{-1}) to corals growing in caves (peak irradiance < 50 µmole quanta m^{-2} s^{-1}). Shade-adapted corals had a lower compensation point and a lower Pn_{max} than their light-adapted conspecifics, but a higher photosynthetic efficiency (Fig 8). In contrast to the study by Titlyanov et al. (2001), Anthony & Hoegh-Guldberg (2003) did not find an increase in the number of zooxanthellae per unit of coral surface, and no increase in the amount of chlorophyll a per cell at lower irradiances. However, the shade-adapted coral had a thinner tissue layer, and may thus have had a higher density of zooxanthellae per unit of tissue volume. Not hampered by secondary effects of high light such as photorespiration, formation of reactive oxygen species, and non-photochemical quenching, low light corals may be optimally adapted to perform photosynthesis with a very high efficiency. Indeed, the quantum yield of nearly 10% estimated by Anthony & Hoegh-Guldberg (2003) in shade-adapted corals is close to the theoretical maximum of 12.5%. In addition, the shape of the corresponding P/E curve closely resembled a solely light-limited process, Pn_{max} already being reached at a quantum irradiance of approximately 100 µmole quanta m^{-2} s^{-1}. The first quantification of the minimum quanta requirements of photosynthesis ($1/\Phi_{max}$) for intact corals, using a correct methodology for the determination of coral absorptance and the amount of energy absorbed by the coral surface was reported by Rodríguez-Román et al. (2006), who quantified an average value of 15.4 ± 2.3 quanta absorbed per O_2 molecule evolved. This represents a quantum efficiency of 6.5% (Φ_{max} = 0.065) for the species *Montastraea faveolata*.

These findings, together with the description of the optical properties of intact coral structures done by Enríquez et al. (2005) confirm that corals are very efficient light collectors and users and can rely they metabolic needs on the symbiotic relationship even at low irradiance values. Thanks to multiple scattering of light by the coral skeletons, effects of self-shading are small. Hence, photosynthesis in coral holobionts or in isolated zooxanthellae will exhibit saturation under relatively low levels quantum irradiance (100-200 µmole quanta m^{-2} s^{-1}, see for example Iglesias-Prieto & Trench 1994; Anthony & Hoegh-Guldberg 2003). Positive effects of higher quantum irradiance values on coral growth and photosynthesis reported for example by Schlacher et al. (2007) and Schutter et al. (2008) may relate to self-shading effects due to the increasing colony size of the growing coral (Fig. 9).

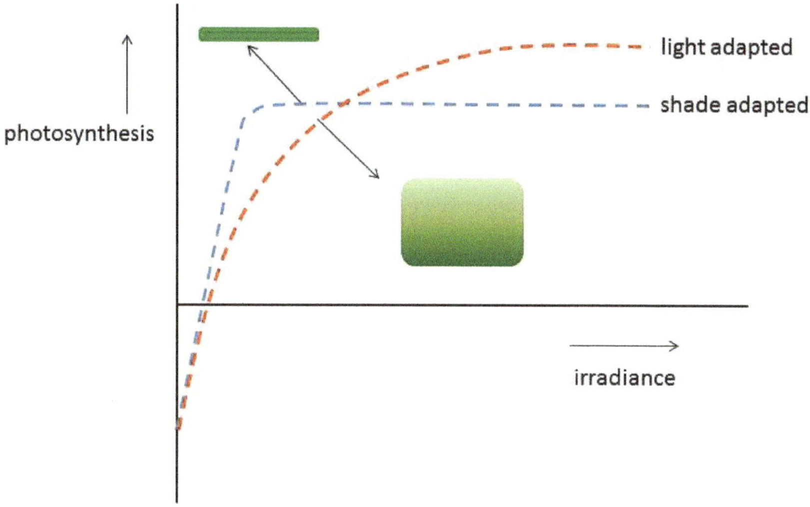

Fig. 8. Schematic representation of the results found by Anthony & Hoegh-Guldberg (2003) on photoadaptation by shade-adapted and light-adapted specimen of *M. monasteriata*. P/E curves for shade-adapted corals (blue line) and light-adapted corals (red line) are shown. The presumed thickness and pigmentation of the coral tissue corresponding to shade-adapted and light-adapted P/E curves is indicated.

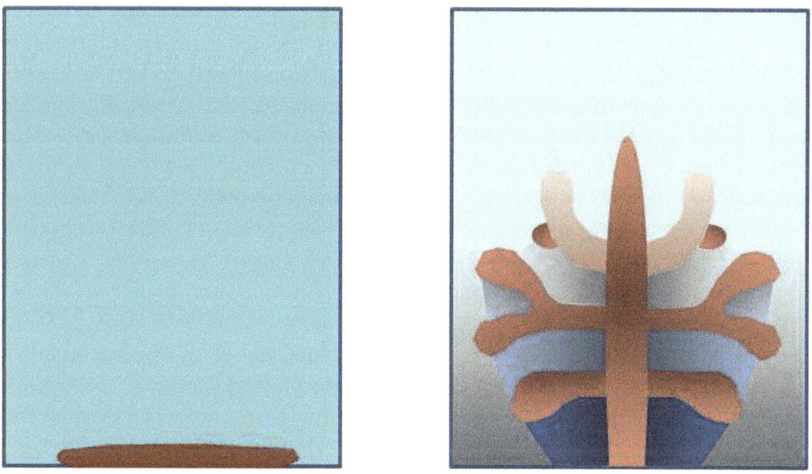

Fig. 9. Schematic representation of different light fields around a shade-adapted, plate-shaped coral (left) and a light adapted branching coral (right).

The strength of the light field is usually measured around the top of the coral colony, but the quantum irradiance at the bottom end of a colony may be tenfold lower. This may cause light limitation of photosynthesis in the lower parts of the colony, and hence, further increasing the quantum irradiance will further enhance the growth of these corals. In agreement to this view is the work of Schutter and coworkers on *Galaxea fascicularis* (Schutter et al. 2010, 2011). Schutter et al. (2010) found very high growth rates of 2.5% per day for small nubbins of *G. fascicularis* that were grown under a quantum irradiance of 90 μmole quanta m^{-2} s^{-1}, which corresponds to the afternoon peak irradiance at a depth of approximately 50 to 60 m (Mass et al. 2007). In another study, which was executed under much higher quantum irradiance levels (300 and 600 μmole quanta m^{-2} s^{-1}), a positive relation between SGR and quantum irradiance was observed, although the growth rates obtained were lower than the 2.5% per day that had been observed for small nubbins grown at 90 μmole quanta m^{-2} s^{-1}.

4.2 Inorganic nutrients

In shallow waters (on average, until a depth of nine metres), light is available in excess and may control coral photosynthesis through inhibitory mechanisms rather than limitation. Under saturating light, the availability of nutrients may as well become a limiting factor for coral photosynthesis. Experimental results suggest that photosynthetic rates of the zooxanthellae are limited by the availability of DIC, whereas the size of the symbiotic population and pigmentation per cell (i.e. the thickness of the coral tissue, the number of zooxanthellae per cm^2 of coral surface and the number of zooxanthellae cells per coral cell) is determined by the availability of inorganic nitrogen (Falkowski et al. 1993). Addition of DIC (e.g. Marubini & Thake 1999; Marubini et al. 2003) enhances photosynthesis and calcification, by increasing the rate of photosynthesis per cell. An increased availability of DIN without concurrent increased supply of DIC apparently disrupts the delicate balance of the symbiosis: additions of ammonium and nitrate increase the biomass of the zooxanthellae and lead to lower rates of calcification (Stambler et al. 1991; Marubini & Davies 1996), whereas concurrent addition of bicarbonate restores the calcification (Marubini & Thake 1999). Addition of planktonic food (i.e. carbon, nitrogen and phosphorous in a natural ratio) enhances tissue growth in corals and increases the size of the symbiotic population and pigmentation per cell leading to thicker tissue (cf Trench & Fisher 1983) and higher photosynthetic rates per unit of coral surface under increasing light than their food-limited conspecifics (Houlbrèque et al. 2004; Fig. 10A). This enables highly fed corals to calcify faster under high light than food-limited corals (Osinga et al. 2011). In contrast to these findings, corals exposed to increased DIN levels displayed a higher P_{max} without a corresponding increase in calcification, despite the denser population of zooxanthellae residing in those corals (Marubini & Davies 1996; Fig 10B). There are two possible explanations for this apparent paradox. First, high DIN loadings reduce translocation of photosynthetic products from the algae to their host. Second, DIN-enriched corals respond to the enrichment by increasing the number of zooxanthellae per cm^2 without a concurrent increase in coral tissue and calcification capacity. Hence, DIN fed corals exhibit a higher zooxanthellae to coral tissue ratio than DON enriched corals, in which both the thickness of the coral tissue and the number of zooxanthellae per cm^2 increase. This implies that the algae are perhaps suboptimally packed in a more dense concentration in DIN-enriched corals.

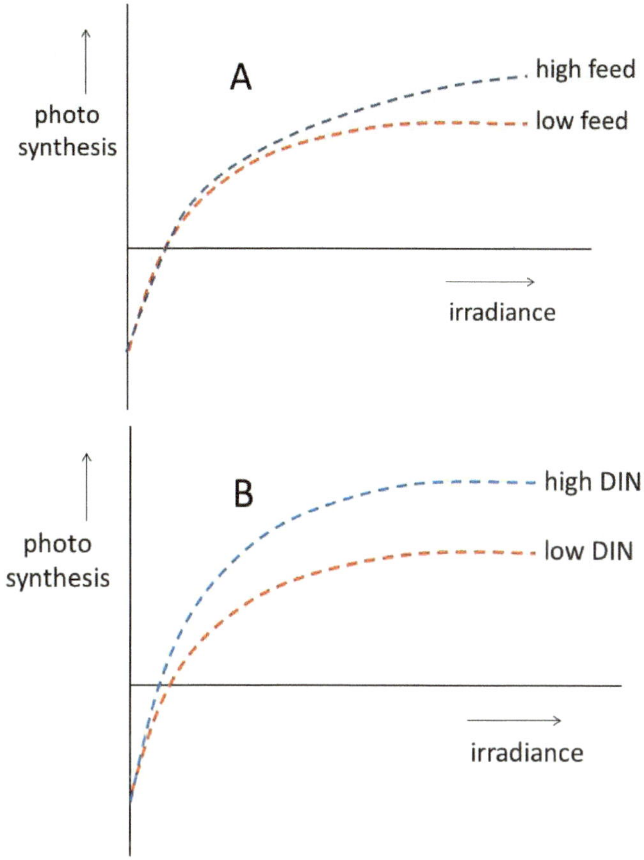

Fig. 10. Schematic representation of the findings of Houlbrèque et al. (2004) on the effects of feeding on photosynthesis in *Stylophora pistillata* (A) and the findings of Marubini & Davies (1996) on the effects of DIN enrichment on photosynthesis in *Montastrea annularis* and *Porites porites* (B).

4.3 The role of water flow and oxygen

Water flow around a sessile organism regulates the rate of gas exchange between the organism and the surrounding water by affecting the thickness of the diffusive boundary layer - a thin, stagnant layer of water around the surface of the organism that determines the rate of mass transfer via diffusion (Patterson 1992). In flume experiments allowing a controlled variation in flow, Dennison & Barnes (1988) found that flow stimulated photosynthesis in corals, which was attributed to an improved influx of DIC, thus relieving DIC limitation of photosynthesis. Recent studies (Mass et al. 2010; Schutter et al. 2011) have related the enhancement of photosynthesis by flow to an increased efflux of oxygen. Under low flow, photosynthetic produced oxygen accumulates in the coral tissue (Gardella & Edmunds 1999; Mass et al. 2010). A concurrent flow-limited influx of DIC will lead to a high oxygen / DIC ratio in the zooxanthellate cell, which will induce high rates of photorespiration, in particular in organisms that contain Type II Rubisco. Photorespiration reduces the efficiency of carbon fixation, but allows maintaining a minimum of carbon fixation under conditions of carbon limitation and high oxygen availability. The impact of photorespiration may be reduced by increasing the flow (Mass et al. 2010; Schutter et al. 2011) or by increasing the availability of DIC, thus reducing carbon limitation of coral photosynthesis. Indeed, several authors reported on increased photosynthetic activity in corals under elevated DIC levels (e.g. Herfort et al. 2008; Marubini et al. 2008; see next subsection).

Fig. 11 depicts a predicted change in P/E curve following an increase in flow. Under high flow, P_n will linearly increase until Pn_{max} has been reached, whereas under low flow, inhibition of net oxygen evolution due to photorespiration will increase with increasing irradiance, thus resulting in a lower Pn_{max} and a quantum yield (slope) that decreases with increasing irradiance.

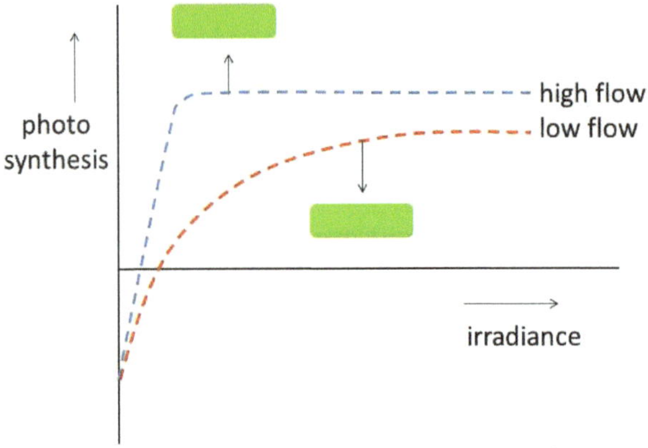

Fig. 11. Predicted effect of flow on photosynthesis in corals. The presumed thickness and pigmentation of the coral tissue corresponding to the P/E curves is equal in both conditions as indicated.

4.4 pH and dissolved inorganic carbon

Due to increasing atmospheric carbon dioxide concentrations, the ocean pH is decreasing. Almost 30% of the anthropogenic CO_2 emissions have been already removed from the atmosphere by the oceans. An increased availability of CO_2 may stimulate photosynthesis in DIC-limited corals (i.e. corals receiving non-limiting quantities of light), because CO_2 diffuses freely through the cell membrane and does not require carbonic anhydrase activity (the conversion of bicarbonate HCO_3^- into CO_2, Langdon & Atkinson 2005). However, doubling the concentration of dissolved CO_2 slightly reduced net photosynthesis in colonies of *Stylophora pistillata* cultured under a quantum irradiance of 380 µmole quanta m^{-2} s^{-1}, even though the number of zooxanthellae per coral cell increased (Reynaud et al. 2003). Several other studies also failed to demonstrate an effect of pH on photosynthesis (Marubini et al. 2008; Goiran et al 1996; Schneider & Erez 2006). It should be noted that these studies all evaluated the effect of pH on net photosynthesis. Hence, it cannot be excluded that the higher [CO_2] simultaneously stimulated to the same extent both algal photosynthesis and coral respiration, so that no increase in Pn could be detected. Indeed, Langdon & Atkinson

(2005) found an increase in net photosynthetic carbon fixation in an experimental coral community upon an increase in [CO_2] without a concurrent increase in the net production of oxygen. Several studies found that calcification was impaired by high [CO_2] (Reynaud et al. 2003; Langdon & Atkinson 2005; Marubini et al. 2008), although Reynaud et al. (2003) only observed such a negative response under elevated temperature. The studies by Langdon & Atkinson (2005), Schneider & Erez (2006 and Marubini et al. (2008) showed that photosynthesis and calcification are uncoupled under high CO_2 availability (a higher photosynthetic production concurrent with a lower calcification), which raises the question for what alternative purpose the excess photosynthetic carbon is being utilized within the holobiont under high [CO_2] conditions. We hypothesize that part of the excess carbon is being respired to provide additional energy needed to maintain a high pH in the calcifying fluid.

Contrasting results have also found with respect to the effect of total [DIC] on coral photosynthesis. Marubini et al. (2008) found that a doubling of the ambient seawater [DIC] nearly doubled the rate of P_n in *Stylophora pistillata*. In agreement with these results, Herfort et al. (2008) found an optimal [DIC] of 6 mM for photosynthesis in *Porites porites* and *Acropora* sp. This concentration is well above the ambient seawater [DIC], which is approximately 2 mM. Other studies failed to show such an effect (Goiran et al. 1996; Schneider & Erez 2006). The contrasting findings may have been caused by differences between species, but also by differences in experimental approaches, such as pre-incubation time (Marubini et al. 2008), effects of limited mass transfer of gases due to low flow (e.g. Dennison & Barnes 1998; see also Section 4.3) and changes in Pg that are not reflected in Pn. Further studies in this field should focus on these aspects and should elucidate the relative importance of CO_2 and HCO_3^- as substrates for photosynthesis in corals.

4.5 Temperature; its role in coral bleaching

Bleaching of zooxanthellate corals, the partial or total expelling by a coral of its zooxanthellae population, is beyond doubt the most intensively studied subject within coral science (see reviews by Lesser 2007; Weis 2008). Whereas bleaching is mainly associated with thermal anomalies (Hoegh-Guldberg 1999; Lesser & Farrell 2004), it must be noted here that with respect to coral bleaching, light and temperature act as partners in crime: they both induce the same type of stress (Iglesias-Prieto 2006). For example, high light and high temperature both have a negative effect on Photosystem II. Re-oxidation of the first quinone electron acceptor Q_A is considered as the rate limiting step in the electron transport reactions in PSII. Over-excitation of Q_A under high light can cause a double reduction of Q_A. This over-reduction induces the formation of reactive oxygen species (ROS), which can cause damage in PSII (Smith et al 2005). High temperature reduces the rate at which Q_A is re-oxidized (Lesser & Farrell 2004; Suggett et al. 2008). Hence, by slowing down re-oxidation of Q_A, high temperature increases the probability that a Q_A molecule becomes over-reduced under high light. Moreover, both high light and high temperature cause an increase in the respiratory demand of a coral, which implies that the proportion of translocated photosynthetic carbon that can be used by the coral for tissue growth becomes smaller. The growing respiratory demand thus causes the coral tissue to become thinner, which makes the zooxanthellae more vulnerable to bleaching (Enriquez et al. 2005), so that the photosynthetic activity is even further reduced. In this way, the bleaching process accelerates rapidly, ultimately resulting in a complete loss of zooxanthellae from the tissue.

Some corals may survive the current era of rapid climate change. The combination of high light and high temperature is restricted to the upper zone of the coral reef environment. Corals usually remain safe from temperature-induced bleaching at depths below 10m. In addition, corals have the potential to adapt to chronic thermal stress, which makes them less vulnerable to acute thermal stress (Mumby et al. 2011). Zooxanthellae play an important role in the responses of corals to thermal stress (Rowan et al. 1997; Ulstrup et al. 2006), although symbiont identity does not fully explain the variation in coral tolerance to thermal stress (Fitt et al. 2009). The different phylotypes (clades) of *Symbiodinium* residing in corals exhibit strong differences in thermo-tolerance (Bhagooli & Hidaka 2003; Baker et al. 2004; Robison & Warner 2006; Suggett et al. 2008). Berkelmans & Van Oppen (2006) were the first to describe a shift in abundance of different *Symbiodinium* types upon transplantation of corals to different light/temperature habitats. Previous work had not documented such shifts in *Symbiodinium* types after coral transplantation to contrasting light environments (Iglesias-Prieto et al. 2004). Shifts in relative abundance in *Symbiodinium* types have been attributed to changes in the relative abundance of the genetic varieties already present in the coral rather than to uptake of new types from the environment. It is not clear yet if corals hosting a mixed population of Symbiodinium would have higher potential for thermal acclimation than corals hosting a single type. The effect of symbiosis plasticity on symbiosis robustness and specifically on holobiont capacity to cope with thermal stress has not been fully elucidated yet. Using real-time PCR techniques, Mieog et al. (2007) found that minute numbers of different types were present in 78% of the samples taken from species that were previously believed to host only one *Symbiodinium* type. Coffroth et al. (2010) recently reported that some corals are capable of taking up *Symbiodinium* from the environment, thus increasing their potential for acclimation through symbiont shuffling. Coffroth et al. (2010) were also the first to note that this uptake of foreign zooxanthellae may not necessarily cause a stable thermo-tolerant symbiosis, and hence, may not prevent the occurrence of bleaching.

Another recent study suggest that the thermo-tolerant *Symbiodinium* type D1a is a rather selfish organism that hardly translocates photosynthetic carbon to its host (Smith et al. 2010), which puts another constraint to symbiont shuffling as a mechanism of thermo-adaptation.

5. Summary and conclusions

1. Photosynthesis in corals is affected by many, often interacting factors.
2. Zooxanthellate corals possess a myriad of mechanisms to adjust their photophysiology to changing environmental conditions, including symbiont shuffling.
3. P/E curves represent a useful characterization of the photosynthetic responses of zooxanthellate corals, the shape and the associated parameters P_{max}, α, E_c and E_k being good indicators for photo-acclimation.
4. Oxygen evolution measurements are the easiest and most reliable way to obtain information on net photosynthesis and dark respiration.

6. Acknowledgment

The authors of this review acknowledge the funding by the FP7 program of the European Commission, project FORCE.

7. References

Abdo, D., Seager, J.W., Harvey, E.S., McDonald, J.I.. Kendrick, G.A. & Shortis, M.R. (2006). Efficiently measuring complex sessile epibenthic organisms using a novel photogrammetric technique. *Journal of Experimental Marine Biology and Ecology* 339, pp. 120-133

Al-Horani, F.A., Al-Moghrabi, S.M., & de Beer, D. (2003). The mechanism of calcification and its relation to photosynthesis and respiration in the scleractinian coral Galaxea fascicularis. *Mar Biol* 142:419–426.

Anthony, K.R.N. & Fabricius, K.E. (2000). Shifting roles of heterotrophy and autotrophy in coral energetics under varying turbidity. *Journal of Experimental Marine Biology and Ecology* 252, pp. 221–253.

Anthony, K.R.N. & Hoegh-Guldberg, O. (2003). Variation in coral photosynthesis, respiration and growth characteristics in contrasting light microhabitats: an analogue to plants in forest gaps and understoreys? *Functional Ecology* 17, pp. 246-259.

Asada, K. (2000). The water-water cycle as alternative photon and electron sinks. *Phil. Trans. R. Soc. Lond. B*, pp. 1419-1431.

Baker, A.C. (2003). Flexibility and specificity in coral-algal symbioses: diversity, ecology and biogeography of Symbiodinium. *Annual Review of Ecology and Systematics* 34, pp. 661-689.

Baker, A.C., Starger, C.J., McClanahan, T.R., & Glynn, P.W. (2004). Corals' adaptive response to climate change. *Nature* 430 p. 741.

Barnes, D.J., & Chalker, B.E. (1990). Calcification and photosynthesis in reef-building corals and algae. Vol. 25. Amsterdam, PAYS-BAS: Elsevier.

Beer, S., Björk, M., Gademan, R., & Ralph, P. (2001). Measurements of photosynthesis in seagrass. In: Short FT, Coles R (eds) Global seagrass research methods. Elsevier, Amsterdam, pp. 183–198

Berkelmans, R. & van Oppen, M. (2006). The role of zooxanthellae in thermal tolerance of corals: a nugget of hope for coral reefs in an era of climate change. *Proc. R. Soc. Lond. B Biol. Sci.* 237, pp. 2305–12.

Bhagooli R, & Hidaka M. (2003). Comparison of stress susceptibility of in hospite and isolated zooxanthellae among five coral species. *J. Exp. Mar. Biol. Ecol.* 291, pp. 181-197.

Boschma, H. (1925). The Nature of the Association between Anthozoa and Zooxanthellae. *Proc. Nat. Ac. Sci.* Vol. 11.

Chalker, B.E. & Taylor, D.L. (1978). Rhythmic variation in calcification and photosynthesis associated with the coral, Acropora cervicornis (Lamarck). *Proc. R. Soc. (Set. B)* 201, pp. 179-189.

Coffroth, M.A., Poland, D.M., Petrou, E.L., Brazeau, D.A., & Holmberg, J.C. (2010). Environmental Symbiont Acquisition May Not Be the Solution to Warming Seas for Reef-Building Corals. *PLoS ONE* 5(10), e13258. doi:10.1371 /journal.pone.0013258

Colombo-Palotta, M.F., Rodriguez-Roman, A., & Iglesias-Prieto, R. (2010). Calcification in bleached and unbleached Montastraea faveolata: evaluating the role of oxygen and glycerol. *Coral Reefs* 29, pp. 899-907.

Davies, P. S. (1984). The role of zooxanthellae in the nutritional energy requirements of Pocillapora eydouxi. *Coral Reefs* 2, pp. 181-186.

Davies, P.S. (1989). Short-term growth measurements of corals using an accurate buoyant weighing technique. *Mar. Biol.* 101, pp. 389–395

Dennison, W.C. & Barnes, D.J. (1988). Effect of water motion on coral photosynthesis and calcification. *J Exp Mar Biol Ecol* 115, pp. 67–77

Dove, S., Lovell, C., Fine, M., Deckenback, J., Hoegh-Guldberg, O., Iglesias-Prieto, R., & Anthony, K.R.N. (2008). Host pigments: potential facilitators of photosynthesis in coral symbioses. *Plant Cell Environment* 31, pp. 1523-1533

Dubinsky, Z. & Jokiel, P.L. 1(994). Ratio of energy and nutrient fluxes regulates symbiosis between zooxanthellae and corals. *Pacific Sciences* 48, pp. 313-324

Dustan, P. (1982). Depth-dependent photoadaptation by zooxanthellae of the reef coral Montastrea annularis. *Marine Biology* 68, pp. 253-264

Falkowski, P.G. & Dubinsky, Z. (1981). Light-shade adaptation of Stylophora pistillata, a hermatypic coral from the Gulf of Eilat. *Nature* 289, pp. 172-174

Edmunds, P.J. & Davies, P.S. (1986). An energy budget for Porites porites (Scleractinia). *Mar Biol* 92, pp. 339-347

Enríquez, S. (2005). Light adsorption efficiency and the package effect in the leaves of the seagrass Thalassia testudinum. *Marine Ecology Progress Series* 289, pp. 141-150

Enríquez, S., Méndez, E.R., & Iglesias-Prieto, R. (2005) Multiple Scattering on Coral Skeletons Enhances Light Absorption by Symbiotic Algae. *Limnology and Oceanography* 50, pp. 1025-1032

Enríquez, S., Borowitzka, M. (2009). The Use of the Fluorescence Signal in Studies of Seagrasses and Macroalgae. In: Sugget, D.J. et al. (eds). Chlorophyll a fluorescence in aquatic sciences: methods and applications. Developments in Applied Phycology 4, Springer, pp. 187-208

Falkowski, P.G., Dubinsky, Z., Muscatine, L., & McCloskey L. (1993). Population control in symbiotic corals. *Bioscience* 43, pp. 606-611

Falkowski, P. G., Z. Dubinsky, L. Muscatine, and J. W. Porter, 1984. Light and the bioenergetics of a symbiotic coral. BioScience 34: 705-709

Fautin, D.G., Buddemaier, R.W. (2004). Adaptive bleaching: a general phenomenon. *Hydrobiologia* 530/531, pp. 459-467

Fitt, W.K., Gates, R.D., Hoegh-Guldberg, O., Bythell, J.C., Jatkar, A., Grottoli, A.G., Gómez, M., Fisher, P., LaJeunesse, T.C., Pantos, O., Iglesias-Prieto, R., Franklin, D.J., Rodrigues, L.J., Torregiani, J.M., van Woesik, R., & Lesser, M.P. (2009). Response of two species of Indo-Pacific corals, Porites cylindrical, and Stylophora postillata, to short-term thermal stress: The host does matter in determining the tolerance of corals to bleaching. *J Exp Mar Biol Ecol.* 373, pp. 102-110

Gardella, D.J. & Edmunds, P.J. (1999). The oxygen microenvironment adjacent to the tissue of the scleractinian Dichocoenia stokesii and its effects on symbiont metabolism. *Mar Biol* 135, pp. 289-295

Gattuso, J.P., Allemand, D., & Frankignoulle M. (1999) Photosynthesis and calcification at cellular, organismal and community levels in coral reefs: a review on interactions and control by carbonate chemistry. *American Zoologist* 39 (1), pp. 160–183

Genty, B., Briantais, J., & Baker, N.R. (1989). The relationship between the quantum yield of photosynthetic electron transport and quenching of chlorophyll fluorescence. *Biochim Biophys Acta* 990, pp. 87-92

Goiran, C., Al-Moghrabi, S., Allemand, D., & Jaubert, J. (1996). Inorganic carbon uptake for photosynthesis by the symbiotic coral-dinoflagellate association I. Photosynthetic performances of symbionts and dependence on sea water bicarbonate. *J. Exp. Mar. Biol. Ecol.* 199, pp. 207–225

Gorbunov, M.Y., Kolber, Z.S., Lesser, M.P., & Falkowski, P.G. (2001). Photosynthesis and photoprotection in symbiotic corals. *Limnol. Oceanogr.* 46, pp. 75–85

Goreau, T.F. & Goreau, N.I. (1960). Distribution of labeled carbon in reef-building corals with and without zooxanthellae. *Science*, 131, pp. 668-669

Hennige, S.J., Suggett, D.J., Warner, M.E., McDougall, K.E., & Smith, D.J. (2008). Photobiology of Symbiodinium revisited: bio-physical and bio-optical signatures. *Coral Reefs* 28, pp. 179-195

Herfort, L., Thake, B., & Taubner, I. (2008). Bicarbonate stimulation of calcification and photosynthesis in two hermatypic corals. *J. Phycol* 44, pp. 91-98

Hoegh-Guldberg, O. (1999). Climate change, coral bleaching and the future of the world's coral reefs. *Marine and Freshwater Research* 50, pp. 839–866

Houlbrèque, F., Tambutté, E., Allemand, D., &. Ferrier-Pagès, C. (2004). Interactions between zooplankton feeding, photosynthesis and skeletal growth in the scleractinian coral Stylophora pistillata. *Journal of Experimental Biology* 207, pp. 1461-1469

Houlbrèque, F., Tambutté, E., & Ferrier-Pagès, C. (2003). Effect of zooplankton availability on the rates of photosynthesis, tissue and skeletal growth of the scleractinian coral Stylophora pistillata *J. Exp. Mar. Biol.Ecol.* 296, pp. 145-166

Iglesias-Prieto, R. (2006). Thermal stress mechanisms in corals. *Comp Biochem Physiol A-Mol Integr Physiol* 143: S132

Iglesias-Prieto, R., Beltrán, V.H., LaJeunesse, T.C., Reyes-Bonilla, H., & Thome, P.E. (2004). The presence of specific algal symbionts explains the vertical distribution patterns of two dominant hermatypic corals. *Proc. R. Soc. Lond. B* 271, pp. 1757-1763.

Iglesias-Prieto, R. & Trench, R.K. (1997a). Acclimation and adaptation to irradiance in symbiotic dinoflagellates. II. Responses of chlorophyll protein complexes to different light regimes. *Mar Biol* 130, pp. 23-33

Iglesias-Prieto, R. & Trench, R.K. (1997b). Photoadaptation, photoacclimation and niche diversification in invertebrate-dinoflagellate symbioses. *Proc. 8th International Coral Reefs Symposium* 2, pp. 1319-1324

Iglesias-Prieto, R., & Trench, R.K. (1994). Acclimation and adaptation to irradiance in symbiotic dinoflagellates. I. Responses of the photosynthetic unit to changes in photon flux density. *Mar Ecol Prog Ser* 113, pp. 163-175

Ishikura, M., Adachi, K., & Maruyama, T. (1999). Zooxanthellae release glucose in the tissue of a giant clam, Tridacna crocea. *Marine Biology* 133, pp. 665–673

Jones, R.J. & Hoegh-Guldberg, O. (2001). Diurnal changes in the photochemical efficiency of the symbiotic dinoflagellates (Dinophyceae) of corals: photoprotection, photoinactivation and the relationship to coral bleaching. *Plant Cell Environment* 24, pp. 89-99

Kana, T.M. (1990). Light-dependent oxygen cycling measured by an O-18 isotope dilution technique. *Mar. Ecol. Prog. Ser.* 64, pp. 293–300

Kawaguti, S.& Sakumoto. D. (1948). The effects of light on the calcium deposition of corals. *Bull. Oceanogr. Inst. Taiwan* 4, pp. 65-70

Kuhl, M., Cohen, Y., Dalsgaard, T., Jørgensen, B.B., & Revsbech, N.P. (1995). Microenvironment and photosynthesis of zooxanthellae in scleractinian corals studies with microsensors for O2, pH and light. *Mar Ecol Prog Ser* 117, pp. 159–172

Langdon, C. & Atkinson, M.J. (2005) Effect of elevated pCO2 on photosynthesis and calcification of corals and interactions with seasonal change in temperature/irradiance and nutrient enrichment. *Journal of Geophysical Research-Oceans* 110, pp. 1-54

Lesser, M.P. (2007). Coral reefs bleaching and global climate change: can corals survive the next century? *Proc. Natl. Acad. Sci. U. S. A.* 104, pp. 5259–60

Lesser, M.P. & Farrell, J.H. (2004). Exposure to solar radiation increases damage to both host tissues and algal symbionts of corals during thermal stress. *Coral Reefs* 23, pp. 367–77

Lesser, M.P., Slattery, M., & Sleichter, J.J. (2009). Ecology of mesophotic coral reefs. *J Exp Mar Biol Ecol* 375, pp. 1-8

Levy, O., Dubinsky, Z., & Achituv, Y. (2003). Photobehavior of stony corals, responses to light spectra and intensity. *Journal of Experimental Biology* 206, pp. 4041-4049

Levy, O., Dubinsky, Z., Achituv, Y., & Erez, J. (2006). Diurnal polyp expansion behavior in stony corals may enhance carbon availability for symbionts photosynthesis. *Journal of Experimental Marine Biology and Ecology* 333, pp 1-11

Maragos, J.E. & Jokiel, P.L. (1986). Reef corals of Johnston Atoll: one of the world's most isolated reefs. *Coral Reefs* 4, pp. 141–150

Marubini, F. & Davies, P.S. (1996). Nitrate increases zooxanthellae population density and reduces skeletogenesis in corals. *Marine Biology* 127(2), pp. 319-328

Marubini, F. & Thake, B. (1999). Bicarbonate addition promotes coral growth. *Limnology and Oceanography* 44(3), pp. 716-720

Marubini, F., Ferrier-Pagès, C., Furla, P., & Allemand, D. (2008). Coral calcification responds to seawater acidification: a working hypothesis towards a physiological mechanism. *Coral Reefs* 27(3), pp. 491-499

Marubini, F., Ferrier-Pagès, C., & Cuif, J.P. (2003). Suppression of growth of scleractinian corals by decreasing ambient carbonate ion concentration: a cross-family comparison. *Proc R Soc Lond B* 270, pp. 179-184

Mass, T., Genin, A., Shavit, U., Grinstein, M., & Tchernov, D. (2010). Flow enhances photosynthesis in marine benthic autotrophs by increasing the efflux of oxygen from the organism to the water. *Proceedings of the National Academy of Sciences* 107(6), pp. 2527-2531

Mass, T., Einbinder, S., Brokovich, E., Shashar, N., Vago, R., Erze, J., & Dubinsky, Z. (2007). Photoacclimation of Stylophora pistillata to light extremes: metabolism and calcification. *Mar Ecol Prog Ser* 334, pp. 93-102

Mieog, J.C., van Oppen, M.J.H., Cantin. N.E., Stam, W.T., & Olsen, J.L. (2007). Real-time PCR reveals a high incidence of Symbiodinium clade D at low levels in four scleractinian corals across the Great Barrier Reef: implications for symbiont shuffling. *Coral Reefs* 26, pp. 449–457

Mumby, P., Elliott, I.A., Eakin, C.M., Skirving, W., Paris, C.B., Edwards, H.J., Enríquez, S., Iglesias-Prieto, R., Cherubin, L,M., & Stevens, J.R. (2011). Reserve design for uncertain responses of coral reefs to climate change. *Ecol Lett* 14, pp. 132-140

Muscatine, L. & Cernichiari, E. (1969). Assimilation of Photosynthetic Products of Zooxanthellae by a Reef Coral. *Biol Bull* 137, pp. 106-123

Muscatine, L., Falkowski, P.G., Porter, J.W., & Dubinsky, Z. (1984). Fate of photosynthetically-fixed carbon in light and shade-adapted colonies of the symbiotic coral Stylophora pistillata. *Proc. R. Soc. Lond. B* 222, pp. 181-202

Niyogi, K.K. (2000). Safety valves for photosynthesis. *Current Opinion in Plant Biology* 3 (6), pp. 455-460

Osinga, R., Janssen, M., & Janse, M. (2008). The role of light in coral physiology and its implications for coral husbandry. In Leewis R.J. and Janse M. (eds). Public Aquarium Husbandry Series, Volume 2: Advances in Coral Husbandry in Public aquariums. Arnhem: Burgers' Zoo, pp. 173-183

Osinga, R., Schutter, M., Griffioen, B., Wijffels, R.H., Verreth, J.A.J., Shafir, S., Henard, S., Taruffi, M., Lavorano, S. & Gili, C. (2011). The biology and economics of coral growth. *Marine Biotechnology* 13, pp. 658-671

Patterson, M.R. (1992). A chemical engineering view of cnidarian symbioses. *Am Zool* 32, pp. 566-582

Ralph, P.J. & Gademann, R. (2005). Rapid light curves: a powerful tool to assess photosynthetic activity. *Aquat Bot* 82, pp. 222–237

Revsbech, N.P. & Jørgensen, B.B. (1983). Photosynthesis of benthic microflora measured with high spatial resolution by the oxygen microprofile method: capabilities and limitations of the method. *Limnol Oceanogr* 28, pp. 749–756

Reynaud, S., Leclercq, N., Romaine-Lioud, S., Ferrier-Pagés, C., Jaubert, J., & Gattuso J.-P. (2003). Interacting effects of CO2 partial pressure and temperature on photosynthesis and calcification in a scleractinian coral. *Global Change Biology* 9(11), pp. 1660-1668

Rinkevich, B. & Loya, Y. (1983). Short term fate photosynthetic products in a hermatypic coral. *Journal of Experimental Marine Biology and Ecology* 73, pp. 175–184

Ritchie, R.J., Eltringham, K., & Hinde, R. (1993). Glycerol Uptake by Zooxanthellae of the Temperate Hard Coral, Plesiastrea versipora (Lamarck). *Proc R Soc Lond B* 253, pp. 189-195.

Robison, J.D. & Warner, M.E. (2006). Differential impacts of photoacclimation and thermal stress on the photobiology of four different phylotypes of Symbiodinium (Pyrrhophyta). *J Phycol* 42, pp. 568–579

Rodríguez-Román, A., Hernández-Pech, X., Thomé, P.T., Enríquez, S., & Iglesias-Prieto, R. (2006). Photosynthesis and light utilization in the Caribbean coral Montastraea faveolata recovering from a bleaching event. *Limnology & Oceanography* 51 (6), pp. 2702-2710

Rowan, R., Knowlton, N., Baker, A.C., & Jara, J. (1997). Landscape ecology of algal symbiont communities explains variation in episodes of coral bleaching. *Nature* 388, pp. 265–69

Schlacher, T.A., Stark, J., & Fischer, A.B.P. (2007). Evaluation of artificial light regimes and substrate types for aquaria propagation of the staghorn coral Acropora solitaryensis. *Aquaculture* 269, pp. 278–289

Schneider, K., Erez, J. (2006). The effect of carbonate chemistry on calcification and photosynthesis in the hermatypic coral Acropora eurystoma. *Limnol Oceanogr* 51, pp. 1284-1293

Schreiber, U., Schliwa, U., & Bilger, W. (1986). Continuous recording of photochemical and non-photochemical chlorophyll quenching with a new type of modulation fluorometer. *Photosynthesis Res* 10, pp. 51-62

Schutter, M., Van Velthoven, B., Janse, M., Osinga, R., Janssen, M., Wijffels, R.H., & Verreth, J.A.J. (2008). The effect of irradiance on long-term skeletal growth and net photosynthesis in Galaxea fascicularis under four light conditions. *Journal of Experimental Marine Biology and Ecology* 367, pp. 75-80

Schutter, M., Crocker, J., Paijmans, A., Janse, M., Osinga, R., Verreth, J.A.J., & Wijffels, R.H. (2010). The effect of different flow regimes on the growth and metabolic rates of the scleractinian coral Galaxea fascicularis. *Coral Reefs* 29, pp. 737-748

Schutter, M., Kranenbarg, S., Wijffels, R.H., Verreth, J.A.J., & Osinga, R. (2011a). Modification of light utilization for skeletal growth by water flow in the scleractinian coral Galaxea fascicularis. *Marine Biology* 158, pp. 769-777

Schutter, M., Van der Ven, R., Janse, M., Verreth, J.A.J., Wijffels, R.H., & Osinga R. (2011b). Light intensity, light period and coral growth in an aquarium setting: A matter of photons? *Journal of the Marine Biological Association of the UK*, DOI: 10.1017/S0025315411000920

Smith, D.J., Suggett, D.J., & Baker, N.R. (2005). Is photoinhibition of zooxanthellae photosynthesis the primary cause of thermal bleaching in corals? *Global Change Biology* 11, pp. 1-11

Smith, R.T. & Iglesias-Prieto, R. (2010). A significant physiological cost to Caribbean corals infected with an opportunistic Symbiodinium. In: Kaandorp JA, Meesters E, Osinga R, Verreth JAJ, Wijgerde THM (Eds). ISRS2010 Reefs in a changing environment, Book of Abstracts, p. 102. Wageningen University, The Netherlands

Stambler, N., Popper, N., Dubinsky, Z., & Stimson, J. (1991). Effects of nutrient enrichment and water motion on the coral Pocillopora damicornis. *Pacific Science* 45(3), pp. 299-307

Stat, M., Carter, D., & Hoegh-Guldberg, O. (2006). The evolutionary history of Symbiodinium and scleractinian hosts—Symbiosis, diversity, and the effect of climate change. *Persp Plant Ecol Evol Syst* 8, pp. 23-43

Steemann Nielsen, E. (1975). Marine photosynthesis with special emphasis on the ecological aspects, 141 pp. Amsterdam: Elsevier Scientific Publishing Co

Suggett, D.J., Warner, M.E., Smith, D.J., Davey, P., Hennige, S., & Baker, N.R. (2008). Photosynthesis and production of hydrogen peroxide by symbiodinium (Pyrrophyta) phylotupes with different thermal tolerances. *J Phycol* 44, pp. 948-956

Teran, E., Méndez, E.R., Enríquez, S., & Iglesias-Prieto, R. (2010). Multiple light scattering and absorption in reef-building corals. *Applied Optics* 49, pp. 532-542

Titlyanov, E.A., Titlyanova, T.V., Yamazato, K., & van Woesik, R. (2001a). Photoacclimation dynamics of the coral Stylophora pistillata to low and extremely low light. *J. Exp. Mar. Biol. Ecol.* 263, pp. 211-225

Ulstrup, K.E., Berkelmans, R., Ralph, P.J., & van Oppen, M.J.H. (2006). Variation in bleaching sensitivity of two coral species across a latitudinal gradient on the Great Barrier Reef: the role of zooxanthellae. *Mar Ecol Prog Ser* 314, pp. 135–148

Van Kooten, O. & Snel, J.F.H. (1990). The use of chlorophyll fluorescence nomenclature in plant stress physiology. *Photosynth Res* 25, pp. 147–150

Vaughan, T.W. (1919). Corals and the Formation of Coral Reefs. Smithsonian Institution. Annual Report for 1917, pp. 189-238

Venn, A., Loram, J.E., & Douglas, A.E. (2008). Photosynthetic symbioses in animals. *J Exp Bot* 59, pp. 1069-1080

Weis, V. (2008). Cellular mechanisms of Cnidarian bleaching: stress causes the collapse of symbiosis. *J Exp Biol* 221, pp. 3059-3066

Wilkinson, C. (1983). Net primary production in coral reef sponges. *Science* 219, pp. 410-412

Yonge, C.M. & Nicholls, A.G. (1931a). Studies on the physiology of corals. IV. The structure, distribution and physiology of zooxanthella. *Gt Barrier Reef Exped Sci Rep* 1, pp. 135-176

Yonge, C.M. & Nicholls, A.G. (1931b). Studies on the physiology of corals. V. The effect of starvation in light and darkness on the relationship between corals and zooxanthellae. *Gt Barrier Reef Exped Sci Rep* 1, pp. 179-211

11

The Xanthophyll Cycle in Aquatic Phototrophs and Its Role in the Mitigation of Photoinhibition and Photodynamic Damage

Miriam Zigman[1], Zvy Dubinsky[1] and David Iluz[1,2]
[1]*The Mina and Everard Goodman Faculty of Life Sciences, Bar-Ilan University, Ramat-Gan*
[2]*The Department of Geography and Environment, Bar-Ilan University, Ramat-Gan Israel*

1. Introduction

Solar energy is the initial power of photosynthesis. Plants and algae cannot proceed in the absence of light, and limited light conditions will limit photosynthesis. However, the conversion of solar energy into chemical energy is a potentially hazardous business that photosynthetic organisms expertly master. Whenever sunlight can actually be converted to chemical energy, there is minimal potential for problems. However, no leaf or algal cell can utilize all the light absorbed by the antenna system during exposure to full sunlight. Excessive light may be potentially dangerous to phototrophic organisms because it has the potential to be transferred to the formation of reactive oxygen species (ROS), which can result in cell damage (Ledford &Niyogi, 2005). It can also inhibit photosynthesis and lead to photooxidative destruction of the photosynthetic apparatus - photoinhibition (Demmig-Adams &Adams, 2006; Lu &Vonshak, 1999). It is known that photosynthesis is the basis of crop yield in plants and primary production in algae, and photoinhibition has an obvious adverse effect on photosynthesis and the accumulation of dry weight, which could lead to a decrease of carbon assimilation by about 10%. Thus, the ability of plants and algae to dissipate excessive light energy in order to resist photoinhibition, would significantly affect plant and alga yield and primary production.

In the first step of the photosynthetic process, light is intercepted by a variety of light-absorbing substances, the photosynthetic pigments. These pigments are associated with proteins forming light-harvesting 'antennae' that have a large optical cross-section for absorbing photons whose energy is efficiently transmitted to reaction centers (Dubinsky, 1992; Emerson &Arnold, 1932; Kirk, 1994).

The light energy absorbed by the chlorophyll of photosynthetic organisms drives photosynthesis and is also dissipated as heat and fluorescence.

To avoid massive ROS accumulation, phytoplankton and plants employ a host of protective mechanisms (Kanervo et al., 2005; Lavaud et al., 2002) – including various alternative energy-dissipation pathways (Adams et al., 2006) and multiple antioxidant systems

(Mullineaux & Rausch, 2005; Noctor & Foyer, 1998). Furthermore, both short- and long-term changes in positioning, stoichiometry, and/or activity of the components of photosystem cores and light-harvesting antennae can occur (Adir et al., 2003; Durnford et al., 2003; Kanervo et al., 2005; Matsubara et al., 2002).

The focus of the present review is photoprotection by the xanthophyll cycle, whereby excess light energy is safely dissipated as heat rather than being transferred to oxygen and, thus, result in ROS production. In this key photoprotective process, potentially damaging energy absorbed by chlorophyll and other light harvesting pigments is dissipated by the carotenoids violaxanthin or diadinoxanthin via the xanthophyll cycle.

In 1962, Yamamoto demonstrated a reversible epoxidation of hydrophilic carotenoids in higher plant leaves (Yamamoto et al., 1962). This class of reactions was subsequently demonstrated in all non-phycobilisomes containing oxygenic photoautotrophs, including algae.

In 1988, Demmig-Adams showed a remarkable correlation between the de-epoxidation reaction and chlorophyll fluorescence quenching. Exposure to high irradiance causes the photosynthetic rates to drop since the harvested light energy cannot be utilized fast enough and has to be dissipated to avoid photodynamic damage due to the formation of free radicals. A significant part of this excess energy turns to heat. This process, which appears to be wasteful, acts as a protective mechanism (Demmig et al., 1988; Krause & Weis, 1991).

2. The evolutionary inheritance of algal pigments in the oceans

The apparatus responsible for the photochemical production of oxygen in photosynthetic organisms is contained within distinct organelles called plastids. Based on small subunit ribosomal RNA sequences, it would appear that all plastids are derived from a single common ancestor that was closely related to extant cyanobacteria (Bhattacharya & Medlin, 1995; Palmer, 2003); however, early in the evolution of eukaryotic photoautotrophs, major schisms occurred that gave rise to two major clades, a 'green lineage' and a 'red lineage', from which all eukaryotic photoautotrophs descended (Delwiche, 1999) (Fig. 1). While all eukaryotic photoautotrophs contain chlorophyll *a* as a primary photosynthetic pigment, one group utilizes chlorophyll *c* and the other appropriated chlorophyll *b* as primary accessory pigments. No extant chloroplast contains all three pigments (Falkowski et al., 2004a, 2004b).

The chlorophyll *c*-containing plastid lineage, which is widely distributed among at least six major groups (i.e., phyla or divisions) of aquatic photoautotrophs, with the exception of some soil-dwelling diatoms and xanthophytes, is not present in any extent terrestrial photoautotroph. In contrast, the chlorophyll *b*-containing plastid lineage is in three groups of eukaryotic aquatic photoautotrophs and in all terrestrial plants. Because additional accessory pigments (carotenoids) found in the chlorophyll *c*-containing group have yellow, red, and orange reflectance spectra (i.e., they absorb blue and green light), the ensemble of organisms in this group are referred to, in the vernacular, as the 'red lineage'. The chlorophyll *b*-containing group contains a much more limited set of carotenoids in the chloroplast, and members of this group generally have a green color. Thus, in effect, the ensemble of organisms responsible for primary production on land is green, while the ecologically dominant groups of eukaryotic photoautotrophs in the contemporary oceans are red.

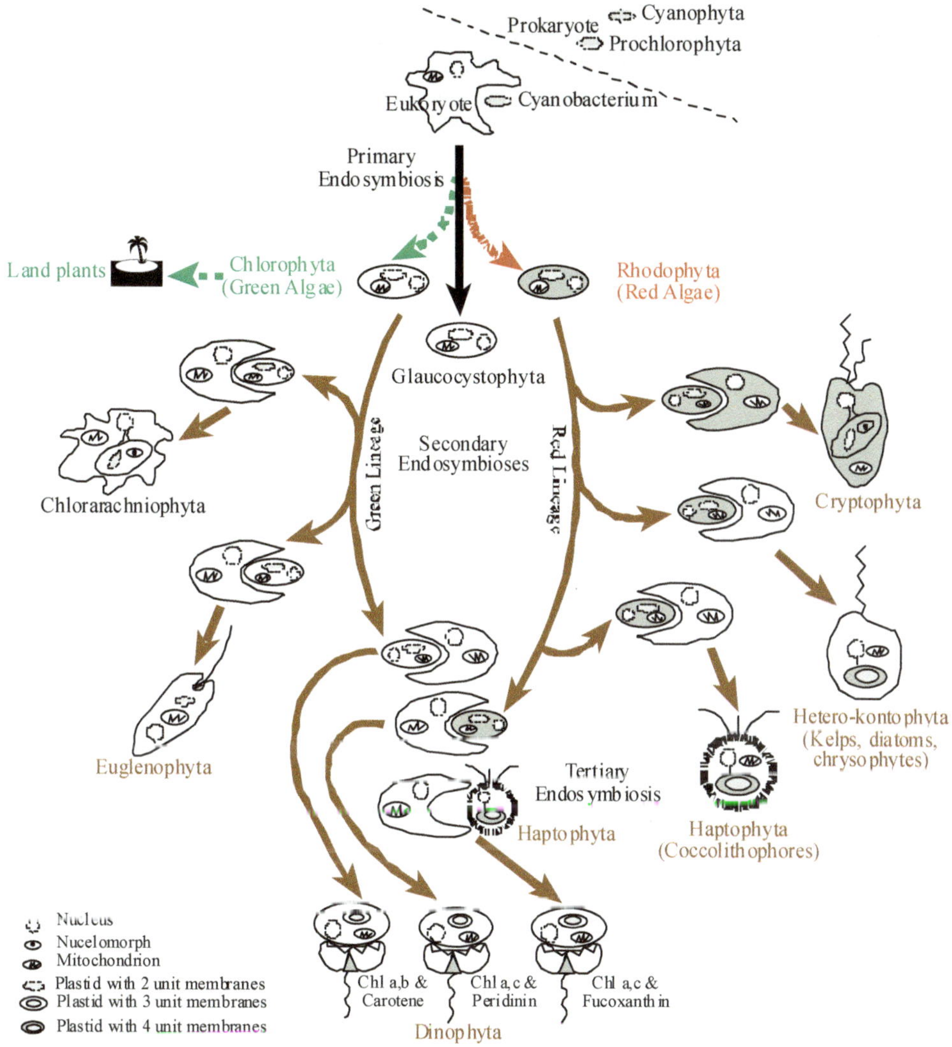

Fig. 1. The evolutionary inheritance of red and green plastids in eukaryotic algae. The ancestral eukaryotic host cell appropriated a cyanobacterium to form a primary photosynthetic symbiont. Two main groups split from this primary association: one formed a 'green' line  and one a 'red' line  (after Falkowski, 2004b).

One potential selection mechanism for red and green plastids is spectral irradiance. Compared to land plants, the majority of the phytoplankton biomass in the ocean is light-limited for growth and photosynthesis. On land, competition for light within a canopy is based on total irradiance, not primarily on the spectral distribution of irradiance. On the average, 85-90% of total incident photosynthetically available radiation on a leaf is absorbed. In contrast, in the oceans, absorption of light by seawater itself is critical to the spectral distribution of irradiance. The spectral irradiance is further modified by dissolved organic matter, sediments, and the spectral properties of the phytoplankton themselves. Hence, it is not surprising that phytoplankton have evolved an extensive array of accessory pigments, including carotenoids and chlorophylls, that permit light absorption throughout a wide range of the visible spectrum (Falkowski et al., 2004a; Jeffrey et al., 1997).

The 'red' and 'green' algal lineages differ, in addition to several cellular and life-cycle characteristics, in the evolution of their photosynthetic pigments, hence, it is not surprising that they also differ in the carotenoids of which their xanthophyll cycle consists. In accordance with their evolutionary linkeage, no xanthophyll cycle was found in cyanophyceae and rhodophyceae (Fig. 2).

From the investigation of fossil evidence regarding the evolution of the eukaryotic phytoplankton taxa, we can consider and develop some hypothesis that may account for the origin and ecological success of the red line in the oceans, while the green line maintained genetic hegemony on land.

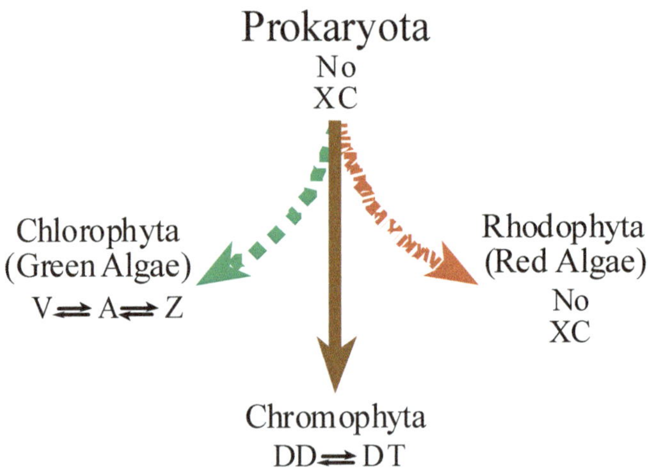

Fig. 2. The three major groups split from the primary prokaryotic prototroph and secondary endosymbiosis. This is characteristic of their xanthophyll cycle. Green line ▪▪▪➧ and red line ➧.

3. Carotenoids in the xanthophyll cycle

Carotenoids containing one or more oxygen atoms have a number of functions in photosynthetic systems. They act as: (i) accessory light-harvesting pigments, such as peridinin and fucoxanthin; (ii) photoprotective pigments, such as carotene and astaxanthin quenchers of triplet-state chlorophyll (chl); and (iii) xanthophyll cycle components, quenchers of singlet oxygen (O_2). Studies on plastid biogenesis and *in vitro* reconstitution have also identified a key role for carotenoids in the structure/organization of the photosynthetic apparatus. Xanthophylls are yellow pigments from the carotenoid group. Some xanthophylls have further been implicated in the non-photochemical quenching of chlorophyll fluorescence in plants and some algae, an important photoprotective process. The role of xanthophylls in this process, resulting in dissipation of excess excitation energy via quenching of chlorophyll fluorescence, is a feature of the interconversion of carotenoids due to the xanthophyll cycle. The xanthophyll cycle involves only 5 carotenoids out of many hundreds of carotenoids found in all phyla of algae and land plants (Tables 1, 2).

Pigment	Solvent	Spectra	Ref.
Diadinoxanthin	acetone	426, 447.5, 478	Johnsen et al. (1974)
Diatoxanthin	Acetone	429, 454, 482	Berger et al. (1977)
Zeaxanthin	acetone	425, 450, 478	Withers et al. (1981)
Antheraxanthin	ethanol	422, 444, 472	Stransky & Hager (1970)
Violaxanthin	acetone	417, 440, 470	Renstrom et al. (1981)

Table 1. Absorbance peaks of xanthophyll-cycle pigment

The xanthophyll cycle involves the enzymatic removal of epoxy groups from xanthophylls (violaxanthin, antheraxanthin, diadinoxanthin) to create so-called de-epoxidized xanthophylls (diatoxanthin, zeaxanthin).The interconversion of violaxanthin to zeaxanthin and of diadinoxanthin to diatoxanthin alters the extent of the conjugated double-bond system as a result of the epoxidation and de-epoxidation reactions (Figs. 3, 4).

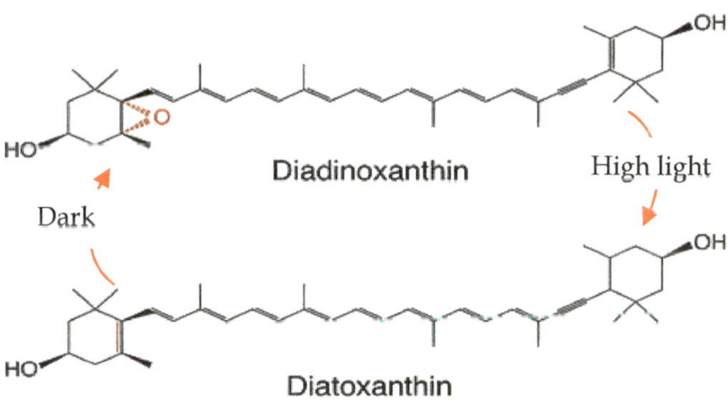

Fig. 3. The xanthophyll biosynthetic pathway in chromophyta algae. Under high light, diadinoxanthin converts to diatoxanthin reverting to diadinoxanthin under dim light or darkness

In chlorophyll *b*-containing organisms (higher plants and green algae), the carotenoid pigment structures that are active in the xanthophyll cycle are: violaxanthin ((3S, 5R, 6S, 3'S, 5'R, 6'S)-5,6;5',6'-diepoxy-5,6,5',6'-tetrahydro-l3,!3-carotene-3,3'-diol), antheraxanthin ((3S, 5R, 6S, 3'R)-5,6-epoxy-5,6-dihydro-B,I3-carotene-3,3'-diol), and zeaxanthin ((3R, 3'R)-l3,Rcarotene-3,3'-diol) (Fig. 4) During light stress, violaxanthin is converted to zeaxanthin via the intermediate antheraxanthin, which plays a direct photoprotective role acting as a lipid-protective antioxidant and by stimulating non-photochemical quenching within light-harvesting proteins. This conversion of violaxanthin to zeaxanthin is done by the enzyme violaxanthin de-epoxidase, while the reverse reaction is performed by zeaxanthin epoxidase. In chlorophyll *c*-containing organisms (some algae groups), the xanthophyll cycle consists of the pigment diadinoxanthin ((3S,5R, *6S*,3 'R)-5,6-epoxy-7',8'-didehydro-5,6-dihydroS,R-carotene-3,3'-diaonld), which is transformed into diatoxanthin ((3R, 3'R)-7,8- didehydro-13,13-carotene-3,3'-diol)) (Fig. 3). Some of chlorophyll *c* algae groups use both cycles (Lohr & Wilhelm, 2001).

The implications of this are: (i) the extent of the conjugated system in carotenoids affects both the energies and lifetimes of their excited states, from 9 to 11 C-C double bonds in violaxanthin and zeaxanthin, respectively; (ii) for carotenoids in which the end-groups are in conjugation with the main polyene chain, a coplanar conformation is energetically favored. In zeaxanthin, steric hindrance prevents it from being fully coplanar and a near-planar conformation is adopted. In contrast, in carotenoids such as violaxanthin, in which the conjugation is removed (by the presence of epoxide groups in the C5,6 position) the end-group occupies a perpendicular position relative to the main chain. Such conformational changes in the carotenoid molecule may, in turn, affect the organization of the light-harvesting complex (LHC).

4. Operation and characteristics of the xanthophyll cycle

The xanthophyll cycle is one of the main processes regulating excessive photon flux in the light-harvesting complexes of photosystems since it is responsible for most of the non-photochemical quenching of chlorophyll fluorescence (NPQ) (Demmig-Adams & Adams, 2006; Goss & Jakob, 2010; Lavaud, 2007).

This photoprotective cycle in its generally recognized form occurs in most eukaryotic algae and in higher plants. While much work has been carried out on higher plants and green algae such as *Chlamydomonas* (Baroli & Melis, 1996; Demmig-Adams & Adams, 1993), much less has been carried out on the other algal taxa. In microalgae, two main groups can be distinguished with regard to pigments involved in the xanthophyll cycle. The first group is characterized by the two-step de-epoxidation of violaxanthin into zeaxanthin via antheraxanthin at high light, which is reversed in the dark (Fig. 3). One molecule of oxygen is released (de-epoxidation) or taken up (epoxidation) for a complete transition. A truncated vilaxanthin-to-antheraxanthin version also occurs in some primitive green algae. The second group exhibits a simpler conversion, with the single step de-epoxidation of diadinoxanthin to diatoxanthin (Fig. 4). Many specificities of the diadinoxanthin cycle relative to the violaxanthin (Goss & Jakob, 2010; Wilhelm et al., 2006) might explain the rapid and effective synthesis of diadinoxanthin in large amounts (Lavaud et al., 2004; Lavaud et al., 2002). Lohr and Wilhelm (Lohr & Wilhelm, 1999) showed that some algae display the diadinoxanthin type of xanthophyll cycle, as well as features of the violaxanthin-based cycle.

In other groups of phytoplankton, there is no (cyanobacteria) or a questionable (red algae) xanthophyll cycle (Goss & Jakob, 2010) but the involvement of various de-epoxidized forms of xanthophylls (including zeaxanthin) in NPQ (Table 3) occurs by reaction center (RC) down-regulation (Kirilovsky, 2007; Stransky & Hager, 1970).

In general, as light intensity increases, the level of violaxanthin/diadinoxanthin decreases, reaching a steady state and, conversely, the level of zeaxanthin/diatoxanthin increases to asymptote (Demmig-Adams & Adams, 1993; Yamamoto, 1979).

	Algal Division/Class												
Pigment	Cyanophyta	Prochlorophyta	Rhodophyta	Cryptophyta	Chlorophyceae	Prasinophyceae	Euglenophyta	Eustigmatophyta	Bacillariophyta	Dinophyta	Prymnesiophyceae	Chrysophyceae	Raphidophyceae
Chlorophyll *a*	+		+	+	+	+	+	+	+	+	+	+	+
b					+	+	+						
c_1									+		+		+
c_2				+					+	+	+	+	+
c_3											+	+	
Carotenes- β,ε		+	+	+	+	+					+		
β,β	+	+			+	+	+	+	+	+	+	+	+
β,ψ					+								
ε,ε				+								+	
ψ, ψ				+									
Xanthophylls Alloxanthin				+									
Antheraxanthin					+	+	+						
Diadinoxanthin							+		+	+	+	+	+
Diatoxanthin							+		+	+	+	+	+
Zeaxanthin	+	+	+		+			+					
Violaxanthin					+	+		+					
Astaxanthin											+	+	
Dinoxanthin										+			
Fucoxanthin									+		+	+	+
Lutein					+	+							
Monadoxanthin				+									
Neoxanthin					+	+	+						
Peridinin										+			
Peridininol										+			

	Algal Division/Class												
Pigment	Cyanophyta	Prochlorophyta	Rhodophyta	Cryptophyta	Chlorophyceae	Prasinophyceae	Euglenophyta	Eustigmatophyta	Bacillariophyta	Dinophyta	Prymnesiophyceae	Chrysophyceae	Raphidophyceae
Prasinoxanthin						+							
Pyrrhoxanthin										+			
Biliproteins Allophycocyanin	+		+										
Phycocyanin	+		+	+									
Phycoerythrin	+		+	+									

Table 2. Distribution of major and taxonomically significant pigments in algal divisions/classes

Group 1	**Group 2**	**Group 3**
Zeaxanthin *No xanthophyll cycle*	*Zeaxanthin* *Violaxanthin*	*Diadinoxanthin* *Diatoxanthin*
Cyanobacteria Rodophyceae Cryptophyceae ? Glaueystophyceae	Phacophyceae Chlorophyceae Chrysophyceae Xanthophyceae Mosses Ferns Gymnosperms Angiosperms	Diatoms Chrysophyceae Xanthophycea Chloromonads Dinoflagellates Euglenophyceae

Table 3. The three groups of algal phyla according to their xanthophylls. Group 1 does not show a reversible epoxidation reaction.

The photoprotective NPQ process takes place in the light-harvesting complex of PSII. When irradiance exceeds the photosynthetic capacity of the cell, NPQ dissipates part of the excessively absorbed light energy, thus decreasing the excitation pressure on PSII (Li et al., 2009). NPQ is composed of three components – qE, qT, and qI, whose respective importance varies among photosynthetic lineages, qE being essential for most of them. qE is the energy-dependent quenching that is regulated by the build-up of a transthylakoid ΔpH and the operation of the xanthophyll cycle. The qT refers to the part of the quenching resulting from state transitions, while qI is due to photoinhibition. qT is relevant in phycobilisome-containing organisms (cyanobacteria and red algae) and green microalgae, but it is not really significant in high light (Ruban & Johnson, 2009). The origin of qI is not clearly defined except for some higher plants, and it requires special conditions (Demmig-Adams & Adams, 2006). Although the relationship between qE and the accumulation of de-epoxidized xanthophylls has been reported in many algal groups (Lavaud et al., 2007), there is still no clear picture of the functioning of qE in microalgae, although models have been proposed (Goss & Jakob, 2010; Lavaud, 2007), with the exception of *Chlamydomonas* (Peers et al., 2009). Regarding cyanobacteria and red algae, although there is a qE quenching that is supported by the presence of xanthophylls and a ΔpH, the composition and organization of the antenna obviously support another type of qE mechanism (Bailey & Grossman, 2008; Kirilovsky, 2007). Nevertheless, qE in cyanobacteria is not as powerful as in other phytoplankton taxa (Lavaud, 2007), possibly because of the lack of a finely regulated xanthophyll cycle. When necessary, cyanobacteria favor other photoprotective processes such as qT (described above) and the rapid repair of the D1 protein of the PS II reaction center (Six et al., 2007; Wilson et al., 2006).

The influence of the size and shape of cells on the capacity for regulation of photosynthesis and, in particular, via the xanthophyll-cycle operation, merits more attention (Key et al., 2010). Cell size could significantly affect xanthophyll-cycle functioning (Dimier et al., 2007b, 2009b; Lavaud et al., 2004). Indeed, physiological acclimation to light changes is a costly process. Cell size determines the structure of the PS II antenna and, therefore, pigment content, which constrains the use of the light resource, hence limiting the energy available for physiological responses to light fluctuations (Key et al., 2010; Litchman & Klausmeier, 2008; Raven & Kübler, 2002) . The influence of cell size/shape on metabolism, coupled with the metabolic theory of ecology (Brown et al., 2004) applied to the fast regulation of photosynthesis versus light, would bring interesting insights for studying photoadaptative strategies versus niche properties in microalgae. This would provide a background to understand how the environmental conditions affect photoregulatory capacity and efficiency and what their impact is on cell metabolism. Picoeukaryotes turned out to be interesting models to further explore this hypothesis (Dimier et al., 2007a, 2009a; Six et al., 2008, 2009). Dimier (2009a) suggested that the energy cost of enhanced photoregulation due to high-light fluctuation could be responsible for the decrease of growth rate in the shade adapted picoeukaryote *Pelagomonas*. Additionally, *in situ* studies showed, in agreement with laboratory experiments, that picoeukaryotes have high plasticity of PS II photoregulatory responses. This is probably related to the fact that the main limiting resource for these organisms is light (Timmermans et al., 2005), since nutrient availability does not seem to significantly determine their rate of primary productivity.

Recently, a specific role for the chlorophyll-binding, 22 kDa protein, psbS, has been shown (Goss & Jakob, 2010). This protein might be located in an intermediate position between LHCII and the inner antenna of RCII (Nield et al., 2000). The evidence suggests that energy-dependent quenching qE (which is defined as that component of the total non-photochemical quenching qN' directly attributable to the energization of the thylakoid membrane and, therefore, the rapidly entrained composition of qN) occurs when: i) there is ΔpH across the thylakoid membrane; and ii) zeaxanthin/diadinoxanthin is at high concentration (and violaxanthin/diadinoxanthin at low concentration) as a result of de-epoxidation of violaxanthin/diatoxanthin. Horton and Ruban (1994) suggested that there is a pocket extending from the intrathylakoid lumen into the membrane, by which low pH in the thylakoid lumen can influence a critical site in the thylakoid membrane. Since psbS is essential for qE to occur, it may be the protein which senses the low pH and binds zeaxanthin or it may play a crucial structural role in energy transfer/dissipation (Li et al., 2000; Raven & Beardall, 2003).

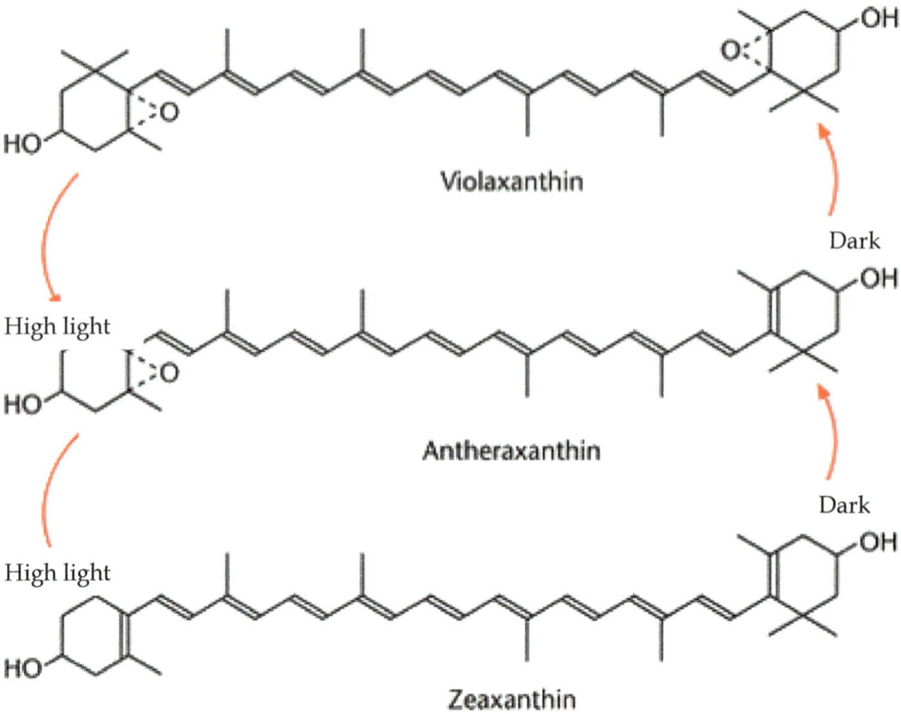

Fig. 4. The xanthophyll biosynthetic pathway in green algae and plants. When exposed to high light, violaxanthin converts to zeaxanthin, turning back into violaxanthin under low light or darkness

5. Xanthophyll cycle in higher plants and algae

Photosynthetic organisms have developed strategies to optimize light harvesting at low intensities while minimizing photoinhibitory damage due to excess energy at high-light intensity. They regulate the quantity and composition of the light-harvesting complexes (LHCs) and a number of other components of their photosynthetic machinery (Anderson et al., 1995; Falkowski & LaRoche, 1991). On shorter time scales, they react to an imbalance between light intensity and photosynthetic capacity (e.g., due to a change in light intensity, temperature, or nutrient supply) by rapid structural modification within the LHC of PSII (Bassi & Caffarri, 2000; Horton et al., 1996). These modifications lead to a decrease in NPQ. The partitioning of absorbed energy between transfer to the reaction center and photoprotective non-radiative dissipation is controlled by the trans-thylakoid pH gradient (Müller et al., 2001) and by the xanthophyll cycle. The molecular mechanisms of photoprotection have been mostly studied in higher plants (Demmig-Adams & Adams, 2006).

In comparison to higher plants, phytoplankton are well known to flourish in turbulent waters (Harris, 1986), where the amount of light available to phytoplankton unicellular organisms is highly unpredictable. The deep vertical mixing continuously sweeps them up and down, exposing the cell to very large short-term changes in light intensity on a time scale of minutes to hours. Higher plants, even though they are attached to the ground, are also exposed to light fluctuations due to a flicker effect caused by leaf movement in forests (Leakey et al., 2002, 2005).

The organization of the photosynthetic apparatus in diatoms differs in many respects from that of green algae and higher plants. The thylakoid membranes are loosely appressed and organized in extended layers of three without grana stacking, and the PSI and PSII are not segregated in different domains. The LHCs, which contain Chl *a*, Chl *b*, fucoxanthin, and the xanthophyll-cycle pigment diadinoxanthin, are equally distributed among appressed and nonappressed regions (Pyszniak & Gibbs, 1992) and there is no evidence of any state 1 to state 2 transitions (Owens, 1986). The xanthophyll concentration relative to chl can be two to four times more than in a higher plant.

The LHC subunits are made of several highly homologous proteins encoded by a multigene family (Bhaya & Grossman, 1993). The CP26 and CP29 subunits present in higher plants are not found in diatoms (Müller et al., 2001). When the cells are suddenly exposed to high-light intensity, an NPQ is rapidly developed. NPQ is associated with a xanthophyll cycle, the diadinoxanthin cycle, which differs from that of higher plants. The diadinoxanthin cycle converts the mono-epoxide carotenoid diadinoxanthin into the de-epoxide form diatoxanthin under high light, and diatoxanthin back into diadinoxanthin under low light or darkness (Arsalane et al., 1994).

In a diatom, the diadinoxanthin content can be modulated by the light regime to which culture is exposed (Willemoes & Monas, 1991). In higher plants, the xanthophyll cycle converts zeaxanthin through asteroxanthin to violaxanthin.

The microalgal xanthophyll-cycle activity shows striking peculiarities with respect to higher plants. This includes a high degree of variation in that cycle's regulation among the different taxa/species (Goss & Jakob, 2010; Lavaud et al., 2004; van de Poll et al., 2010), together with the growth phase (Arsalane et al., 1994; Dimier et al., 2009b; Lavaud et al., 2002, 2003;), the nutrient state (Staehr et al., 2002; Van de Poll et al., 2005), and the light history with both visible and UV radiation (Laurion & Roy, 2009; Lavaud, 2007; van de Poll & Buma, 2009). Also, recent reports demonstrated how xanthophyll-cycle activity and efficiency might be influenced by niche adaptation, and vice versa, in both pelagic (Dimier et al., 2007b, 2009a; Lavaud et al., 2004, 2007; Meyer et al., 2000) and benthic {Serodio, 2005 #3081;van Leeuwe, 2008 #3082} species, and how this could influence species succession {Meyer, 2000 #3080;Serodio, 2005 #3081;van Leeuwe, 2008 #3082}. This functional trait is part of the overall adaptive photophysiological properties of PS II, as shown for the diatoms (Wagner et al., 2006; Wilhelm et al., 2006). It highlights the narrow functional relationship between the niche adaptation and the capacity for photo-regulation/-acclimation, thus elucidating that the fast regulation of photosynthesis might be a crucial functional trait for microalgal ecology. In this respect, diatoms are currently the most studied group, probably because they appear to be the best xanthophyll cycle/NPQ performers among microalgae (Lavaud, 2007). Nevertheless, diatoms show a large interspecies xanthophyll cycle/NPQ diversity (Dimier et al., 2007b; Lavaud et al., 2004, 2007), which might take its source in the special evolution of this group (Armbrust, 2009), leading to its successful adaptation to all aquatic habitats driven by a change from a benthic to a pelagic way of life (Kooistra et al., 2007).The decrease in accessory-pigment diversity in diatoms compared with other microalgal groups would especially be an advantage for an opportunistic strategy (Dimier et al., 2009b) that might be related to the high plasticity of their PS II antenna function, including the xanthophyll cycle (Lavaud, 2007).

6. PSII protection by NPQ against photoinhibition

In higher plants and algae, the capacity for photosynthesis tends to saturate at high light intensities while the absorption of light remains linear. Therefore, the potential exists for the absorption of excess light energy by photosynthetic light-harvesting systems. This excess excitation energy leads to an increase in the lifetime of singlet excited chlorophyll, increasing the chances of the formation of long-lived chlorophyll triplet states by inter-system crossing. Triplet chlorophyll is a potent photosensitizer of molecular oxygen forming singlet oxygen, which can cause oxidative damage to the pigments, lipids, and proteins of the photosynthetic thylakoid membrane. One photoprotective mechanism that exists to counter this problem is the so-called non-photochemical quenching of chlorophyll fluorescence (NPQ), which relies upon the conversion and dissipation of the excess excitation energy into heat (Fig. 5). Excitation energy is, thereby, diverted away from the photosynthetic reaction centers and is no longer available for photochemistry. Although this increase in the rate of radiationless dissipation is associated with a reduction in the

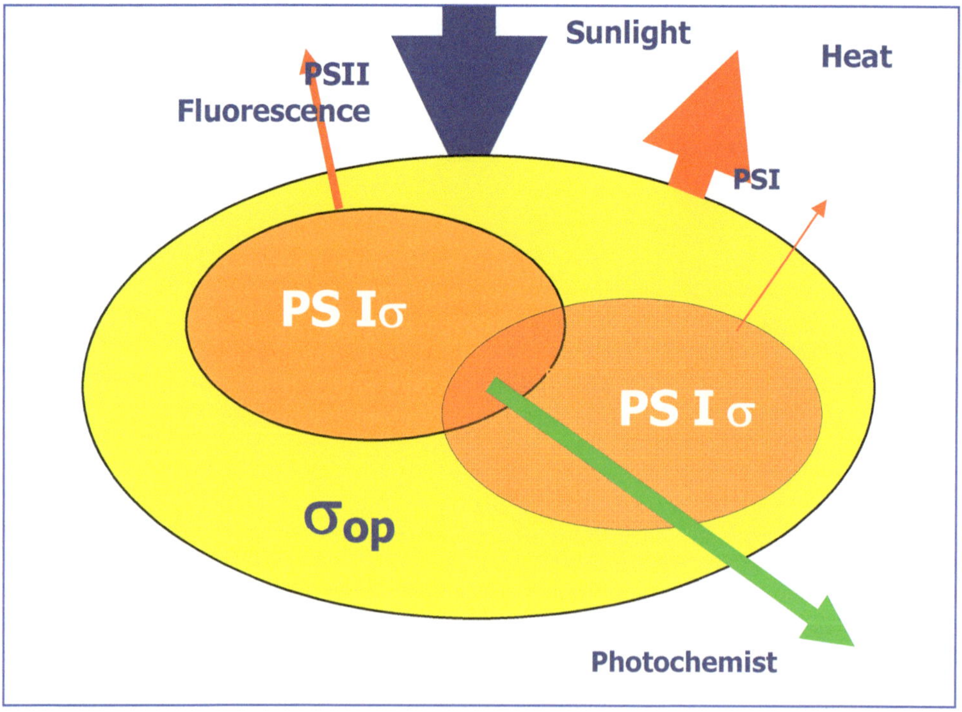

Fig. 5. The paths of energy dissipation and the use of harvested light energy in phototrophs. The σ symbol represents the different optical and functional cross sections of photosynthesis (for definitions, see Dubinsky (1980, 1992). Light energy diverted by the xanthophyll cycle is dissipated as heat, also included in the term NPQ.

efficiency of photosynthesis at low light, this disadvantage is most likely outweighed by the benefits of preventing the accumulation of excess excitation energy at high light, whereby damage to the reaction centers is avoided. In higher plants, the quenching could be as high as 80% and was induced by exposure to high irradiance (Demmig et al., 1988), and from one half to nearly all of the absorbed energy in algae. NPQ involves conformational changes within the light-harvesting proteins of photosystem II that bring about a change in pigment interactions, causing the formation of energy traps. The conformational changes are stimulated by a combination of transmembrane proton gradient, the PsbS subunit of photosystem II, and the enzymatic conversion of the carotenoid violaxanthin to zeaxanthin or diadinoxanthin to diatoxanthin (the xanthophyll cycle).

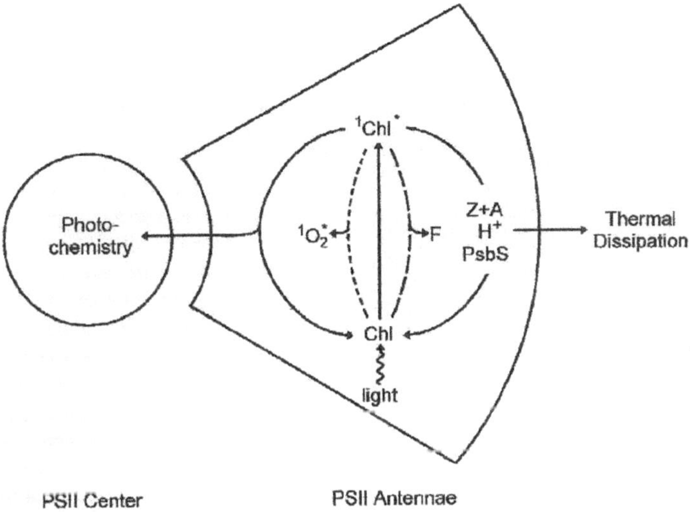

Fig. 6. Four routes of excitation energy in photosystem II light–collecting antennae. After moving chlorophyll electron by light to the (singlet) excited state, this energy can be used either for photochemistry or, alternatively, be dissipated thermally (as heat) in a process facilitated by xanthophylls (via the xanthophyll cycle), an acidic thylakoid, and the PsbS protein. A very small fraction of excitation energy is re-emitted as chlorophyll fluorescence, and can be used to monitor the excitation energy. If excited singlet chlorophyll were allowed to accumulate transiently, energy could also be transformed to oxygen, forming destructive singlet excited oxygen (after Demmig-Adams (2003)).

The xanthophyll cycle allows the fine tuning of the photosynthetic apparatus to ambient light by switching between two or three states of pigment couples constituting the xanthophyll cycle. When exposed to low light, most of its energy is used in the photochemical photolysis of water and the subsequent reduction of CO_2 to high-energy photosynthate. Under high light, the xanthophyll-cycle pigments undergo epoxidation and now divert light energy to harmless heat rather than damaging excess light (Adams et al., 1999; Demmig-Adams, 1998).

The NPQ mechanism depends on the size of the diadinoxanthin pool and can reach much larger values in algae than in higher plants (Lavaud et al., 2002; Li et al., 2002). NPQ and diatoxanthin are directly linearly related and if diatoxanthin is not present, NPQ cannot be formed. A fast diadinoxanthin de-epoxidation and concomitant formation of NPQ occurs within seconds (Lavaud et al., 2004). A linear relationship between zeaxanthin formation and NPQ has also been frequently observed in higher plants (Demmig-Adams & Adams, 1996). On the other hand, a number of reports show poor correlation between the light-induced zeaxanthin accumulation and the quenching of variable chlorophyll fluorescence in higher plants (Lichtenthaler et al., 1992; Schindler & Lichtenthaler, 1994; Schindler & Lichtenthaler, 1996), as well as in green algae (Masojidek et al., 1999). Poor correlation has been also found in green algae *Dunaliella,* which has similar light-harvesting and xanthophyll-cycle pigments to that of higher plants (Casper-Lindley & Bjorkman, 1998).

7. The biophysical mechanism of NPQ

There is a lack of knowledge concerning the exact nature and organization of light-harvesting complex containing fucoxanthin (LHCF) subunits in diatoms, and especially the location of diadinoxanthin and diatoxanthin in the antenna complex. In diatoms, the size of the diadinoxanthin pool increases under intermittent light and a larger fraction of the pool is susceptible to de-epoxidation (diatoxanthin). The diatoxanthin molecules produced under high light were shown to enhance the dissipation of excess energy and were, therefore, likely to be bound to the antenna subunits responsible for the NPQ. One hypothesis about diadinoxanthin enrichment under intermittent light is pigment-pigment 'replacement'. Under such conditions, the LHCF subunit would bind two diadinoxanthin molecules at the same time in continuous light. This is supported by the observation that the diadinoxanthin enrichment is correlated with a stoichiometrically parallel decrease in fucoxanthin content (Lavaud et al., 2003). In higher plants, the possibility that a given site of a LHC protein can bind different xanthophylls has been demonstrated.

The existence of LHCF subunits with different pigment content is very likely. Since, in parallel to the increase in diadinoxanthin content, Chl *c* decreases to the same extent as fucoxanthin, some subunits could be specifically rich in diadinoxanthin while others could mainly bind fucoxanthin and Chl *c* (Lavaud et al., 2003).

Under excess light, a higher degree of de-epoxidation occurs in diadinoxanthin- enriched cells. To be de-epoxidized, xanthophylls have to be accessible to the de-epoxidase localized in the lipid matrix. The fraction of diadinoxanthin that can be transformed to diatoxanthin is thus likely located at the periphery of pigment-protein complexes.

The exact biophysical mechanism by which fluorescence is quenched via the xanthophyll cycle is unclear. One possibility is that physical aggregation of chlorophyll molecules induced by the de-epoxidation of the xanthophylls leads to a reduction in the optical absorption cross section (i.e., a*) of the entire antenna system, and associated quenching of the fluorescence (Horton et al., 1996). The quenching mechanism in this scenario is unclear, however, if this scenario is valid, it should lead to change in both the optical and effective cross sections. An alternative hypothesis is based on direct competition for excitation energy within the LHC between the reaction center and the xanthophylls, upon de-epoxidation from a singlet state that can be populated with excitations emanating from the lowest singlet excited state of chlorophyll *a* (Frank & Cogdell, 1996; Owens, 1994). In effect, this hypothesis suggests that xanthophylls become reversibly activated quenchers within the pigment bed. Upon activation, they increase the probability that absorbed photons will be dissipated as heat through nonradioactive energy transfer to carotenoids. The process should lead to a change in the effective cross section of one or both reaction centers (depending upon which antenna system the xanthophyll cycle is associated with), but not a change in the optical absorption cross section (Falkowski & Chen, 2003).

8. Microalgal response to variable light environments

Phytoplankton species must cope with a highly variable environment that continuously requires energy for maintenance of photosynthetic productivity and growth. This is relevant in such aquatic ecosystems in which biodiversity is high and competition for resources is strong. Indeed, in a few cubic millimeters of water, many phytoplankton species can grow together, sharing and competing for the same energy resources, especially light and nutrients (Liess et al., 2009). To be competitive, phytoplankton must be able to respond quickly to any kind of changes occurring in their habitat. The main abiotic driving forces are temperature, nutrients, and light, the latter showing the highest variations in amplitude and frequency (Dubinsky & Schofield, 2010; MacIntyre et al., 2000; Raven & Geider, 2003). Hence, the response of phytoplankton might be supported by at least one irradiance-dependent physiological process that must be fast, flexible, and efficient (Dubinsky & Schofield, 2010; Li et al., 2009). Huisman et al. (2001) proposed that the diversity of life history and physiological abilities might promote the high biodiversity of phytoplankton. It has been proposed that the variability of physiological responses to light fluctuations would allow competitive exclusion and thus the spatial co-existence and/or the temporal succession of a multitude of species in both pelagic (Dimier et al., 2007b, 2009b; Lavaud et al., 2007) and benthic {van Leeuwe, 2008 #3082} ecosystems. Indeed, growth rate responds to fluctuating light in different ways as a function of groups/species of phytoplankton (Flöder et al., 2002; Litchman, 2000; Mitrovic et al., 2003; Wagner et al., 2006) and of the photoacclimation ability and light history of the cells (Laurion & Roy, 2009; Litchman & Klausmeier, 2001; van de Poll et al., 2007; van Leeuwe et al., 2005; Wagner et al., 2006). The photoresponse ultimately leading to a change in growth rate is thus a matter of both genomic plasticity and time scale (Dubinsky & Schofield, 2010; Grobbelaar, 2006). In the short term (a few seconds/minutes), the light fluctuations are mainly due to cloud movement, surface sunflecks, ripples on water surface, and vertical mixing generating unpredictable changes. These fluctuations, and especially their extremes (darkness and excess light), are generated by a lensing effect that simultaneously focuses and diffuses sunlight in the upper few meters of the water column, producing a constantly moving pattern of interspersed light and shadows on the substrate (Fig. 6). Due to the lensing effect, light intensity in shallow water environments sometimes reaches more than 9,000 µmol quanta m^{-2} s^{-1}, corresponding to 300-500% of the surface light intensity. For review of the "flicker effect" see Alexandrovich et al. (this volume).

These fluctuations in light can be harmful for the photosynthetic productivity of microalgae by promoting an imbalance between the harvesting of light energy and its use for photochemical processes and carbon fixation (Dubinsky & Schofield, 2010; Long et al., 1994; Raven & Geider, 2003). In order to regulate photosynthesis versus rapid light fluctuations, phytoplankton have evolved a number of physiological photoprotective mechanisms such as the photosystem II (PS II) and PS I electron cycles, the state-transitions, the fast repair of the D1 protein of the PS II reaction center, and the scavenging of reactive oxygen species (see (Falkowski & Raven, 1997; Lavaud, 2007; Ruban & Johnson, 2009)). Among these processes, the xanthophyll cycle and the dependent thermal dissipation of the excess light energy (NPQ for non-photochemical fluorescence quenching) play a central role. At longer time scales (hours to seasons), acclimation processes are supported by gene regulation, which modifies the architecture of the photosynthetic apparatus as well as enzymatic reactions involved in photochemistry and metabolism (Dubinsky & Schofield, 2010; Grobbelaar, 2006). These two types of responses, regulative and acclimative, are not mutually exclusive (Lavaud et al., 2007). For instance, long-term acclimation to a prolonged light regime (low or high intensity, or intermittent light) modifies the amount of pigments involved in the xanthophyll cycle, leading to a modulation of the high light response via the kinetics and amplitude of NPQ (Dimier et al., 2009b; Gundermann & Büchel, 2008; Lavaud et al., 2003; van de Poll et al., 2007).

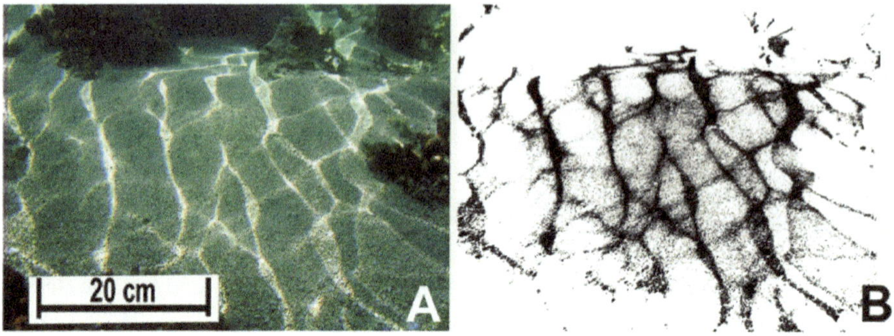

Fig. 7. Flicker light: (A) on a sandy bottom substrate; (B) digital processing of the original photo to enhance the high light regimes. Photos were taken at Bise, Okinawa, Japan (Yamasaki & Nakamura, 2008).

9. Conclusions

1. Phytoplankton evolution results in a taxonomic linkage between microalgae groups and xanthophyll-cycle development. In the ancestral cyanobacteria and the red algae, no xanthophyll-cycle mechanism is found. In the chlorophyta and land plants, the violaxanthin cycle evolved, whereas in the chromophyta, the diadinoxanthin variant has become the rule.
2. In two of the phototrophic phytoplankton groups (cyanobacteria and red algae), the composition of the antennae does not support a xanthophyll-cycle mechanism. In these taxa, there is another type of photoprotective process, a quenching mechanism that is supported by the xanthophylls and a pH gradient.

10. Acknowledgments

This study was supported by NATO SfP 981883. We thank Sharon Victor for the English editing.

11. References

Adams, W.W., Demmig-Adams, B., Logan, B.A., Barker, D.H. & Osmond, C.B. (1999). Rapid changes in xanthophyll cycle-dependent energy dissipation and photosystem II efficiency in two vines, Stephania japonica and Smilax australis, growing in the understory of an open Eucalyptus forest. *Plant Cell and Environment,* Vol.22, No.2, pp. 125-136, ISSN 0140-7791

Adams, W.W. III, Zarter, C.R., Mueh, K.E., Amiard, V. & Demmig-Adams, B. (2006). Energy dissipation and photoinhibition: a continuum of photoprotection. In: *Photoprotection, photoinhibition, gene regulation, and environment. Advances in photosynthesis and respiration,*Vol. 21, W.W. Demmig-Adams III & A.K. Mattoo (Eds.), 49-64, Springer, Dordrecht, The Netherlands

Adir, N., Zer, H., Shochat, S. & Ohad, I. (2003). Photoinhibition - a historical perspective. *Photosynthesis Research,* Vol.76, No.1-3, pp. 343-370, ISSN 0166-8595

Anderson, J.M., Chow, W.S. & Park, Y.I. (1995). The grand design of photosynthesis: Acclimation of the photosynthetic apparatus to environmental cues. *Photosynthesis Research,* Vol.46, No.1-2, pp. 129-139, ISSN 0166-8595

Armbrust, E.V. (2009). The life of diatoms in the world's oceans. *Nature,* Vol.459, No.7244, pp. 185-192, ISSN 0028-0836

Arsalane, W., Rousseau, B. & Duval, J.C. (1994). Influence of the pool size of the xanthophyll cycle on the effects of light stress in a diatom - Competition between photoprotection and photoinhibition. *Photochemistry and Photobiology,* Vol.60, No.3, pp. 237-243, ISSN 0031-8655

Bailey, S. & Grossman, A. (2008). Photoprotection in cyanobacteria: Regulation of light harvesting. *Photochemistry and Photobiology,* Vol.84, No.6, pp. 1410-1420, ISSN 0031-8655

Baroli, I. and Melis, A. (1996). Photoinhibition and repair in *Dunaliella salina* acclimated to different growth irradiances. *Planta,* Vol.198, No.4, pp. 640-646, ISSN 0032-0935

Bassi, R. & Caffarri, S. (2000). LHC proteins and the regulation of photosynthetic light harvesting function by xanthophylls. *Photosynthesis Research,* Vol.64, No.2-3, pp. 243-256, ISSN 0166-8595

Berger, R., Liaaen-Jensen, S., McAlister, V. & Guillard, R.R.L. (1977). Carotenoids of the Prymneiophyceae (Haptophyceae). *Biochemical Systematicas and Ecology,* Vol.5, pp. 71-75

Bhattacharya, D. & Medlin ,L. (1995). The phylogeny of plastids - a review based on comparisons of small-subunit ribosomal-RNA coding regions. *Journal of Phycology,* Vol.31, No.4, pp. 489-498, ISSN 0022-3646

Bhaya, D. & Grossman, A.R. (1993). Characterization of gene clusters encoding the fucoxanthin chlorophyll proteins of the diatom *Phaeodactylum tricornutum. Nucleic Acids Research,* Vol.21, No.19, pp. 4458-4466, ISSN 0305-1048

Brown, J.H., Gillooly, J.F., Allen, A.P., Savage, V.M. & West, G.B. (2004). Toward a metabolic theory of ecology. *Ecology,* Vol.85, No.7, pp. 1771-1789, ISSN 0012-9658

Casper-Lindley, C. & Bjorkman, O. (1998). Fluorescence quenching in four unicellular algae with different light-harvesting and xanthophyll-cycle pigments. *Photosynthesis Research,* Vol.56, No.3, pp. 277-289, ISSN 0166-8595

Delwiche, C.F. (1999). Tracing the thread of plastid diversity through the tapestry of life. *American Naturalist,* Vol.154, pp. S164-S177, ISSN 0003-0147

Demmig-Adams, B. (1998). Survey of thermal energy dissipation and pigment composition in sun and shade leaves. *Plant and Cell Physiology,* Vol.39, No.5, pp. 474-482, ISSN 0032-0781

Demmig-Adams, B. (2003). Linking the xanthophyll cycle with thermal energy dissipation. *Photosynthesis Research*, Vol.76, No.1-3, pp. 73-80, ISSN 0166-8595

Demmig-Adams, B. & Adams, W.W. (1993). The xanthophyll cycle, protein turnover, and the high light tolerance of sun-acclimated leaves. *Plant Physiology*, Vol.103, No.4, pp. 1413-1420, ISSN 0032-0889

Demmig-Adams, B. & Adams, W.W. (1996). The role of xanthophyll cycle carotenoids in the protection of photosynthesis. *Trends in Plant Science*, Vol.1, No.1, pp. 21-26, ISSN 1360-1385

Demmig-Adams, B. & Adams, W.W. (2006). Photoprotection in an ecological context: the remarkable complexity of thermal energy dissipation. *New Phytologist*, Vol.172, No.1, pp. 11-21, ISSN 0028-646X

Demmig, B., Winter, K., Kruger, A. & Czygan, F.C. (1988). Zeaxanthin and the heat dissipation of excess light energy in *Nerium oleander* exposed to a combination of high light and water stress. *Plant Physiology*, Vol.87, No.1, pp. 17-24, ISSN 0032-0889

Dimier, C., Corato, F., Saviello, G. & Brunet, C. (2007a). Photophysiological properties of the marine picoeukaryote Picochlorum RCC 237 (Trebouxiophyceae, Chlorophyta). *Journal of Phycology*, Vol.43, No.2, pp. 275-283, ISSN 0022-3646

Dimier, C., Corato, F., Tramontano, F. & Brunet, C. (2007b). Photoprotection and xanthophyll-cycle activity in three marine diatoms. *Journal of Phycology*, Vol.43, pp. 937-947, ISSN 0022-3646

Dimier, C., Brunet, C., Geider, R. & Raven, J. (2009a). Growth and photoregulation dynamics of the picoeukaryote Pelagomonas calceolata in fluctuating light. *Limnology and Oceanography*, Vol.54, No.3, pp. 823-836, ISSN 0024-3590

Dimier, C., Saviello, G. & Tramontano, F. (2009b). Comparative ecophysiology of the xanthophyll cycle in six marine phytoplanktonic species. *Protist*, Vol.160, pp. 397-411

Dubinsky, Z. (1980). Light utilization efficiency in natural phytoplankton communities. In: *Primary Productivity in the Sea*, P.G. Falkowski (Ed.), 83-97, Plenum Press, New York

Dubinsky, Z. (1992). The optical and functional cross-sections of phytoplankton photosynthesis. In: *Primary Productivity and Biogeochemical Cycles in the Sea*, P.G. Falkowski & A.D. Woodhead (Eds.), 31-45, Plenum Press, New York

Dubinsky, Z. & Schofield, O. (2010). From the light to the darkness: thriving at the light extremes in the oceans. *Hydrobiologia* ,Vol.639, No.1, pp. 153-171, ISSN 0018-8158

Durnford, D.G., Price, J.A., McKim, S.M. & Sarchfield, M.L. (2003). Light-harvesting complex gene expression is controlled by both transcriptional and post-transcriptional mechanisms during photoacclimation in Chlamydomonas reinhardtii. *Physiologia Plantarum*, Vol.118, No.2, pp. 193-205, ISSN 0031-9317

Emerson, R. & Arnold, W. (1932). The photochemical reaction in photosynthesis. *Journal of General Physiology*, Vol.16, pp. 191-205

Falkowski, P.G. & LaRoche, J. (1991). Acclimation to spectral irradiance in algae. *Journal of Phycology*, Vol.27, pp. 8-14

Falkowski, P.G. & Raven, J.A., (1997). *Aquatic Photosynthesis*. Blackwell Science Inc., Malden, USA.

Falkowski, P.G. & Chen, Y.B. (2003). Photoacclimation of light harvesting systems in eukaryotic algae. In: *Light-harvesting Antennas in Photosynthesis*, B.R. Green and W.W. Parson (Eds.), Kluwer Academic Publishers, Dordrecht, Boston

Falkowski, P.G., Katz, M., van Schootenbrugge, B., Schofield, O. & Knoll, A.H. (2004a). Why is the land green and the ocean red? In: *Coccolithophores: From Molecular Processes to Global Impact*, H.R. Thierstein & J. Young (Eds.), Springer-Verlag, Berlin

Falkowski, P.G., Katz, M.E., Knoll, A.H., Quigg, A., Raven, J.A., Schofield, O. & Taylor, F.J. (2004b). The evolution of modern eukaryotic phytoplankton. *Science*, Vol.305, pp. 354-360

Flöder, S., Urabe, J. & Kawabata, Z. (2002). The influence of fluctuating light intensities on species composition and diversity of natural phytoplankton communities. *Oecologia*, Vol.133, No.3, pp. 395-401, ISSN 0029-8549

Frank, H.A. & Cogdell, R.J. (1996). Carotenoids in photosynthesis. *Photochemistry and Photobiology*, Vol.63, No.3, pp. 257-264, ISSN 0031-8655

Goss, R. & Jakob, T. (2010). Regulation and function of xanthophyll cycle-dependent photoprotection in algae. *Photosynthesis Research*, Vol.106, No.1-2, pp. 103-122, ISSN 0166-8595

Grobbelaar, J.U. (2006). Photosynthetic response and acclimation of microalgae to light fluctuations In: *Algal Cultures Analogues of Blooms and Applications*, D.V. Subba Rao (Ed.), 671-683, Science Publishers, Enfield (NH), USA, Plymouth, UK

Gundermann, K. & Büchel, C. (2008). The fluorescence yield of the trimeric fucoxanthin-chlorophyll-protein FCPa in the diatom Cyclotella meneghiniana is dependent on the amount of bound diatoxanthin. *Photosynthesis Research*, Vol.95, No.2-3, pp. 229-235, ISSN 0166-8595

Harris, G.P. (1986). *Phytoplankton Ecology: Structure, Function and Fluctuation*. Chapman & Hall, London.

Horton, P. and Ruban, A.V. (1994). The role of LHCII in energy quenching. In: *Photoinhibition of photosynthesis - from molecular mechanisms to the field (monograph)*, N.R. Baker & J.R.G. Bowyer (Eds.), 111-128, Bios Scientific Publishers, Oxford

Horton, P., Ruban, A.V. & Walters, R.G. (1996). Regulation of light harvesting in green plants. *Annual Review of Plant Physiology and Plant Molecular Biology*, Vol.47, pp. 655-684, ISSN 0066-4294

Huisman, J., Johansson, A.M., Folmer, E.O. & Weissing, F.J. (2001). Towards a solution of the plankton paradox: the importance of physiology and life history. *Ecology Letters*, Vol.4, No.5, pp. 408-411, ISSN 1461-023X

Jeffrey, S.W., Mantoura, R.F.C. & Wright, S.W. (1997). *Phytoplankton Pigments in Oceanography*. UNESCO Publishing, Paris

Johansen, J.E., Svec, W.A., Liaaenje. S & Haxo, F.T. (1974). Algal carotenoids. 10. Carotenoids of Dinophyceae. Phytochemistry, Vol.13, No.10, pp. 2261-2271, ISSN 0031-9422

Kanervo, E., Suorsa, M. & Aro, E.M. (2005). Functional flexibility and acclimation of the thylakoid membrane. *Photochemical & Photobiological Sciences*, Vol.4, No.12, pp. 1072-1080, ISSN 1474-905X

Key, T., McCarthy, A., Campbell, D.A., Six, C., Roy, S. & Finkel, Z.V. (2010). Cell size trade-offs govern light exploitation strategies in marine phytoplankton. *Environmental Microbiology*, Vol.12, No.1, pp. 95-104, ISSN 1462-2912

Kirilovsky, P. (2007). Photoprotection in cyanobacteria, the orange carotenoid protein (OCP)-related non-photochemical quenching mechanism. *Photosynthesis Research*, Vol.93, pp. 7-16

Kirk, J.T.O. (1994). *Light and Photosynthesis in Aquatic Ecosystems*, 2nd ed. Cambridge University Press, London, New York

Kooistra, W.H.C.F., Gersonde, R., Medlin, L.K. & Mann, D.G. (2007). The origin and evolution of the diatoms: their adaptation to a planktonic existence. In: *Evolution of planktonic photoautotrophs*, P.G. Falkowski & A.H. Knoll (Eds.), 207-249, Academic Press, Inc., Burlington, MA

Krause, G.H. & Weis, E. (1991). Chlorophyll fluorescence and photosynthesis - the basics. *Annual Review of Plant Physiology and Plant Molecular Biology*, Vol.42, pp. 313-349

Laurion, I. & Roy, S. (2009). Growth and photoprotection in three dinoflagellates (including two strains of Alexandrium tamarense) and one diatom exposed to four weeks of natural and enhanced ultraviolet-B radiation. *Journal of Phycology*, Vol.45, No.1, pp. 16-33, ISSN 0022-3646

Lavaud, J. (2007). Fast regulation of photosynthesis in diatoms: mechanisms, evolution and ecophysiology. *Plant Science and Biotechnology*, Vol.1, pp. 267-287

Lavaud, J., Rousseau, B., van Gorkom, H.J. & Etienne, A.L. (2002). Influence of the diadinoxanthin pool size on photoprotection in the marine planktonic diatom *Phaeodactylum tricornutum. Plant Physiology*, Vol.129, No.3, pp. 1398-1406, ISSN 0032-0889

Lavaud, J., Rousseau, B. & Etienne, A.L. (2003). Enrichment of the light-harvesting complex in diadinoxanthin and implications for the nonphotochemical fluorescence quenching in diatoms. *Biochemistry*, Vol.42, pp. 5802-5808

Lavaud, J., Rousseau, B. & Etienne, A.L. (2004). General features of photoprotection by energy dissipation in planktonic diatoms (Bacillariophyceae). *Journal of Phycology*, Vol.40, No.1, pp. 130-137, ISSN 0022-3646

Lavaud, J., Strzepek, R.F. & Kroth, P.G. (2007). Photoprotection capacity differs among diatoms: Possible consequences on the spatial distribution of diatoms related to fluctuations in the underwater light climate. *Limnology and Oceanography*, Vol.52, No.3, pp. 1188-1194, ISSN 0024-3590

Leakey, A.D.B., Press, M.C., Scholes, J.D. & Watling, J.R. (2002). Relative enhancement of photosynthesis and growth at elevated CO_2 is greater under sunflecks than uniform irradiance in a tropical rain forest tree seedling. *Plant Cell and Environment*, Vol.25, No.12, pp. 1701-1714, ISSN 0140-7791

Leakey, A.D.B., Scholes, J.D. & Press, M.C. (2005). Physiological and ecological significance of sunflecks for dipterocarp seedlings. *Journal of Experimental Botany*, Vol.56, No.411, pp. 469-482, ISSN 0022-0957

Ledford, H.K. & Niyogi, K.K. (2005). Singlet oxygen and photo-oxidative stress management in plants and algae. *Plant Cell and Environment*, Vol.28, No.8, pp. 1037-1045, ISSN 0140-7791

Li, X.P., Bjorkman, O., Shih, C., Grossman, A.R., Rosenquist, M., Jansson, S. & Niyogi, K.K. (2000). A pigment-binding protein essential for regulation of photosynthetic light harvesting. *Nature*, Vol.403, No.6768, pp. 391-395, ISSN 0028-0836

Li, X.P., Gilmore, A.M. & Niyogi, K.K. (2002). Molecular and global time-resolved analysis of a psbS gene dosage effect on pH- and xanthophyll cycle-dependent nonphotochemical quenching in photosystem II. *Journal of Biological Chemistry*, Vol.277, No.37, pp. 33590-33597, ISSN 0021-9258

Li, Z.R., Wakao, S., Fischer, B.B. & Niyogi, K.K. (2009). Sensing and responding to excess light. *Annual Review of Plant Biology*, Vol.60, pp. 239-260, ISSN 1543-5008

Lichtenthaler, H.K., Burkart, S., Schindler, C. & Stober, F. (1992). Changes in photosynthetic pigments and *in-vivo* chlorophyll fluorescence parameters under photoinhibitory growth conditions *Photosynthetica*, Vol.27, No.3, pp. 343-353, ISSN 0300-3604

Liess, A., Lange, K., Schulz, F., Piggott, J.J., Matthaei, C.D. & Townsend, C.R. (2009). Light, nutrients and grazing interact to determine diatom species richness via changes to productivity, nutrient state and grazer activity. *Journal of Ecology*, Vol.97, No.2, pp. 326-336, ISSN 0022-0477

Litchman, E. (2000). Growth rates of phytoplankton under fluctuating light. *Freshwater Biology*, Vol.44, No.2, pp. 223-235, ISSN 0046-5070

Litchman, E. & Klausmeier, C.A. (2001). Competition of phytoplankton under fluctuating light. *American Naturalist*, Vol.157, No.2, pp. 170-187, ISSN 0003-0147

Litchman, E. & Klausmeier, C.A. (2008). Trait-based community ecology of phytoplankton. *Annual Review of Ecology Evolution and Systematics*, Vol.39 , pp. 615-639, ISSN 1543-592X

Lohr, M. & Wilhelm, C. (1999). Algae displaying the diadinoxanthin cycle also possess the violaxanthin cycle. *Proceedings of the National Academy of Sciences of the United States of America*, Vol.96, No.15, pp. 8784-8789, ISSN 0027-8424

Lohr, M. & Wilhelm, C. (2001). Xanthophyll synthesis in diatoms: quantification of putative intermediates and comparison of pigment conversion kinetics with rate constants derived from a model. *Planta*, Vol.212, No.3, pp. 382-391, ISSN 0032-0935-

Long, S.P., Humphries, S. & Falkowski, P.G. (1994). Photoinhibition of photosynthesis in nature. *Annual Review of Plant Physiology and Plant Molecular Biology*, Vol.45, pp. 633-662

Lu, C. & Vonshak, A. (1999). Photoinhibition in outdoor *Spirulina platensis* cultures assessed by polyphasic chlorophyll fluorescence transients. *Journal of Applied Phycology*, Vol.11, No.4, pp.355-359, ISSN 0921-8971

MacIntyre, H.L., Kana, T.M. & Geider, J.R. (2000). The effect of water motion on the short-term rates of photosynthesis by marine phytoplankton. *Trends in Plant Science*, Vol.5, pp. 12-17

Masojidek, J., Torzillo, G., Koblizek, M., Kopecky, J., Bernardini, P., Sacchi, A. & Komenda, J. (1999). Photoadaptation of two members of the Chlorophyta (Scenedesmus and Chlorella) in laboratory and outdoor cultures: changes in chlorophyll fluorescence quenching and the xanthophyll cycle. *Planta*, Vol.209, pp. 126-135

Matsubara, S., Gilmore, A.M., Ball, M.C., Anderson, J.M. & Osmond, C.B. (2002). Sustained downregulation of photosystem II in mistletoes during winter depression of photosynthesis. *Functional Plant Biology*, Vol.29, No.10, pp. 1157-1169, ISSN 1445-4408

Meyer, A.A., Tackx, M. & Daro, N. (2000). Xanthophyll cycling in *Phaeocystis globosa* and *Thalassiosira* sp.: a possible mechanism for species succession. *Journal of Sea Research*, Vol.43, No.3-4, pp. 373-384, ISSN 1385-1101

Mitrovic, S.M., Howden, C.G., Bowling, L.C. & Buckney, R.T. (2003). Unusual allometry between in situ growth of freshwater phytoplankton under static and fluctuating light environments: possible implications for dominance. *Journal of Plankton Research*, Vol.25, No.5, pp. 517-526, ISSN 0142-7873

Müller, P., Li, X.P. & Niyogi, K.K. (2001). Non-photochemical quenching. A response to excess light energy. *Plant Physiology*, Vol.125, No.4, pp.1558-1566, ISSN 0032-0889

Mullineaux, P.M. & Rausch, T. (2005). Glutathione, photosynthesis and the redox regulation of stress-responsive gene expression. *Photosynthesis Research*, Vol.86, No.3, pp. 459-474, ISSN 0166-8595

Nield, J., Orlova, E.V., Morris, E. P., Gowen, B., van Heel, M. & Barber, J. (2000). 3D map of the plant photosystem II supercomplex obtained by cryoelectron microscopy and single particle analysis. *Nature Structural Biology*, Vol.7, No.1, pp. 44-47, ISSN 1072-8368

Noctor, G. & Foyer, C.H. (1998). Ascorbate and glutathione: Keeping active oxygen under control. *Annual Review of Plant Physiology and Plant Molecular Biology*, Vol.49, pp. 249-279, ISSN 1040-2519

Owens, T.G. (1986). Light-harvesting function in the diatom *Phaeodactylum tricornutum*. 2. Distribution of excitation-energy between the photosystems. *Plant Physiology*, Vol.80, No.3, pp. 739-746, ISSN 0032-0889

Owens, T.G. (1994). In vivo chlorophyll fluorescence as a probe of photosynthetic physiology. In: *Plant Responses to the Gaseous Environment*, R.G. Alscher & A.R. Wellburn (Eds.), 195-217, Chapman & Hall, London

Palmer, J.D. (2003). The symbiotic birth and spread of plastids: How many times and whodunit? *Journal of Phycology*, Vol.39, No.1, pp. 4-11, ISSN 0022-3646

Peers, G., Truong, T.B., Ostendorf, E., Busch, A., Elrad, D., Grossman, A.R., Hippler, M. & Niyogi, K.K. (2009). An ancient light-harvesting protein is critical for the regulation of algal photosynthesis. *Nature*, Vol.462, No.7272, pp. 518-U215, ISSN 0028-0836

Pyszniak, A.M. & Gibbs, S.P. (1992). Immunocytochemical localization of photosystem-I and the fucoxanthin-chlorophyll-a/c light-harvesting complex in the diatom *Phaeodactylum tricornutum*. *Protoplasma*, Vol.166, No.3-4, pp. 208-217, ISSN 0033-183X

Raven, J.A. & Kübler, J.E. (2002). New light on the scaling of metabolic rate with the size of algae. *Journal of Phycology*, Vol.38, pp. 11-16

Raven, J.A. & Beardall, J. (2003). CO_2 acquisition mechanisms in algae: Carbon dioxide diffusion and carbon dioxide concentrating mechanisms. In: *Photosynthesis in the algae*, A. Larkum, J.A. Raven & S. Douglas (Eds.), 225-244, Kluwer

Raven, J.A. & Geider, J.R. (2003). Adaptation, acclimation and regulation in algal photosynthesis. In: *Photosynthesis in Algae*, W.D. Larkum, S.E. Douglas & J.A. Raven (Eds.), 385-412, Kluweer Academic Publishers, Dordrecht, The Netherlands

Renstrom, B., Borch, G., Skulberg, O.M. & Liaaenjensen, S. (1981). Natural occurrence of enantiometric and meso-astaxanthin. 3. Optical purity of (3s,3's)-astaxanthin from *Haematococcus pluvialis*. *Phytochemistry*, Vol.20, No.11, pp. 2561-2564, ISSN 0031-9422

Ruban, A.V. & Johnson, M.P. (2009). Dynamics of higher plant photosystem cross-section associated with state transitions. *Photosynthesis Research*, Vol.99, No.3, pp. 173-183, ISSN 0166-8595

Schindler, C. & Lichtenthaler, H.K. (1994). Is there a correlation between light-induced zeaxanthin accumulation and quenching of variable chlorophyll a fluorescence. *Plant Physiology and Biochemistry*, Vol.32, No.6, pp. 813-823, ISSN 0981-9428

Schindler, C. & Lichtenthaler, H.K. (1996). Photosynthetic CO_2 assimilation, chlorophyll fluorescence and zeaxanthin accumulation in field grown maple trees in the course of a sunny and a cloudy day. *Journal of Plant Physiology*, Vol.148, No.3-4, pp. 399-412, ISSN 0176-1617

Serodio, J., Cruz, S ,.Vieira, S. & Brotas, V. (2005). Non-photochemical quenching of chlorophyll fluorescence and operation of the xanthophyll cycle in estuarine microphytobenthos. *Journal of Experimental Marine Biology and Ecology*, Vol.326, No.2, pp. 157-169, ISSN 0022-0981

Six, C., Finkel, Z.V., Irwin, A.J. & Campbell, D.A. (2007). Light variability illuminates niche-partitioning among marine picocyanobacteria. *PLoS One*, Vol.2, No.12, pp. e1341, ISSN 1932-6203

Six, C., Finkel, Z.V., Rodriguez, F., Marie, D., Partensky, F. & Campbell, D.A. (2008). Contrasting photoacclimation costs in ecotypes of the marine eukaryotic picoplankter Ostreococcus. *Limnology and Oceanography*, Vol.53, No.1, pp. 255-265, ISSN 0024-3590

Six, C., Sherrard, R., Lionard, M., Roy, S. & Campbell, D.A. (2009). Photosystem II and pigment dynamics among ecotypes of the Green Alga Ostreococcus. *Plant Physiology*, Vol.151, No.1, pp. 379-390, ISSN 0032-0889

Staehr, P.A., Henriksen, P. & Markager, S. (2002). Photoacclimation of four marine phytoplankton species to irradiance and nutrient availability. *Marine Ecology Progress Series*, Vol.238, pp. 47-59

Stransky, H. & Hager, A. (1970). Carotenoid pattern and occurrence of light induced xanthophyll cycle in various classes of algae. 2. Xanthophyceae. *Archiv Fur Mikrobiologie*, Vol.71, No.2, pp. 164-190, ISSN 0003-9276

Timmermans, K.R., van der Wagt, B., Veldhuis, M.J.W., Maatman, A. & de Baar, H.J.W. (2005). Physiological responses of three species of marine pico-phytoplankton to ammonium, phosphate, iron and light limitation. *Journal of Sea Research*, Vol.53, pp. 109-120

Van de Poll, W.H., Van Leeuwe, M.A., Roggeveld, J. & Buma, A.G.J. (2005). Nutrient limitation and high irradiance acclimation reduce PAR and UV-induced viability loss in the Antarctic diatom Chaetoceros brevis (Bacillariophyceae). *Journal of Phycology*, Vol.41, pp. 840-850

van de Poll, W.H., Visser, R.J.W. & Buma, A.G.J. (2007). Acclimation to a dynamic irradiance regime changes excessive irradiance sensitivity of *Emiliania huxleyi* and *Thalassiosira weissflogii*. *Limnology and Oceanography*, Vol.52, No.4, pp. 1430-1438, ISSN 0024-3590

van de Poll, W.H. & Buma, A.G.J. (2009). Does ultraviolet radiation affect the xanthophyll cycle in marine phytoplankton? *Photochemical & Photobiological Sciences*, Vol.8, No.9, pp. 1295-1301, ISSN 1474-905X

van de Poll, W.H., Buma, A.G.J., Visser, R.J.W., Janknegt, P.J., Villafane, V.E. & Helbling, E.W. (2010). Xanthophyll cycle activity and photosynthesis of *Dunaliella tertiolecta* (Chlorophyceae) and *Thalassiosira weissflogii* (Bacillariophyceae) during fluctuating solar radiation. *Phycologia*, Vol.49, No.3, pp. 249-259, ISSN 0031-8884

van Leeuwe, M.A., van Sikkelerus, B., Gieskes, W.W.C. & Stefels, J. (2005). Taxon-specific differences in photoacclimation to fluctuating irradiance in an Antarctic diatom and a green flagellate. *Marine Ecology-Progress Series*, Vol.288, pp. 9-19, ISSN 0171-8630 van Leeuwe, M.A., Brotas, V., Consalvey, M., Forster, R.M., Gillespie, D., Jesus, B., Roggeveld, J. & Gieskes, W.W.C. (2008). Photoacclimation in microphytobenthos and the role of xanthophyll pigments. *European Journal of Phycology*, Vol.43, No.2, pp. 123-132, ISSN 0967-0262

Wagner, H., Jakob, T. & Wilhelm, C. (2006). Balancing the energy flow from captured light to biomass under fluctuating light conditions. *New Phytologist*, Vol.169, No.1, pp. 95-108, ISSN 0028-646X

Wilhelm, C., Büchel, C., Fisahn, J. Goss, R., Jakob, T., LaRoche, J., Lavaud, J., Lohr, M., Riebesell, U., Stehfest, K., Valentin, K. & Kroth, P.G. (2006). The regulation of carbon and nutrient assimilation in diatoms is significantly different from green algae. *Protist*, Vol.157, No.2, pp. 91-124, ISSN 1434-4610

Willemoes, M. & Monas, E. (1991). Relationship between growth irradiance and the xanthophyll cycle pool in the diatom *Nitzschia palea*. *Physiologia Plantarum*, Vol.83, No.3, pp. 449-456, ISSN 0031-9317

Wilson, A., Ajlani, G., Verbavatz, J.M., Vass, I., Kerfeld, C.A. & Kirilovsky, D. (2006). A soluble carotenoid protein involved in phycobilisome-related energy dissipation in cyanobacteria. *Plant Cell*, Vol.18, No.4, pp. 992-1007, ISSN 1040-4651

Withers, N.W., Fiksdahl, A., Tuttle, R.C. & Liaaenjensen, S. (1981). Carotenoids of the Chrysophyceae. *Comparative Biochemistry and Physiology B-Biochemistry & Molecular Biology*, Vol.68, No.2, pp. 345-349, ISSN 0305-0491

Yamamoto, H.Y. (1979). Biochemistry of the violaxanthin cycle in higher plants. *Pure and Applied Chemistry*, Vol.51, pp. 639-648

Yamamoto, H.Y., Nakayama, T.O.M. & Chichester, C.O. (1962). Studies on the light and dark interconversion of leaf xanthophylls. *Archives of Biochemistry*, Vol.97, pp. 168-173

Yamasaki, H. & Nakamura, T. (2008). Flicker light effects on photosynthesis of symbiotic algae in the reef-building coral *Acropora digitifera* (Cnidaria: Anthozoa: Scleractinia). *Pacific Science*, Vol.62, No.3, pp. 341-350, ISSN 0030-8870

12

Physiology of Crops and Weeds Under Biotic and Abiotic Stresses

Germani Concenço[1], Ignacio Aspiazú[2], Evander A. Ferreira[3], Leandro Galon[4] and Alexandre F. da Silva[5]
[1]*Embrapa Agropecuária Oeste, Dourados-MS*
[2]*Universidade Estadual de Montes Claros-MG*
[3]*UFVJM, Diamantina-MG*
[4]*Universidade Federal do Pampa, Itaqui-RS*
[5]*Embrapa Milho e Sorgo, Sete Lagoas-MG*
Brazil

1. Introduction

The application of knowledge with strong physiological basis of crop yield, allied to genetic and environmental factors, is essential in developing proper practices for crop management aiming high yields (Floss, 2008). Several aspects determine the performance of a particular crop plant in a given environment, such as temperature, water availability, incidence of pests, plant genetics and management applied. Although it is virtually impossible to control all these factors, plant behavior can be assessed when submitted to different levels of these factors to understand how the responses of the plant to that given stress are formed (Radosevich et al., 2007; Gurevitch et al., 2009).

Population growth leads to an increasing demand for food, fibers and energy, and requires both expansion of the area for crops as well as higher crop yields. In both cases, one of the limiting factors is the occurrence of weed species. The high cost of human labor led farmers to choose weed control practices which allow reduction of the production costs (Silva et al., 2007). In the past, chemical management was used as the only method of weed control, but several problems led to the development of cultural methods as effective tools for lasting and low cost weed management.

More recently the physiology and ecology of crops and weed species gained increasing importance in the development of methods of weed control (Radosevich et al., 2007; Gurevitch et al., 2009). Several studies of competition between crops and weed species were conducted and the results of these studies are being applied at planning of integrated management practices such as crop rotation, succession, crop-livestock integration and winter crops as tools for suppressing weed occurrence (Severino, 2005; Ceccon, 2007). Most of these studies allow modeling the dynamics of weeds infestation in certain crops and optimizing the system as a whole, based in dry mass accumulation, plants height, number of tillers or branches, number of inflorescences and other directly measured variables (Galon et al., 2007; Fleck et al., 2008; Bianchi et al., 2010).

On the other hand, there is still a big gap between physiological, high specialized studies and application of these results for practical everyday weed management inside crops. Weed biologists, mainly from under developed countries, often do not use physiological parameters in association to the directly measured variables as tools for supporting their findings. Basic research materials which support applied studies (Radosevich et al., 2007; Larcher, 2004; Gurevitch et al., 2009; Aliyev, 2010), propose changes to this scenario.
This chapter is proposed to present options involving the application of physiological parameters in studies of crop-weed competition, whether the chemical tool of weed control is present or not, highlighting examples of studies of the types of findings which could result from the use of physiological parameters in ecophysiological and competition studies.

2. Weed species

Elaborate a definition for "weed species" in agricultural terms was never easy. All the current concepts are based in the not desired occurrence of a plant species in a given time and place (Silva et al., 2007). A plant species can be considered a weed if its presence interferes in some way in a given situation. Therefore, no specific plant species can be considered essentially a weed, since this will depend on the place and moment this species is present for it to be considered harmful to the system - a weed. Even a crop plant, like corn, can be considered a weed if it is growing inside a soybean field, for example.
Weed species can be classified in terms of vegetative cycle, habitat or growth habit. But despite this, general features of these species are (1) high capacity of producing seeds and other reproductive structures; (2) seeds keep viability even under unfavorable conditions; (3) seeds capable of germinating and emerging from deep soil layers; (4) seeds dormancy, which allows a continuous and low percentage of germination; (5) alternative methods of propagation (rhizomes, stolons, bulbs, underground seeds); (6) mechanisms of seed dispersion (seeds with wings or hooks); (7) fast initial growth and development; and (8) seeds longevity (Silva et al., 2007).

3. Chemical weed control

The wide acceptance of chemical weed control with herbicides can be attributed to: (1) less demand of human labor; (2) efficient even under rainy seasons; (3) efficient in controlling weeds at the crop row with no damage to crop root system; (4) essential tool for no-till planting systems; (5) efficient in controlling vegetatively propagated weed species ; and (6) allows free decision about planting system (in rows, sowing) and crop row spacing (Silva et al., 2007).
It is important to consider, however, that a herbicide is a chemical molecule that should be correctly managed to avoid human intoxication as well as environmental contamination (Silva et al., 2007). The knowledge in plant physiology, chemical herbicide groups and technology of pesticide application is essential for the success of the chemical weed control (Floss, 2008). Surely there are risks involved at this method, but if they are known they can be avoided and controlled.
Chemical weed control should be applied as an auxiliary method. Efforts should be focused on the cultural method of weed management once it allows the best conditions for the development of crops while at the same time creating barriers for the proper development of seedlings of weed species (Silva et al., 2007; Floss, 2008).

In addition, the application of herbicides will, as mandatory, cause some level of harm over the crops - from almost no harm to near total plant death. These impacts are sometimes not easily visible externally at the plant, but the physiological parameters would be imbalanced resulting in lower plant performance. Because of that, more susceptible parameters like the ones associated to the photosynthesis and water use of plants are essential tools for monitoring herbicide safety for crops.

4. Competition between plant species

Among several interpretations, "plant competition" essentially means a reduction in performance of a given plant species of importance, due to shared use of a limited available resource (Gurevitch et al., 2009). Competition between plants is different from the competition between animals. Due to the lack of mobility, the competition among plants apparently is passive, not being visible at the beginning of the development (Floss, 2008). It is known, however, that crops in general terms do not present high competitive ability against weed species, due to the genetic refinement they were submitted to increase the occurrence of desired productive features in detriment of aggressiveness (Silva et al., 2007).

According to Grime (1979), as cited in Silva et al. (2007), competition is established when neighboring plants use the same resources, and success in competition is strongly determined by the *plant capacity to capture these resources*. Thus, a good competitor has a high relative growth rate and can use the available resources quickly. However, Tilman (1980), cited in Silva et al. (2007), claims that competitive success is the ability to extract scarce resources and to tolerate this lack of resources - essentially to be *more efficient in extracting and using a given resource*. Therefore, in theory, a good competitor could be the *species with least resource requirement* (Radosevich et al., 2007).

In agricultural systems, both the crop and weeds grow together in the same area. As both groups usually demand similar environmental factors as water, light, nutrients and CO_2, and usually these resources are not enough even for the crops, the competition is established. Under this situation, any strange plant which emerges at this area will share these limited resources, causing a reduction both in the volume produced by the crops, as well as in the quality of the harvested product (Floss, 2008). Radosevich et al. (2007) classified the environmental factors which determine plant growth in "resources" and "conditions".

Resources are the consumed factors such as water, CO_2, nutrients and light, and the response of plants usually follows a standard curve: it is small if the resource is less available and maximum at the saturation point, usually declining again in case of excessive availability of the resource (e.g. toxicity due to excessive zinc availability in the soil). Conditions are factors not directly consumed, such as pH and soil density, although they influence directly plant ability in exploring the resources. However, plant competition will only be established when the demand of a given resource by a plant community surpasses the ability of the environment in supplying the demanded level of the given resource (Floss, 2008).

The competition between crops and weed species is critical for the crop in cases where the weed is established together or before the crop (Radosevich, 2007). However, if the crop presents similar competitive ability to the weeds and is capable of establishing itself first, it will cover the soil, preventing access of weeds to light (Silva et al., 2007) - which is one of the most determinant factors for plant establishment (Floss, 2008; Gurevitch et al., 2009).

The competition can be established both among individuals of the same species (intra-specific competition), or among distinct plants (inter-specific competition). There is also the intra-plant competition, where distinct parts of the same plant (leaves, roots, flower buds) vie for photo-assimilates. Based on the previously exposed, in general terms the competitive process among plant species should be faced as follows (Silva et al., 2007):

- Competition is more serious in younger stages of development of the crop, e.g. at the first eight weeks for annual crops;
- Weed species morpho-physiologically similar to the crop are usually the most competitive in comparison to those which differ greatly from the crops;
- A moderate weed infestation in crop fields can be as harmful as a heavy infestation, depending on the moment these weeds are established;
- The competition is established for water, light, CO_2, nutrients and physical space. Weed species can also exsudate to soil allelochemicals capable of inhibiting germination and/or growth of other plant species.

This chapter will be focused on competition for light, but competition for water and its related parameters will also be addressed as competition for these factors are related to photosynthesis rate or efficiency.

5. The physiology of competition

When plants are subjected to strong competition in the plant community, the physiological characteristics of growth and development are usually changed. This results in differences in the use of environmental resources, especially the water, which directly affects the availability of CO_2 in leaf mesophyll and leaf temperature, therefore, the photosynthetic efficiency (Procópio et al., 2004b).

5.1 Competition for light

For some authors, competition for light is not as important as competition for water and nutrients. However, it should be considered that there is an interrelation among these factors (Silva et al., 2007). In fact, researchers are only starting to understand how the plant physiology is related to conditions of competition (Larcher, 2004). When the crop shades completely the soil, there is no competition for light between crops and weed species. For other authors, the genetic improvement of crops allowed these plants to be more efficient in intercepting and using light. As a consequence, plants of crop species present high Light Use Efficiency (LUE) when evaluated alone (Floss, 2008). Probably because of this, competition for light is often not considered in studies of plant competition.

Santos et al. (2003) evaluated the LUE of bean and soybean plants and of weed species *Euphorbia heterophylla*, *Bidens pilosa* and *Desmodium tortuosum*, and concluded that crops accumulated more dry mass per unit of light intercepted than any of the weeds studied. These authors also reported that, although the weeds were less efficient than crops in using light, they present high competitive ability in field conditions due to be more efficient in the extraction and use of other resources, like water and nutrients.

It is known that the competition for light is complex and its amplitude is influenced by the plant species, e.g. if the species is native to shaded or sunny environments and if it presents carbon metabolism of the type C_3, C_4 or CAM. The differences between these plant groups are based on the reactions that take place at the dark phase of photosynthesis (Floss, 2008; Gurevitch et al., 2009).

It is common to imagine that C_4 plants are always more efficient than C_3 plants; however, this is true only under certain conditions (Silva et al., 2007). The C_4 plants demand higher levels of energy for producing photoassimilates, because they present two carboxilative systems, and thus need to recover two enzymes for a new photosynthetic cycle. It is known that the relation CO_2 fixed/ATP/NADPH is 1:3:2 for C_3 species and 1:5:2 for C_4 species. This remarks the higher need of energy for photosynthesis in C_4 plants. As all this energy comes from light, if the access to light is reduced, C_4 plants will be less competitive than C_3 species.

On the other hand, the enzyme responsible for carboxilation in C_4 species presents some characteristics like high affinity for CO_2; no oxygenase function; optimal performance at higher temperatures; and no saturation under high light availability. As a function of these and other features, when plants are under high temperatures, light availability and also temporary water deficit, C_4 species are capable of completely overcoming C_3 species, being able to accumulate twice the dry mass per unit of leaf area in the same time interval (Silva et al., 2007).

5.2 Competition for water

Plants are powerful pumps extracting water from the soil, and because of this in hot days it is common to see crops submitted to water deficit presenting some degree of wilting, while plants of some weed species are still completely turgid. Usually, the competition for water causes the plant to compete also for light and nutrients (Silva et al., 2007). Several factors influence the competitive ability of a plant in competing for water, highlighting the volume of soil explored (proportional to the volume of the root system), physiological characteristics of the plant, stomatal regulation, osmotic adjustment in roots and hydraulic conductivity capacity of the roots (Floss, 2008).

Some plant species are capable of using less water per unit of dry mass accumulated than others, because they are more efficient in the use of the water (Water Use Efficiency - WUE – amount of dry mass accumulated as a function of water used at the same period). It is possible to infer that plants with higher WUE (more efficient in the use of water) are more productive when submitted to periods of limited water availability, as well as more competitive (Radosevich et al., 2007). However, some weed species may present distinct values of WUE throughout the cycle, being more competitive for water in certain stages of their development (Silva et al., 2007).

Differences in WUE are important in plant aggressiveness, although this is not the only mechanism allowing survival to water competition. The stomatal self-regulation, in terms of stomatal conductance, plays an important role in overcoming water deficit periods.

5.3 Competition for CO_2

In relation to CO_2, competitive aspects involving the availability of this gas are usually not considered. However, when the distinct carbon cycles presented by crops and weed species are detailed, it is possible to observe that the CO_2 concentration in the leaf mesophyll, necessary for a given species to properly accumulate dry mass, is distinct. As the efficiency in capturing CO_2 from the air is distinct between C_3 and C_4 species, and also the concentration of CO_2 may vary inside a given mixed plant community, the availability of CO_2 may be limiting for photosynthesis under competition, mainly for C_3 plant species (Silva et al., 2007).

6. Characteristics related to the photosynthesis and water use efficiency

The photosynthesis rate surely is one of the main processes responsible for high crop yields, but the liquid photosynthesis is a result of interaction among several processes, and each one of these processes, alone or in sets, may limit plant gain in terms of photoassimilates (Floss, 2008). The genetic variation among species and even among biotypes of the same species may shift the enzimatic mechanism and make a given species more capable than other in extracting or using efficiently a given environmental resource aiming to maximize its photosynthetic rate. Until recently, it was widely accepted that light affected indirectly the stomatal opening through the CO_2 assimilation dependent on light - i.e., light increased the photosynthesis rate, which would reduce the internal CO_2 concentration in the leaf and as a consequence the stomata would open. However, more accurate studies concluded that stomatal response is less connected to the internal CO_2 concentration of the leaf than anticipated; most of the response to light in stomatal opening is direct, not mediated by CO_2 (Sharkey & Raschke, 1981).

Distinct light regimes, both in terms of quantity and composition, influence almost all physiological processes like photosynthesis and respiration rates, affecting also variables like plant height, fresh and dry mass and water content of the plant (Pystina & Danilov, 2001). Water content, on the other hand, shifts both the stem length and leaf area of the plant, in a way to adapt the plant to the amount or quality of light intercepted (Aspiazú et al., 2008).

Interspecific and intraspecific plant competition affect the amount and the quality of the final product, as well as its efficiency in utilization of environmental resources (VanderZee & Kennedy, 1983; Melo et al., 2006). This is noted when assessing physiological characteristics associated to photosynthesis, such as concentration of internal and external gases (Kirschbaum & Pearcy, 1988), light composition and intensity (Merotto Jr. et al., 2009) and mass accumulation by plants under different conditions.

Although gas exchange capability by stomata is considered a main limitation for photosynthetic CO_2 assimilation (Hutmacher & Krieg, 1983), it is unlikely that gas exchange will limit the photosynthesis rate when interacting with other factors. However, photosynthetic rate is directly related to the photosynthetically active radiation (composition of light), to water availability and gas exchange (Naves-Barbiero et al., 2000). Plants have specific needs for light, predominantly in bands of red and blue (Messinger et al., 2006). When plants do not receive these wave lengths in a satisfactory manner, they need to adapt themselves in order to survive (Attridge, 1990). When under competition for light, the red and far-red ratio affected by shading is also important (Merotto Jr. et al., 2009) and influences photosynthetic efficiency (Da Matta et al., 2001).

The physiological parameters associated to photosynthesis rate and dynamics of water use in plant species will be presented, and on the following the application of these parameters in studies of plant competition (both in crops and weed species), and studies of herbicide toxicity to crops and forest trees will be discussed.

6.1 Physiological parameters

Table 1 presents the main physiological parameters evaluated by the equipment called *Infra Red Gas Analyzer* (IRGA). Details on the principles of measurements by this equipment, as well as cares to be taken to avoid reading errors, can be found in Dutton et al. (1988), Long & Bernacchi (2003) and in the technical manual of the equipment. Please note that the parameters available vary among equipments from different manufacturers, as well as

Parameter	Usual Unit	Name and Description
A	µmol m^{-2} s^{-1}	*Photosynthesis rate* – Rate of incorporation of carbon molecules from the air into biomass. Supplied by equipment.
E	mol H_2O m^{-2} s^{-1}	*Transpiration* – Rate of water loss through stomata. Supplied by equipment.
Gs	mol m^{-1} s^{-1}	*Stomatal conductance* - Rate of passage of either water vapor or carbon dioxide through the stomata. Supplied by equipment.
Ci	µmol mol^{-1}	*Internal CO_2 concentration* – Concentration of CO_2 in the leaf mesophyll . Supplied by equipment.
E_{an}	mBar	*Vapor pressure at sub-stomatal chamber* – Water pressure at the sub-stomatal zone within the leaf. Not supplied by some equipments.
ΔC	µmol mol^{-1}	*Carbon gradient* – Gradient of CO_2 between the interior and the exterior of the leaf. Usually not supplied by equipment. May be calculated by the difference between the CO_2 of reference (supplied by equipment) and its concentration at the mesophyll (Ci – supplied by equipment).
T_{leaf}	°C	*Leaf temperature* – Supplied by equipment in °C or °F.
ΔT	Δ °C	*Temperature gradient* – Not supplied by equipment. May be calculated by the difference between the temperature of the leaf (supplied by equipment) and the environmental temperature (supplied by equipment).
WUE	µmol CO_2 mol H_2O^{-1}	*Water use efficiency* – Describes the relation between the rate of incorporation of CO_2 into biomass and the amount of water lost at the same time interval. Usually not supplied. May be calculated by dividing photosynthesis (A) by transpiration (E). May be presented in several distinct units.
A/Gs	Curve*	*Intrinsic water use efficiency* – Not widely used, but describes a relation between the actual photosynthesis rate and the stomatal conductance. The original units of each parameter is maintained.
A/Ci	Curve*	*Photosynthesis / CO_2 relation* – Describes the curve of photosynthesis rate as the concentration of CO_2 within the leaf is increased. The original units of each parameter is maintained.

*These data should be represented as a graph showing the relation between variables as the concentration of one of them is increased; thus, the original units are maintained.

Table 1. Physiological parameters usually available when using an Infra Red Gas Analyzer (IRGA). Parameters available vary among manufacturers as well as among models of the same manufacturer. Some parameters are supplied by the equipment while others have to be calculated.

among models of the same manufacturer. Some of them (ΔC, T_{leaf}, ΔT and WUE) are usually not automatically supplied, but may be easily calculated based on parameters supplied by the equipment. The measuring units of the parameters also vary, and the most common units were adopted at Table 1. Water use efficiency (WUE) have several interpretations and may be presented in several different units, for distinct purposes. For a more comprehensive overview of this parameter, please consult Tambuci et al. (2011).

6.2 Photosynthesis

The photosynthesis (A) and thus the respiration, depend upon a constant flux of CO_2 and O_2 in and out of the cell; this free flux is a function of the concentration of CO_2 (Ci) and O_2 at the intercellular spaces, which depend on the stomatal opening, major controller of the gas flux through stomata (Taylor Jr. & Gunderson, 1986; Messinger et al., 2006). This is mainly controlled by the turgescence both of the guard cells (which control stomatal opening) as well as by the epidermic cells at the stomata (Humble & Hsiao, 1970). A low water potential will promote reduction in stomatal opening and reduce the leaf conductance, inhibiting photosynthesis and also the respiration (Attridge, 1990), and increasing the gradient of CO_2 concentration between the leaf mesophyll and the exterior of the leaf (ΔC).

6.3 Water use efficiency

When plants are studied in terms of the efficiency they present when using water, the parameters stomatal conductance of water vapor (Gs), vapor pressure at the sub-stomatal chamber (E_{an}), transpiratory rate (E) and water use efficiency (WUE) should be considered. The WUE is obtained by the relation between CO_2 incorporated in the plant and the amount of water lost by transpiration during the same period (Gurevitch et al., 2009). The more efficient water use is directly related to the photosynthetic efficiency as well as the dynamics of stomatal opening, because while the plant absorbs CO_2 for the photosynthesis, it also loses water to the atmosphere by transpiration, in rates that depend on the potential gradient between the interior and the exterior of the leaf (Floss, 2008). Water exchange also allows the plant to keep adequate temperature levels, which can be evaluated by the leaf temperature (T_{leaf}), as well as by the difference between the leaf temperature and the temperature of the air surrounding the leaf (ΔT).

7. Physiology of crops

7.1 Sugarcane

Control of weed species is a mandatory practice which should be applied in sugarcane fields, and among the control methods available, the chemical is usually the most widely adopted on this crop. The reasons for that are (1) the big size of the sugarcane fields; (2) higher cost of the other control methods; (3) high efficiency of the chemical method; and (4) speed of the method.

There is a very limited knowledge about the impact of the application of herbicides on the physiology of crops. According to Azania et al. (2005), the application of late post-emergence herbicides in sugarcane fields may result in high toxicity to the crop, limiting the yield. These authors attribute this to physiological changes in sugarcane plants which would result in negative effects, also on the quality of the harvest.

Galon et al. (2010) studied the following herbicides, applied over several sugarcane genotypes: ametryn - 2.000 g ha^{-1}; trifloxysulfuron-sodium - 22,5 g ha^{-1}; and a commercial mixture containing ametryn + trifloxysulfuron-sodium at 1.463 + 37,0 g ha^{-1}, respectively. Treatments were compared against a check with no herbicide. The results are summarized in Table 2.

Treatment	Sugarcane Genotype					
	RB72454	RB835486	RB855113	RB867515	RB947520	SP801816
	Shoot Dry Mass (g $plant^{-1}$)					
TC	A 5,77 a	A 5,46 a	A 4,93 a	A 5,02 a	A 5,02 a	B 2,46 a
HA	B 2,83 b	B 3,69 ab	B 2,89 b	A 5,97 a	B 3,38 ab	B 3,63 a
HB	A 5,21 ab	B 2,81 b	B 3,81 ab	B 2,99 b	B 2,03 b	B 2,30 a
HC	A 2,90 b	A 1,99 b	A 2,46 ab	A 2,09 b	A 1,66 b	A 1,30 a
	Consumed CO_2 - ΔC (μmol mol^{-1})					
TC	A 124 ab	AB 149 a	B 120 a	AB 139 a	A 177 a	B 117 a
HA	A 109 b	B 75 c	AB 105 b	AB 91 b	AB 102 b	A 110 a
HB	AB 119 ab	B 106 b	AB 114 ab	AB 118 ab	A 144 ab	B 108 a
HC	A 131 a	A 107 b	AB 112 ab	A 95 b	A 118 b	A 111 a
	Internal CO_2 Concentration – Ci (μmol mol^{-1})					
TC	AB 102 a	A 177 a	AB 104 b	AB 123 b	B 68 c	AB 120 b
HA	A 165 a	A 136 ab	A 169 a	A 178 a	A 134 a	A 157 a
HB	AB 126 ab	AB 127 ab	AB 128 ab	AB 114 b	B 88 b	A 137 ab
HC	AB 146 ab	B 87,7 b	AB 125 ab	A 179 a	AB 145 a	A 158 a
	Photosynthesis Rate (μmol m^{-2} s^{-1})					
TC	AB 45,1 a	AB 51,2 a	B 41,3 a	AB 47,9 a	A 60,7 a	AB 48,0 a
HA	A 37,5 b	B 25,8 b	A 36,1 a	B 28,9 b	A 37,0 b	A 37,3 b
HB	B 41,1 ab	B 36,6 ab	B 38,9 a	B 40,3 ab	A 49,5 ab	B 38,8 b
HC	A 42,1 ab	A 36,4 ab	A 38,4 a	B 32,8 b	A 40,5 b	A 40,1 b

Table 2. Physiological variables evaluated in sugarcane genotypes as a function of herbicide treatment. TC: control with no herbicide; HA: ametryn at 2.000 g ha^{-1} a.i.; HB: trifloxysulfuron-sodium at 22,5 g ha^{-1} a.i.; HC: ametryn + trifloxysulfuron-sodium at 1.463 + 37,0 g ha^{-1} a.i. Means followed by the same letter at the column, inside each variable, are not different by the DMRT test at 5% probability. Source: adapted from Galon et al. (2010).

In general terms, the CO_2 consumed by photosynthesis (ΔC) was smaller in treatments including the herbicide ametryn. There were also remarkable differences between genotypes. The ΔC is directly related to the photosynthesis rate of the plant by the time of the evaluation. In this situation, it is possible to observe that the genotypes RB72454 and SP80-1816 were less susceptible to ametryn than the other genotypes.

The concentration of CO_2 within the leaf (Ci) was affected by the herbicide treatments, being also observed once more differences between genotypes. As expected, this parameter presented, in general terms, opposite behavior in comparison to ΔC. The application of the herbicide ametryn, a photosynthesis II (PSII) inhibitor, resulted in higher concentrations of CO_2 within the leaf, once the photosynthesis of the genotypes under application of this herbicide was more severely affected. The CO_2 concentration within the leaf was around 50% higher in treatments involving ametryn in comparison to the control treatment with no herbicide. Trifloxysulfuron-sodium also caused changes in Ci, but not at the same magnitude of ametryn.

In general terms, the photosynthesis rate (A) observed at the treatment with trifloxysulfuron alone was similar to the control with no herbicide. On the same way treatments involving the PSII inhibitor presented photosynthesis rate inferior to rates observed at the control treatment. When considering the treatment containing ametryn + trifloxysulfuron, it was possible to highlight the genotype RB947520.

The authors highlight that, even the damages caused by ametryn being more easily identified by evaluating parameters associated to the photosynthesis, variations due to the application of trifloxysulfuron-sodium were also detectable by changes in these parameters by using an Infra Red Gas Analyzer (IRGA). In other words, herbicidal damage on crops can be effectively quantified by evaluating direct and indirect damage to the photosynthetic route. Furthermore, the accumulation of dry mass did not correlate directly with most of the studied physiological parameters, because plant growth is a result of biomass accumulation since the emergence until the moment of the evaluation. In this way, the authors remark the importance of evaluating both types of variables, physiological and biomass/growth - related, before concluding about the efficacy or impact of a given herbicide treatment. In addition, the authors remark the existence of differences among genotypes in terms of susceptibility to herbicides, which were effectively identified by physiological parameters.

7.2 Cassava

Several factors have contributed to the low productivity of cassava in under-developed countries, being the inadequate management of weeds one of the most important. Usually, cassava producers believe that this crop is rustic and there is no need to worry too much about weed control (Albuquerque, 2008). However, the competition between weeds and cassava can affect its production in quantitative and qualitative ways (Aspiazú et al., 2010a, 2010b). This competition alters the efficiency of use of environmental resources such as water, light, nutrients and space among species that occupy the same ecological niche (Melo et al., 2006; Floss, 2008).

Aspiazú et al. (2010b) studied the physiological interactions of cassava with three weed species in order to determine the mechanism the crop plants used to overcome the stress imposed by weed competition. One cassava plant was submitted to competition with one of these weed species: *Bidens pilosa* (three plants m^{-2}), *Brachiaria plantaginea* (six plants m^{-2}) or *Commelina benghalensis* (three plants m^{-2}).

Cassava plants grown free of competition showed greater leaf internal CO_2 concentration (Ci) than when competing with weeds (Table 3). Lower CO_2 concentrations in leaves were observed when cassava plants were grown under competition with *C. benghalensis*. The highest values for Ci observed in cassava plants were when under competition with *B. pilosa* or *B. plantaginea*, if compared to *C. benghalensis*. That can be attributed to the accelerated plant metabolism as a way to increase growth rate and escape shading caused by weeds, as will be further detailed. It is believed that under these conditions, where lower Ci occurs, CO_2 was consumed due to the increase at metabolic rate.

Studies conducted by some researchers showed that responses to changes in Red:Far Red (R:Fr) ratios occur before the mutual shading among neighbors. Based on these studies, it was proposed that the Far Red radiation, which is reflected by adjacent leaves, is a means of early detection that signals the imminence of competition during canopy development

(Ballaré et al., 1990; Merotto Jr. et al., 2009; Aspiazú et al., 2010b). In addition, some species are capable of recognizing others by the amount of radiation reflected in each wave length - essentially, the composition of the light reflected (Larcher, 2004; Aspiazú et al., 2010b).

Treatment	Ci (µmol mol-1)	ΔC (µmol mol-1)	T_{leaf} (°C)	A (µmol m-2 s-1)
Cassava	295,7 a	14,43 b	32,1 b	5,71 b
Cassava x *B. pilosa*	256,4 b	12,73 b	33,6 ab	5,17 b
Cassava x *C. benghalensis*	232,3 c	19,60 a	35,1 a	7,32 a
Cassava x *B. plantaginea*	266,9 b	14,50 b	33,6 ab	5,93 b
CV (%)	**10,0**	**19,4**	**3,6**	**15,2**

Table 3. Parameters associated to the photosynthesis of cassava plants 60 days after emergence, as a function of the weed species with whom cassava plants were under competition. Means followed by the same letter at the column, are not different by the DMRT test at 5% probability. Source: Aspiazú et al. (2010b).

It was probably what happened when cassava plants competed with *C. benghalensis*; the presence of this slow growth, poor competitive weed species might have been simply enough to increase cassava metabolism aiming to avoid the imposition of future competition.

The authors also remark that the consumption of CO_2 (ΔC) increased proportionally to the decrease of Ci. When the three weeds were compared, *C. benghalensis* was the one that most increased the ΔC of cassava, because there was a higher gradient between the internal and the external sides of the cassava leaves - this may indicate accelerated plant metabolism.

Cassava leaves temperature was higher when under competition with *C. benghalensis*, mainly due to the stimulus in metabolism caused by the presence of the weed, and due to the low competition exerted, or simply due to the changing of the quality of light, which allowed cassava plants to recognize the species of weed (Radosevich et al., 2007; Larcher, 2004). The competition-free control indicates the metabolic rate usual to cassava plants when free of competition. In this situation, plant growth becomes more balanced, distributing photoassimilates proportionally between shoots and roots, which is not the case when under competition (Radosevich et al., 2007).

The difference between leaf temperature and the air around it (ΔT) is commonly only 1 or 2 °C, but in extreme cases it may exceed 5 °C (Attridge, 1990). Cassava plants under competition with *B. pilosa* or *B. plantaginea* stayed half way compared to the ones observed in weed-free control, and at treatment under competition with *C. benghalensis*. This indicates that these weeds were able to prevent cassava from reacting adequately to the imposition of competition, probably by limiting crop access to appropriate levels of a given resource, such as light or water, for example.

The photosynthetic rate was also higher for cassava plants under competition with *C. benghalensis*, when compared to other treatments. Photosynthetic rate of cassava under competition with *C. benghalensis* was 7,32 µmol m^{-2} s^{-1} CO_2, while for the average of the remaining treatments was 5,50 µmol m^{-2} s^{-1} CO_2. The radiation balance and composition on the plant when in competition or shading, combined with carbohydrate level in leaves, may increase respiratory rate directly or through alternative pathways associated with the respiratory chain (Pystina & Danilov, 2001). This could make photosynthesis balance even smaller and reduce the ability of the plant to accumulate mass. The interaction between measured parameters is presented at Table 4.

	Ci (μmol mol^{-1})	ΔC (μmol mol^{-1})	T_{leaf} (°C)	A (μmol m^{-2} s^{-1})
Ci	1	-	-	-
ΔC	-0,60 *	1	-	-
T_{leaf}	-0,53	0,34	1	-
A	-0,60 *	0,96 *	0,37	1

* = significant at 5% of probability.

Table 4. Pearson linear correlation matrix as a function of the parameters associated to the photosynthesis of cassava plants grown alone or under competition with weed species, evaluated 60 days after emergence. Source: Aspiazú et al. (2010b).

In terms of parameters associated to the water use efficiency, the stomatal conductance (Gs) of cassava plants in coexistence with *B. pilosa* was inferior only to the treatment where cassava was grown alone (Table 5).

Treatment	E (mol H_2O m^{-2} s^{-1})	Gs (mol m^{-1} s^{-1})	E_{an} (mbar)	WUE (μmol CO_2 mol^{-1} H_2O)
Cassava	3,57 a	0,13 a	22,40 a	1,81 b
Cassava x *B. pilosa*	2,31 c	0,07 b	21,03 a	2,26 a
Cassava x *C. benghalensis*	3,69 a	0,11 ab	24,57 a	1,98 ab
Cassava x *B. plantaginea*	3,15 b	0,12 ab	23,47 a	1,90 ab
CV (%)	**19,7**	**24,40**	**6,6**	**9,8**

Table 5. Parameters associated to the water use of cassava plants 60 days after emergence, as a function of the weed species with whom cassava plants were under competition.
Means followed by the same letter at the column, are not different by the DMRT test at 5% probability. Source: Aspiazú et al. (2010a).

Vapor pressure in the substomatal cavity of the cassava leaves (E_{an}) did not change in function of the competition between cassava and any of the weed species (Table 5). In Table 6, a high correlation between E and E_{an} can be observed, although the latter was not modified in function of the species with which cassava competed. In this case, the changes in Gs can be attributed to factors not related to cassava plants, such as availability of soil water or wind speed, factors that can be changed by the presence of weeds.

The transpiration rate (E) was dependent on the weed species with which the cassava plant competed, being superior for the control free of weeds and for the cassava plant which competed with *C. benghalensis* (Table 5). This supports the considered hypothesis by the authors that *C. benghalensis* does not show a good competitive ability for light, because the crop had higher leaf temperature (Table 3) and higher transpiration rate (Table 5) when competing with this species, which indicates high metabolic rate and accelerated growth due to recognition of imposition of competition before it was actually established. The interactions between parameters associated to the water use are presented at Table 6.

	E ($mol\ H_2O\ m^{-2}\ s^{-1}$)	Gs ($mol\ m^{-1}\ s^{-1}$)	E_{an} (mbar)	WUE ($\mu mol\ CO_2\ mol\ H_2O^{-1}$)
E	1	0,72 *	0,87 *	-0,26
Gs	-	1	0,49	-0,30
Ean	-	-	1	-0,09
WUE	-	-	-	1

* = significant at 5% of probability.

Table 6. Pearson linear correlation matrix as a function of the parameters associated to the water use of cassava plants grown alone or under competition with weed species, evaluated 60 days after emergence. Source: Aspiazú et al. (2010a).

The photosynthetic characteristics of cassava were influenced by its competition with *B. pilosa* and *B. plantaginea*. The authors noticed that cassava plants are affected by these species especially concerning competition for light and water. However, there were no negative effects when cassava plants competed with *C. benghalensis*. This species seems to rather affect the composition of light by reflecting peculiar spectrum of specific wavelengths, which allows cassava plants to anticipate the imposition of competition even before they get to harmful levels.

7.3 Eucalyptus

The management of weed species in *Eucalyptus* plantations is based both on mechanical and chemical methods (Machado et al., 2010). When using the chemical method, the main active ingredient applied is the glyphosate, due to several advantages (Tuffi Santos et al., 2010). As this is a non selective herbicide, it is applied between plant rows, avoiding contact with the *Eucalyptus* plants, which could result in lower growth rates and eventually plant death (Tuffi Santos et al., 2010).

The possibility of severe damage for *Eucalyptus* by the application of glyphosate instigated the conduction of several researches involving the concept of "simulated drift". However, there are just a few researches in Brazil aiming at the physiological implications of glyphosate application in *Eucalyptus* plantations. In a pioneer study, Machado et al. (2010) modeled the impact of the herbicide glyphosate over clones of *Eucalyptus grandis* and *E. urophylla*. These authors were not able to identify differences in susceptibility among clones, but were able to characterize relatively well the behavior of physiological parameters in *Eucaliptus* under application of glyphosate. Four doses of glyphosate (43,2; 86,2; 129,6; and 172,8 g ha^{-1} of the commercial formulation containing 360 g L^{-1} of the acid equivalent of glyphosate) were applied on two clones of each species when plants were around 40 cm height. The main results are synthesized in Figure 1.

When the photosynthesis rate was analyzed as a function of herbicide doses, there was no change seven days after application (18,49 $\mu mol\ m^{-2}\ s^{-1}$). However, 21 days after application (DAA) the photosynthesis was reduced as the dose of the herbicide increased (Figure 1). At the higher herbicide doses, there was a proportional increasing in toxicity to the plants of *Eucalyptus* which caused leaf abscission, thus reducing the leaf area available for photosynthesis. The stomatal conductance (Gs) did not differ among herbicide doses 7 DAA, but it was reduced as the dose was increased at the evaluation of 21 DAA. According to the authors, this is due to the slow action of the glyphosate which usually causes most of the damage between 9 and 15 DAA.

In the same way of A and Gs, differences in transpiration (E) were not observed 7 DAA with increasing at the dose of glyphosate, but these differences were reported 21 DAA of the herbicide. The water use efficiency (WUE), however, was affected by increasing doses of the herbicide at the two evaluations (Figure 1), although these differences were more drastic 21 DAA. According to the authors, these differences in WUE are due at least in part to the reduction observed at the photosynthesis rate, which also resulted in smaller dry mass accumulation.

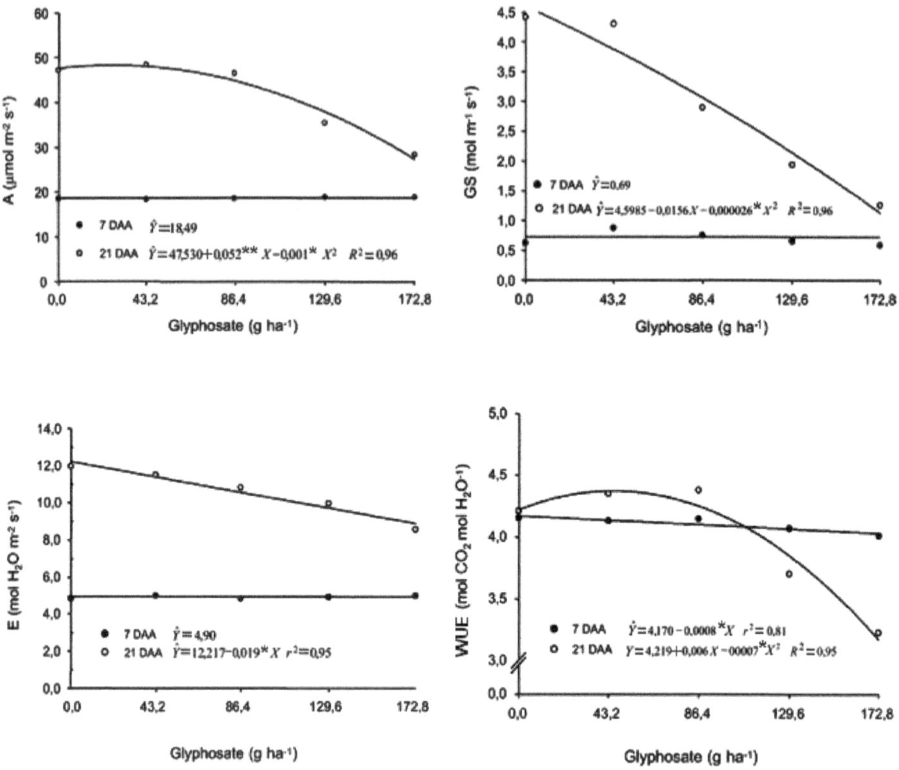

Fig. 1. Physiological parameters of plants of *Eucaliptus* spp. submitted to increasing doses of glyphosate, 7 and 21 days after application (DAA) of the treatments. Doses are supplied based on commercial formulation containing 360 g L^{-1} of acid equivalent of glyphosate. Source: Machado et al. (2010).

The authors suggest that the smaller dry mass accumulated in treatments under high doses of glyphosate (Figure 2) are due to the higher toxicity caused to plants which resulted in high rates of necrosis and foliar abscission, associated to the lower photosynthesis rate and water use efficiency observed in these treatments 21 DAA of the herbicide. In these terms the authors recommend maximum care when using glyphosate in *Eucalyptus* plantations aiming to avoid toxicity to the crop and possible plant death, and concluded that the physiological parameters are an essential tool to determine herbicide injury to tree plants, when the evaluations are conducted at the right time after herbicide application.

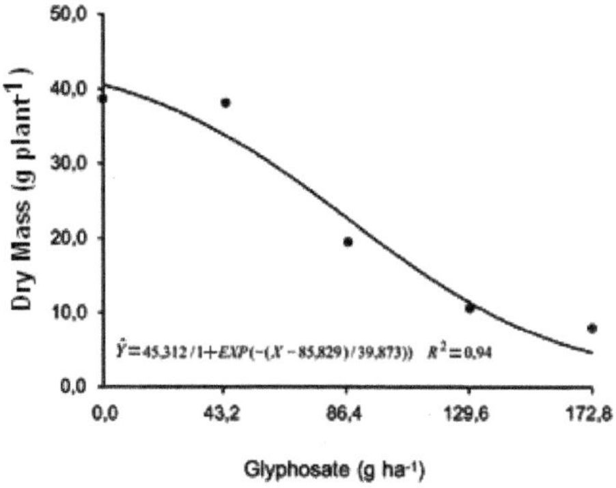

Fig. 2. Dry mass (g plant^{-1}) of plants of *Eucaliptus* spp. submitted to increasing doses of glyphosate, 50 days after application (DAA) of the treatments. Doses are supplied based on commercial formulation containing 360 g L^{-1} of acid equivalent of glyphosate. Source: Machado et al. (2010).

8. Physiology of weed species

8.1 *Bidens pilosa, Commelina benghalensis, Brachiaria plantaginea*

B. pilosa, C. benghalensis and *B. plantaginea* usually are associated in the same community in areas where cassava is grown. There are several implications in controlling these weed species in cassava plantations, and the main one is the lack of herbicides registered to be used in this crop - only four herbicides are registered to be applied in cassava in Brazil. Based on this, it is important to determine the extent of the damage by competition these species are capable of causing to crops, and understand the physiological differences among these weed species.

These weed species were grown alone at densities of 3 plants m^{-2} (*B. pilosa* and *C. benghalensis*) and 6 plants m^{-2} (*B. plantaginea*), in soil with moisture constantly maintained at $^{2}/_{3}$ of the field capacity. Sixty days after emergence, plants were evaluated in terms of physiological characteristics associated to photosynthesis and water use efficiency.

The stomatal conductance (Gs) did not differ among species, with values around 0,06 mol m^{-1} s^{-1}. The stomatal conductance is composed by the small cuticular conductance of the epidermis, and by the stomatal conductance when stomata are open. Because of that, Gs is proportional to stomata number and size. The water vapor exchange between the interior and exterior of the leaf was similar for all species (Table 7). The control of stomatal opening depends upon light availability, CO_2 levels in the mesophyll, relative air humidity and water potential inside the plant, as well as other less impacting factors like wind speed, application of growth substances and endogenous rhythms proper of a given species.

Similarly to the Gs, the E_{an} was also equal among species. This variable is directly related to the water status of the plant and to the dynamics of the water vapor. Even in a leaf with high transpiration rate the relative humidity at the sub-stomatal chamber (E_{an}) may be superior to 95% and the resultant water potential may be around 0%, increasing the vapor exchange with the external environment, which presents water potential highly negative. Under these conditions the instantaneous vapor pressure is the saturation vapor pressure of the temperature of the leaf. In this way the E_{an} is controlled by the leaf humidity level and temperature, causing changes over both the stomatal conductance of water vapor (Gs) and over the transpiration rate (E).

Water Use				
Weed Species	**Gs ($mol\ m^{-1}\ s^{-1}$)**	**E_{an} (mbar)**	**E ($mol\ H_2O\ m^{-2}\ s^{-1}$)**	**WUE ($mol\ CO_2\ mol\ H_2O^{-1}$)**
B. pilosa	0,08 a	19,90 a	2,17 a	0,98 b
C. benghalensis	0,07 a	21,57 a	2,40 a	1,26 b
B. plantaginea	0,04 a	20,30 a	1,71 b	4,85 a
CV (%)	**32,9**	**4,2**	**16,8**	**91,3**
Photosynthesis				
Weed Species	**Ci ($\mu mol\ mol^{-1}$)**	**ΔC ($\mu mol\ mol^{-1}$)**	**T_{leaf} (°C)**	**A ($\mu mol\ m^{-2}\ s^{-1}$)**
B. pilosa	303,6 a	8,30 b	30,9 b	2,26 b
C. benghalensis	282,1 b	8,17 b	34,3 a	3,07 b
B. plantaginea	31,2 c	21,83 a	35,1 a	8,44 a
CV (%)	**73,6**	**61,5**	**7,4**	**73,2**

Table 7. Parameters associated to the photosynthesis and water use of weed species. In the same variable, means followed by the same letter at the column are not different by the Least Significant Difference (LSD) test at 5% probability. Source: Aspiazú et al. (2010c).

Bidens pilosa is known by its wide capacity of extracting water from the soil, although it is not highly efficient in water use. This species is capable of keeping high growth rates under soil water potentials in which most of the crops and weed species reached the permanent wilting point, and in some cases this species is benefited by the occurrence of water deficit (Procópio et al., 2004b). *B. plantaginea,* on the other hand, has a distinct strategy in overcoming severe water stress. While *B. pilosa* is highly efficient in extracting water from soil, *B. plantaginea* is highly efficient in using the amount of water extracted, in part because it is a C_4 species. As *B. plantaginea* is capable of keeping high photosynthesis rate even with CO_2 concentration in the leaf mesophyll very close to 0 ppm, it will allow this species to keep stomata closed for a longer period of time and increase the water use efficiency (WUE). *B. plantaginea* was superior to the others in terms of water loss as it was able to keep a tighter stomatal opening control. It is probably related to the C_4 metabolism of this species and to the lower compensation point of CO_2 concentration. As stomata of this species are closed for more time in relation to other species, the amount of water lost by transpiration is reduced. The parameters Gs, E_{an} and E are connected to a cost-benefit ratio because the processes of transpiration and CO_2 interception from the outside of the leaf only occur when stomata are open. In addition, E_{an} is usually reduced in periods of stomatal opening (Aspiazú et al., 2010a).

The higher WUE observed for *B. plantaginea* is directly related to the C_4 carbon cycle in this species and the consequent smaller period of stomatal opening (Table 7). The Ci of the leaf differed among species, and *B. pilosa* showed higher levels of CO_2, followed by *C. benghalensis* and *B. plantaginea*. This is considered a physiological characteristic affected by environmental factors such as water and light availability (Gurevitch et al., 2009). The higher the photosynthesis rate, the higher the amount of CO_2 consumed (Table 7), and smaller will be its concentration at the interior of the leaf supposing the stomata are closed (Aspiazú et al., 2010c). The CO_2 consumption increases the differences in concentration of this gas between the interior and the exterior of the leaf, and in general terms as higher is this gradient (ΔC), faster the CO_2 will enter the leaf in moments of stomatal opening. Because of this, it was also observed that a smaller Ci also resulted in higher ΔC.

The leaf temperature (T_{leaf}) is affected by the metabolic rate of the leaf, and in proper conditions this temperature is always 1 - 3 °C superior to the environmental temperature around it. In these terms, plant metabolism may be indirectly estimated by the leaf temperature (Attridge, 1990). According to the observed for Ci and ΔC, the leaf temperature was lower for *B. pilosa*, because this plant consumed less CO_2 per unit of time than the other species, and presented higher Ci and lower ΔC than *B. plantaginea* (Table 7).

According to the Ci, ΔC and T_{leaf}, the photosynthesis rate was higher for *B. plantaginea* in comparison to the other weed species. The photosynthesis and transpiration rates depend upon a constant flux of both CO_2 and O_2 in and out of the leaves (Messinger et al., 2006).

The authors emphasize that, in general terms, under lower water availability, *B. plantaginea* tend to be more competitive than *B. pilosa* and *C. benghalensis* due to its superiority at photosynthesis rate and the other physiological parameters studied. On the other hand, *B. pilosa* is capable of keeping high photosynthesis rate even under moderate water stress due to its higher capacity of extracting water from soil - in other words, this species is specialized in avoiding the stress. While *B. plantaginea* is more efficient in characteristics related to the photosynthesis, i.e. more efficient in use of light, *B. pilosa* was more efficient in characteristics related to water use.

8.2 *Lolium multiflorum*

Ryegrass (*Lolium multiflorum*) is an annual winter forage, also extensively used as winter crop, aiming to supply mass for the no-till planting system. This species is distributed in all temperate climate regions of Brazil, especially at the Southern region. By the time of planting, this plant is usually desiccated with the herbicide glyphosate, and recently this species became resistant to this herbicide (Ferreira et al., 2009).

Ferreira et al. (2009) conducted some studies on growth and development of the ryegrass biotype resistant to glyphosate in comparison to the standard susceptible biotype, and concluded that the susceptible biotype accumulates more dry mass than the resistant one. These authors attributed this to the smaller number of tillers at the resistant biotype: while the susceptible biotype presented around 7,2 tillers per plant, the resistant one presented in average 4,4 tillers per plant under the same conditions. Due to this, the number of inflorescences - and as a consequence the number of seeds produced - is smaller at the resistant biotype. Another fact reported by this group of researchers was that the susceptible biotype shifted from the vegetative to the reproductive period 19 days before the resistant, and completed the cycle around 25 days before.

The parameters photosynthesis (A) and water use efficiency (WUE) are usually the main determiners of the predominance of a given species or biotype over the others at the same area, mainly as a function of the environmental stresses to which these plants are subjected to.

As the factors causing this behavior in the resistant biotype were not clear, further work by the same authors went deeper at the physiological causes of these differences. Characteristics related with both the photosynthesis and the water use efficiency at the resistant biotype differed from the susceptible biotype, probably as a consequence of the mechanism conferring resistance to the herbicide glyphosate. The competitive ability of the plant affects directly its capacity of using environmental resources, mainly those related to the photosynthesis rate (VanderZee & Kennedy, 1983; Melo et al., 2006).

When the resistant biotype was put under competition with the susceptible one, the photosynthesis rate was greatly affected. As a consequence, the internal concentration of CO_2 in the leaf is increased because its consumption by photosynthesis was reduced (Figure 3). Considering the previously discussed fact that the stomatal opening is related with both the internal CO_2 concentration of the leaf (in smaller degree) and to the level of light available (in higher degree), it is hypothesized by the authors that under field conditions there would be a smaller period of stomatal opening for the resistant biotype, when under competition with the susceptible one, resulting in smaller dry mass accumulation.

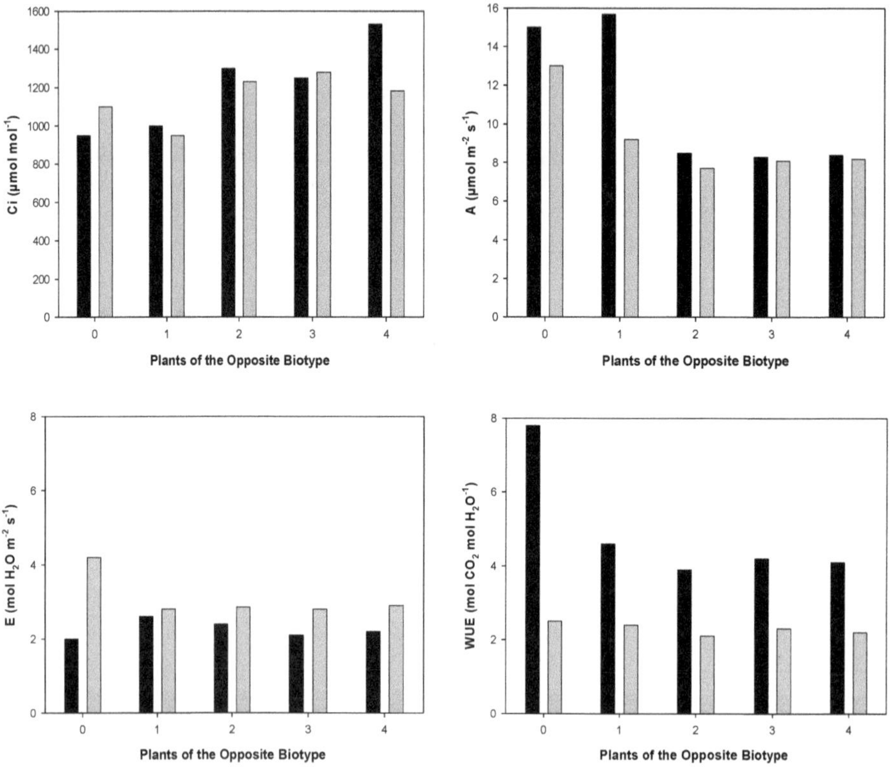

Fig. 3. Physiological parameters measured at biotypes of *Lolium multiflorum* resistant (■) or susceptible (░) to the herbicide glyphosate. Source: Ferreira et al. (2009).

Under low competition levels, the resistant biotype is more efficient in water use than the susceptible one (Figure 3). However, this biotype becomes less efficient in water use as the competition level is increased. Some studies show that the resistant biotype accumulates less dry mass because the period of stomatal opening is smaller to avoid losing excessive water, and as a consequence small amounts of CO_2 move from the external environment to the interior of the leaf. On the other hand, a possible smaller speed of water movement from roots to shoots would limit the water loss to that contained at the sub-stomatal chamber during the period of stomatal opening, which would make the resistant biotype more efficient in water use at the expense of smaller dry mass accumulation capabilities.

The most probable reason for the resistant biotype to be more efficient at the water use is a supposed smaller efficiency in water absorption or translocation, as well as the smaller period of stomatal opening. It is highlighted, however, that a higher WUE does not guarantee that this biotype has any physiological advantage over the susceptible one under low competition levels. As previously presented, plants of *B. pilosa* are highly efficient in extracting water from soil, but present relatively low WUE (Procópio et al., 2004b).

8.3 Echinochloa crusgalli

Barnyardgrass (*Echinochloa crusgalli*) infest rice fields worldwide being rarely found in shaded places. This species is adapted to anaerobic environments, being capable of germinating even under 10 cm of water (Concenço et al., 2009). The high number of seeds produced increases both its importance as a weed and the survival of the species. In addition, this species present C_4 carbon cycle while rice is a C_3 species.

The competition for light between rice and barnyardgrass plants occur when the latter grow more than the former, shading the plants of the crop and reducing both the quantity and the quality of the light received by the rice plants (Concenço et al., 2009). Several studies showed that taller plants present superior ability of causing interference in crops, mainly in terms of amount of light intercepted. Barnyardgrass plants are usually twice the size of the current rice varieties when mature.

Although relatively easy to control, barnyardgrass plants are becoming a more serious problem because there are several biotypes resistant to herbicides around the world. In Brazil the main problem is the occurrence of several biotypes resistant to the herbicide quinclorac. As the chemical control method becomes inefficient in controlling this species, the physiology of the plant should be explored and understood to allow application of management practices which will allow a cultural suppression of the occurrence of this weed species.

Preliminary works showed that there are variations among biotypes in terms of growth rate, and hypothesized that this was probably related to the physiology of the plant. Later studies by Concenço et al. (2009) supplied some light on the physiological behavior of these biotypes in terms of characteristics associated to the photosynthesis and growth rates. At Table 8 is presented a comparison between a susceptible and a resistant biotype to the herbicide quinclorac, in terms of variables associated to the photosynthesis.

In relation to the variables associated to the photosynthesis, differences between biotypes were not observed (Table 8). In relation to the increasing at the competition intensity, it is possible to conclude that both biotypes are similar but the susceptible biotype was generally more affected than the resistant one under high competition levels, in comparison to the control with no competition. More vigorous plants are capable to cover the available space faster and avoid access of the less vigorous plants to light.

At Table 9 it is supplied a comparison between a susceptible and a resistant biotype to the herbicide quinclorac in terms of variables associated to the water use efficiency. Also, differences were not found. The authors concluded that unlike *Lolium multiflorum* biotypes which differ greatly in terms of physiologic parameters and competitive ability, barnyardgrass biotypes resistant and susceptible to the herbicide quinclorac present similar performance in terms of physiology and biomass accumulation, and the probable mutation responsible for conferring resistance to the herbicide did not affect the environmental fitness of the biotype at significant levels.

Plants under Competition	Resistant Biotype	Susceptible Biotype	Difference between Biotypes
	Sub-Stomatal CO_2 Concentration – Ci (μmol mol^{-1})		
1	23,4 b	18,1 c	+5,3 ns
2	29,7 b	29,3 bc	+0,4 ns
3	36,5 ab	33,6 bc	+2,9 ns
4	43,2 ab	48,5 ab	-5,3 ns
5	55,3 a	60,2 a	-4,9 ns
	Photosynthesis Rate – A (μmol m^{-2} s^{-1})		
1	23,2 a	23,0 a	+0,2 ns
2	19,4 ab	22,2 ab	-2,8 ns
3	21,0 ab	20,8 abc	+0,2 ns
4	21,5 ab	17,9 bc	+3,6 ns
5	18,1 b	16,8 c	+1,3 ns

Table 8. Physiological parameters associated to the photosynthesis in barnyardgrass (*Echinochloa crusgalli*) biotypes, resistant or susceptible to the herbicide quinclorac. Means followed by the same letter at the column, inside each variable, are not significantly different by the DMRT test at 5% of probability. Source: Concenço et al. (2009).

Plants under Competition	Resistant Biotype	Susceptible Biotype	Difference between Biotypes
	Transpiratory Rate – E (mol H_2O m^{-2} s^{-1})		
1	1,86 a	1,97 a	-0,11 ns
2	1,84 a	1,78 ab	+0,06 ns
3	1,65 a	1,70 ab	-0,05 ns
4	1,72 a	1,35 b	+0,37 ns
5	1,39 a	1,29 b	+0,10 ns
	Water Use Efficiency – WUE (μmol CO_2 mol H_2O^{-1})		
1	13,1 a	11,7 a	+1,4 ns
2	10,5 a	12,5 a	-2,0 ns
3	12,7 a	12,2 a	+0,50 ns
4	12,5 a	13,2 a	-0,70 ns
5	13,0 a	13,0 a	0 ns

Table 9. Physiological parameters associated to the water use in barnyardgrass (*Echinochloa crusgalli*) biotypes, resistant or susceptible to the herbicide quinclorac. Means followed by the same letter at the column, inside each variable, are not significantly different by the DMRT test at 5% of probability. Source: Concenço et al. (2009).

9. Final comments and conclusions

The weed species present distinct strategies to avoid stresses aiming to keep high rates of physiological metabolism. As example, *B. plantaginea* is highly efficient at the use of water;

B. pilosa is not efficient in using water but is capable of extracting this resource from soil at water potentials where most other weeds are reaching the permanent wilting point, and *C. benghalensis* produces aerial and subterranean seeds, being also capable of propagating vegetatively.

Biotypes of weeds resistant to herbicides, like *Echinochloa crusgalli* resistant to quinclorac and *Lolium multiflorum* resistant to glyphosate, may present different abilities to those at the susceptible biotype. These differences are related to morphophysiological changes caused by the mechanism which confers resistance to the herbicide, and the physiological parameters associated to the photosynthesis rate and water use efficiency are highly effective in pointing out these discrepancies among plant biotypes when they are present, and probably also useful for differentiating plant ecotypes of the same species. Thus, the measurement of physiological parameters associated to photosynthesis and water use present a wide range of applicabilities, and many of them are currently not fully explored.

Some weed species or biotypes of weed species are so poor competitors against certain crops, that it is unnecessary to eliminate them from the field when they occur at low densities, if all other developmental factors (light, water, CO_2) are supplied at minimum required levels for crop development. In fact, some low competitive weed species are recognized by some crops before the competition starts, and these crops are able to increase their metabolism as an answer to the presence of that given weed species, growing faster while trying to avoid the competition with that weed species. The presence of low infestation of such weed species is hypothesized to result in positive effect over the crop competitiveness; maybe not increasing yield, but increasing crop capacity to compete against other weeds.

In relation to the application of chemicals for weed management inside crop fields and tree plantations, it is advised to observe the optimal time after application when symptoms of phytotoxicity are present to avoid mis-determining a lower impact than the real from the herbicide and dose over the crops. In this scenario, physiological parameters are more efficient than directly quantified variables, also supplying results faster.

In general terms, more attention should be given to the behavior of physiological parameters in crops and weed species when they are submitted to some type of stress. At this chapter the possibility of using physiological parameters in determining the effects of competition between plant species over each one of the species involved, as well as a tool to describe the impact of a given herbicide treatment, were illustrated. However, these parameters are suitable for inferences in many other types of applications not discussed at this chapter.

10. References

Albuquerque, J.A.A., Sediyama, T., Silva, A.A., Carneiro, J.E.S., Cecon, P.R. & Alves, J.M.A. (2008). Interferência de plantas daninhas sobre a produtividade da mandioca (*Manihot esculenta*). *Planta Daninha*, Vol.26, No.2, (April/June 2008), pp.279-289, ISSN 0100-8358.

Aliyev, J.A. (2010). Photosynthesis, photorespiration and productivity of wheat and soybean genotypes. *Proceedings of ANAS (Biological Sciences)*, Vol.65, No.5/6, pp.7-48.

Aspiazú, I., Concenço, G., Galon, L., Ferreira, E.A. & Silva, A.F. (2008). Relação colmos/folhas de biótipos de capim-arroz em condição de competição. *Revista Trópica*, Vol.2, No.1, pp.22-30, ISSN 1982-4831.

Aspiazú, I., Sediyama, T., Ribeiro Jr., J.I., Silva, A.A., Concenço, G., Ferreira, E.A., Galon, L., Silva, A.F., Borges, E.T. & Araujo, W.F. (2010a). Water use efficiency of cassava plants under competition conditions. *Planta Daninha*, Vol.28, No.4, (December 2010), pp.699-703, ISSN 0100-8358.

Aspiazú, I., Sediyama, T., Ribeiro Jr., J.I., Silva, A.A., Concenço, G., Ferreira, E.A., Galon, L., Silva, A.F., Borges, E.T. & Araujo, W.F. (2010b). Photosynthetic activity of cassava plants under weed competition. *Planta Daninha*, Vol.28, pp.963-968, ISSN 0100-8358. Special Number.

Aspiazú, I., Sediyama, T., Ribeiro Jr., J.I., Silva, A.A., Concenço, G., Galon, L.V., Ferreira, E.A., Silva, A.F., Borges, E.T. & Araujo, W.F. (2010c). Eficiência fotosintética y de uso del agua por malezas. *Planta Daninha*, Vol.28, No.1, pp.87-92, ISSN 0100-8358.

Attridge, T.H. (1990). The natural environment, In: *Light and Plant Responses*, Attridge, T.H. (Ed.), pp.1-5, Edward Arnold, ISBN 978-052-1427-48-7, London, England.

Azania, C.A.M., Rolim, J.C., Casagrande, A.A., Lavorenti, N.A. & Azania, A.A.P.M. (2005). Seletividade de herbicidas. II - Aplicação de herbicidas em pós-emergência inicial e tardia da cana-de-açúcar na época das chuvas. *Planta Daninha*, Vol.23, No.4, (October/December 2005), pp. 669-675, ISSN 0100-8358.

Ballaré, C.L., Scopel, A.L. & Sánchez, R.A. (1990). Far-red radiation reflected from adjacent leaves an early signal of competition in plant canopies. *Science*, Vol. 247, No.4940, (January 1990), pp. 329-332, ISSN 0036-8075.

Bianchi, M.A., Fleck, N.G., Lamego, F.P. & Agostinetto, D. (2010). Papéis do arranjo de plantas e do cultivar de soja no resultado da interferência com plantas competidoras. *Planta Daninha*, Vol.28, pp.979-991, ISSN 0100-8358. Special Number.

Ceccon, G. (2007. Milho safrinha com solo protegido e retorno econômico em Mato Grosso do Sul. *Revista Plantio Direto*, Ano16, No.97, (Janeiro/Fevereiro 2007), pp.17-20, ISSN 16778081.

Concenço, G., Noldin, J.A., Eberhardt, D.S. & Galon, L. (2009). Resistência de *Echinochloa* sp. ao herbicida quinclorac, In: *Resistência de Plantas Daninhas a Herbicidas no Brasil*, Agostinetto, D. & Vargas, L. (Eds.), pp.309-350, Berthier, ISBN 978-858-9873-92-5, Passo Fundo, Brazil.

Da Matta, F.M., Loos, R.A., Rodrigues, R. & Barros, R. (2001). Actual and potential photosynthetic rates of tropical crop species. *Revista Brasileira de Fisiologia Vegetal*, Vol.13, No.1, (Abril 2001), pp.24-32, ISSN 0103-3131.

Dutton, R.G., Jiao, J., Tsujita, J. & Grodzinski, B. (1988). Whole plant CO_2 exchange measurements for non destructive estimation of growth. *Plant Physiology*, Vol.86, No.2, (February 1988), pp.355-358, ISSN 0032-0889.

Ferreira, E.A., Concenço, G., Vargas, L., Silva, A.A. & Galon, L. (2009). Resistência de *Lolium multiflorum* ao glyphosate, In: *Resistência de Plantas Daninhas a Herbicidas no Brasil*, Agostinetto, D. & Vargas, L. (Eds.), pp.271-289, Berthier, ISBN 978-858-9873-92-5, Passo Fundo, Brazil.

Fleck, N.G., Agostinetto, D., Galon, L. & Schaedler, C.E. (2008). Competitividade relativa entre cultivares de arroz irrigado e biótipo de arroz-vermelho. *Planta Daninha*, Vol.26, No.1, (January/March 2008), pp.101-111, ISSN 0100-8358.

Floss, E.L. (2008). *Fisiologia das Plantas Cultivadas* (4th Ed.), Universidade de Passo Fundo, ISBN 978-857-5156-41-4, Passo Fundo, Brazil.

Galon, L., Agostinetto, L., Moraes, P.V.D., Dal Magro, T., Panozzo, L.E., Brandolt, R.R. & Santos, L.S. (2007). Níveis de dano econômico para decisão de controle de capim-arroz (*Echinochloa* spp.) em arroz irrigado (*Oryza sativa*). *Planta Daninha*, Vol.25, No.4, (October/December 2007), pp.709-718, ISSN 0100-8358.

Galon, L., Ferreira, F.A., Silva, A.A., Concenço, G., Ferreira, E.A., Barbosa, M.H.P., Silva, A.F., Aspiazú. I., França, A.C. & Tironi, S.P. (2010). Influência de herbicidas na atividade fotossintética de genótipos de cana-de-açúcar. *Planta Daninha*, Vol.28, No.3, pp.591-597, ISSN 0100-8358.

Gurevitch, J., Scheiner, S.M. & Fox, G.A. (2009). *Ecologia Vegetal* (2nd Ed.), Artmed, ISBN 978-853-6319-18-6, Porto Alegre, Brazil.

Humble, G.D. & Hsiao, T.C. (1970). Light-dependent influx and efflux of potassium of guard cells during stomatal opening and closing. *Plant Physiology*, Vol.46, No.3, (September 1970), pp.483-487, ISSN 0032-0889.

Hutmacher, R.B. & Krieg, D.R. (1983). Photosynthetic rate control in cotton. *Plant Physiology*, Vol.73, No.3, (November 1983), pp.658-661, ISSN 0032-0889.

Kirschbaum, M.U.F. & Pearcy, R.W. (1988). Gas exchange analysis of the relative importance of stomatal and biochemical factors in photosynthetic induction in *Alocasia macrorrhiza*. *Plant Physiology*, Vol.86, No.3, (March 1988), pp.782-785, ISSN 0032-0889.

Larcher, W. (2004). *Ecofisiologia Vegetal*, Rima, ISBN 858-655-2038, São Paulo, Brazil.

Long, S.P. & Bernacchi, C.J. (2003). Gas exchange measurements, what can they tell us about the underlying limitations to photosynthesis? Procedures and sources of error. *Journal of Experimental Botany*, Vol.54, No.392, (November 2003), pp.2393-2401, ISSN 0022-0957.

Machado, A.F.L., Ferreira, L.R., Santos, L.D.T., Ferreira, F.A., Viana, R.G., Machado, M.S.V. & Freitas, F.C.L. (2010). Eficiência fotossintética e uso da água em plantas de eucalipto pulverizadas com glyphosate. *Planta Daninha*, Vol.28, No.2, (April/June 2010), pp.319-327, ISSN 0100-8358.

Melo, P.T.B.S., Schuch, L.O.B., Assis, F. & Concenço, G. (2006). Comportamento de populações de arroz irrigado em função das proporções de plantas originadas de sementes de alta e baixa qualidade fisiológica. *Revista Brasileira de Sementes*, Vol.12, No.1, pp.37-43, ISSN 0101-3122.

Merotto Jr., A., Fischer, A.J. & Vidal, R.A. (2009). Perspectives for using light quality knowledge as an advanced ecophysiological weed management tool. *Planta Daninha*, Vol.27, No.2, (April/June 2009), pp.407-419, ISSN 0100-8358.

Messinger, S.M., Buckley, T.N. & Mott, K.A. (2006). Evidence for involvement of photosynthetic processes in the stomatal response to CO_2. *Plant Physiology*, Vol.140, No.2, (February 2006), pp.771-778, ISSN 0032-0889.

Naves-Barbiero, C.C., Franco, A.C., Bucci, S.J. & Goldstein, G. (2000). Fluxo de seiva e condutância estomática de duas espécies lenhosas sempre-verdes no campo sujo e cerradão. *Revista Brasileira de Fisiologia Vegetal*, Vol.12, No.1, (Abril 2000), pp.119-134, ISSN 0103-3131.

Procópio, S.O., Santos, J.B., Silva, A.A., Martinez, C.A. & Werlang, R.C. (2004a). Características fisiológicas das culturas de soja e feijão e de três espécies de plantas daninhas. *Planta Daninha*, Vol.22, No.2, (April/June 2004), pp.211-216, ISSN 0100-8358.

Procópio, S.O., Santos, J.B., Silva, A.A., Donagemma, G.K. & Mendonça, E.S.V. (2004b). Ponto de murcha permanente de soja, feijão e plantas daninhas. *Planta Daninha*, Vol.22, No.1, (January/March 2004), pp.35-41, ISSN 0100-8358.

Pystina, N.V. & Danilov, R.A. (2001). Influence of light regimes on respiration, activity of alternative respiratory pathway and carbohydrates content in mature leaves of *Ajuga reptans* L. *Revista Brasileira de Fisiologia Vegetal*, Vol.13, No.3, (Dezembro 2001), pp. 285-292, ISSN 0103-3131.

Radosevich, S.R., Holt, J.S. & Ghersa, C.M. (2007). *Ecology of Weeds and Invasive Plants: Relationship to Agriculture and Natural Resource Management* (3rd Ed.), John Wiley & Sons, ISBN 978-047-1767-79-4, Hoboken, USA.

Santos, J.B., Procópio, S.O., Silva, A.A. & Costa, L.C. (2003). Captação e aproveitamento da radiação solar pelas culturas da soja e do feijão e por plantas daninhas. *Bragantia*, Vol.62, No.1, (Janeiro/Abril 2003), pp.147-153, ISSN 0006-8705.

Severino, F.J. (2005). *Supressão da Infestação de Plantas Daninhas pelo Sistema de Produção de Integração Lavoura-Pecuária*. 113p. Thesis (Doctorate) - Escola Superior de Agricultura "Luiz de Queiroz", Universidade de São Paulo, Piracicaba.

Sharkey, T.D. & Raschke, K. (1981). Effect of light quality on stimatal opening in leaves of *Xanthium strumarium*. *Plant Physiology*, Vol.68, No.5, (November 1981), pp.1170-1174, ISSN 0032-0889.

Silva, A.A., Ferreira, F.A., Ferreira, L.R. & Santos, J.B. (2007). Biologia de plantas daninhas, In: *Tópicos em Manejo de Plantas Daninhas*, Silva, A.A. & Silva, J.F., pp.17-61, Universidade Federal de Viçosa, ISBN 978-857-2692-75-5, Viçosa, Brazil.

Tambuci, E.A., Bort, J. & Araus, J.L. (2011). Water use efficiency in C3 cereals under mediterranean conditions: a review of some physiological aspects. *Options Méditerranéennes*, Series B, No.57, pp.189-203, ISSN 1016-1228.

Taylor Jr., G.E. & Gunderson, C.A. (1986). The response of foliar gas exchange to exogenously applied ethylene. *Plant Physiology*, Vol.82, No.3, (November 1986), pp.653-657, ISSN 0032-0889.

Tuffi Santos, L.D.T., Ferreira, F.A., Machado, A.F.L., Ferreira, L.R. & Santos, B.F.S. (2010). Glyphosate em eucalipto: formas de contato e efeito do herbicida sobre a cultura, In: *Manejo Integrado de Plantas Daninhas na Cultura do Eucalipto*, Ferreira, L.R., Machado, A.F.L., Ferreira, F.A. & Santos, L.D.T. (Eds.), pp.91-116, Universidade Federal de Viçosa, ISBN 978-857-2693-76-9, Viçosa, Brazil.

Vanderzee, D. & Kennedy, R.A. (1983). Development of photosynthetic activity following anaerobic germination in rice-mimic grass (*Echinochloa crus-galli* var. *oryzicola*). *Plant Physiology*, Vol.73, No.2, (October 1983), pp.332-339, ISSN 0032-0889.

Part 2

Applied Aspects

13

Phloem Feeding Insect Stress and Photosynthetic Gene Expression

Anna-Maria Botha[1], Leon van Eck[2], Carlo S. Jackson[3], N. Francois V. Burger[1,3] and Thia Schultz[1]
[1]*Department of Genetics, Stellenbosch University, Stellenbosch,*
[2]*Department of Soil and Crop Sciences, Colorado State University, Ft Collins, CO,*
[3]*Department of Genetics, Forestry and Agriculture Biotechnology Institute, University of Pretoria, Hillcrest, Pretoria,*
[1,3]*South Africa*
[2]*USA*

1. Introduction

The ability to photosynthesise (*i.e.*, to utilize solar energy for conversion into chemical energy) is a distinguishing characteristic unique to plants, algae and photoautotrophic bacteria. It is believed that photosynthesis was already well established at least 3.5 billion (Gyr) years ago in ancient organisms, with similar capabilities as that of modern cyanobacteria (Schidlowski, 1984, 1988; Blakenship, 1992). However, it is only much later (*i.e.*, between 2.3 to 2.7 Gyr ago), with the advent of oxygen-evolving photosynthesis, that advanced life became possible (Buick, 1992; Björn & Govindjee, 2009).

For sunlight to be converted into chemical energy, it must first be absorbed by organisms through the use of pigments. The primary light absorbing pigments, located in the thylakoid membrane of chloroplasts of eukaryotic cells, are Chlorophyll *a* (Chl *a*) and Chlorophyll *b* (Chl *b*). These pigments are located in the thylakoid membrane of the chloroplast, and absorb different light wavelengths so as to accumulate energy in the form of excited electrons. Secondary pigments, such as carotenoids (carotenes and xanthophyll), are located in the chloroplast membrane and outer membrane in order to absorb residual light wavelengths not efficiently absorbed by the primary pigments (Blankenship, 1992; Nelson & Yocum, 2006; Björn & Govindjee, 2009). This conversion of solar to chemical energy is a complex process and involves a large number of pigments and electron transfer proteins, collectively known as a photosynthetic unit (*i.e.*, photosynthetic reaction centre) (Buttner et al., 1992). In a photosynthetic system the pigments serve as an antenna, collecting light and transferring the energy to the reaction centre, where the reactions leading to chemical energy conversion take place.

The photosynthetic reaction centres, or the cores of light harvesting systems, consist of special protein-chlorophyll complexes which play a major role in the energy conversion process (Buttner et al., 1992). Oxygenic photosynthesis of chloroplasts involves two photosystems: the oxygen-evolving photosystem II (PSII) that originated from purple bacteria and the ferredoxin reducing photosystem I (PSI) that originated from the green sulphur bacteria (Figure 1) (Xiong et al., 2000; Dent et al., 2001).

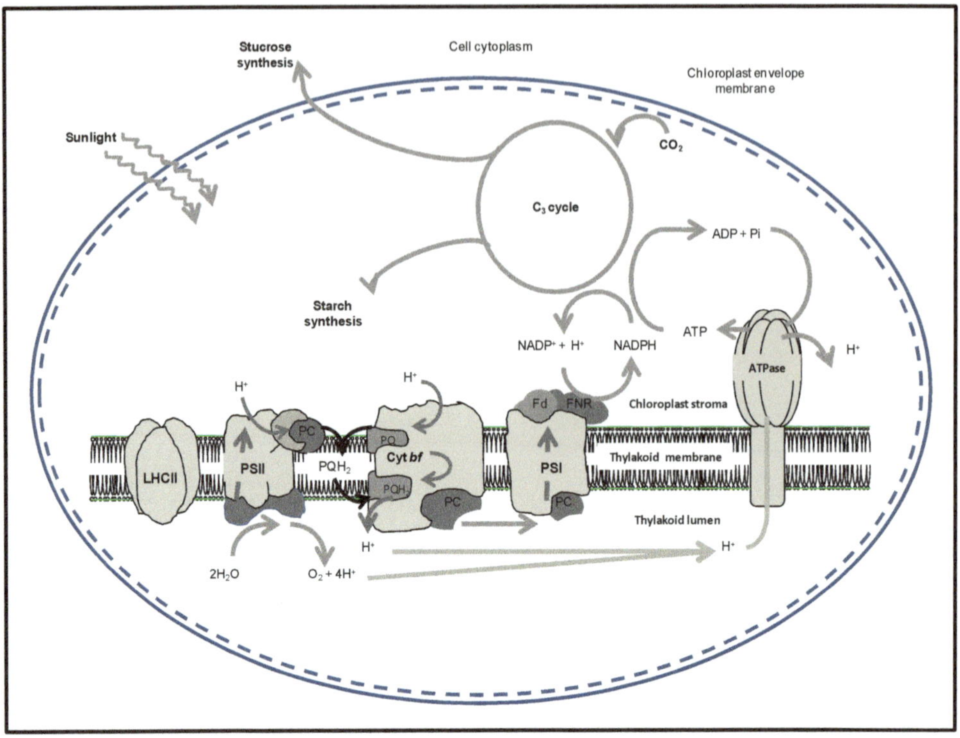

Fig. 1. Indicated are Photosystems I and II's location and their respective functions. The thylakoid membrane with PSI and PSII are indicated with the energy flow through the Calvin cycle (Modified from Dent et al., 2001).

Photosystem I (PSI) reaction centre complex consists of 6 polypeptides containing two of subunit I, which associates with P700, subunit PSI-D, subunit PSI-E, quinones and fluorenones. TMP14 thylakoid membrane phosphoprotein (14 kDa), a novel subunit of the plant PSI (Khrouchtchova et al., 2005) is found second, after PSI-D, as phosphorylation subunits of PSI (Hansson & Vener, 2003).). It is probably involved in the interaction with LHCII and together with PSI-D ensures PSI's function by accepting electrons from PSII (Khrouchtchova et al., 2005). Photosystem I P700 is bound by PsaA and PsaB in PSI and function as the primary electron donor. PSI converts photonic excitation into a charge separation, which transfers an electron from the donor P700 chlorophyll pair to the spectroscopically characterized acceptors A0, A1, FX, FA and FB in turn. Photosystem I P700 induction ensures electron excitation and reduction might force the synthesis of reactive oxygen intermediates (ROIs) for the hypersensitive response (*i.e.*, oxidative burst during plant defence) (Grotjohann & Fromme, 2005). Each PSI P700 antenna molecule consists of twenty chlorophyll *a* molecules and a cytochrome 522 heme (Bengis & Nelson, 1977). PSI utilises photons at 700 nm wavelength to excite electrons collected from its antenna molecule P700. The electrons produced by PSI are transferred to PSII, where it is excited, captured by ferredoxin and finally used to reduce $NADP^+$ to NADPH. ATP is produced from ADP and pyrophosphate via chemiosmosis. The energy for this process is produced by three hydrogen ions, which supply the energy by passing from the thylakoid to the stroma of the chloroplast. Both ATP and NADPH are subsequently used in the light-independent reactions of the PSII complex, to convert carbon dioxide into glucose using the hydrogen atom extracted from water by PSII, and releasing oxygen as a by-product (Fromme, 1996; Nelson & Yocum, 2006).

Photosystem II (PSII) reaction centre complex on the other hand consists of D-1 and D-2 polypeptides, five chlorophyll *a*, two pheophytin *a*, one *B*-carotene, and one or two cytochrome *b*-559 heme- molecules (Nanba & Satoh, 1987). In PSII, the P680 reaction centre captures photons, and the light energy is used for oxidation (splitting) of water molecules. Upon electron release, the water molecule is broken into oxygen gas and released into the atmosphere. The resulting hydrogen ions are then used to power ATP synthesis. The electrons, excited at the antenna molecule P680, are passed down a chain of electron-transport proteins while receiving extra electrons from PSI. More hydrogen ions are pumped across the membrane as these electrons flow down the chain providing more protons for ATP synthesis. Chloroplast ATP synthase (cpATPase) is found to be essential for photosynthesis (Maiwald et al., 2003) by playing a direct role in the translocation of protons across the membrane as a key component of the proton channel. Nine different polypeptides make up the cpATPase, which consists of intrinsic CF_0 and extrinsic CF_1 segments. The F-type ATPase CF_1 segment functions as the catalytic core and the CF_0 segment functions as the membrane proton channel (Cramer et al., 1991; Groth & Strotmann, 1999). The electrons are then transported as NADPH molecules to enzymes that build sugar from water and carbon dioxide (Nanba & Satoh, 1987).

2. Photosynthetic genes respond to biotic stressors

Plants are constantly locked in an evolutionary arms race with their biological attackers, whether they be viral, bacterial or fungal pathogens, parasitic plants or herbivorous insects, therefore imposing the need to evolve defensive strategies to overcome this onslaught. Although there are a few examples of compensatory stimulation of photosynthesis (Trumble et al., 1993), most reports suggest that a decline in photosynthetic capacity is inevitable and this may represent the "hidden fitness costs" to defence (Fouché et al., 1984; Zangerl et al., 2002, 2003; Heng-Moss et al., 2003; Bilgin et al., 2008, 2010; Nabity et al., 2009). A recent study by Bilgin et al. (2010), reported that photosynthesis associated genes were down-regulated in seven different dicotyledonous and one coniferous plant species upon exposure to twenty different forms of biotic damage, regardless of the type of biotic attack. The hosts seem able to down-regulate these genes as an adaptive response to biotic attack, since a reduction in gene expression does not necessarily translate into loss of function.

Once invasion by an attacker has been recognized, through the detection of various effectors (either pathogen-associated molecular patterns (PAMPs), microbe-associated molecular patterns (MAMPs), viral coat proteins or insect salivary elicitors), the host must balance competing demands for metabolic resources between either supporting defence versus sustaining cellular maintenance, growth and reproduction (Berger et al., 2007a,b). Plant defence can be costly in terms of plant growth and fitness (Tian et al., 2003; Zavala &

Baldwin, 2004), as in addition to the mobilization of an array of defensive strategies (*i.e.*, up-regulation of a suite of defence response genes and production of chemical defence responses), the plant usually also has to cope with a reduction in effective biomass, and a decline in photosynthetic capacity in the remaining leaf tissue (Zangerl et al., 2002; Bilgin et al., 2008, 2010; Nabity et al., 2009).

2.1 Phloem-feeding insects

Phloem-feeding insects (PFI), on the other hand selectively down-regulate the expression of only certain photosynthesis-related genes (Heidel & Baldwin, 2004; Voelckel et al., 2004; Zhu-Salzman et al., 2004; Qubbaj et al., 2005; Yuan et al., 2005; Van Eck, 2007) (Table 1), and by manipulating the host carbohydrate metabolism, induce a change in carbon flux to their own advantage (Zhu-Salzman et al., 2004).

Regulated gene	Host species	Aphid species	Platform[1]	Ref[2]
PSI P700 apoprotein	*Triticum aestivum*	*Diuraphis noxia*	SSH libraries/cDNA arrays	1
PSI reaction centre SU2/SU IV	*Apium graviolens*	*Myzus persicae*	Oligonucleotide arrays	2
	Sorghum bicolor	*Schizaphis graminum*	SSH libraries/cDNA arrays	3
	Arabidopsis thaliana	*Bemisia tabaci*	Oligonucleotide arrays	4
PSI antenna & assembly proteins	*Apium graviolens*	*Myzus persicae*	Oligonucleotide arrays	2
	Triticum aestivum	*Diuraphis noxia*	SSH libraries	5
PSI chain D precursor	*Sorghum bicolor*	*Schizaphis graminum*	SSH libraries/cDNA arrays	3
PSII 5 kD protein	*Arabidopsis thaliana*	*Bemisia tabaci*	Oligonucleotide arrays	4
PSII 10 kD protein	*Triticum aestivum*	*Diuraphis noxia*	cDNA-AFLP analysis	6
	Apium graviolens	*Myzus persicae*	Oligonucleotide arrays	2
PSII protein DI	*Triticum aestivum*	*Diuraphis noxia*	SSH libraries	5
PSII LS1 protein	*Triticum aestivum*	*Diuraphis noxia*	SSH libraries	7
	Apium graviolens	*Myzus persicae*	Oligonucleotide arrays	2

[1] Where cDNA-AFLPs = cDNA-amplified fragment length polymorphism and SSH = suppression subtractive hybridization

[2] 1, Botha et al., 2006; 2, Divol et al., 2005; 3, Park et al., 2005; 4, Kempema et al., 2007; 5, Boyko et al., 2006; 6, Schultz, 2010; 7, Lacock et al., 2003; 8, Voelckel & Baldwin, 2004; 9, Zhu-Salzman et al., 2004; 10, Botha et al., 2010; 11, Qubbaj et al., 2005; 12, De Vos et al., 2005; 13, Van Eck, 2007; 14, Kuśniersczyk et al., 2008; 15, Zaayman et al., 2009.

PSII O_2 evolving complex peptide	*N. attenuata*	*Myzus nitotianae*	cDNA microarrays	8
	Sorghum bicolor	*Schizaphis graminum*	Oligonucleotide arrays	9
	Triticum aestivum	*Diuraphis noxia*	SSH libraries	5
PSII type I chlorophyll *a/b* binding protein	*Sorghum bicolor*	*Schizaphis graminum*	Oligonucleotide arrays	9
	Apium graviolens	*Myzus persicae*	Oligonucleotide arrays	2
	Sorghum bicolor	*Schizaphis graminum*	SSH libraries/cDNA arrays	3
	Triticum aestivum	*Diuraphis noxia*	Oligonucleotide arrays	10
	Triticum aestivum	*Diuraphis noxia*	SSH libraries	5
PSII chlorophyll *a* binding protein psbB	*Triticum aestivum*	*Diuraphis noxia*	SSH libraries	5
ATP synthase δ subunit	*Triticum aestivum*	*Diuraphis noxia*	SSH libraries	7
	Malus domestica	*Dysaphis plantaginea*	cDNA-AFLP analysis	6, 11
	Arabidopsis thaliana	*Myzus persicae*	Oligonucleotide arrays	12
	Triticum aestivum	*Diuraphis noxia*	SSH libraries/cDNA arrays	1
	Triticum aestivum	*Diuraphis noxia*	Oligonucleotide arrays	10
Aconitate hydratase	*Triticum aestivum*	*Diuraphis noxia*	cDNA-AFLP analysis	13
Beta-glucosidase	*Triticum aestivum*	*Diuraphis noxia*	cDNA-AFLP analysis	6
	Apium graviolens	*Myzus persicae*	Oligonucleotide arrays	2
Chloroplast carbonic anhydrase	*Triticum aestivum*	*Diuraphis noxia*	cDNA-AFLP analysis	6
	Sorghum bicolor	*Schizaphis graminum*	Oligonucleotide arrays	9
Chloroplast genome DNA	*Triticum aestivum*	*Diuraphis noxia*	SSH libraries	7
Chloroplast 23S ribosomal RNA gene	*Triticum aestivum*	*Diuraphis noxia*	cDNA-AFLP analysis	6
Chloroplast 50S ribosomal protein L28	*Triticum aestivum*	*Diuraphis noxia*	Oligonucleotide arrays	10

Chloroplast precursor CH1C_ARATH 20 kDa chaperonin,	*Triticum aestivum*	*Diuraphis noxia*	Oligonucleotide arrays	10
29 kD ribonucleoprotein chloroplast precursor	*Sorghum bicolor*	*Schizaphis graminum*	SSH libraries/cDNA arrays	3
Ribosomal protein chloroplast-like	*Sorghum bicolor*	*Schizaphis graminum*	SSH libraries/cDNA arrays	3
Chlorophyllase	*Apium graviolens*	*Myzus persicae*	Oligonucleotide arrays	2
Chlorophyll synthetase	*Sorghum bicolor*	*Schizaphis graminum*	Oligonucleotide arrays	9
	Triticum aestivum	*Diuraphis noxia*	Oligonucleotide arrays	10
	Triticum aestivum	*Diuraphis noxia*	Oligonucleotide arrays	10
Chloroplast 50S ribosomal protein L28	*Arabidopsis thaliana*	*Bemisia tabaci*	Oligonucleotide arrays	4
Chlorophyll A oxygenase (CAO)	*Apium graviolens*	*Myzus persicae*	Oligonucleotide arrays	2
Carbonic anhydrase	*Triticum aestivum*	*Diuraphis noxia*	Oligonucleotide array	10
	Triticum aestivum	*Diuraphis noxia*	cDNA-AFLP analysis	6
Cytochrome P450	*Arabidopsis thaliana*	*Brevicoryne brassicae*	Oligonucleotide arrays	14
	Triticum aestivum	*Diuraphis noxia*	Oligonucleotide arrays	10
	Apium graviolens	*Myzus persicae*	Oligonucleotide arrays	2
	Triticum aestivum	*Diuraphis noxia*	cDNA-AFLP analysis	6
Cytochrome *c*1, *c*6 (ATC6) and/or B6	*Sorghum bicolor*	*Schizaphis graminum*	SSH libraries/cDNA arrays	3
	Arabidopsis thaliana	*Bemisia tabaci*	Oligonucleotide arrays	4
	Sorghum bicolor	*Schizaphis graminum*	SSH libraries/cDNA arrays	3
Ferredoxin	*Sorghum bicolor*	*Schizaphis graminum*	Oligonucleotide arrays	9
	Sorghum bicolor	*Schizaphis graminum*	Oligonucleotide arrays	9
Ferredoxin-thioredoxin reductase	*Triticum aestivum*	*Diuraphis noxia*	Oligonucleotide arrays	10

	Triticum aestivum	*Diuraphis noxia*	Oligonucleotide arrays	1
	Triticum aestivum	*Diuraphis noxia*	cDNA-AFLP analysis	6, 13, 15
Ferrochelatase	*Sorghum bicolor*	*Schizaphis graminum*	Oligonucleotide arrays	9
Fructose-1,6-bisphosphatase	*Triticum aestivum*	*Diuraphis noxia*	Oligonucleotide arrays	10
	Triticum aestivum	*Diuraphis noxia*	Oligonucleotide arrays	10
	Triticum aestivum	*Diuraphis noxia*	cDNA-AFLP analysis	6
Malate dehydrogenase [NADP], chloroplast precursor	*Triticum aestivum*	*Diuraphis noxia*	Oligonucleotide arrays	10
Mg-chelate subunit chlH	*Triticum aestivum*	*Diuraphis noxia*	Oligonucleotide arrays	10
Monooxygenase 2	*Triticum aestivum*	*Diuraphis noxia*	cDNA-AFLP analysis	6
NADPH:quinone oxidoreductase	*Triticum aestivum*	*Diuraphis noxia*	Oligonucleotide arrays	10
Nonphototrophic hypocotyl 1b	*Apium graviolens*	*Myzus persicae*	Oligonucleotide arrays	2
Non-green plastid inner envelope membrane protein	*Triticum aestivum*	*Diuraphis noxia*	Oligonucleotide arrays	10
Phytochrome association *PAP2*	*Triticum aestivum*	*Diuraphis noxia*	Oligonucleotide arrays	10
Photolyaseblue-light receptor	*Triticum aestivum*	*Diuraphis noxia*	SSH libraries/cDNA arrays	1
Quinone oxidoreductase	*Triticum aestivum*	*Diuraphis noxia*	SSH Libraries	7
Red chlorophyll catabolic reductase gene	*Triticum aestivum*	*Diuraphis noxia*	cDNA-AFLP analysis	6, 13, 15
Ribulose-1,5-bisphosphate carboxylase /oxygenase LSU	*Malus domestica*	*Dysaphis plantaginea*	cDNA-AFLP analysis	11
	Nicotiana attenuata	*Myzus nicotianae*	Oligonucleotide arrays	8
	Triticum aestivum	*Diuraphis noxia*	SSH libraries/cDNA arrays	1

	Apium graviolens	*Myzus persicae*	Oligonucleotide arrays	2
	Triticum aestivum	*Diuraphis noxia*	Oligonucleotide arrays	10
	Triticum aestivum	*Diuraphis noxia*	SSH libraries	7
	Triticum aestivum	*Diuraphis noxia*	cDNA-AFLP analysis	6
Ribulose-1,5-bisphosphate carboxylase/oxygenase SSU	*Nicotiana attenuata*	*Myzus nicotianae*	Oligonucleotide arrays	8
	Sorghum bicolor	*Schizaphis graminum*	Oligonucleotide arrays	9
	Apium graviolens	*Myzus persicae*	Oligonucleotide arrays	2
	Triticum aestivum	*Diuraphis noxia*	SSH libraries	7
	Triticum aestivum	*Diuraphis noxia*	SSH libraries/cDNA arrays	1
T51328 transcription initiation factor sigma5, plastid -specific	*Triticum aestivum*	*Diuraphis noxia*	Oligonucleotide arrays	10
Thioredoxin	*Triticum aestivum*	*Diuraphis noxia*	Oligonucleotide arrays	10
TMP 14 kDa thylakoid membrane phosphoprotein	*Triticum aestivum*	*Diuraphis noxia*	cDNA-AFLP analysis	13
	Arabidopsis thaliana	*Bemisia tabaci*	Oligonucleotide arrays	4
Thylakoid luminal 15 kD protein	*Triticum aestivum*	*Diuraphis noxia*	Oligonucleotide arrays	10
Thylakoid luminal protein-related	*Triticum aestivum*	*Diuraphis noxia*	Oligonucleotide arrays	10
Ubiquinol--cytochrome-*c* reductase	*Triticum aestivum*	*Diuraphis noxia*	Oligonucleotide arrays	10
	Triticum aestivum	*Diuraphis noxia*	cDNA-AFLP analysis	6

Table 1. Genes involved in photosynthesis under regulation after aphid feeding. Underlined genes are down-regulated, non-underlined genes are mostly up-regulated.

PFIs achieve these benefits, at some cost to their host by inducing genes, involved in carbon assimilation and mobilization, so as to increase their own sugar uptake, whilst at the same time depleting sugars and creating localized metabolic sinks (Moran & Thompson, 2001; Zhu-Salzman et al., 2004). PFIs also modify nitrogen allocation in their hosts by up-regulating genes involved in nitrogen assimilation. In particular, genes encoding enzymes required for the synthesis of tryptophan and other essential amino acids are up-regulated to fulfil to the dietary requirements of the PFI (Heidel & Baldwin, 2004; Zhu-Salzman et al., 2004; Botha et al., 2010).

2.2 Linking photosynthesis and plant defence

The linkage between photosynthesis and host defence was recently demonstrated by silencing two central photosynthetic proteins, *i.e.*, ribulose-1,5-bisphosphate carboxylase/oxygenase (Rubisco) and Rubisco activase in *Nicotiana attenuata* using virus-induced gene silencing (VIGS) (Mitra & Baldwin, 2008). Silencing of these genes improved the performance of a native generalist (*Spodoptera littoralis*) and specialist (*Manduca sexta*) herbivorous larvae on transformed host plants (Mitra & Baldwin, 2008). Similarly, it was shown that independent silencing of the TMP 14 kDa thylakoid membrane phosphoprotein, PSI P700 apoprotein, and Fructose-1,6-bisphosphatase in near-isogenic (NILs) *Triticum aestivum* lines using VIGS, also affected host resistance to *Diuraphis noxia* in varying degrees (Jackson, 2010). In the latter study, no significant decrease in aphid fecundity was observed in the susceptible Tugela plants after silencing with any of the genes when compared with the uninfested Tugela plants (Table 2). Silencing with BSMV:TMP14 caused a significant increase in aphid fecundity when aphids were fed on the resistant Tugela-*Dn1* plants, while silencing with BSMV:FBPase caused an increase in aphid fecundity in resistant Tugela-*Dn2* plants. A significant increased aphid fecundity was also observed upon silencing of Tugela-*Dn2* plants with BSMV:P700, but in Tugela-*Dn1* plants no significant increase in aphid fecundity was observed (Table 2).

Treatment	Susceptible NIL	Resistant NILs	
	Tugela	Tugela-Dn1[a]	Tugela-Dn2[b]
Control	31.0 ± 1.82	14.3 ± 2.08	21.0 ± 2.64
BSMV:00	30.25± 2.62	15.0 ± 1.00	20.3 ± 2.08
BSMV:TMP14	34.4 ± 3.58	21.8 ± 2.63*	23.6 ± 2.06
BSMV:FBPase	33.8 ± 2.93	17.1 ± 2.04	26.2 ± 2.78*
BSMV:P700	34.0 ± 1.54	12.2 ± 1.17	24.8 ± 2.31*

*Significantly different from control ($P< 0.05$)
[a]SA1684/6* 'Tugela' (*Dn1*+) – confers antibiosis to *Diuraphis noxia* (Wang et al., 2004)
[b]SA2199/6* 'Tugela' (*Dn2*+) – confers tolerance to *Diuraphis noxia* (Wang et al., 2004)

Table 2. Summary of *Diuraphis noxia* fecundities when feeding on three near isogenic wheat lines before and after gene silencing. Aphid fecundity is indicated as number of aphids per plant 10 d.p.i. (n=3)(From Jackson, 2010).

2.3 Does the photosynthetic compensation of the wheat host form part of *Diuraphis noxia*'s defence strategy?

Diuraphis noxia feeding on susceptible wheat causes chlorosis (*i.e.*, longitudinal chlorotic streaking) and leaf rolling in the leaves of susceptible wheat (Figure 2). Leaf chlorophyll content is reduced by *D. noxia* infestation (Heng-Moss et al., 2003; Botha et al., 2006). This results in decreased photosynthetic potential and the eventual collapse of the plant (Burd & Burton, 1992).

Fig. 2. Wheat expressing the longitudinal chlorotic streaking phenotype associated with *Diuraphis noxia* susceptibility.

Aphid damage has historically been ascribed to a phytotoxin injected during feeding, which is responsible for chloroplast disintegration (Fouché et al., 1984). Although, such a phytotoxin has never been described or isolated, ultrastructural studies revealed limited chloroplast breakdown in the leaves of resistant cultivars after aphid feeding (Van der Westhuizen et al., 1998). Since cell fluorescence data has shown that *D. noxia* feeding causes reduced photosynthetic capacity even in intact chloroplasts (Haile et al., 1999), this chloroplast rupture mechanism seems unlikely. *D. noxia* feeding probably induces malfunctioning of the photosynthetic apparatus in the stacked region of the thylakoid membrane, but the exact site of interference has not been determined (Burd & Elliott, 1996; Heng-Moss et al., 2003). Chlorosis induced by *D. noxia* differs significantly from normal chlorophyll degradation during leaf senescence (Ni et al., 2001). *D. noxia* feeding stimulates an increase in the activity of Mg-dechelatase, a catabolic enzyme that converts chlorophyllide *a* to pheophorbide *a* as the final step in the chlorophyllase pathway (Ni et al., 2001; Wang et al., 2004)

Total chlorophyll concentration assays indicate that *D. noxia* feeding causes a marked decrease in chlorophyll levels in Tugela, but that the reduction in the antibiotic near-isogenic line (NIL) Tugela-*Dn1* is much less severe (Botha et al., 2006). Since the phenotypes afforded by different *Dn* genes vary — *Dn1* confers antibiosis, *Dn2* tolerance and *Dn5* a combination of antibiosis and antixenosis (Wang et al., 2004) — it appears that the presence of these genes activate transcription of defence-related genes differently (Botha et al., 2008). Antibiotic Betta-*Dn1* plants are also unable to compensate for chlorophyll loss, which has been attributed to an increase in defence compound production. Tolerant Betta-*Dn2* plants have very stable chlorophyll content during *D. noxia* feeding, suggesting that they can compensate for chlorophyll loss in some way (Heng-Moss et al., 2003).

Chlorosis due to *D. noxia* infestation is thought to originate from interference with electron transport (Burd & Elliott, 1996; Haile et al., 1999; Heng-Moss et al., 2003; Botha et al., 2006). Susceptible wheat shows decreased levels of chlorophyll *a* upon infestation by *D. noxia* (Burd & Elliott, 1996; Ni et al., 2001; Wang et al., 2004) which indicates damage to PSI (Botha et al., 2006). If this is indeed the case, it has serious implications for susceptible wheat under aphid attack. PSI catalyzes the electron transport from plastocyanin to ferredoxin (Haldrup et al., 2003). This reduced ferredoxin pool is mostly employed in generating NADPH for CO_2 assimilation, but is also used in regulating the activity of, among others, CF_1-ATPase and several enzymes in the Calvin cycle (Ruelland & Miginiac-Maslow, 1999). Under-reduced ferredoxin directly diminishes the plant's ability to synthesize ATP and carbohydrates. Studies by Van Eck (2007) using qRT-PCR analysis of the CF_1-ATPase response to aphid feeding indicated an increased demand for ATPase transcripts as infestation progressed (Figure 3).

Since damaged PSI can no longer act as electron acceptor from PSII via the cytochrome $b_{6\text{-}f}$ complex, inefficient reoxidation of the reduced plastoquinone occurs, halting state transitions and resulting in an over-reduction of PSII (Burd & Elliott, 1996). This leads to photoinactivation of PSII, and thus the irreversible decline in functional PSII complexes, because the absorbed light energy exceeds the amount that can be employed in electron transport (Kornyeyev et al., 2006). An acute induction of a TMP 14 kDa thylakoid membrane phosphoprotein, a putative component of PSI, was observed in Tugela-*Dn2* after infestation with RWA (Table 1, Figure 3) which indicates transcriptionally regulated photosynthetic compensation (Van Eck, 2007).

An induction of TMP14 could be a strategy to overcome pest attack in order to keep PSI stable and energy production going, while a reduction of TMP14 might force energy to flow in a different direction. Thus, up-regulation of PSI complexes would ensure the integrity of electron transport from PSII during state 2 as well as increased levels of NADPH and possibly increased CO_2 assimilation. In growth tolerance experiments, the tolerant PI 262660 line containing the *Dn2* gene maintained vigorous growth during aphid infestation when compared to the susceptible Arapahoe and antibiotic PI 137739 line (Haile et al., 1999). It is suggested that increased photosynthetic capacity via up-regulation of photosystem components may provide a mechanism for passive resistance against *D. noxia* feeding (Botha et al., 2006).

2.4 Regulating plant homeostasis

The production of ROIs is a by-product of normal cellular processes, such as photosynthesis and respiration, but can also be produced in response to a variety of environmental conditions, *i.e.*, light, cold, drought, as well as pathogen and pest attack. The latter event is known as the hypersensitive response and has proven to be effective against sedentary insects, such as PFIs that target a specific tissue. However, for an effective hypersensitive response-based programmed cell death to occur, a cascade of events have to occur, including production of ROIs and associated downstream defensive responses. Indeed, increases in the activity of oxidative enzymes such as peroxidases, polyphenol oxidases and lipoxygenases were observed after *Diuraphis no*xia feeding (Van der Westhuizen et al., 1998; Ni et al., 2000, 2001; Ni & Quisenberry, 2003). This increase occurs not only at the site of feeding, but it also spreads systemically (Van der Westhuizen et al. 1998). Thus, if this spreading is not kept at bay, it can be lethal to the host. ROIs are partially reduced forms of atmospheric oxygen (O_2), and typically result from the excitation of O_2 to form singlet oxygen O_2^1 or from the transfer of one, two or three electrons to O_2 to form, respectively, a superoxide radical (O_2^-), hydrogen peroxide (H_2O_2) or a hydroxyl radical (HO^-). Unlike atmospheric oxygen, ROIs are capable of unrestricted oxidation of various cellular components and can lead to the oxidative destruction of the cell (Asada, 1999). A variety of mechanisms exist for the dissipation of excess excitation that may give rise to the generation of reductants and the production of ROIs, act as signalling agents, or to serve as alternative

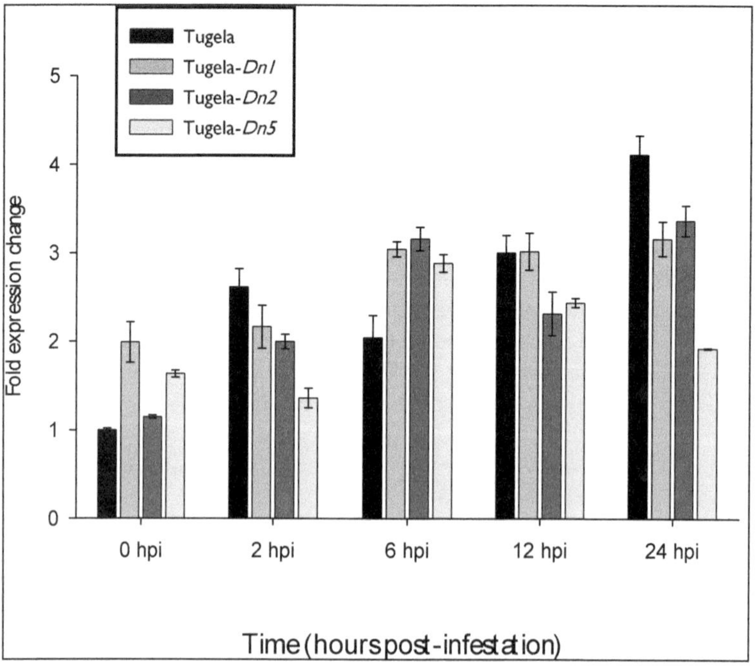

Fig. 3. qRT-PCR expression profiles of ATPase, where the expression level is calculated relative to the expression level of the uninfested, susceptible Tugela (at 0 h.p.i.) sample and is normalized to the expression of the unregulated chloroplast 16S rRNA transcript.

electron acceptors to avoid over-reduction and potentially the generation of toxic intermediates (Avenson et al., 2005; Mullineaux & Karpinski, 2002; Foyer & Noctor, 2005). It is thus suggested that cellular homeostasis will be maintained as long as the mechanisms for redox poising are in place, otherwise uncontrolled cellular damage will follow leading to death of the host (Schelbe et al., 2005).

3. Acknowledgements

This work was partially funded by the Winter Cereal Trust, the National Research Foundation, and THRIP-DTI of South Africa.

4. References

Asada, K. (1999). The water-water cycle in chloroplasts scavenging of active oxygen and dissipation of excess photons. *Annual Review of Plant Physiology and Plant Molecular Biology*, Vol. 50, pp. 601-69

Avenson, T.J., Kanazawa, A., Cruz, J.A., Takizawa, K., Ettinger, W.E. & Kramer, D.M. (2005). Integrating the proton circuit into photosynthesis: progress and challenges. *Plant, Cell and Environment*, Vol. 28, pp. 97-109

Bengis, C. & Nelson, N. (1977). Subunit structure of chloroplast photosystem I reaction Center. *The Journal of Biological Chemistry*, Vol. 252, No. 13, pp. 4564-4569

Berger, S., Sinha, A.K. & Roitsch, T. (2007a). Plant physiology meets phytopathology: plant primary metabolism and plant-pathogen interactions. *Journal of Experimental Botany*, Vol. 58, pp. 4019–4026

Bilgin, D.D., Aldea, M., O'Neill, B.F., Benitez, M., Li, M., Clough, S.J. & DeLucia, E.H. (2008) Elevated ozone alters soybean-virus interaction. *Molecular Plant-Microbe Interactions*, Vol. 21, pp. 1297–1308

Bilgin, D.D., Zavala, J.A., Zhu, J., Clough, S. J., Ort, D.R. & DeLucia, E.H. (2010). Biotic stress globally downregulates photosynthesis genes. *Plant, Cell and Environment*, Vol. 33, pp. 1597–1613

Björn, L.O. & Govindjee. (2009). The evolution of photosynthesis and chloroplasts. *Current Science*, Vol. 96, No. 11, pp. 1466-1474

Blankenship, R.E. (1992). Origin and early photosynthesis. *Photosynthesis Research*, Vol. 33, pp. 91-111

Botha, A-M., Lacock, L., van Niekerk, C., Matsioloko, M.T., du Preez, F.B., Loots, S., Venter, E., Kunert, K.J. & Cullis, C.A. (2006). Is photosynthetic transcriptional regulation in *Triticum aestivum* L. cv. 'TugelaDN' a contributing factor for tolerance to *Diuraphis noxia* (Homoptera: Aphididae)? *Plant Cell Reports*, Vol. 25, pp. 41–54

Botha, A-M., Swanevelder, Z.H., Shultz, T., Van Eck, L. & Lapitan, N.L.V. (2008). Deciphering defense strategies that are elucidated in wheat containing different *Dn* resistance genes. Proceedings of the 11th Wheat Genetics Symposium, Brisbane, Australia, August 24-29, 2008. p. O29 Edited by Rudi Appels Russell Eastwood Evans Lagudah Peter Langridge Michael Mackay Lynne. Sydney University Press. ISBN: 9781920899141 (http://hdl.handle.net/2123/3514)

Botha, A-M., Swanevelder, Z.H. & Lapitan, N.L.V. (2010). Transcript profiling of wheat genes expressed during feeding by two different biotypes of *Diuraphis noxia*. *Journal of Environmental Entomology*, Vol. 39, No. 4, pp. 1206-1231

Boyko, E.V., Smith, C.M., Thara, V.K., Bruno, J.M., Deng, Y., Strakey, S.R. & Klaahsen, D.L. (2006). Molecular basis of plant gene expression during aphid invasion: wheat Pto - and Pti-like sequences areinvolved in interactions between wheat and Russian wheat aphid (Homoptera: Aphididae). *Journal of Economic Entomology*, Vol. 99, pp. 1340-1445

Buick, R. (1992). The antiquity of oxygenic photosynthesis: Evidence from stromatolites in sulphate-deficient archean lakes. *Science*, Vol. 255, pp. 74-77

Burd, J.D. & Burton, R.L. (1992). Characterization of plant damage caused by Russian wheat aphid (Homoptera: Aphididae). *Journal of Economic Entomology*, Vol. 85, pp. 2015-2022

Burd, J.D. & Elliott, N.C. (1996). Changes in chlorophyll *a* fluorescence induction kinetics in cereals infested with Russian wheat aphid (Homoptera: Aphididae). *Journal of Economic Entomology*, Vol. 89, pp. 1332-1337

Buttner, M., Xie, D.-L., Nelson, H., Pinther, W., Hauska, G. & Nelson, N. (1992). Photosynthetic reaction center genes in green sulfur bacteria and in photosystem 1 are related. *Proceedings of the National Academy of Science USA*, Vol. 89, pp. 8135-8139

Cramer, W.A., Furbacher, P.N., Szczepaniak, A. & Tae, G-S. (1991). Electron transport between photosystem II and photosystem I. In: *Current Topics in Bioenergetics*, Lee, C.P. ed., Vol. 16. pp 179–222, Academic Press, San Diego

Dent, R.M., Han, M. & Niyogi, K.N. (2001). Functional genomics of plant photosynthesis in the fast lane using *Chlamydomonas reinhardtii*. *Trends in Plant Science*, Vol. 6, No. 8, pp. 384-371

De Vos, M., Van Oosten, V.R., Van Poecke, R.M.P., Van Pelt, J.A., Pozo, M.J., Mueller, M.J., Buchala, A.J., Métraux, J.-P., Van Loon, L.C., Dicke, M. & Pieterse, C.M.J. (2005). Signal signature and transcriptome changes of *Arabidopsis* during pathogen and insect attack. *MPMI*, Vol. 18, No. 9, pp. 923–937

Divol, F., Viliaine, F., Thibivilliers, S., Amselem, J., Palauqui, JC., et al. (2005). Systemic response to aphid infestation by *Myzus persicae* in the phloem of *Apium graveolens*. Plant Molecular Biology, Vol. 57, pp. 517-540

Fouché, A., Verhoeven, R. L., Hewitt, P. H., Walters, M. C., Kriel, C. F. & De Jager, J. (1984). Russian wheat aphid (*Diuraphis noxia*) feeding damage on wheat, related cereals and a Bromus grass species. In: *Technical communication Department of Agriculture Republic of South Africa*, Vol. 191, pp. 22-33, Walters, M.C. ed., Bloemfontein

Foyer, C.H. & Noctor, G. (2005). Redox homeostasis and antioxidant signalling: A metabolic interface between stress perception and physiological responses. *The Plant Cell*, Vol. 17, pp. 1866-1875

Fromme, P. (1996). Structure and function of photosystem I. *Current Opinion in Structural Biology*, Vol. 6, pp. 473-484

Grotjohann, I. & Fromme, P. (2005). Structure of cyanobacterial Photosystem I. *Photosynthesis Research*, Vol. 85, pp. 51-72

Groth, G. & Strotmann, H. (1999). New results about structure, function and regulation of the chloroplast ATP synthase (CF_0CF_1). *Physiologia Plantarum*, Vol. 106, pp. 142–148

Haile, F.J., Higley, L.G., Ni, X. & Quisenberry, S.S. (1999). Physiological and growth tolerance in wheat to Russian wheat aphid (Homoptera: Aphididae) injury. *Environmental Entomology*, Vol. 28, pp. 787-794

Haldrup, A., Lunde, C. & Scheller, H.V. (2003). *Arabidopsis thaliana* plants lacking the PSI-D subunit of ph,otosystem I suffer severe photoinhibition, have unstable photosystem I complexes, and altered redox homeostasis in the chloroplast stroma. *Journal of Biological Chemistry*, Vol. 278, pp. 33276-33283

Hansson, M. & Vener, A.V. (2003). Identification of three previously unknown in vivo phosphorylation sites in thylakoid membranes of *Arabidopsis thaliana. Molecular & Cellular Proteomics*, Vol. 2, pp. 550-559

Heidel, A. J. & Baldwin, I. T. (2004). Microarray analysis of salicylic acidand jasmonic acid-signalling in responses of *Nicotiana attenuate* to attack by insects from multiple feeding guilds. *Plant, Cell and Environment*, Vol. 27, pp. 1362-1373

Heng-Moss, T.M., Ni, X., Macedo, T., Markwell, J.P., Baxendale, F.P., Quisenberry, S.S. & Tolmay, V. (2003). Comparison of chlorophyll and carotenoid concentrations among Russian wheat aphid (Homoptera: Aphididae)-infested wheat isolines. *Journal of Economic Entomology*, Vol. 96, No. 2, pp. 475-481

Jackson, C.S. (2010). Significance of photosynthesis and the photosynthesis related genes TMP14, FBPase and P700 in Russian wheat aphid resistant wheat. M.Sc. Dissertation, University of Pretoria, Pretoria, pp. 191

Kempema, L.A., Cui, X., Holzer, F.M. & Walling, L.L. (2007). Arabidopsis transcriptome changes in response to phloem-feeding silverleaf whitefly nymphs. Similarities and distinctions in responses to aphids. *Plant Physiology*, Vol. 143, pp. 849-865

Khrouchtchova, A., Hansson, M., Paakkarinen, V., Vainonen, J. P., Zhang, S., Jensen, P. E., Scheller, H. V., Vener, A. V., Aro, E. M. & Haldrup, A. (2005). A previously found thylakoid membrane protein of 14 kDa (TMP14) is a novel subunit of plant photosystem I and is designated PSI-P. *Federation of European Biochemical Societies*, Vol. 579, pp. 4808–4812

Kornyeyev, D., Logan, B.A., Tissue, D.T., Allen, R.D. & Holaday, A.S. (2006). Compensation for PSII photoinactivation by regulated non-photochemical dissipation influences the impact of photoinactivation on electron transport and CO_2 assimilation. *Plant and Cell Physiology*, Vol. 47, pp. 437-446

Kuśnierszczyk, A., Per Winge, Jørstad, T.S., Troczyńska, J. T., Rossiter, J.T. & Bones, A.M. (2008). Towards global understanding of plant defence against aphids – timing and dynamics of early *Arabidopsis* defence responses to cabbage aphid (*Brevicoryne brassicae*) attack. *Plant Cell Environ*. Vol. 31, pp. 1097-1115

Lacock, L., van Niekerk, C., Loots, S., du Preez, F. & Botha, A-M. (2004). Functional and comparative analysis of expressed sequences from *Diuraphis noxia* infested wheat obtained utilizing the conserved Nucleotide Binding Site. *African Journal of Biotechnology*, Vol. 2, No. 4, pp. 75-81, ISSN 1684-5315

Maiwald, D., Dietzmann, A., Jahns, P., Pesaresi, P., Joliot, P., Joliot, A., Levin, J.Z., Salamini, F. & Leister, D. (2003). Knock-out of the genes coding for the Rieske protein and the ATP-Synthase δ-subunit of Arabidopsis. Effects on photosynthesis, thylakoid protein composition, and nuclear chloroplast gene expression. *Plant Physiology*, Vol. 133, pp.191-202

Mitra, S. & Baldwin, I.T. (2008). Independently silencing two photosynthetic proteins in *Nicotiana attenuata* has different effects on herbivore resistance. *Plant Physiology*, 148, pp. 1128-1138

Moran, P. J. & Thompson, G.A. (2001). Molecular responses to aphid feeding in Arabidopsis in relation to plant defense pathways. *Plant Physiology*, Vol. 125, pp. 1074-1085

Mullineaux, P. & Karpinski, S. (2002). Signal transductiojn in response to excess light: getting out of the chloroplast. *Current Opinion in Plant Biology*, Vol. 5, pp. 43-48

Nabity, P.D., Zavala, J.A. & DeLucia, E.H. (2009). Indirect suppression of photosynthesis on individual leaves by arthropod herbivory. *Annals of Botany*, Vol. 103, pp. 655-663

Nanba, O. & Satoh, K. (1987). Isolation of a photosystem II reaction center consisting of D-1 and D-2 polypeptides and cytochrome b-559. *Proceedings of the National Academy of Sciences, USA*, Vol. 84, No. 1, pp. 109-112

Nelson, N. & Yocum, C. F. (2006). Structure and function of photosystems I and II. *Annual Review of Plant Biology*, Vol. 57, pp. 521-565

Ni, X. & Quisenberry, S.S. (2003). Possible roles of esterase, glutathione S-transferase, and superoxide dismutase activities in understanding aphid-cereal interactions. *Entomologia Experimentalis et Applicata*, Vol. 108, pp. 187-195

Ni, X., Quisenberry, S.S., Pornkulwat, S., Figarola, J.L., Skoda, S.R. & Roster, J.E. (2000). Hydrolase and oxido-reductase activity in *Diuraphis noxia* and *Rhopalosiphum padi* (Hemiptera: Aphididae). *Journal of Economic Entomology*, Vol. 93, pp. 595-601

Ni, X., Quisenberry, S.S., Markwell, J., Heng-Moss, T., Higley, L., Baxendale, F., Sarath, G. & Klucas, R. (2001). *In vitro* enzymatic chlorophyll catabolism in wheat elicited by cereal aphid feeding. *Entomologia Experimentalis et Applicata*, Vol. 101, pp. 159-166

Park, S-L., Huang, Y. & Ayoubi, P. (2005). Identification of expression profiles of sorghum genes in response to greenbug phloem-feeding using cDNA subtraction and microarray analysis. *Planta*, Vol. 223, pp. 932-947

Qubbaj, T., Reineke, A. & Zebitz, C.P.W. (2005). Molecular interactions between rosy apple aphids, *Dysaphis plantaginea*, and resistant and susceptible cultivars of its primary host *Malus domestica*. *Entomologia Experimentalis et Applicata*, Vol. 115, pp. 145–152

Ruelland, E. & Miginiac-Maslow, M. (1999). Regulation of chloroplast enzyme activities by thioredoxins: activation of relief or inhibition? *Trends Plant Sci*, Vol. 4, pp. 136–141

Schelbe, R., Backhausen, J.E., Emmerlich, V. & Hollgrete, S. (2005). Strategies to maintain redox homeostatis during photosynthesis nunder changing condictions. *Journal of Experimental Botany*, Vol. 56, No. 416, pp. 1481-1489

Schidlowski, M. (1984). Early atmospheric oxygen levels: constraints from Archaean photoautotrophy. *J. Geol. Soc. London*, Vol. 141, pp. 243-250

Schidlowski, M. (1988). A 3800-million-year isotopic record of life from carbon in sedimentary rocks. *Nature*, Vol. 333, pp. 313-318

Schultz, T. (2010). Transcript profiling of *Diuraphis noxia* elicited responses in Betta NILs. M.Sc. Dissertation, University of Pretoria, South Africa, 190 p

Tian, D., Traw, M.B., Chen, J.Q., Kreitman, M. & Bergelson, J. (2003). Fitness costs of R-gene-mediated resistance in *Arabidopsis thaliana*. *Nature,* Vol. 423, pp. 74–77

Trumble, J.T., Kolondy-Hirsch, D.M. & Ting, I.P. (1993). Plant compensation for arthropod herbivory. *Annual Review Entomology,*Vol. 38, pp. 93–119

Van der Westhuizen, A.J., Qian, X-M. & Botha, A-M. (1998). Differential induction of apoplastic peroxidase and chitinase activities in susceptible and resistant wheat cultivars by Russian wheat aphid infestation. *Plant Cell Reports,* Vol. 18, pp. 132-137

Van Eck, L. (2007). Aphid-induced transcriptional regulation in near-isogenic wheat. M.Sc. Dissertation, University of Pretoria, Pretoria, South Africa, 145 p

Voelckel, C., Weisser, W.W. & Baldwin, I.T. (2004). An analysis of plant-aphid interactions by different microarray hybridization strategies. *Molecular Ecology,* Vol. 13, pp. 3187-3195

Voelckel, C. & Baldwin, I.T. (2004). Herbivore-induced plant vaccination. Part II. Array-studies reveal the transience of herbivorespecific transcriptional imprints and a distinct imprint from stress combinations. *The Plant Journal,* Vol. 38, pp. 650–663

Wang, T., Quisenberry, S.S., Ni, X. & Tolmay, V. (2004). Aphid (Hemiptera: Aphididae) resistance in wheat near isogenic lines. *Journal of Economic Entomology,* Vol. 97, pp. 646-653

Xiong, J., Fischer, W.M., Inoue, K., Nakahara, M. & Bauer, C.E. (2000). Molecular evidence for the early evolution of photosynthesis. *Science,* Vol. 289, pp. 1724-1730

Yuan, H. Y., Chen, X. P., Zhu, L. L. & He, G. C. 2005. Identification of genes responsive to brown planthopper *Nilaparvata lugens* Stal (Homoptera, Delphacidae) feeding in rice. *Planta,* Vol. 221, pp. 105-112

Zaayman, D., Lapitan, N.L.V., & Botha, A-.M. (2009). Dissimilar molecular defense responses are elicited in *Triticum aestivum* L. after infestation by different *Diuraphis noxia* (Kurdjumov) biotypes. *Physiologia Plantarum,* Vol. 136, pp. 209-222

Zangerl, A.R., Hamilton, J.G., Miller, T.J., Crofts, A.R., Oxborough, K., Berenbaum, M.R. & DeLucia, E.H. (2002). Impact of folivory on photosynthesis is greater than the sum of its holes. *Proceedings of the National Academy of Sciences USA,* Vol. 99, pp. 1088–1091

Zangerl, A.R. & Berenbaum, M.R. (2003). Phenotype matching in wild parsnip and parsnip webworms: causes and consequences. *Evolution,* Vol. 57, pp. 806–815

Zavala, J.A. & Baldwin I.T. (2004). Fitness benefits of trypsin proteinase inhibitor expression in *Nicotiana attenuata* are greater than their costs when plants are attacked. *BMC Ecology*, Vol. 4, pp. 11

Zhu-Salzman, K., Salzman, T.A., Ahn, J-.E. & Koiwa, H. (2004). Transcriptional regulation of sorghum defense determinants against a phloem-feeding aphid. *Plant Physiology*, Vol. 134, pp. 420-431

14

Relation Between Characteristics of Plant Bioelectric Potential and Purification Function Under LED Light

Shin-ichi Shibata[1], Haruhiko Kimura[1] and Takashi Oyabu[2]
[1]*Division of Electrical Engineering and Computer Science, Graduate School of Natural Science & Technology, Kanazawa University,*
[2]*Graduate School of Strategic Management, Kanazawa Seiryo University, Japan*

1. Introduction

Recently, awareness for food safety and environmental issues due to climate change is increasing. The plant factory is effective in the issues and food self-sufficiency. The pick-time and the production of the food (for example vegetable, rice and fruits) can be controlled artificially by controlling the growing environment, for example light, temperature and amount of carbon dioxide. Agricultural chemical is unnecessary in a fully-closed factory. Secure and safe production is also possible in the factory. However, high efficiency-production method is not still established and the growing control for environmental factors is based on the experience of researcher and practical farmer. Meanwhile, plant (vegetation) transmits various kinds of information with their growing and environmental variation. One of them is bioelectric potential and it is very complex. The plant-potential characteristic includes the physiological and environmental information and the growing factors can be controlled using the characteristic. It will make important contributions for the growing and ingathering control. Light (frequency and intensity) plays important roles for the growing of leafy and fruit vegetables. Growing part of plant and the picking season can be controlled by light frequency and intensity in detail. Namely, production control of the vegetation becomes possible. The control which is adjusted for social situation (consumer needs and food self-sufficiency) is possible. The study aims to construct the production control system for improving food self-sufficiency ratio. The system should give due consideration to the nature social environments. Plant bioelectric potential is a kind of internally-occurred electric signal depending on the environmental factors. The potential changes due to variety of factors, namely light, temperature, humidity and atmospheric pressure. If the relationship among the potential characteristic, environmental factors and plant growth can become obvious, the growth could be controlled using the self-generating potential. The relationship between the potential and light frequency is investigated as a preliminary step in this study. And purification capability for an airborne chemical (ethyl alcohol) under LED is also investigated. Plant can intake the chemical as nutrition. The capability is depend on environmental factors, especially light frequency.

2. Experoment

A pothos (Epipremnum aureum) was adopted as the plant in this study. The plant is familiar to humans and it is easy to control to grow in this plant. The pothos is suitable for experiment because environmental factor to grow easily. It was kept in a pot with an internal diameter of 10 cm and a height of 14 cm. the plant height from the bottom of the pot was about 55 cm. The plant was installed in an experimental chamber (Inner volume: 575×510×1000mm: 300litters) and the experiment was carried out about characteristics of plant bioelectric potential and purification function under LED Light. The photograph of experimental and diagram of the system is shown in Fig.1. The experimental system was composed with environmental sensor, LED panel (CCS: ISL-150×150), LED units (CCS: ISC-201-2), bioelectric code, A-D converter, and portable computer. There were four types of LED panel, namely, Blue, Green, Red, and White. The Specification of these LED panels is shown in table.1. Bioelectrical sensor, tin oxide gas sensor (Figaro: TGS#800, Oska) for sensing air purification , temperature•humidity•atmospheric pressure (T&D: TR-73U, Nagano), illuminance (T&D: PHR-51, Nagano) were adopted. And the chamber was set up to protect from light outside. The bioelectric potential's sampling interval is 0.1 s and the others are 1 m. Two electrodes (material: aluminum) are attached to two leaves of plant, which are near the surface of the pot soil and are adjacent. The difference of potential among the electrodes is amplified 100-fold because the original potential of the plant is the order of millivolts. And the signal inputs into a portable computer through an A-D converter and data logger (KEYENCE: NR-350, Osaka). The ground is connected to the soil of pots in this system. A measured characteristic of original data for 1m (600 data) is indicated in Fig.2. The bioelectric potential changes rapidly. It is difficult to derive from relevant to environmental factors. In this study, summation values of the absolute in original data (vh1) were adopted. The equation is shown in Eq. (1).

$$v_{h1} = \sum_{i=1}^{n} |S_i| \tag{1}$$

A Si means that the sampled bioelectric potential input every 0.1 s. The study used summation per 1 h, namely, i = 36,000. This summation can observe a trend in the bioelectrical potential responses to environmental factors. In this study, two types of experiment were examined.

1. Characteristics for bioelectric potential under LED
2. Characteristics for purification function under LED

This experience was set at the same time each from 15:00 to 09:00, and measured bioelectric potential to observe relation to environmental factors. Each measurement item was examined in the same condition more three times. In experiment of air purification, Ethyl alcohol was used as a sample of pollution material. The purification experiment was set up as 8ppm in the chamber and the characteristics for decrease in concentration by purification capability of plants were measured. The chemical (Ethyl alcohol) was injected into the chamber using a micro syringe after setting the plant 3h (Offset level). And the purification characteristics were developed. If the plants would largely have a purification capability and be used as food, this study could lead to a high added value. In the feature, these added-value vegetables would be made in plant factory and home garden. It could contribute to more human society. Also it will be considered that a amount of production increases. It is thought that our findings contribute to growth environment and control of production in plant factory.

(a) Photograph of experimental system

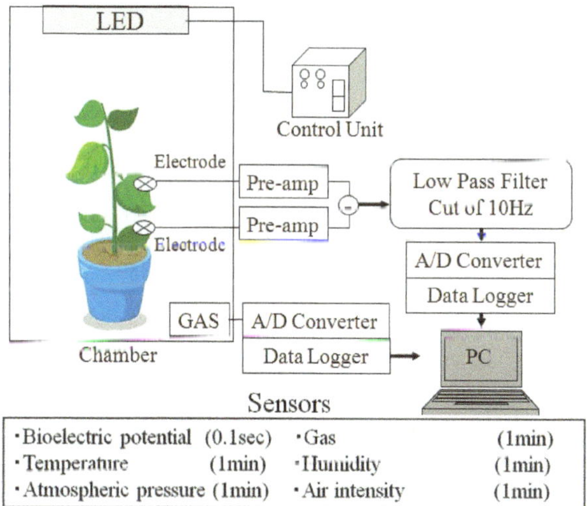

(b) Schematic diagram of the system

Fig. 1. Experimental system

	uE	*lux*	*w/m*^{2}	*nm*
White	100	2835	23	---
Blue	100	929	18	475
Green	100	4755	21	525
Red	100	394	25	660

Table 1. Specification of LED panels

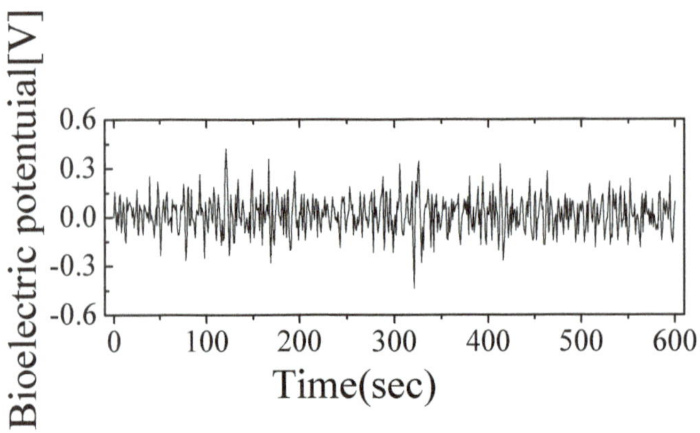

Fig. 2. Original characteristic of bioelectric potential for pothos

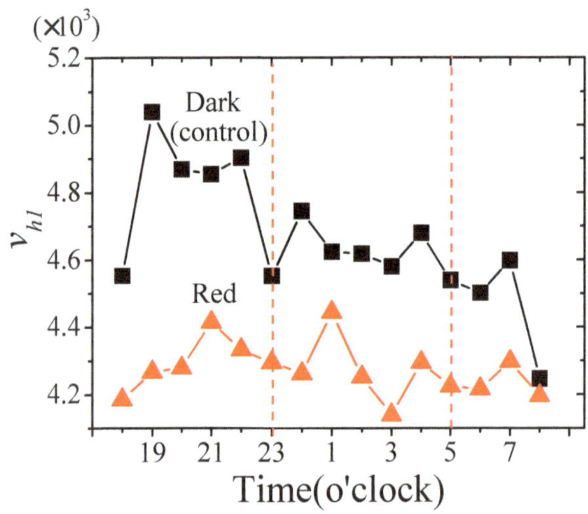

Fig. 3. Bioelectric characteristic (v_{h1}) in Red LED and under darkness

3. Experoment result

3.1 Characteristics for bioelectric potential under LED

3.1.1 Radiation bioelectric potential under lights

LED panels can shed light on the plant with four types of light, namely, Blue, Green, Red, and White. By photosynthesis, all plants including fruit vegetable and leaf vegetable grow with various parts formed. It is expected that the circadian rhythm of plants is disturbed by difference time and fluctuation and times are slightly different about physiological effect. Day length time is also largely related to photosynthesis rate. It is considered that time dilation differs due to frequency of irradiating light.

In this study, first, a bioelectric potential under dark as control was measured in a row. And then those under LED lights were measured. Measured characteristic under red light and dark is shown in Fig.3. This figure shows similar trends. In these characteristics, the characteristic under the LED lights are given by the 1 h than one in darkness and the correlation coefficient (R_{xy}: Cross Correlation Function) between shifted characteristic and control is introduced. It is used to express the similarity and the time lag between two signals ($x(n)$, $y(n)$). The equation is indicated in Eq.2. The similarity becomes the highest in $R_{xy} = |1|$ and the more lower under the condition that the value becomes close to 0.

$$R_{xy}(k) = \frac{1}{N} \sum_{n=0}^{N-1-k} x(n) \cdot y(n+k) \quad \{n = 0, 1, 2, \ldots, N-1\} \tag{2}$$

(n: the number of data, k: shifted time)

The scatter graph is indicated in Fig.4. It was recognized that the R_{xy} has a maximum value between control v_{h1Dark} and $v_{h1Redshifted3}$ at the shifted period of 3 h under red light. The characteristic has time lag for 3 h. Equally, characteristics under Blue and Green lights and these results are shown in Fig.5. A horizontal axis means wave length, and a vertical axis means maximum correlation coefficient to v_{h1LED}. The experimental results denoted the tendency of disturbing a circadian rhythm in plants. In fact, a wave length is longer light, a control characteristic more significantly delay, according to Fig.5.

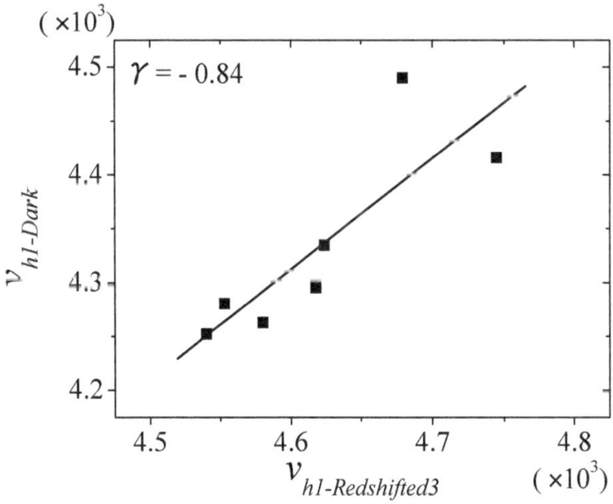

Fig. 4. Scattering Diagram (v_{h1Dark}-$v_{h1Redshifted3}$) for the shifted characteristic for 3 hours in red light and the control characteristic

3.1.2 Relation in between bioelectric potential and light frequency

The fluctuation of bioelectric potential was examined by light frequency. Histograms of these bioelectric potential for control (Dark), Blue, Green, and Red was examined. By comparison, the average of control (v_{h1Dark}) is the highest among that of v_{h1LED}. It is thought that the plants have effect to substantial stress under dark. This characteristic was indicated in Fig.6 (a). A lot of data in dark was measured for comparison. The average of the bioelectric potential (Ave) is the most largely value under Green light and followed by one under Blue light. And one under Red light means the smallest. It is not clear yet that the large or small bioelectric potential stress associate with plant's stress. By investigating these functions, it is thought that the relationship between the bioelectric potential and the stress become clear. In general, the sunlight includes various kinds of frequency light. In this experiment, the panel of White light is close to the sunlight. The data under white, Blue, Green, and Red light are marshaled and these histograms are shown in Fig.6 (b). The average (Ave, unit: 10^3[V]) of 4.27 under White light and the ones of 4.25 among Blue, Green, and Res light is derived. This result becomes pretty much the same. In the Fig6 (a), the control (Dark) was considered as a standard of comparison. The averages of the bioelectric potential ($difv_{h1}$) were derived and plotted as Fig.7. The difvh1 is indicated in Eq. (3). As is clear from the figure, difference value is higher under red light and bioelectric potential becomes small as compared with that under dark. The red light also the most important contributes to the photonic synthesis It is said that the photonic synthesis is the most effective to the photosynthesis rate in the case of rate ten to one between red light and blue light. Namely, as indicated in Fig.6 (a), the bioelectric potential becomes high under dark and green light.

$$difv_{h1} = ave\ (v_{h1\text{-}control}) - ave\ (v_{h1}) \tag{3}$$

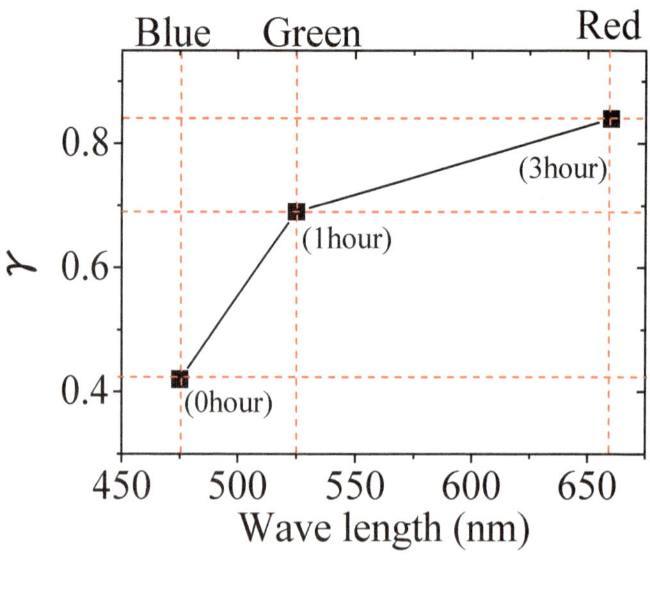

Fig. 5. Illuminated light frequency and the maximum correlative relationship

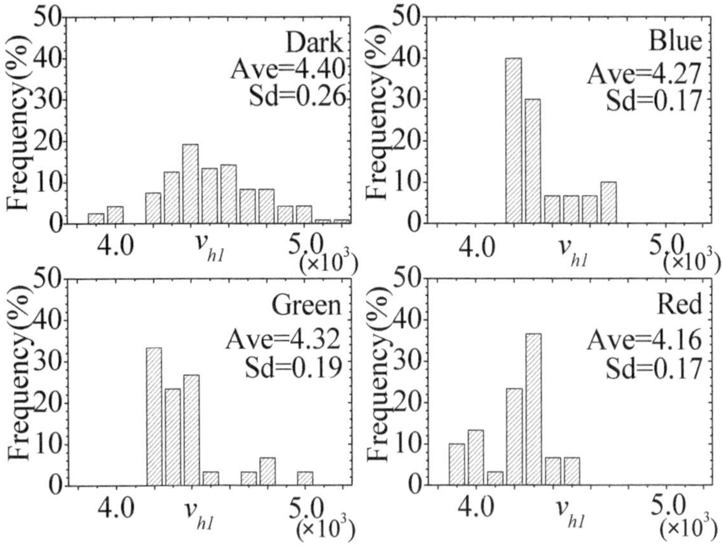

(a) Histograms of vh1 for Dark, Blue, Green and Red

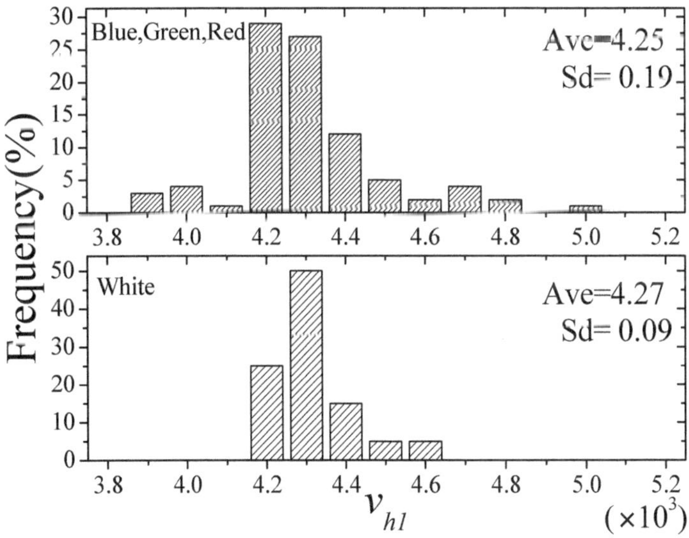

(b) Histograms for three lights and white

Fig. 6. Histograms of v_{h1} for each LED panel

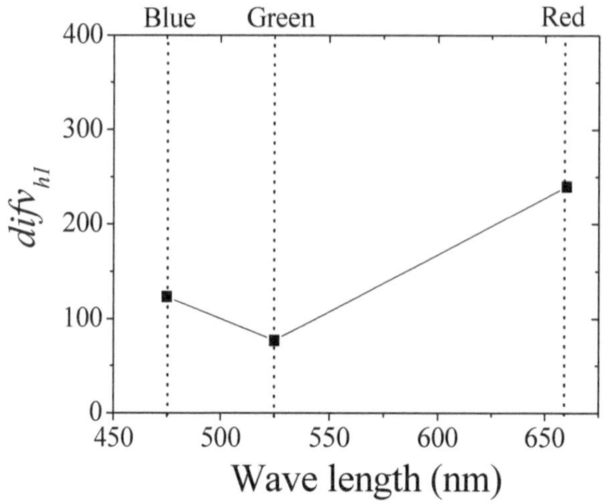

Fig. 7. Relationship of light frequency and $difv_{h1}$

3.2 Characteristics for purification function under

3.2.1 Relation between purification function *Pa* and light frequency

A foliage plants as photos process a capability to absorb air-polluting substance and break down it in physiological tissue. Especially, in indoor environment, the plants have a capability to purify Volatile organic compound (VOCs: Volatile Organic Compound), which is a causative substance for sick-building syndrome. And it is common knowledge that the plants fix carbon dioxide from the atmosphere as factor contributing to global warming. Light frequency causes these functions to ability difference. The irradiated plants are affected a differences in these functions per frequency. Additionally, it is thought that the plants are consuming air-pollutant substances as nitroxide and sulfur oxide. For this reason, the plants growth rate is high in area where car-exhaust pollution partway becomes advanced. It remains possible that these functions can be ready applied to plant factories. At all, the plants are food and system of environmental purification. In this study, a relationship between a characteristic air-purification and wavelength was examined. After subjective plant was placed in the experimental chamber, a pollutant was injected into the chamber using microsyringe and a change in concentration was investigated. The pollutant concentration was estimated by an output from tin gas sensor. Change in concentration was used as the baseline, namely, it means offset level before injecting the pollutant into the chamber. The pollutant disperse into the chamber and a sensor detected the increase in the concentration of the pollutant in the chamber when the chemical when the pollutant was injected. Then the sensor output [V] increases in response to the concentration. After the sensor output reached a peak-level (maximum concentration), it gradually became lower because the polluting chemical was purified by capability of the plant. Eventually, the concentration drop to a lower value by offset level at the time of injection. Then, the *h* [V] means peak-value from the offset level and the *tw* means the half-value width, namely, the time in which the *h* becomes half value. The purification function *(Pa)* was detected with the use of these values. So, the *Pa* is indicated as Eq. (4). A schematic diagram is shown n Fig.8. In this study, ethyl alcohol was adopted as the pollution substance.

$$Pa = h \,/\, tw \times 100 \tag{4}$$

An effect to purification capability of the plant by light frequency was examined. Under four types of frequency, these *Pa* values were indicated in Fig.9. From this figure, white light including any wave length means the most effective and the average of *Pa* is 15.8. While on the other hand, the capability under Dark means the lowest value and the average of *Pa* is 9.9. It is considered that the purification capability would be lower due to the photosynthesis. That is, it is thought that the carbon constituent in the pollution substance decreases at the same time various kinds of functions of the plants also decreases. The values of *Pa* under Red, Blue lights are higher following the one under White. The value of *Pa* under Green light was indicated the lowest among three of monochromatic light. The *Pa* under Blue, Green, and Red was located during Dark and White. This result in purification capability agreed with the effectual wave length to the photosynthesis. The characteristic graph shown in Fig.7 is remarkably similar to that Fig.9. That is, a characteristic of *V configuration* was indicated. Scatter diagram of the relationship between Fig.7 and Fig.9 was shown in Fig.10. The more effectual wave length for photosynthesis becomes higher purification capability of the Pa. The difference in bioelectric potential of the $difv_{h1}$ becomes larger as the purification capability of the *Pa* becomes smaller. This result deeply relates to the photosynthesis function. Because the plant grows by photosynthesis, this growth relates to the purification capability, which is one of the plant functions. A characteristic of purification under Red light advance 14% compared to Green light.

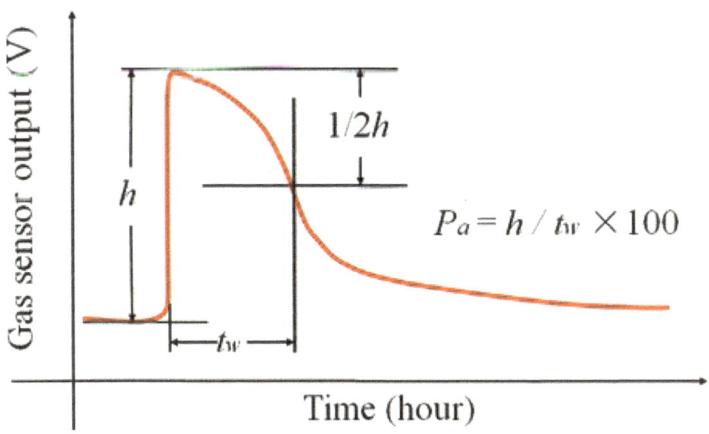

Fig. 8. Conceptual diagram for air purification capability *Pa* of plant

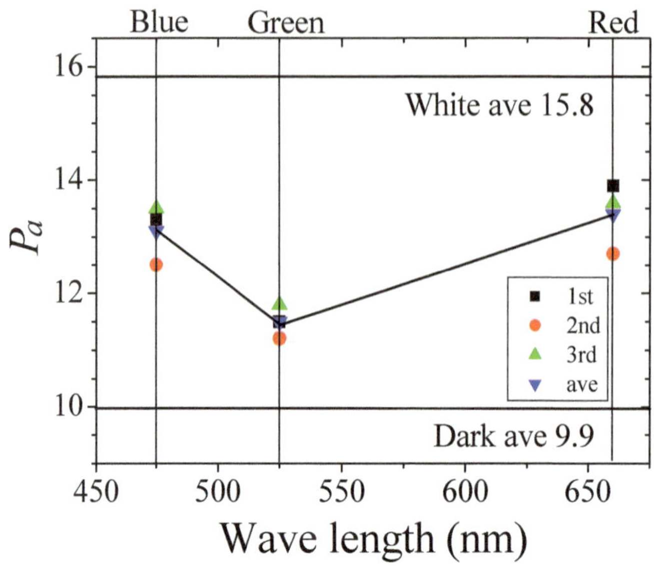

Fig. 9. *Pa* as a function of illuminating light

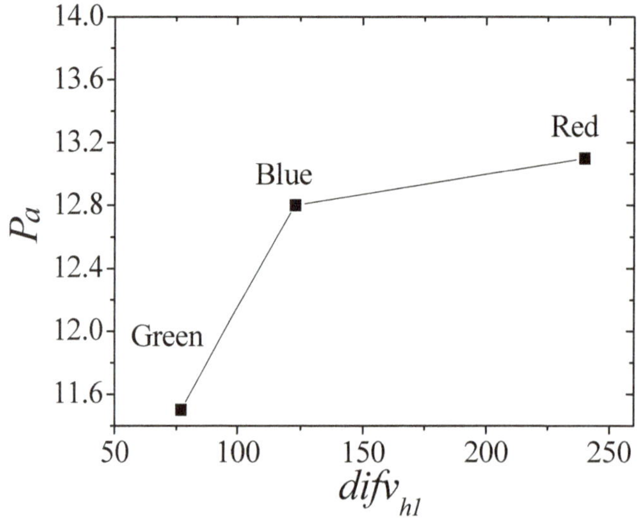

Fig. 10. *Pa* as a function of $difv_{h1}$

3.2.2 Derivation to *Pa* by multiple liner regression analysis

In the plant factories, it is necessary to control the growth, picking season, and yield amount of the plants. Factors of light, temperature, carbon dioxide, and nutrition (for example, carbon, air-pollutant, etc) effect the plant growth. It is also important to effect the plant growth by the three major nutrients as nitrogen, phosphorus, and potassium. By using this experimental result, a multiple regression equation was derived from the environmental factors as explanatory variable. Especially, a degree to affect *Pa* was introduced by statistical validation. A kind of LED lights, average of the temperature, humidity, and atmosphere were used as explanatory variable. The equation is indicated in Eq. (5).

$$y = 4.769\, W_{hite} + 2.960\, B_{lue} + 2.164\, R_{ed} + 0.783\, G_{reen} +$$

$$0.242\, T_{emp} - 0.022\, H_{umid} + 0.002\, A_{toms} + 4.446 \quad (5)$$

A value of the White is the largely compared to the coefficient about the explanatory variable. Next, the values of Blue and Red are large in a row. This three were tested by variable hypothesis test. White was dismissed as the level of 1% and Blue, Red was as one of 5%. Namely, these coefficients have largely effect to purification capability of the Pa. From this result, it is thought that the *Pa* required can be configured by controlling the environmental condition for example kinds of lights, temperature, humidity, and atmosphere. But, in this study the environmental conditions were not controlled and the coefficient to temperature was very small. After this, it is necessary to analysis with changing all environmental factors. Deriving from Eq. (4), Scatter diagram between measured value *Pa* and estimated value *Pa'* is shown in Fig.11. The correlation coefficient was the acceptable value of 0.89. By the category for types of the plant, it is easy to manage a production line deriving these equations. Also it is able to apply to production simulation as yield amount and picking season.

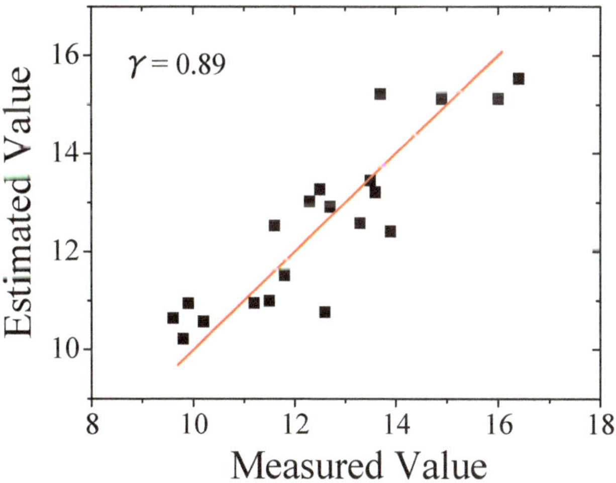

Fig. 11. Scattering diagram for *Pa* and *Pa'*

4. Conclusion

In Japan, the environmental problem and the food problem are urgent issue. It is obvious that the plants assume important role to solve these problems. Meanwhile, it is necessary to reduce the effects on the environment in transportation as food mileage by local production for local consumption. From these social environments, the plant factories of energy saving type using solar panel are proposed and close to practical use. These factories are being controlled based on the practical farmer's hunch and experience. In the future, the maximum cost benefit performance will be achieved if these plants could be controlled by the information that the plants send by itself, for example the bioelectric potential. In this study, as the part of the basic data, the characteristic for the bioelectric potential and purification capability under LED were examined. As the results, it was obvious that the characteristics under Blue and Red lights, which are photosynthetically active wave length, became significantly high. Especially, the Red light's effect has a significant impact on the characteristics. The characteristic for air purification was examined due to carry out as one element of a carbon fix in growing plants. As a consequence, it is thought that effects on light frequency are need to be utilized. For the future, the experiment as temperature, humidity, intermittent light will be examined and give a thorough report about establish a system of the full-control to the plant factories.

5. Acknowledgment

The authors would like to express my appreciation to associate professor Yuki Hasegawa (Saitama University Graduate School of Science and Engineering) for her kind advices. They also would like to thank Kazuto Sumita (Kanazawa Seiryo University Faculty of Economics) for meaningful discussions about statistical processing.

6. References

M. Takatsuji. (1979). Plant Factory, ISBN 978-406-1180-10-9, Kodansya, Tokyo, Japan

M. Takatsuji. (2007). Introduction to Plant Factories and Examples, Syokabou, Tokyo, Japan

C. Backster. (2003). Primary Perception, White Rose Millennium Press, ISBN 978-096-6435-43-6, California, United States

T. Oyabu, Y. Hasegawa, T. Katsube & H. Nanto. (August 2006). Plant Bioelectric-Characteristics and Profitable Merchandising, Vol.11, No.4, pp.239-244, An International Journal of Asia Pacific Management Review, ISSN 1029-3132

T. Shimbo, H. Kimura, & T. Oyabu. (2008). Prediction of plant bioelectric potential based on environmental information in house space, EICA, Vol13, No.1, pp.27-33, Japan

N. Takahashi & J. Maeda. (2000). Relation between electric potential difference and color change on surface of leaf, IEICE, D-II, Vol.J38-D-II, No.9, pp.1946-1951, Japan

K. Baosheng, S. Shibata, A. Sawada, T. Oyabu & H. Kimura. (2009). Air Purification Capability of Potted Phoenix Roebelenii and Its Installation Effect in Indoor Space, Sensors and Materials, Vol.21, No.8, pp.445-455, Japan

M. Takatsuji.(2010). Completely-Controlled Plant Factories, Ohmsya, Tokyo, Japan

M. Takatsuji. (2010). Present status of completely controlled plant factories, SHITA, Vol.22, No.1 pp.2-7, Japan

S.Tani, H. Kuroda, A. Sawada, & T. Oyabu. (2006). Pirification grades of interior plants for ammonia and VOC，EICA，Vol.11，No.1 pp.29-34, Japan

T. Oyabu & T. Katube, (2009). Plant bioelectric potential and its communication, kaibundo, Tokyo, Japan

F.Baluska, S.Mancuso & D.Volkman (Eds.). (2000). Communication in Plants: Neuronal Aspects of Plant Life, Springer, pp.291-320, 333-349

E. Baena-Gonzalez, F. Rolland, J.M. Thevelein & J. Sheen. (2007). A central integrator of transcription networks in plant stress and energy signaling, Nature, Vol.448, No.7156, pp.938-942

15

The Distribution of Dry Matter in Bean Seedlings in Light and Darkness Conditions

Ramón Díaz-Ruiz
Colegio de Postgraduados, Campus Puebla
México

1. Introduction

Plants express different stages of development to complete their life cycle, where in each phase have different demands on nutrition and the distribution of photosynthates among the structures. Thus, during growth, the plant accumulates and allocates different proportions of dry matter to their structures and root.

In the seed is the embryo formed by the radicle and plumule, and cotyledons. They have stored nutrient reserves that proportion to the embryo that originate the root and plumule forms the stem and leaves of the seedling. The seeds can contain one cotyledon as monocotyledons such as maize, two in dicotyledonous plants such as beans and many cotyledons in gymnosperms. The seedling formed the hypocotyl located between the root and cotyledons and the epicotyl which includes the stem and plumule formed above the cotyledons.

In a bean seedling formed are defined the root system which consists of the main root, adventitious roots, primary roots, secondary and tertiary roots and stem consists of the hypocotyl, cotyledons, epicotyl, simple leaves, first trifoliate leaf and second trifoliate leaf. Each structure complies with specific functions to maintain growth and seedling development. The roots absorb water and minerals and the vegetative part absorbs the solar radiation to produce the photosynthate that nourish the seedling.

In the germination of the seed resumes growth by activating the metabolism of the embryo, which occurs when the seed absorbs water available around them and ends when specifically embryonic radicle elongates (Bidwell, 1990; Mayer and Poljakoff, 1989). From a practical standpoint, it is considered that germination is complete when the radicle has emerged from the shell and from this moment is considered the growth of the seedling. During germination different biochemical and morphological processes perform and it marks the beginning of the decline in reserves in the cotyledons and their exportation, which is accompanied by an increase in respiration (Bathellier et al., 2008). The first structure that appears in the seed is the radicle, which grows down, then hook plumule emerges as part of the hypocotyl and first structure that comes into contact with the light so soon chlorophyll actives and triggers photosynthesis.

After the germination, growth begins, which leads to establish the seedling ranging from when the seedling emerges from the ground until it becomes autotrophic (Holman and Robbins, 1982). In light, the seedlings quickly initiate the synthesis of photosynthates and

ensure their establishment (Mayer and Poljakoff, 1989). Seedlings emerge in darkness, they continue to grow until senescence and die without being autotrophic (Bidwell, 1990).
The nutrition of the seedling is done by the translocation of photosynthates and nutrients from germination and in all development. There are two concepts that explain the translocation of nutrients called "source" conceived as the regions of nutrient export and "demand" that are the places of import (Wolswinkel, 1992). Thus establishing a nutrient concentration gradient between the two regions, which move through the vascular system (Ho et al., 1989). During the germination process is initiated the mobilization of seed reserves from the cotyledons to the embryonic axis. This process continues until the seedling stage, in which the primary sources are exhausted and the seedling becomes autotrophic to form leaves that carry out photosynthesis, which are the most important source of photosynthate of the plant (Bewley and Black, 1985; Ho et al., 1989). The first regions of demand are the apical meristems of the plumule and radicle. Later, during the plant development, the diverse meristems of stem and root apical meristems are the sites of most intense demand (Eschrich, 1989).
The accumulation of dry matter in seedlings developed in light and darkness is different because in each condition, the amount of nutrients available to the seedling is different. Seedlings show marked differences in morphology and proportions of distribution and allocation of dry matter in the structures and root that are formed. Therefore, this chapter describes the dynamic allocation of dry matter in bean seedlings developed in conditions of light and darkness.

2. Materials and methods

2.1 Establishment of experiments

Two experiments were installed in light conditions, one in greenhouse and the other in a growth chamber at constant temperature of 25°C. In darkness, the seeds were sown in a dark room and in a growth chamber at 25°C.
The planting of bean cultivar Cacahuet-72 habit type I (Debouck and Hidalgo, 1985) was realized in transparent polyethylene cylindrical tubes of 14 cm of diameter and 70 cm of length. The tubes were filled with sand washed with water. The seed was placed at 2.5 cm depth after irrigation to field capacity. The seed was placed with the micropyle toward up and the lens down, in order to facilitate the growth of the radicle. Seeds sown weighed between 290 and 300 mg per seed.

2.2 Management of seedlings

In the light experiment there were five samples at 8, 13, 18, 21 and 25 days after sowing. The sample size was of two seedlings in the first and second sampling, and three in the remaining. In dark conditions some samples were obtained at 2, 5, 8, 11, 13, 18, 21, 25, 29, 32 days.
Seedlings in dark were observed with green light by a flashlight covered with three layers of green cellophane paper. The light is commonly used as a security light in physiology experiments mainly where phytochrome is involved because it keeps the photostationary state between Pr and Pfr forms because these pigments are not effective at absorbing green light. In addition, our eyes are sensitive to this light and allows us to visual perception (Smith, 1975; Salisbury and Ross, 1994).

Seedlings were removed from the tubes getting the whole root tubes cut longitudinally and the roots were freed from the sand with water. To avoid desiccation of the seedlings they were placed on paper newspaper. Each seedling was dissected into its various structures, which were analyzed for fresh weight and then dried to constant weight (80ºC) for 48 hours to obtain dry weight. Both the dry and fresh weight were obtained in an analytical balance. The roots are separated by category, taproot, adventitious, secondary and tertiary.

2.3 Shoot structures

Structure was called to each of the parts of the stem, which include cotyledons, epicotyl, hypocotyl and radicle. The last three structures form the embryonic axis. The plumule are all parts located above the cotyledons that including apical meristems. Epicotyl was named to the portion of the stem between the cotyledons and the first simple leaves, hypocotyl to the structure located between the cotyledons and the neck of the root differentiated by a concave line in the root observed between these two parties. The simple leaves are formed in the second node in the opposite way.

2.4 Classification of roots

The bean root system is characterized by a taproot. Roots arising from it are called secondary roots, those formed from them is called tertiary roots. Adventitious roots were formed at the base of hypocotyl.

2.5 Use of reserves

In order to know the initial distribution of dry matter in the seed, 20 of them were used from 290-300 mg in weight, which were soaked. They were then separated the shell, cotyledons and embryonic axis were placed in petri dishes and dried to a constant weight at temperature of 80ºC for 48 hours to get weight on analytical balance.

The dry matter contained in the cotyledons is metabolizable dry matter (which is the reserve material) and non metabolizable (which is what constitutes the cell walls). The amount of reserves of the cotyledons was considered its initial dry weight minus the dry weight when they reach their constant weight after drying representing the structural matter of the cotyledons (non-metabolizable dry matter).

2.6 Determinación of the root/shoot

In each sample were estimated the root/shoot dry weight relation by dividing the dry weigh of root and the dry weight of shoot.

3. Results and discussion

The embryonic axis consists of the plumule and radicle, which demands nutrients from the seed when germination begins, which are provided by the cotyledons. The nutrients are sufficient until the hook plumule emerges and makes contact with sunlight to start photosynthesis. As the cotyledons have nutrient reserves, the seedling can live without performing photosynthesis. However, when the seedling is not exposed to sunlight show etiolation and dies soon at the time that deplete nutrients in the cotyledons. This establishes the relationships between growth of the shoot, root and provides nutrients from the cotyledons (McDonald, 1994), which can be expressed by dry matter accumulation in the stem and root and to lose dry matter of cotyledons.

3.1 Accumulation of dry matter

Bean seedlings planted under light conditions accumulate dry matter during their development (Figure 1). The same occurs with the shoot and root. However, the cotyledons decreased dry weight. This is because these structures provide nutrients to the shoot and root. The dry matter accumulation in the seedling and the different structures forming are the product of the translocation of nutrients from the cotyledons and the photosynthesis performed. For this reason the cotyledons tend to lose dry weight during the seedling growth and become shoot structures dying first (Díaz-Ruiz et al., 1999). According to Bathellier et al. (2008), during imbibition of the seed, cotyledons mass decrease slowly but after three days it was found that accelerated markedly. Shoot accumulates more dry matter than the root which is possibly due to greater number of structures that require significant amounts of nutrients. This occurs from the beginning of germination, this tendency has been reported in cotton plants developed under light (De Souza and Da Silva, 1987).

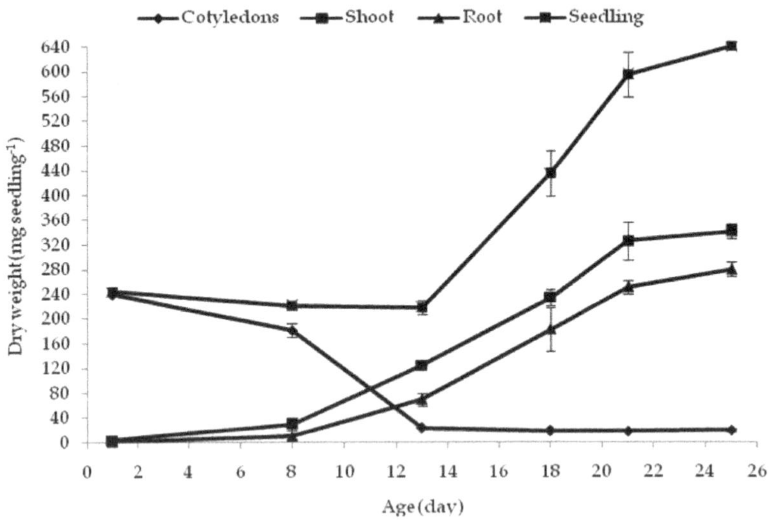

Fig. 1. Dry matter accumulation in bean seedlings developed in light. The bars represent the standard deviation.

Seedlings grown in darkness tend to decrease their dry matter during development up to senescence (Figure 2). The dry matter in shoot and root increased from emergence to 13 days and then tend to decrease their dry weight. Cotyledons reduced its dry weight because they are the only structures that provide nutrients to the seedling. The shoot and root depends on the amount of nutrients stored in the cotyledons for their growth. Both in darkness and in light, the shoot accumulate greater quantity of dry matter than the root with the difference being greater the accumulation in darkness than in light. Under darkness the only source of nutrients are the cotyledons, when they run out the shoot and root dry weight decreased and die (Diaz-Ruiz et al., 1999).

By comparing the growth of bean seedlings in light and darkness conditions, it is clear that the dry matter accumulation in darkness depends exclusively on the reserves of the cotyledons which were reflected in the correlation between weight gain in shoot and root and dry weight loss of the cotyledons. Under light conditions, the cotyledons perform photosynthesis, but their contribution is minimal compared to the first leaf blade which was determined by Harris et al. (1986) in soybeans. This indicates that their main function is to provide photosynthates that have stored to the seedling. However, photosynthesis of cotyledons helps balance the energy loss through respiration until the first leaf performs full photosynthesis (Harris et al., 1986). Even if the seedlings grown in darkness were exposed to light, the content chlorophyll increases and activates. Maricle (2010) reported that etiolated seedlings by exposing to light for four days increased the chlorophyll. In general, beans and wheat are species when exposed to light rapidly develop chlorophyll and increase the activity of catalase (Maricle, 2010). The early declines in dry matter of the cotyledons of bean seedlings have been reported by Barthellier et al. (2008).

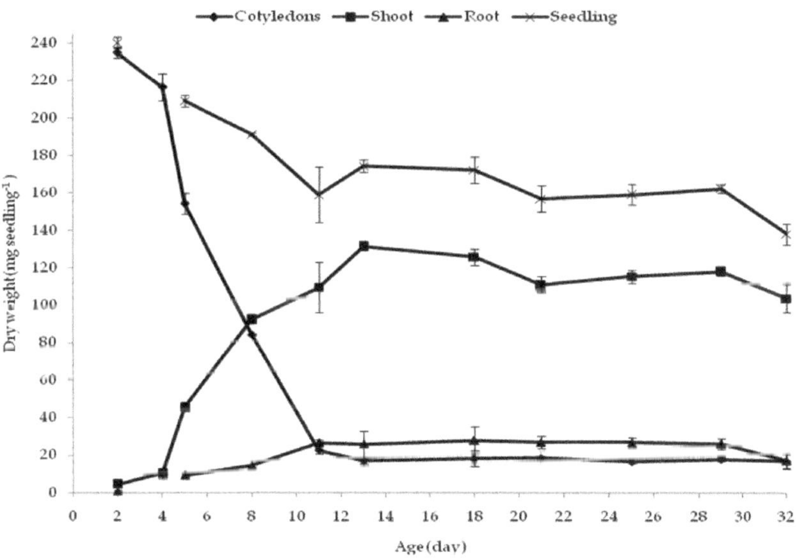

Fig. 2. Dry matter accumulation in bean seedlings developed in darkness. The bars represent the standard deviation.

3.2 Distribution of dry matter

The dry matter in seedlings developed with light is distributed in different proportions to be tangible in the early days, in this case at 8 days after planting (Figure 3). When the cotyledons reach their maximum nutrients, shoot and root tend to stabilize the distribution of photosynthates, where the shoot becomes more dry matter than the root which is greater than 50%. Although the shoot is composed of more structures such as leaves and stem, the root is channeled a significant amount of dry matter. The cotyledons are the only structures of seedling that stabilize dry matter reaching around 3%. These structures serve the function of supplying photosynthate to the shoot and root and themselves. While performing photosynthesis their life is short and so are the first structures to die and slough of the shoot. Dry matter detected in them to reach senescence is not metabolizable. The transfer of more dry matter to shoot was also recorded by Metivier and Paulilo (1980) in bean, they also found that storage proteins were transferred faster than dry matter.

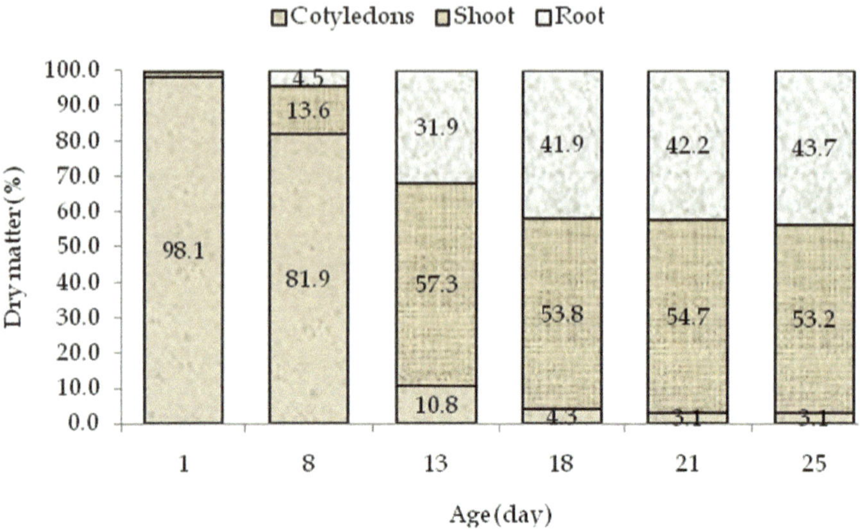

Fig. 3. Distribution of dry matter in bean seedlings developed in light.

The distribution of dry matter in seedlings developed in darkness is different compared to seedlings in light. In this condition, the cotyledons have a similar trend with the decisive role of supplying nutrients to the shoot and root because it is the only source of reserves, in this case do not perform photosynthesis, so its depletion is faster. Thus, 8 days after sowing cotyledons in darkness contain 44.1% dry matter (Figure 4) and in light conditions have 81.9% (Figure 3) which is twice the dry matter. The shoot is the structure which holds as much dry matter in relation to the root as happens in light conditions, the difference is the amount that accumulates in darkness is over 70% and around of 53% in light. In contrast, the accumulation of matter in the root is significantly lower, in the darkness is about 16% and around of 42% in light. Thus, in light of dry matter distribution is more equitable between the shoot and root compared with the distribution occurred in darkness.

3.3 Water content

The seedlings developed under light, mg of water presented an upward trend, the same happen in the stem and root (Table 1). The cotyledons expressed a slight ascent to 8 days and then tended to decline. The stem was the structure with more water content after 8 days. The seedling was a maximum of 2390.2 mg. In mg of water showed a significant difference both in structures and in the seedling, but the differences in water content were not very different, indicating a trend towards a stable proportion of water in the range of 85 to 90% mainly in the new tissue, which go forward age, the water decreases and the dry matter increases.

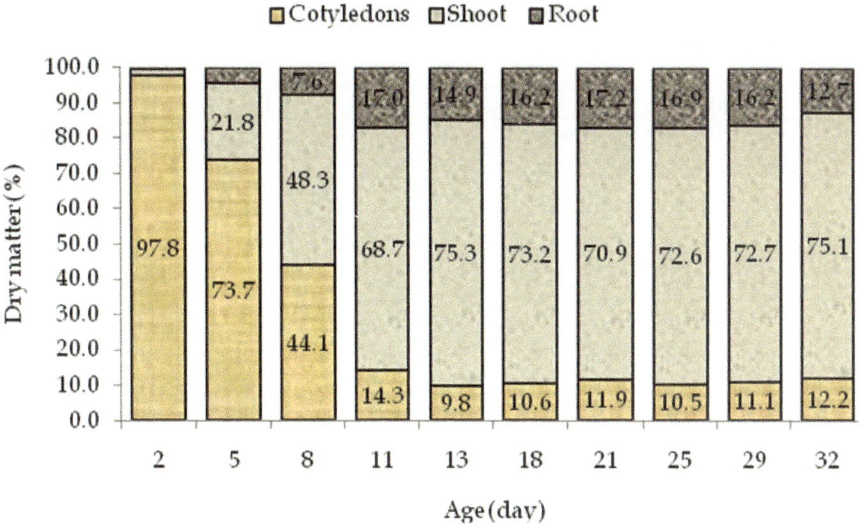

Fig. 4. Distribution of dry matter in bean seedlings developed in darkness.

Age (day)	Cotyledons		Shoot		Root		Seedling	
	mg	%	mg	%	mg	%	mg	%
2	279.0 ±3.6	54.4 ±0.7	20.6 ±2.7	86.0 ±0.6	2.3 ±0.5	85.9 ±1.8	301.9 ±2.2	56.0 ±1.0
8	289.3 ±45.3	75.4 ±1.9	728.7 ±132.6	92.2 ±0.7	207.9 ±61.4	91.4 ±0.8	1225.9 ±79.7	87.4 ±1.1
13	120.7 ±20.6	85.8 ±1.7	1288.3 ±93.0	90.6 ±9.5	581.2 ±158.2	89.7 ±1.6	1990.1 ±90.6	90.1 ±4.2
18	109.3 ±13.0	87.4 ±1.0	1593.0 ±82.0	86.5 ±0.5	687.9 ±159.1	84.7 ±4.3	2390.2 ±84.6	89.3 ±1.9

Table 1. Water content in bean seedlings developed in light.

In dark conditions, the seedling, cotyledons, shoot and root showed upward trends and then down in water content (Table 2), which was more tangible in the cotyledons. Seedlings reached a maximum water content equal to 3379.7 mg (13 days). The seedling, shoot and root showed a water content greater than 90%. Cotyledons water accounted for more than 80% but less than 90%. At roots, the water content tended to decrease (13 days) first than in the shoot (22 days). In the cotyledons, the increase in water content at the beginning is attributable to imbibition process and then the fact that the water content drops more slowly than dry weight, coupled with these, there is little loss of water by transpiration because the substrate in which plants were sown always remained wet.

Seedlings more water stored in darkness than in light, however in both conditions the amount of water tended to decrease. In darkness is attributable to the decrease of dry matter which coincides with the senescence of seedlings, shoot and root, in the light, is attributable to the tissue becomes more fibrous and accumulation in higher dry matter, of thus the water becomes less retained. Both in light and in darkness, the cotyledons did not exceed more than 90% water and was markedly decreased between 13 and 18 days in darkness in relation to the occurred in light

Age (day)	Cotyledons		Shoot		Root		Seedling	
	mg	%	mg	%	mg	%	mg	%
2	318.0 ±19.1	57.5 ±1.4	33.6 ±4.1	88.5 ±0.7	8.1 ±1.4	89.6 ±1.7	359.7 ±8.2	59.9 ±1.3
5	360.3 ±17.5	70.0 ±0.8	730.3 ±82.6	94.1 ±0.4	164.0 ±12.2	94.6 ±0.5	1254.6 ±37.4	85.6 ±0.6
8	212.0 ±8.0	82.8 ±1.9	2124.0 ±48.2	95.6 ±0.2	274.3 ±9.6	95.0 ±0.2	2610.3 ±21.9	94.4 ±0.8
13	97.7 ±8.3	83.9 ±1.5	2916.7 ±63.2	96.3 ±0.2	365.3 ±57.5	94.6 ±1.7	3379.7 ±43.0	95.6 ±1.1
18	29.6 ±12.4	60.9 ±6.8	2662.7 ±141.2	96.1 ±0.3	121.2 ±50.0	85.4 ±4.7	2813.5 ±67.9	95.0 ±3.9
22	3.0 ±1.0	15.5 ±4.9	1648.7 ±127.1	95.0 ±0.3	79.8 ±11.0	79.8 ±3.3	1731.5 ±46.4	93.2 ±2.8

Table 2. Water content in bean seedlings developed in darkness.

3.4 Dry matter ratio root/shoot

In light-grown seedlings, the ratio derived from the root/shoot tended to increase as was occurring seedling growth (Table 3). Up to 18 days, the ratio significantly increased from 0.32 to 0.78, then remained in stable values. It is likely that the trend expressed is due to the lack of fertility in the substrate. Initially, the increase in rates due to the instability of the dynamics of dry matter allocation to shoot and root, then the proportions of dry matter were more defined, which was reflected in a disparity in rates.

Age (day)	Light	Darkness
8	0.32 ±0.04	0.16 ±0.01
13	0.55 ±0.08	0.20 ±0.06
18	0.78 ±0.16	0.22 ±0.06
21	0.77 ±0.07	0.24 ±0.02
25	0.82 ±0.06	0.23 ±0.02
29	-	0.22 ±0.03

Table 3. Ratio dry matter of root/shoot in bean seedlings developed in darkness and light.

In dark conditions the root/shoot had its largest increase between 8 and 13 days, after the increases were minimal. The maximum value of root/shoot was 21 days. The indexes were having minimal increases from 13 days, when the stem presented slight increases of dry matter and the root remained more constant. Velazquez-Mendoza (1989) found in beans under drought increased root/stem indexes, which were attributed to greater distribution of nutrients to the root at the expense of stem and leaves. In our case, we observed a similar trend after the cotyledons contributed the most assimilated.

In light conditions higher rates of root/shoot were obtained than in dark conditions at all ages sampled. This indicates that the root/shoot may increase with the supply of nutrients to the seedlings, in light were supplied of photosynthates by the cotyledons and leaves and in darkness were consumed the reserves of the cotyledons only.

3.5 Dry matter in shoot structures

In the seedlings developed under light all the structures of both the shoot and the roots tended to increase dry matter accumulation (Figure 5). The leaves accumulated more dry matter in all samples except at 8 days. Adventitious and secondary roots had higher dry matter than the other structures of the stem and taproot after 18 days except for simple leaves. Similarly at 18 days dry weight was higher in the hypocotyl than the epicotyl and main root. In the seedling stage the leaves become the main source of photosynthates replacing the cotyledons reach senescence. This indicates that the flow of nutrients begins in the cotyledons and continues in simple leaves, so it takes the largest proportion of nutrients from the cotyledons during the early days, from germination until the leaves are coming almost half of growth in dry matter accumulation. The transition from simple leaves of source organ to of demand organ is associated with its ability to form photoassimilates and maintain a balance between their synthesis and use by itself (Loescher et al., 1982). Thus, after senescence of cotyledons, simple leaves senescence start, when this happens, the first compound leaf begins to provide photosynthate to the seedling. In this moment is important to apply fertilizer to the seedling to meet the demand for nutrients (Díaz-Ruiz et al., 2008).

Adventitious and secondary roots are most important at this stage, obtain more dry weight which is marked at 18 days. Thus, the water absorption in the seedling depends on these two types of roots. In addition to the above, the increased demands for nutrients in the roots are the adventitious roots and secondary roots. For his part in the shoot are simple leaves and the hypocotyl.

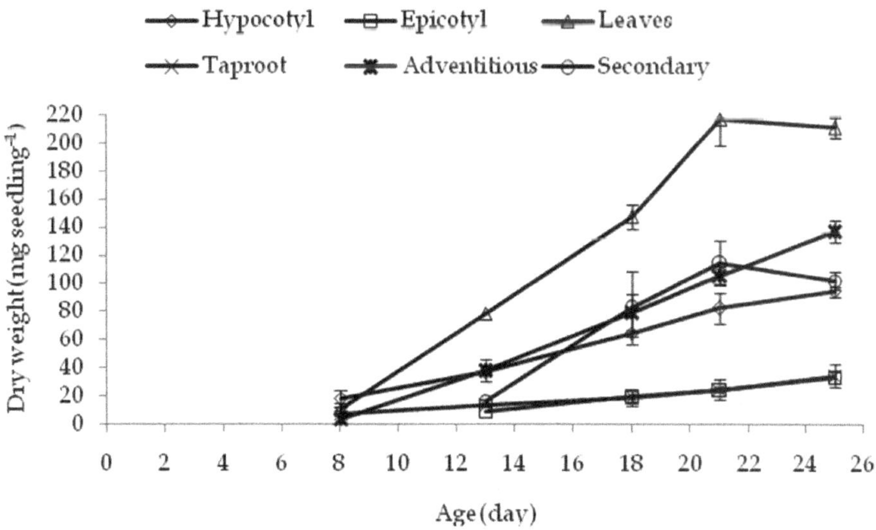

Fig. 5. Dry matter accumulation in the structures of shoot and root of bean seedlings developed in light. The bars represent the standard deviation.

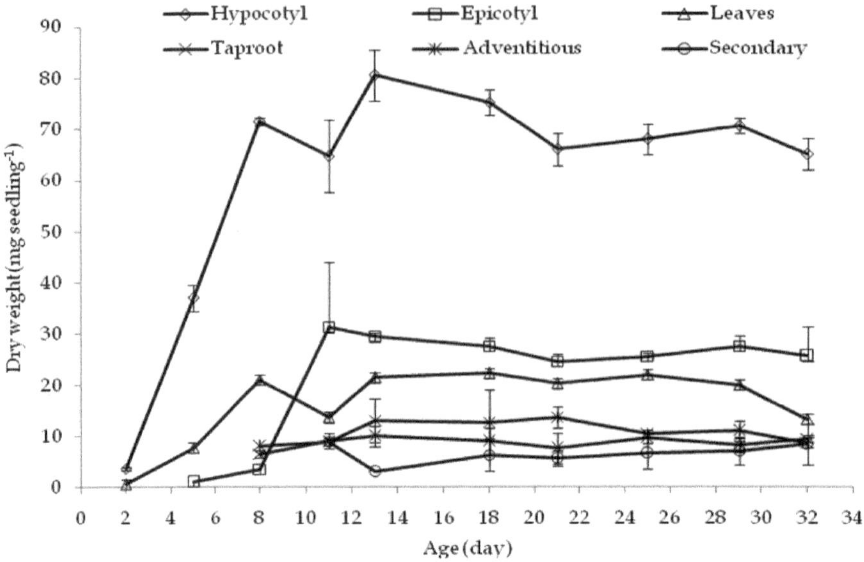

Fig. 6. Dry matter accumulation in the structures of shoot and roots of bean seedlings developed in darkness. The bars represent the standard deviation.

In darkness conditions, dry matter accumulation was different to that observed in seedlings with light. Shoot structures and the different types of roots showed a sigmoid trend (Figure 6), except in secondary roots was not observed clearly. Up to 13 days the dry matter accumulation in the root and structures was evident, after being stable and tended to decrease. The dry matter accumulation in the hypocotyl was significant compared to other structures and roots. In general stem structures accumulated more dry matter than roots. The order from highest to lowest accumulation was as follows: hypocotyl, epicotyl, leaves, adventitious roots, taproot and secondary roots. The decrease in dry matter was strongest in the hypocotyl, after the cotyledons exhausted most of its reserves. This could indicate the translocation of nutrients from other structures, as well, the hypocotyl was commissioned to provide nutrients to other structures to keep them alive, so the decrease in dry weight of roots and other structures is less pronounced, however, the nutrients provided are not sufficient to continue increasing in dry weight.

Adventitious roots formed first secondary roots extract nutrients from the cotyledons first and then the base of the hypocotyl.

3.6 Distribution of dry matter in shoot structures

In the seedlings developed under light gives greater amount of dry matter to leaves, which occurred from 13 days (Figure 7). At 8 days, more dry matter was assigned to the hypocotyl, followed by the leaves. The formation of new organs influences the distribution of dry matter in the seedling, thus, after 13 days the dry matter distribution tends to be constant. Under these conditions, the dry matter production of seedlings is a function of the photosynthate produced in primary leaves, so if there is damage to them, the growth of the seedling is affected. Thus, Hodgkinson and Baas-Becking (1977) indicate that the defoliation causes death roots and decrease its ramifications. However, maintaining carbohydrate stores that allow them to survive (Buwai and Trlica, 1977). In the shoot, the structures with greater allocation of dry matter are the leaves and the hypocotyl and in the roots are adventitious and secondary roots. It is likely that the roots, the taproot becomes less important and the role of nutrient uptake and soil water is realized by adventitious roots and secondary.

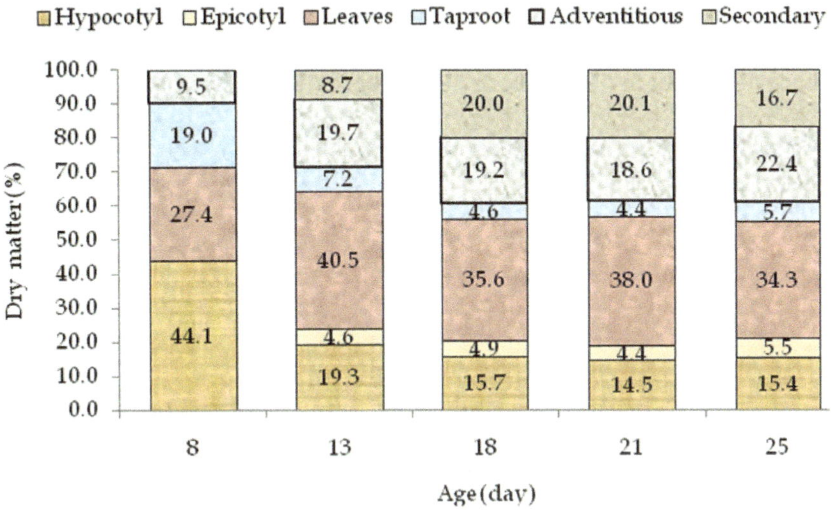

Fig. 7. Distribution of dry matter in bean seedlings developed in light.

In seedlings under darkness was allocated more dry matter to shoot structures (Figure 8), the hypocotyl had higher dry matter, followed the epicotyl and leaves. Unlike seedlings developed in the light, where the leaves accumulated more dry matter, in darkness, it was

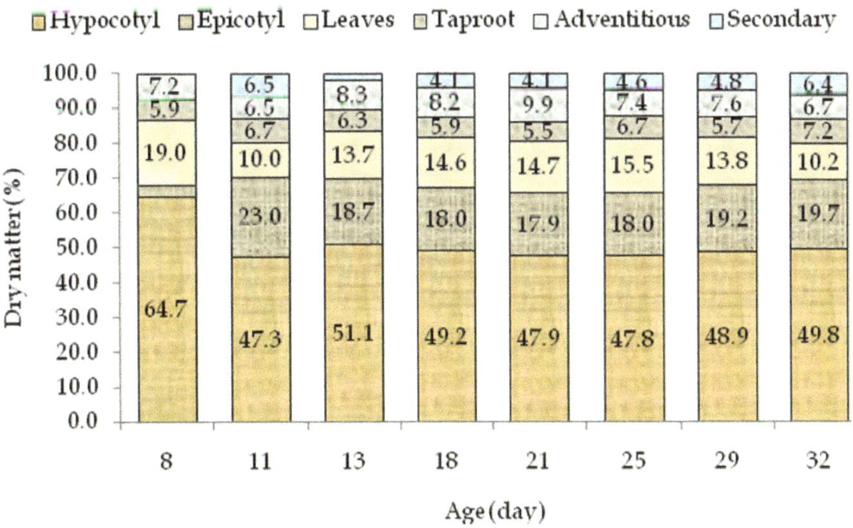

Fig. 8. Distribution of dry matter in bean seedlings developed in darkness.

the hypocotyl. This is evident from the 11 days where the growth is more stable because the 8 days the hypocotyl and the leaves accumulate more dry matter. The distribution of dry matter in roots was more equitable in darkness than in light. However the taproot had less dry matter.

3.7 Water content in shoot structure an roots

The seedlings developed with light, the amount of water upward and downward trends presented in each structure. Hypocotyl water content increased to 8 days, decreased to 13 and increased to 18 days (Table 4), however, the corresponding percentage of water decreased from 8 days. Probably due to their tissue became more fibrous. The epicotyl had a rise in mg of water but a decrease in the percentage of water corresponding. For their part, maintained a rise in leaf water content in both milligrams and percentage. These structures were the most water accumulated at 18 days was the last sampling. It is probably that exist a displacement of water to accumulate as much dry matter.

Age (day)	Hypocotyl		Epicotyl		Leaves	
	mg	%	mg	%	mg	%
2	17.9 ±2.8	86.4 ±0.6	-	-	2.8 ±0.6	83.2 ±3.0
8	525.0 ±95.1	94.9 ±0.9	35.3 ±6.0	77.9 ±1.3	168.3 ±27.5	87.8 ±0.5
13	512.3 ±41.0	92.9 ±0.1	99.7 ±14.6	92.3 ±0.4	676.3 ±32.7	88.7 ±0.3
18	557.3 ±45.3	89.0 ±0.8	128.3 ±10.3	90.6 ±0.5	857.0 ±16.1	93.4 ±0.6

Table 4. Water content in shoot structures of bean seedling developed in light.

In the case of roots, it showed a pattern similar to the structures of the stem with upward and downward trends in milligrams of water in each structure (Table 5). However, the corresponding percentage of water decreased. Adventitious and secondary roots tended to decrease in the percentage of water from 8 to 18 days, however the amount of water recorded in milligrams increased over the same interval. Just as in the structures of the stem, the conduct in the water content in the roots is attributable to tissues become more fibrous and less water.

Age (day)	Taproot		Adventitious		Secondary	
	mg	%	mg	%	mg	%
2	2.3 ±0.5	85.9 ±1.8	-	-	-	-
8	77.8 ±18.6	91.0 ±1.6	93.3 ±28.1	90.4 ±0.9	36.8 ±15.2	95.0 ±2.2
13	114.5 ±43.1	90.5 ±2.5	319.8 ±20.5	90.9 ±0.6	113.5 ±106.4	84.5 ±5.9
18	91.5 ±81.4	84.3 ±7.5	330.0 ±149.5	84.9 ±4.8	155.8 ±109.0	81.9 ±4.4

Table 5. Water content in the roots of bean seedlings developed in light.

In darkness conditions the upward and downward of water content in the shoot structures were more evident than in light. The hypocotyl was the structure that reached the highest water content (2124 mg) at 13 days (Table 6). The epicotyl was the structure with lower water content than the hypocotyl but higher than the leaves, the maximum amount was reached at 18 days. The leaves are structures with less water, it is likely to influence the development precarious that obtain. It is obvious that the seedlings in darkness have higher water content than seedlings developed in the light, but the corresponding percentage was similar, which reached a 90%. In herbaceous plants reported a water content of 80-90% (Kramer, 1983), however, Ehlers and Goss (2003) report between 75 and 95% in stem, leaf and root. Accordingly, the structures of bean seedlings did not affect the percentage of water in the darkness. It may be mentioned that there is a balance in the percentage distribution of water structures in both light and darkness.

Age (day)	Hypocotyl		Epicotyl		Leaves	
	mg	%	mg	%	mg	%
2	30.0 ±2.8	89.7 ±1.3	-	-	-	-
5	670.7 ±86.0	94.9 ±0.3	7.7 ±0.6	88.4 ±0.8	52.0 ±3.0	87.1 ±0.4
8	1981.7 ±59.0	96.2 ±0.1	50.0 ±7.0	92.6 ±1.5	92.3 ±11.0	86.3 ±0.5
13	2124.0 ±87.5	96.7 ±0.3	672.3 ±56.3	96.2 ±0.4	120.4 ±7.0	88.1 ±0.6
18	1817.0 ±201.9	96.4 ±0.5	683.7 ±29.7	96.4 ±0.4	162.0 ±22.8	90.8 ±2.0
22	1104.7 ±173.0	94.8 ±0.3	426.0 ±76.2	95.7 ±0.8	118.0 ±6.7	91.7 ±1.5

Table 6. Water content in structures bean seedlings developed in darkness.

Age (day)	Taproot		Adventitious		Secondary	
	mg	%	mg	%	mg	%
2	8.1 ±2.4	89.6 ±2.1	-	-	-	-
5	87.2 ±9.3	94.4 ±0.3	67.5 ±10.6	94.6 ±0.7	9.3 ±2.1	95.7 ±1.1
8	105.3 ±21.1	94.8 ±0.9	118.4 ±7.0	95.0 ±0.9	50.6 ±9.5	95.4 ±1.0
13	140.0 ±35.3	94.8 ±1.6	164.2 ±21.7	95.0 ±1.1	61.1 ±36.4	91.4 ±6.3
18	59.0 ±47.3	86.2 ±7.8	37.5 ±16.8	79.1 ±3.5	24.6 ±9.8	83.9 ±0.4
22	23.2 ±9.4	81.2 ±2.5	40.3 ±21.0	78.7 ±5.6	16.3 ±5.5	78.7 ±3.9

Table 7. Water content in the roots of bean seedlings developed in darkness.

In the roots appeared similar trends in water content than in the structures of the shoot, with the difference that accumulated less milligrams of water. In all roots reached the highest amount of water at the same time (13 days). Adventitious roots were the more water accumulated (164.2 mg) followed by the taproot (140 mg) (Table 7). The decrease in water content was very dramatic in both milligrams and the percentage of water. This coincided with the decrease in stem dry weight and stability in the roots.

3.8 Distribution and use of nutrients

Considering the weight of the seed represents 100%, it is distributed as follows: 8.9% in shell and 91.15 in the embryo (Figure 9). Of 91.1%, 89.9% corresponds to the cotyledons and 1.2% to the embryonic axis. However, the cotyledons allocated 83.4% to metabolism (metabolic dry matter) and 6.4% as part of its structure (structural dry matter). The embryonic axis use metabolic dry matter available in the cotyledons to resume growth untapped 100% (83.4%), because it spends 32.6% on the respiration. The decrease in dry matter by respiration in darkness conditions has been tested, specifically in leaves (Bathellier et al., 2008). The allocation of dry matter in the seedling is uneven, most are distributed to the shoot than the root. This indicates that the shoot exerts more force on the extraction of nutrients and becomes part of the seedling more demanding but the depletion of reserves in the cotyledons are the first structures to resent the lack of nutrients and tend to lose dry matter. Instead, the roots extract nutrients from the stem to remain. According to Zhang et al. (2010) the seedling hypocotyl elongates significantly by the effect of sucrose stored in the cotyledons, which probably contributes to greater dry matter accumulation in this structure.

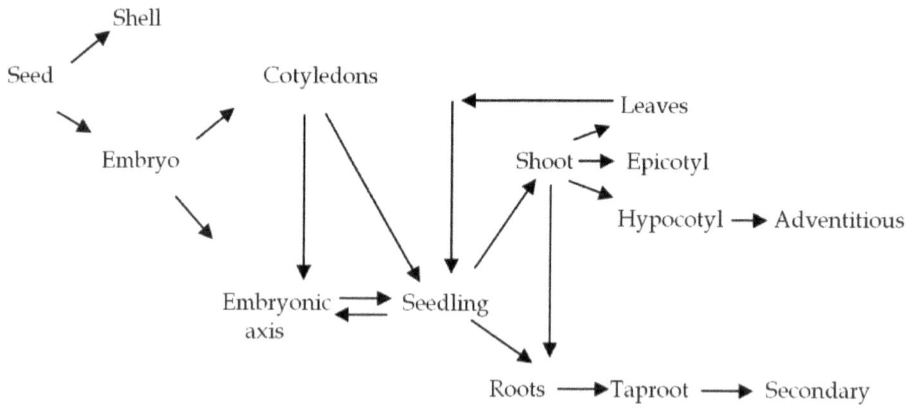

Fig. 9. Allocation and flow of dry matter bean seedlings grown in darkness.

In light conditions, the cotyledons export nutrients to the embryonic axis formed by the stem and root, the seedling until not realize photosynthesis depends on its reserves (Figure 10). However, the cotyledons lose metabolic dry matter and conserved structural dry matter during the development of the seedling so that they become the only structures that die. The embryonic axis uses the nutrients for their development and originates the seedling, in this process is lost dry matter through respiration, which is recovered by the generation of photosynthate through the simple leaves photosynthesis and to a lesser proportion by the cotyledons. In addition to cotyledon, senescence occurs in adventitious roots and secondary roots but do not die it continue to grow and generate more roots. This process culminates with the release of some individual roots or only in the degeneration of the epidermis and barking, the latter feature has been observed in grape (Mapfumo and Aspinall, 1994). After the death of the cotyledons, photosynthates demanding by the structures of the stem and roots are provided by simple leaves that realize photosynthesis. These structures become the main source of nutrients to reach senescence (Yin and Watson, 1990). Finally, establishing the flow of nutrients from leaves to the seedling, which distributes the nutrients in the stem and root, root turn supplies water to the seedling.

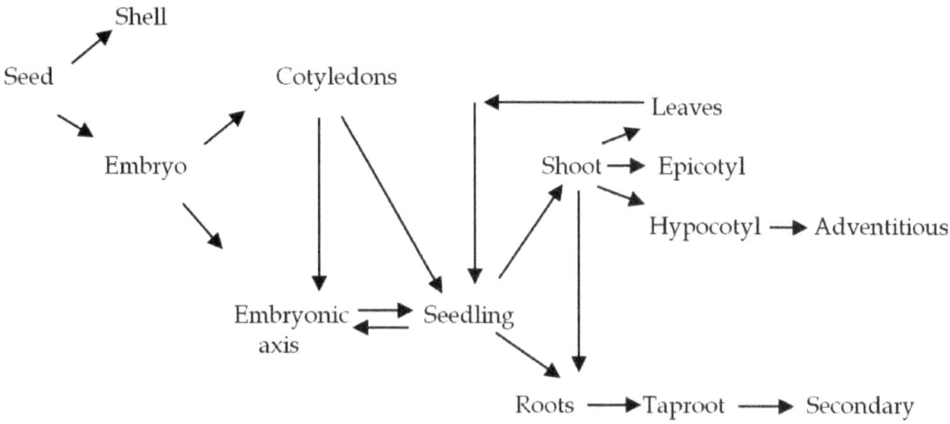

Fig. 10. Flow of nutrients in bean seedlings developed in darkness.

3.9 Relationship of dry matter and seedling senescence

During the development of seedlings in the darkness all the structures of shoot and root dry matter accumulated except for cotyledons. These structures are stored nutrients which are used in the growth of seedlings. As a result, these organs show a decrease in dry weight as nutrients export, although some of the dry weight decrease is attributed to respiration. When the cotyledons die deplete its reserves, the event coincides with the maximum dry matter accumulation in the seedling.

Seedling senescence starts at the tips of the simple leaves (not counting the cotyledons), five days after maximum dry matter accumulation, specifically in the shoot, when the roots reach their maximum dry weight (Díaz-Ruiz et al., 1999). Likewise, adventitious and secondary maintained a slight increase in dry matter. This indicates that at the beginning of senescence, the leaves provide nutrients as well as the epicotyl and hypocotyl to the roots (Figure 11). The decrease in dry matter in the adventitious roots was noted after the start of senescence at the tips of both of them as secondary root. Adventitious roots before the secondary present the senescence as a result of the death of the hypocotyl that becomes their main source of nutrients after the cotyledons die. The main or primary root decreased their dry weight for the following reasons: nutrient intake by respiration, the existence of detachment of the cortex in the apical region of the root that cause death secondary root located in that region and export nutrients to the secondary roots and their use for maintenance of herself. As the nutrient reserves were decreasing, seedling allocated less amount of nutrients to the roots to be zero allocation, so the secondary roots are affected and die first than taproot (Figure 11). According to Klepper (1991), the way in which grows the stem is different from the root but the relationship between them is undeniable.

Initially, the cotyledons exported nutrients to the shoot and root first, the root supplies to shoot the extracted soil water. At death the cotyledons, shoot sends nutrients to the root, thus kept alive longer than the shoot, which die for lack of nutrient supply. In the shoot, hypocotyl dies last and roots are the taproot. The allocation of more dry matter to the shoot is growing faster than the root and has increased demand for nutrients. Because they do not perform photosynthesis, shoot structures supplement for a short period of time the demand for nutrients that makes the root, this allows to extend the time of accumulation of dry matter. Thus, by failing to supply photosynthate to shoot dies first than the root. In general, all structures of the shoot and roots, senescence started after reaching its maximum dry matter accumulation

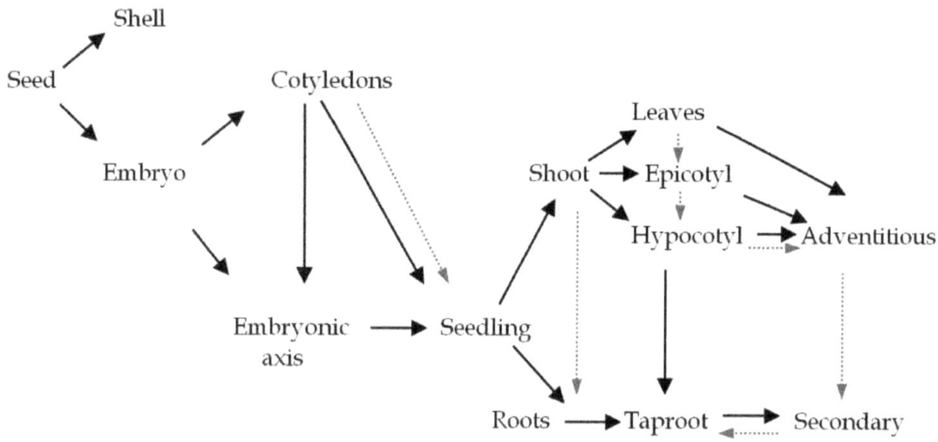

Fig. 11. Nutrient flow (→) and progress of senescence (⋯→) in bean seedlings developed in darkness.

4. Conclusions

The seedlings developed in light use nutrients from the cotyledons and the photosynthate formed by photosynthesis by the leaves. In darkness the seedlings grow with nutrients from the cotyledons only. Thus, seedlings become autotrophic light and dark heterotrophic only because they do not perform photosynthesis.

The dry matter accumulation in bean seedlings developed in darkness and under light is different because in each condition, the amount of nutrients available to seedlings is different. Seedlings show marked differences in morphology and proportions of distribution and allocation of dry matter in the structures shoot and root formed.

The dry matter decreases in the cotyledons because provide nutrients to the shoot and root, which increases its content of dry matter. Shoot accumulates more dry matter than the root. The senescence of the cotyledons is shown in light and darkness conditions. In light conditions the leaves are the structures that accumulate more dry matter and in darkness is the hypocotyl.

5. References

Bathellier, C.; Badeck, F. W.; Couzi, P.; Harcoet, S.; Mauve, C. & Ghashghaie, J. (2008). Divergence in &13C of dark respired CO_2 and bulk organic matter occurs during the transition between heterotrophy and autotrophy in *Phaseolus vulgaris* plants. *New Phytologist*. Vol. 177, pp 406-418.

Bidwell, R. G. S. (1990). *Fisiología vegetal*. A. G. T. Editor. ISBN 968-463-015-8, México. 784 p.

Bewley, J. D. & Black, M. (1985). Seeds physiology of development and germination. Plenum Press. ISBN 0-306-41687-5, New York, 367 p.

Buwai, M. & Trlica, M. J. (1977). Defoliation effects on root weights and total nonstructural carbohydrates of blue grama and western wheatgrass. *Crop Sci.* Vol. 17, pp 15-17.

Debouck, D. G & Hidalgo, R. (1985). Morfología de la planta de frijol común, In: *Frijol: Investigación y producción*. CIAT-PNUD, 7-42, Colombia.

De Souza, J. G. & Da Silva, J. V. (1987). Partitioning of carbohydrates in anual and perennial cotton (*Gossypium hirsutum* L.). *J. Exp. Bot.* Vol. 38, pp 211-218.

Díaz-Ruiz, R.; Kohashi-Shibata, J.; Yáñez-Jiménez, P. & Escalante-Estrada, A. (2008). Growth and allocation of dry matter in bean seedlings developed up to the senescence of the cotyledons. *Agric. Conspec. Sci.* Vol. 73, No. 4, pp 203-210.

Díaz-Ruiz, R.; Kohashi-Shibata, J.; Yáñez-Jiménez, P. & Escalante-Estrada, A. (1999). Crecimiento, asignación de materia seca y senescencia de plántulas de frijo común en oscuridad. *Agrociencia*. Vol. 33, pp 313-321.

Ehlers, W. & Goss, M. (2003). *Water dynamics in plant production*. CAB International. ISBN 0-85199-694-9. London, UK. 273 p.

Eschrich, W. (1989). Phloem unloading of photoassimilates. In: *Trnasport of photoassimilates*. Baker, D. A. & Milburn, J. A. (Eds.), 206-263, Longman Scientific & Technical. Great Britain.

Harris, M.; Mackender, R. O. & Smith D. L. (1986). Photosynthesis of cotyledons of soybean seedlings. *New Phytologist*. Vol. 104, No. 3, (Nov., 1986), pp 319-329.

Hodgkinson, K. C. & Baas-Becking, H. G. (1977). Effect of defoliation on root growth of some arid zone perennial plants. *Aust. J. Agric. Res.* Vol. 29, pp 31-42.

Holman, R. M. & Robbins, W. W. (1982). Botánica general. Uteha. ISBN 968-438-168 8, México. 632 p.

Ho, L. C.; Grange, R. I. & Shaw, A. F. (1989). Source/sink regulation. In: *Transport of photoassimilates*. Baker, D. A. & Milburn; J. A. (Eds.), 306-343, Longman Scientific & Technical. Great Britain.

Klepper, B. (1991). Root-shoot relationships. In: *Plant roots*. Waisel, Y.; Eshel, A. & Kafkafi, U. (Eds.), 265-286, The Hidden Half, Marcel Dekker, INC. New York.

Kramer, P. J. (1983). Water relations of plants. Academic Press, Inc. ISBN 0-12-425040-8, United State America. 489 p.

Loescher, W. H.; Marlow, G. C. & Kennedy, R. A. (1982). Sorbitol metabolism and sink-source interconversion in developing apple leaves. *Plant Physiology*. Vol. 70, pp 335-339.

Mapfumo, E. & Aspinall, D. (1994). Anatomical changes of grapevine (*Vitis vinifera* L. cv. Shiraz) roots related to radical resistance to water movement. *Aust. J. Plant Physiol.* Vol. 21, pp 437-447.

Maricle, B. R. (2010). Changes in chlorophyll content and antioxidant capacity during dark to light transitions in etiolated seedlings: comparisons of species and units of enzyme activity. *Transactions of the Kansas Academy of Science*. Vol. 113, (3/4), pp 177-190.

Mayer, A. M. & Poljakoff M.(1989). *The germination of seeds*. Fourth edition. Pargamon Press. Great Britain. 270 p.

McDonald, M. B. (1994). Seed germination and seedling establishment. In: *Physiology and determination of crop yield*. American Society of Agronomy, Crop Science Society of America, Soil Science Society of America, 37-60, Madison, USA.

Metivier, J & Paulilo, M. T. (1980). The utilisation of cotyledonary reserves in *Phaseolus vulgaris* L. cv. Carioca. *Journal of Experimental Botany*. Vol. 31, pp 1257-1270.

Salisbury, F. B. & Ross, C. W. (1994). *Fisología vegetal*. Grupo Editorial Iberoamérica, ISBN 970-625-024-7, México. 759 p.

Smith, H. (1975). *Phytochrome and photomorphogenesis: an introduction to the photocontrol of plant development*. McGraw-Hill, ISBN 0-07-084038-5, Great Britain. 335 p.

Velazquez-Mendoza, J. (1989). Algunos aspectos morfológicos, fisiológicos y bioquímicos de *Phaseolus vulgaris* L. bajo sequía. In: *El agua en las plantas cultivadas*. Lorque-Saavedra, A. (Copilador), 19-26, Centro de Botánica. Colegio de Postgraduados, ISBN 968-839-073-9, Montecillo, México.

Wolswinkel, P. (1992). Transport of nutrients into developing sedes: a review of physiologgical mechanisms. *Seed Science Research*. Vol. 2, pp 59-73.

Ying, L. & Watson M. A. (1990). Leaf senescence in a perennial clonal plant. *American Journal of Botany. Abtracts for Richmond Meeting, Botanical Society of America*. Vol. 77, No. 6, p 57.

Zhang, Y.; Liu, Z.; Wang, L.; Zheng, S.; Xie, J. & Bi, Y. (2010). Sucrose-induced hypocotyl elongation of *Arabidopsis* seedlings in darkness depends on the presence of gibberellins. *Journal of Plant Physiology*. Vol. 167, pp 1130-1136.

16

Salinity Dependent Photosynthetic Response and Regulation of Some Enzymes in Halophytes from Indian Sundarbans

Sauren Das
Agricultural and Ecological Research Unit
Indian Statistical Institute
India

1. Introduction

The Environment and Ecosystem of tropical and subtropical coastal zone is marked with unique geophysical characters like sea surges with tidal waves, upland discharges, rapid sedimentation, substrate erosion and episodic cyclones. Mangroves are representing a genetic adaptation of a large variety of plant community of different families to a typical saline environment and are best developed on shorelines of tropical world particularly in vast areas of tidal influence. The mangroves are specially suited for the inhospitable environmental condition and thus pose a lot of challenging problems to the biologists. The main feature of mangroves is in their ability to successful colonization under constant physiological stress (Chaudhuri 1996). These plants grow by developing some morphological, physiological and reproductive adaptation (Zimmermann 1983; Das 1999). This vegetation provides a multidimensional beneficial impact on coastal ecosystem in the form of production and protection. This vast greenery nurses several estuarine habitats and mitigate the violence of cyclonic effect (Hogarth, 1999). Recently, these economic and ecologically utility plant communities are under severe threats world-wide. Hence, conservation and management of such ecosystem is a front-line issue to the scientific world. It is well established that biodiversity of the mangrove vegetation is getting degraded to a large extent all over the world due to human interference and tectonic activities. Mangrove ecosystems currently cover 146,530 km^2 of the tropical shorelines of the world (FAO 2003). This represents a decline from 198,000 km^2 of mangroves in 1980, and 157,630 km^2 in 1990 (FAO 2003). These losses represent about 2.0% per year since 1980–1990, and 0.7% per year within 1990–2000. These figures show the magnitude of mangrove loss, and hence the potentiality of mangrove restoration programme.

In the Indian subcontinent (extends between 21°31' - 22°30' N and 88°10' - 89°51' E), two important river systems, namely the Ganga and Brahmaputra, constitute the largest delta formation where the vast mangrove vegetation thrives with highest species diversity. Given the marked uniformity of zonation pattern, mangrove communities may be useful in interpreting minor changes in coastal conditions and serving as biological indicators. In the Sundarbans delta, there has been a very slow tilting of the coast due to tectonic uplift in the northwestern part (India) and subsidence in the east (Bangladesh). This has a major impact on mangrove species distribution as increased salinity prevails in the western part (India). This forest area (Indian territory) covers approximately 2195 km^2 (Sanyal, 1996) excluding the anastomosing network of creeks and backwaters. Presently the soil salinity ranges between 15 - 27 PPT and maximum available irradiance was reported as approximately 2000 μmol $m^{-2}s^{-1}$ (Nandy (Datta) et al., 2007). The flora comprises 36 true mangroves, 28 associates and seven obligatory mangrove species representing 29 families and 49 genera (Naskar and Guha Bakshi, 1983). Unfortunately, excessive demographic pressure, over-harvesting for timber and fuel-wood, poaching, reclamation for aquaculture and industrial pollution are being detrimental for these coastal resources. The soil of the Sundarbans is saline due to regular tidal interactions, although the salinity is low compared to soil salinity in other mangrove forests of the world (Karim, 1988). Soil salinity, however, is regulated by a number of other factors including surface runoff and groundwater seepage from adjacent areas, amount and seasonality of rainfall, evaporation, groundwater recharge and depth of impervious subsoil, soil type and topography etc. It is found that, conductivity of subsurface soil is much higher than that of surface soil (Chaffey et al., 1985). Using the salinity scale established by Walter (1971), the forest areas have been divided into three zones based on soil salinity (Karim, 1994). These are:

i. Oligohaline (ormiohaline) Zone: The zone is characterised by the soil containing less than 5 ppt of NaCl salt. The oligohaline zone occupies a small area of the north-eastern part of the forest;
ii. Mesohaline Zone: The zone is characterized by NaCl content within the concentration range of 5 to 10 ppt in soil. This zone covers the north-central to south-central part of the forest; and
iii. Polyhaline Zone: The NaCl content of the soil in this zone is higher than 10 ppt. This zone covers the western portion of the forest (Figure 1).

These manual and environmental adversities proved disastrous for some important plant species like *Heritiera fomes, Nypa fruticans, Xylocarpus granatum* and *X. mekongensis* (Banerjee, 1999; Upadhyay et al., 2002). These species predominate in between the Raimangal and Matla rivers, where fresh water influx from the Ichamati river towards Raimangal is much better (in Bangladesh part). Especially *H. fomes* prefers slightly and/or moderately saline zone and the ridges of higher elevation that are inundated only during spring tide (Alim, 1979). Previously in West Bengal, these trees used to be 2m in girth, but over 1 m girth are no longer common and top dying of *H. fomes* is very frequent in the Sundarbans forest.

High irradiance and elevated salinity of intertidal swamps impose at least two potential restrictions on the photosynthetic rate of mangroves; water deficiency and low stomatal conductance enforce these plants to thrive under considerable stress. Generation of Reactive oxygen species (ROS) such as superoxide, hydroxyl and peroxyl radicals are inevitable under oxidative stress as does the level of ROS-induced oxidative damage to lipids, proteins, and nucleic acids (Meloni et al., 2003). Current strategies for improving salt tolerance rely primarily on the production of low-molecular weight solutes e.g. Flavonoids and polyphenols and radical-scavenging enzyme systems (Tarczynski et al., 1993; Kishor et al., 1995) in order to defend alteration in the cytosolic osmotic potential. Mangroves, being the well recognized halophytes have high ability of salt tolerance and these plants can be used as a potential source of several salt resistant genes and proteins.

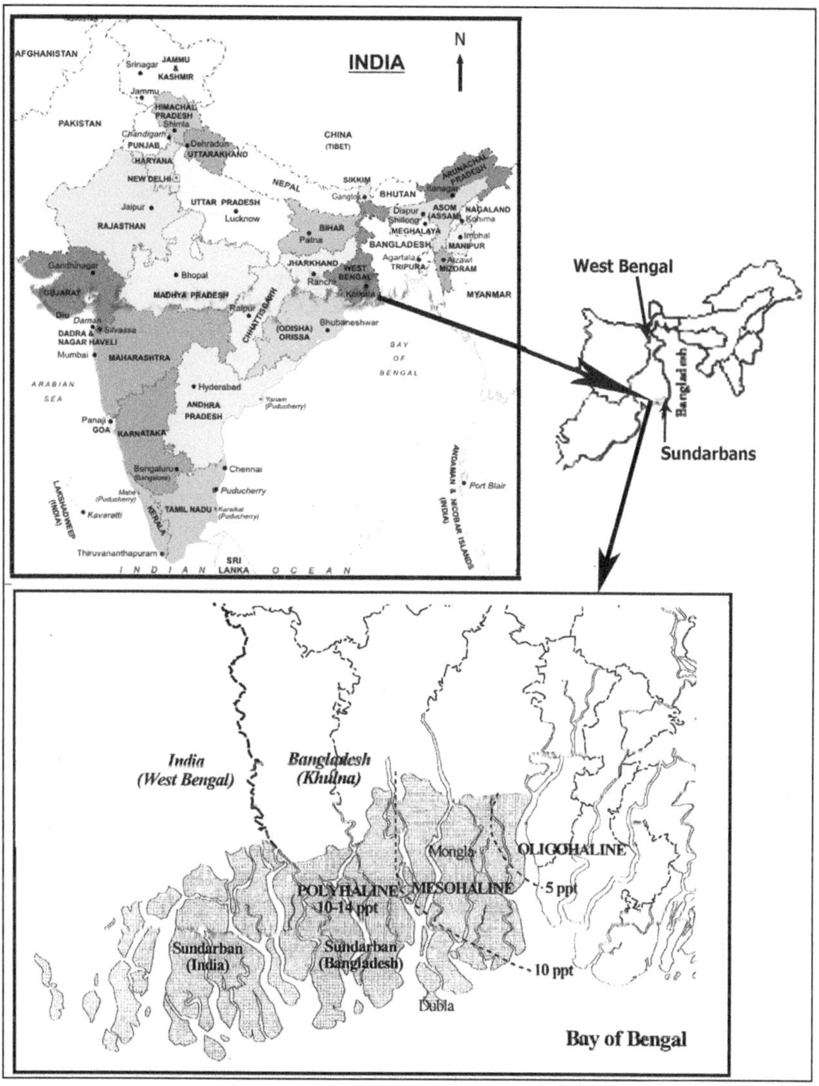

Fig. 1. Map of the study area. Differential salinity zone of undevided Sundarbans (in Bangladesh and India) area (After Karim, 1988).

To mitigate the extent of destruction of cellular components by ROS, a front line defense mechanism is developed in plants with complex antioxidant enzyme mechanisms like peroxidase (PRX), superoxide dismutase (SOD), catalase (CAT) and glutathione reductase (GR). Salinity resistance is improved by elevated regulation of antioxidant enzymes leading to ROS scavenging (Alscher et al. 2002). Salinity imposed up regulation of cellular ROS accumulation leading to destruction of membrane lipids, proteins and nuclic acids have been reported by earlier works (Hernandez et al. 1999, 2000; Mansour et al. 2005; Ben-Amor et al. 2007; Eyidogan and Oz, 2007).

Due to regular tidal inundation, increased saline water makes the substrate physiologically dry and hence, mangroves have to combat with the problem of maintaining turgour pressure and protecting their metabolic activity from high NaCl concentration (Greenway and Munns, 1980). The cumulative effects of extreme microclimate (irradiance and temperature) and high salinity affect the rate of photosynthesis (Ball, 1988). To prevent photoinhibition, mangroves have to maintain considerably low assimilation rate throughout the day (Cowan, 1982; Nandy and Ghose, 2001). From a recent comprehensive study it was revealed that both salt and drought stress led to down-regulation of some photosynthetic genes, with most of the changes being small (ratio threshold lower than 1) possibly reflecting the mild stress imposed and compared with drought, salt stress affected more genes and more intensely, possibly reflecting the combined effects of dehydration and osmotic stress in salt-imposed habitats (Chaves et al., 2009). Desingh and Kanagaraj (2007) pointed out that photosynthetic rate and RuBP carboxylase activity decreased with increasing salinity but significantly increased for some antioxidative enzymes. An important consequence of salt stress is the excessive generation of reactive oxygen species (ROS) such as superoxide anion ($O_2^{\cdot}$), hydrogen peroxide (H_2O_2) and the hydroxyl radicals ($OH^{\bullet}$) particularly in chloroplast and mitochondria (Asada, 1994; Prochazkova and Wilhelmova, 2007).
Unlike morphological markers, molecular markers are not prone to environmental influences and provide some vital information towards the priority areas for conservation strategies. Therefore, the use of molecular markers (enzymes, DNA) might enhance the understanding of such situation. Enzyme analysis is an added tool for detecting this diversity (Zeidler, 2000). The International Union for Protection of New Varieties of Plants (UPOV) have harmonized and adopted test guidelines and procedures for the use of isozyme electrophoresis as a characteristic for establishing uniqueness of plants (UPOV, 1997).
Scientific knowledge on the structural and functional characteristics of mangroves and the natural processes operating in these vulnerable fragile ecosystems is rather poor (Upadhay et al., 2002). Mangroves have to cope with considerably high soil salinity and consequently, a physiologically dry substrate. As such they are confronted with the problem of maintaining adequate turgour pressure within the cell sap because of high salt concentrations in the growth medium and thus protecting their metabolic activity (Flowers et al., 1977). This leads to accumulation and /or synthesis of organic substances in the form of compatible solutes within the vacuole (Hasegawa et al., 2000). Cheesman et al. (1997) experimentally showed that ascorbate peroxidase and SOD synthesis are much higher in field grown mangroves. Superoxide dismutase (SOD) and several antioxidant enzymes are potentially involved in H_2O_2 metabolism leading to photoprotection. Parida et al. (2002) reported that sugar, proline and some polyphenolic compounds accumulate in the cell sap of *Bruguiera parviflora* to restore the water potential more negative. Experimental works reported that in mangroves, the synthesis of these osmolytes, specific proteins and translatable mRNA induced and increased by salt stress (Hurkman et al., 1989; Bray, 1993; Xu et al., 1996; Swire-Clark and Marcotte, 1999; Xu et al., 2001). A Positive linear relationship between peroxidase activity and leaf tissue metal concentrations were reported in *Avicennia marina* (Macfarlane and Burchett, 2001). *In-vitro* experiment on *B. parviflora* resulted the differential changes of the isoforms of antioxidative enzymes in the levels due to NaCl treatment which may be useful as markers for recognizing salt tolerance in mangroves and suggested that the elevated levels of the antioxidant enzymes protect the plants against the reactive oxygen species (ROS) thus avoiding lipid peroxidation during salt stress (Parida et al. 2004a, b). Amirjani (2010) opined that the major ROS scavenging mechanism of plants involve over expression of antioxidant enzymes such as superoxide dismutase, catalase and glutathion peroxidase. The primary scavenger is SOD which convert singlet oxygen ($O_2^{\bullet -}$) to H_2O_2, which is finally scavenged by peroxidase. If this

antioxidative defense mechanism against ROS is somehow disrupted, the cellular homeostasis will be lost (Li, 2009). An increased level of peroxidase and SOD accumulation was reported in water logging stress in *Kandelia candel* and *Bruguiera gymnorrhiza* (Ye et al., 2003).

In obligate halophytes, reverse adaptation often provoke significant metabolic shifts that can be partially characterized by isozyme study. Peroxidase (in different isoforms) is widely distributed throughout the growing phase and has great biological importance. In plants, peroxidase is either bound to cell wall or located in the protoplast (Mader, 1976). Cell wall bound peroxidases are probably involved in lignifications while other isoenzymes have the regulatory role in plant senescence or in the destruction of auxins (Frenkel, 1972; Stonier and Yang, 1973). Beside the morphological adaptation, certain biochemical changes occur in halophytes. Depending on the efficient salt management strategies, mangroves show their differential suitability in elevated salinity level of the present day's Sundarbans forest. Much works on carbon assimilation and its attributes were done in *in vitro* condition under different salinity grdient (Cheeseman, 1994; Cheeseman et al., 1991; Kathiresan and Moorthy, 1993, 1994; Parida et al., 2004).

The literature review reveals that hardly any information is available on the physiological responses of 15-17 years old mangroves grown in fresh water conditions. The present study points to the reverse adaptation in five Indian 'true mangrove' species (*Bruguiera gymnorrhiza, Excoecaria agallocha, Heritiera fomes, Phoenix paludosa* and *Xylocarpus granatum*) with respect to their photosynthetic response in relation to their sustainable existence in Indian Sundarbans. Due to changed ecology, isoforms of these stress related enzymes were differentially expressed. There are hardly any report dealt with a comparative account of quantitative and qualitative analysis of antioxidant and hydrolyzing enzymes in Indian context. In view of above, this work aim to understand the extent of changes of isoforms of two antioxidant enzymes (peroxidase and superoxide dismutase) and two important hydrolyzing enzymes (esterase and acid phosphates) in five true mangrove species grown in the natural field condition (in Sundarbans) and their counterparts grown in the fresh water condition in the garden of ISI Kolkata. The comparative assessment of some physiological response (efficiency of PAR acquisition, Carbon assimilation and stomatal conductance) and biochemical characterization (enzimes and proteins) through both gel electrophoretic study and quantitative estimation of total leaf protein and enzyme, would provide some important clues towards their reverse adaptability to mesophytic condition for postulating proper conservation technique in *ex-situ* condition.

2. Methodology

Five species of true mangroves (*viz. Bruguiera gymnorrhiza, Excoecaria agallocha, Heritiera fomes, Phoenix paludosa* and *Xylocarpus granatum*) were selected for this experiment among which *Heritiera* and *Xylocarpus* are in very much stressed and rest three are profusely grown in western Sundarbans. The youngest leaf buds were collected in ice from properly identified and well matured *in-situ* (from Sundarbans forest, where salinity ranges between 15 - 27 PPT) plants (about 10-12 years old) and their counter parts from *ex-situ* (grown in fresh water condition - in the premises of Indian Statistical Institute, of all most same age, salinity ranges between 2 - 2.5PPT) conditions.

2.1 Measurement of photosynthesis and stomatal conductance

The rate of net photosynthesis and stomatal conductance in different PAR were measured with an infrared CO_2 gas analyzer (PS 301 CID, USA) that uses an electronic mass flow meter to monitor airflow rate. Measurements were taken from the exposed surface of leaves from top, middle and bottom of each plant. The rate of net photosynthesis (P_n) was determined measuring the rate, at which a known leaf area assimilated CO_2 concentration at a given time. The data were taken from randomly 20 plants of almost same age in full sunshine condition. The average data and their standard error bars were presented in the graphs.

$P_n = -W \times (C_o - C_I) = -2005.39 \times \{(V \times P) / (T_a \times A)\} \times (C_o - C_I)$ .... [W = mass flow rate per leaf area (mmol $m^{-2}s^{-1}$); C_o (C_I) = outlet (inlet) CO_2 conc. (μmol $m^{-2}s^{-1}$); P = atm. pressure (bar); and T_a = air temp. (K)].

Stomatal conductance (C_{leaf}) was calculated from the rate of water efflux and leaf surface temperature (^{0}C).

$C_{leaf} = W / [\{e_{leaf} - e_o) / (e_o - e_I)\} \times \{(P - e_o) / P\} - R_b W] \times 1000$.........[$e_{leaf}$ = saturated water vapour at leaf temperature (bar); R_b = leaf boundary layer resistance (m^2s / mol); P = atm. pressure (bar) and W = mass flow rate per leaf area (mmol $m^{-2}s^{-1}$)].

The data were downloaded and computed through RS 232 Port.

2.2 Protein estimation and SDS-PAGE analysis

Total protein estimation was carried out for five mangrove taxa from the both habitat following Lawrey et al. (1951). Extraction of proteins for gel electrophoresis was done from 2g of fresh leaf. Leaf samples were macerated in a mortar-pestle, add 5ml of extraction buffer (containing 10% (w/v) SDS, 10mM β- Marcaptoethanol, 20% (v/v) glycerol, 0.2 M Tris/HCl (pH 6.8) and 0.05% Bromophenol blue). Centrifuge at 10000 rpm for 20 min. Supernatants were used as samples. Protein samples were resolved in 12.5% SDS-PAGE gels following the procedure of Laemmli (1970) and stained with Coomassie Brilliant Blue R-250 (Sigma). Molecular weights of different protein bands were determined with respect to standard protein marker (Bioline Hyper Page pre-stained protein marker, 10kDa - 200kDa) with the Kodak MI software after documentation the gel slab with Gel-Doc system (Biostep GmbH - Germany).

2.3 Extraction of enzymes for native gel electrophoresis

Two grams of young leaf buds were macerated to powder with liquid Nitrogen with a mortar- pestle, then 0.1 g PVP and 5 ml of extraction buffer (consists of 1 M Sucrose, 0.2 M Tris-HCL and 0.056 M β - Marcaptoethanol; pH is adjusted at 8.5) was added to it and homogenized. The extractants were centrifuged at 10000 rpm for 20 minutes at 4°C; supernatants were used as samples for gel electrophoresis. Isozyme analysis of four enzymes *viz.* Peroxidase, Superoxide dismutase, Esterase and Acid phosphatase was done for the investigated five taxa. Equimolar amount of enzymes were loaded in each well. Samples from saline and non-saline environment were loaded side by side for precision of polymorphic band expression. Slab gels were stained for definite enzymes following Das and Mukherjee (1997). Gels were documented with a Gel-Doc System (Biostep GmbH - Germany) and analysis for band intensity and Relative Mobility Factor (R_{mf}) were estimated with Kodak-MI software.

2.4 Enzyme assay

Peroxidase (PRX, E.C.1.11.1.7): 200 mg fresh leaf sample was extracted in 1-1.5 ml 0.9% KCl and centrifuged at 12,000 rpm for 15 min at 4°C; supernatant used as enzyme sample. Absorbances were taken by Helios γ spectrophotometer (Thermo electron Corporation,

USA) at 460 nm in respect to the standard curve prepared following Shannon et al. (1966) with minute modification.

Superoxide dismutase (SOD, E.C.1.15.1.1): Cell sap was extracted from 200mg of leaf and 1-1.5 ml 50 mM Phosphate buffer, ph adjusted to 7.0; centrifuged at 12,000 rpm for 15 min at 4°C. Supernatants were used for enzyme samples. Different aliquots (50, 100, 150, 200, 250 μg/ml) of the standard enzyme samples were also used for preparing the standard curve and absorbance were measured at 550 nm following the protocol described by Keith et al. (1983) with minute modification.

Esterase (EST, E.C.3.1.1.1): Enzyme sample was prepared from 200 mg fresh leaf sample extracted with 1-1.5 ml ice cold 0.1 M Tris/HCl buffer adjusted pH 8.0. Extractants were centrifuged at 12,000 rpm for 15 min at 4°C. The supernatant used as sample. Absorbances were noted at 322 nm with respect to the prepared standard curve following the procedure described by Balen et al. (2004).

Acid Phosphatse (ACP, E.C.3.1.3.2): 200 mg fresh leaf sample was extracted in 1-1.5 ml 40 mM succinic acid /NaOH buffer, pH adjusted to 4.0; centrifuged at 12,000 rpm for 15 min at 4°C. Supernatant was taken for enzyme assay. Prepared a standard curve with the known enzyme samples and absorbances were taken at 322 nm following Huttová et al. (2002).

The data presented was the average of 20 readings for each plant and standard errors were also depicted in the figures. SPSS 12.0 version was used for statistical analysis towards estimating the correlation value, if any, between the total protein amount and quantitatively assayed enzymes. For each enzyme, the pure samples (Sigma chemicals) were used for preparing the standard curves.

3. Results

The salinity ranged between 15-27 ppt throughout the year in Sundarbans, but in garden (mesophytic) soil it never exceeds beyond 2 ppt. The irradiances (PAR) were measured in two ecosystems and range obtained between 428 - 2110 μmol $m^{-2}s^{-1}$ in the saline habitat (Sundarbans) and 600 - 1880 μmol $m^{-2}s^{-1}$ in the mesophytic (ISI garden) environment.

3.1 Photosynthesis and stomatal conductance

The average net photosynthesis was generally higher in mangroves of non-saline habitat than that of the native ones (Fig. 2A), but the PAR acquisition for maximum photosynthesis were greater in most of the Sundarbans species except *H. fomes* and *X. granatum* (Fig. 2A). In *B. gymnorrhiza*, the maximum photosynthesis (10.47μmol $m^{-2}s^{-1}$) was achieved only at 873μmol $m^{-2}s^{-1}$ PAR when grown in non-saline soil, but as high as 1078.5μmol $m^{-2}s^{-1}$ PAR was utilized to obtain the highest assimilation rate (9.19μmol $m^{-2}s^{-1}$) under saline condition (Fig. 2A). In *E. agallocha* the optimum PAR required for maximum photosynthesis was 1445.8 μmol $m^{-2}s^{-1}$ in Sundarbans and 1402.6 μmol $m^{-2}s^{-1}$ in the garden, whereas the highest assimilation rates were 12.27 μmol $m^{-2}s^{-1}$ and 14.69 μmol $m^{-2}s^{-1}$ respectively (Fig.2A). Similarly, in *P. paludosa*, the optimum PAR value was 1662.3 μmol $m^{-2}s^{-1}$ in Sundarbans forest beyond which photosynthesis started declining, whereas in the garden, the highest rate of net photosynthesis (6.92 μmol $m^{-2}s^{-1}$) was recorded at a much lower PAR value (1012.6 μmol $m^{-2}s^{-1}$) (Fig. 2A). On the contrary, under salt stress, the rate of assimilation in *X. granatum* dropped just beyond 827.7 μmol $m^{-2}s^{-1}$ PAR, whereas in non-saline condition, the optimum PAR was as high as 1557.6 μmol $m^{-2}s^{-1}$ (Fig. 2A). Among the studied species, photosynthesis rate was maximal in *H. fomes* under both the environmental conditions (10.63 μmol $m^{-2}s^{-1}$ in Sundarbans and 12.63 μmol $m^{-2}s^{-1}$ in garden) (Fig. 2A).

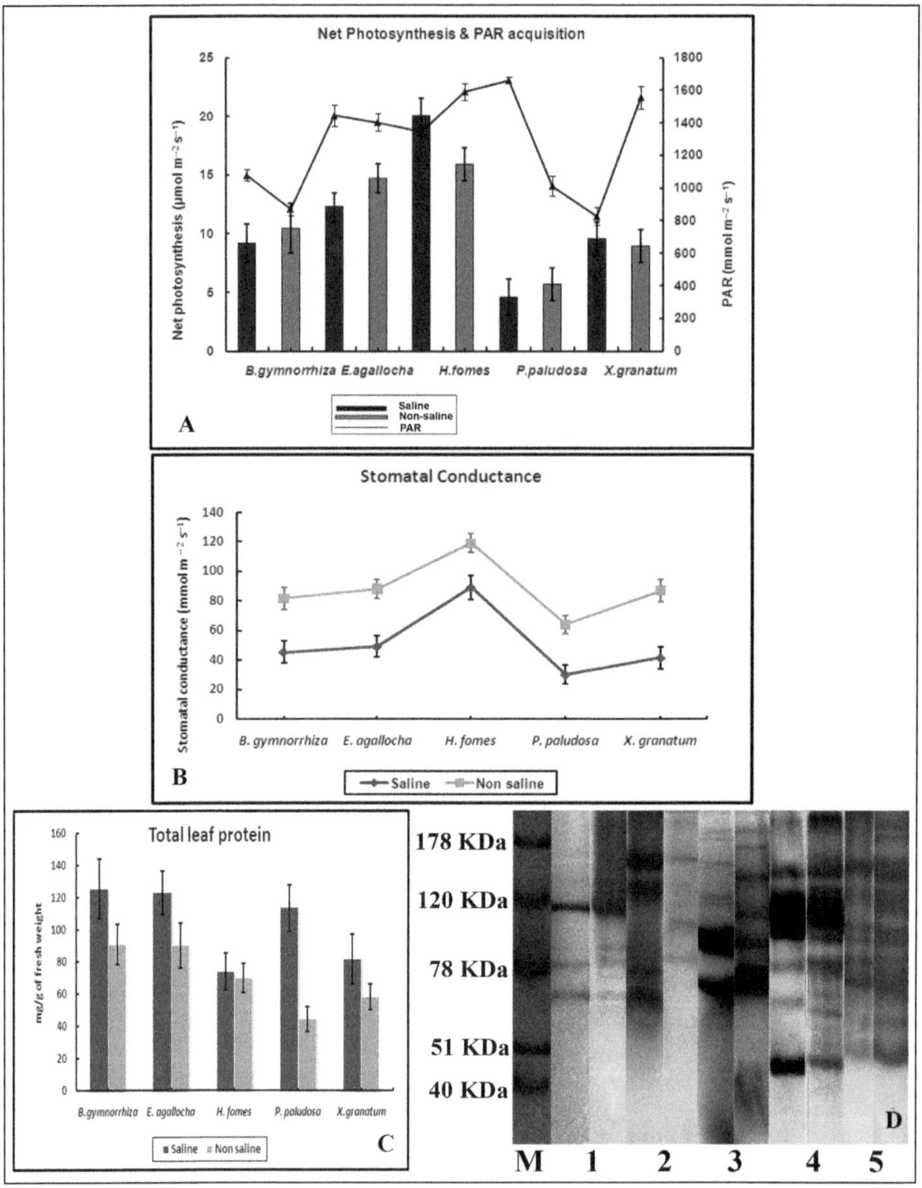

Fig. 2. **A.** Comparative account of net photosynthesis and PAR acquisition of the investigated mangroves in two different salinity zones. **B.** Graphical representation of stomatal conductance of five mangroves from two different environments. **C.** Total leaf protein content along with standard errors. **D.** Documentation of SDS-PAGE analysis of proteins along with standard marker (M - marker, **1.** - *B. gymnorrhiza*, **2.** - *E. agallocha*, **3.** - *H. fomes*, **4.** - *P. paludosa* and **5.** - *X. granatum*). Each consecutive two lanes representing one species; in left - non-saline and in right - saline environment grown plant species.

Stomatal conductance was remarkably decreased under salinity stressed habitats as compared to those of the sweet-water counterparts (Fig. 2B). In *B. gymnorrhiza* and *E. agallocha* the salinity-imposed restriction of stomatal conductance was about 44%, in *P. paludosa* and *X. granatum* it was nearly 52% and in *H. fomes* 25%.

3.2 Total protein

Total leaf protein was estimated from the five enough mature taxa, grown in both saline and fresh water environment. In all five species, the total protein content showed higher amount in fresh water grown plants than that of their Sundarban counterpart (salt stress environment). The highest amount was estimated in *B. gymnorrhiza* (125.82 mg/g fr. wt.) and *E. agallocha* (123.2 mg/g fr. wt.) and minimum was in *X. granatum* (73.96 mg/g fr. wt.) grown in *ex-situ* condition. The increment of total protein was estimated at highest in *P. paludosa* (156%) and lowest in *X. granatum* (5.7%). In *H. fomes*, fresh water habitat showed 57% more protein content than that of the *in-situ* habitat (Fig. 2C).

SDS - PAGE analysis: This analysis revealed that the numbers of protein bands were expressed differentially in the same species from two different habitats. The molecular weights of these bands were calculated with respect to standard marker, run in the same gel. The result revealed that in *Bruguiera*, the saline habitat individual showed one extra band than its non-saline replica and molecular weight ranged between 169.1 - 66.67kDa (non-saline) and 210.7 - 66.11kDa (saline). *Excoecaria* showed the same number of bands in both habitats having molecular weight ranged between 205.8 - 65.55kDa (non-saline) and 213.2 - 77.72kDa (saline). The highest number of protein bands appeared in *Heritiera* from both the environments, nine bands in each having molecular weight 211.2 - 26.71kDa in non-saline and 212.2 - 37.0kDa in saline taxa. One extra band appeared in non-saline *Phoenix* than it saline pair and the molecular weight ranged between 201.3 - 46.43kDa and 213.2 - 46.0kDa respectively. In *Xylocarpus*, one more band was expressed in saline plant, having 202.8 - 50.57kDa (non-saline) and 197.3 - 58.27kDa (saline) (Fig. 2D).

3.3 Native gel electrophoresis: Peroxidase (PRX)

Band expression obtained from gel electrophoresis revealed that in *H. fomes* and *X. granatum* showed the same number of isoforms in two different habitats, where as in *B. gymnorrhiza, P. paludosa* and *E. agallocha* the number of isoforms were higher in Sundarbans species than that of their replica from fresh water condition (Table 1). But the R_{mf} and band intensity were different to a large extent in all the five species. In *Bruguiera*, the saline plant showed eight isoforms with highest OD 163.5 (0.07 R_{mf}) where as the fresh water individual showed five isoforms with highest OD was 51.37 (0.68 R_{mf}). In *Heritiera* and *Xylocarpus*, the numbers of isoforms were same but highest OD obtained 206.0 (0.18 R_{mf}) and 180.0 (0.68 R_{mf}) from saline individual and from fresh water habitats highest OD values were 166.0 (0.07 R_{mf}) and 89.9 (0.07 R_{mf}) respectively. Non-saline *Phoenix* and *Excoecaria* showed three and two isoforms of PRX and saline partners expressed four and three isoforms respectively (Table 1; Fig. 3A).

Superoxide dismutase (SOD)

The experimental data showed that in all five species, isoforms of SOD expressed in less number from the fresh water grown individuals than that of their saline replica. All four species expressed three isoforms in non-saline environment, except *Phoenix*, where it was two. The plants from saline habitat, *Heritiera, Phoenix* and *Xylocarpus* showed five isoforms, *Bruguiera* and *Excoecaria* have two. The Densitometric scanning resulted that the band intensity of each isoforms were much higher in saline habitats. In *Heritiera*, highest intensity

Enzymes / Env. / Plants	Peroxidase				Superoxide Dismutase				Esterase				Acid phosphatase			
	Saline		Non-saline		Saline		Non-saline		Saline		Non-saline		Saline		Non-saline	
	R_{mf}	OD	R_{mf}	OD	R_{mf}	OD	R_{mf}	OD	R_{mf}	OD	R_{mf}	OD	R_{mf}	OD	R_{mf}	OD
B. gymnorrhiza	0.07 0.17 0.24 0.35 0.41 0.5 0.67 0.79	163.5 121.1 99.38 26.12 24.32 21.11 20.64 72.51	0.09 0.18 0.47 0.59 0.68	8.4 23.02 6.76 8.96 51.37	0.31 0.42 0.66 0.73	135.0 134.0 127.0 142.0	0.03 0.11 0.18	16.42 25.75 41.53	0.37 0.48 0.53	150.0 153.0 221.0	0.36 0.48	17.33 20.0	0.32 0.49	148.6 157.0	0.53	194
E. agallocha	0.07 0.15 0.21	80.12 25.3 115.66	0.18 0.26	6.03 8.05	0.02 0.31 0.34 0.57 0.66	129.38 170.07 86.36 90.99 74.39	0.18 0.33 0.49	113.0 163.0 162.0	0.17 0.37 0.53	214.0 212.0 210.0	0.24 0.3 0.56	102.27 30.82 14.62	0.40	196.0	0.53	26.15
H. fomes	0.07 0.17 0.31	166.0 184.0 196.0	0.18 0.26 0.76	206.0 204.0 182.0	0.06 0.2 0.31 0.42 0.78	63.0 26.5 43.45 50.31 138.7	0.33 0.63 0.75	5.43 6.48 5.42	0.37 0.48 0.62 0.71	214.0 226.0 205.0 198.5	0.36 0.48	53.6 32.48	0.17 0.36 0.49	127.4 146.6 168.4	0.19 0.37	189.0 245.0
P. paludosa	0.04 0.09 0.41 0.67	184.94 125.62 139.67 136.81	0.18 0.26 0.76	154.0 192.0 39.0	0.06 0.31 0.34 0.48 0.78	160.0 131.0 128.2 176.0 184.0	0.11 0.26 0.39	40.0 42.0 39.6	0.1 0.37 0.48	178.0 186.0 222.0	0.3 0.48	177.0 174.0	0.12 0.32 0.4 0.49	227.0 195.0 185.0 214.0	0.17 0.46 0.63	96.03 29.31 18.6
X. granatum	0.07 0.31 0.41	89.91 18.0 75.43	0.59 0.68 0.76	168.0 180.0 150.0	0.14 0.42 0.57 0.84 0.87	44.0 104.0 41.0 155.0 147.0	0.24 0.49 0.63	5.5 12.0 5.87	0.10 0.32 0.37 0.48 0.71	190.0 214.0 223.0 174.0 162.0	0.3	59.5	0.17 0.32 0.49	210.0 194.0 247.0 210.0	0.12 0.37 0.46	203.0 248.0 194.0

Table 1. Band intensity (OD) and Relative Mobility Fraction (R_{mf}) of isozymes from mangroves in two environments.

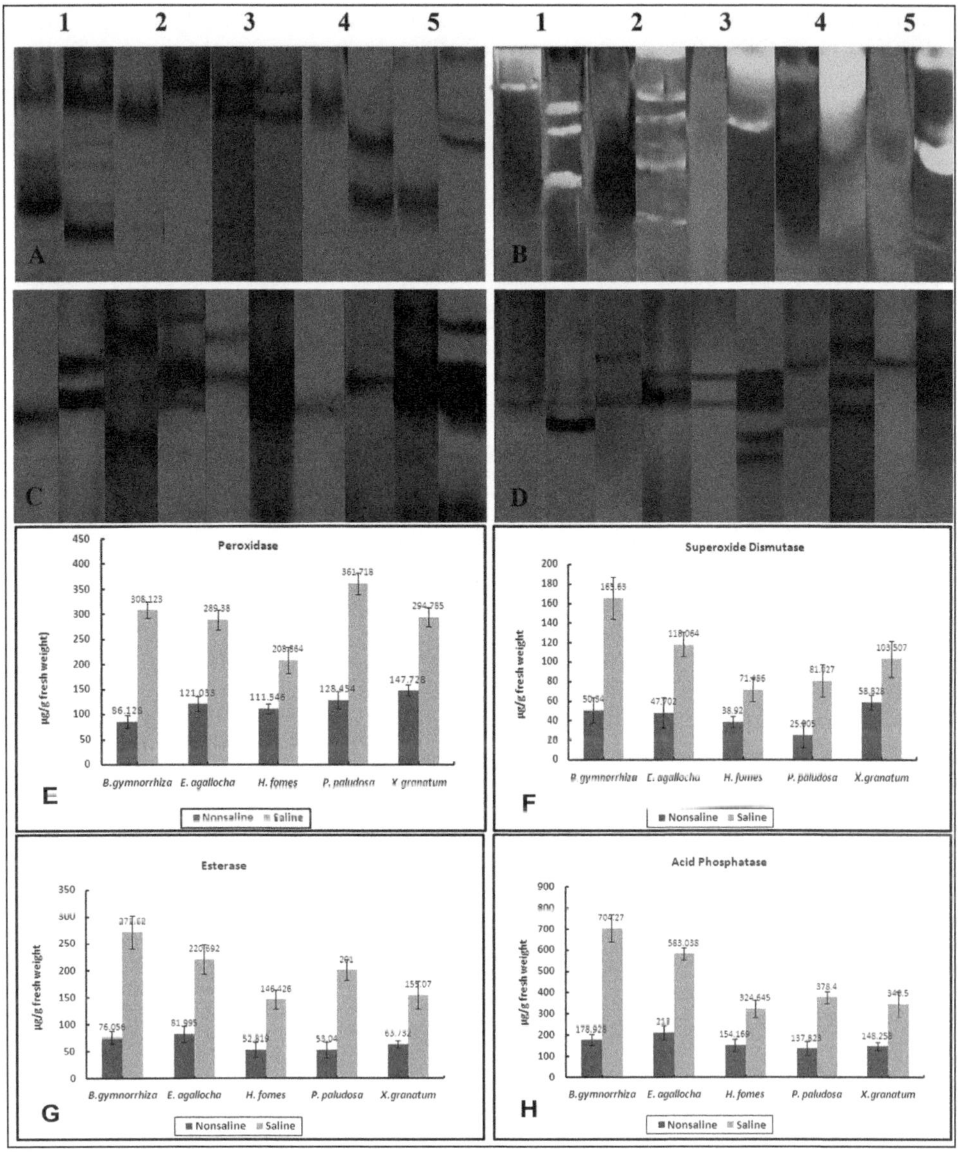

Fig. 3. A - D, Documentation of different enzymes; A - Peroxidase, B - Superoxide dismutase, C - Esterase and D - Acid phosphatase (*1 - B. gymnorrhiza, 2 - E. agallocha, 3 - H. fomes, 4 - P. paludosa and 5 - X. granatum*). Each consecutive two lanes representing one species; in left - non-saline and in right - saline environment grown plant species. E - H, Graphical representation of quantitative estimation of studied enzymes. Standard error bars were showed on each bar diagram.

(138.7 OD) occurred with R_{mf} value 0.78 in saline individual, where in reverse habitat it was much less (6.48 OD and 0.63 R_{mf}). Similarly, in *Bruguiera*, it was 142.0 OD with 0.73 R_{mf} in saline and 41. 53 OD at 0.18 R_{mf} in non-saline habitat. In *Xylocarpus* and *Phoenix*, saline and fresh water condition showed the highest peak as 147.0 OD (0.87 R_{mf}) and 12.0 OD (0.49 R_{mf}) and 184.0 OD (0.78 R_{mf}) and 42.0 OD (0.26 R_{mf}) respectively. In *Excoecaria* highest peak of intensity were observed in saline and non-saline habitats as 170.07 OD (0.31 R_{mf}) and 163.0 OD (0.33 R_{mf}) (Table 1; Fig.3B).

Esterase (EST)

From the gel staining it revealed that EST expression in all species from fresh water habitats were two isoforms, except *Xylocarpus* (single band) and *Excoecaria* (three bands). The comparative band intensity was also remarkably high from all saline habitat taxa except in *Phoenix*, where it was slightly higher (222.0 OD at 0.48 R_{mf} in saline plants and 177.0 OD at 0.3 R_{mf} in non-saline habitat). In *Heritiera*, among the four expressed bands in saline habitat, the highest band intensity occurred at 226.0 OD (0.48 R_{mf}) and it was 53.6 OD (0.36 R_{mf}) in reverse habitat. *Bruguiera* showed as high as 221.0 OD (0.53 R_{mf}) in saline (expressed number of isoforms was three) and 20.0 OD (0.48 R_{mf}). In *Xylocarpus*, out of five isoforms in saline condition, the highest OD was 223.0 (0.37 R_{mf}) in saline and the other side it was 59.5 OD (0.3 R_{mf}). Out of three isoforms, in saline species of *Excoecaria*, highest OD obtained 214.0 (0.17 R_{mf}) and in non-saline it was 102.27 OD (0.24 R_{mf}) (Table 1; Fig3C).

Acid phosphatase (ACP)

Among the five investigated taxa all four species showed excess number isoforms of ACP in saline individual except in *Excoecaria*, where it was single band in both the environment, though the band intensity was higher in saline plants (196.0 OD, 0.4 R_{mf}) than non-saline partner (26.15 OD, 0.53 R_{mf}). In *Bruguiera* the saline habitat expressed two isoforms of ACP with higher intensity of 157.0 (0.49 R_{mf}) and 148.6 OD (0.32 R_{mf}) but the freshwater plant have only one band with 124.0 OD (0.53 R_{mf}). In both Xylocarpus and Phoenix, saline environment expressed one more isoforms than that of their reverse habitat (three isoforms were expressed in fresh water habitat in each). The highest band intensity in *in-situ Xylocarpus* occurred with 247.0 OD (0.49 R_{mf}) and in reverse condition the highest band intensity and Rmf value were almost same (248.0 OD and 0.46). In *ex-situ* plant of *Phoenix* the highest intensity observed at 96.03 OD (0.17 R_{mf}) and in counterpart it was 227.0 at 0.12 R_{mf}. Among the three expressed bands, highest OD value occurred as 168.4 (0.49 R_{mf}) in *Heritiera* (saline) and 145.0 OD (0.37 R_{mf}) in non-saline plant (Table 1; Fig. 3D).

3.4 Quantitative assay of enzymes

The plant species from saline environment showed the all four (PRX, SOD, EST and ACP) investigated enzymes were in higher quantities than that of their fresh water grown individual. Increase in PRX quantity (µg/g) was highest in *Bruguiera* (257%), then *Xylocarpus* (209%), *Phoenix* (181%) and *Heritiera* (176%) while the increment was 139% in *Excoecaria* (Fig. 3E). In case of SOD, the highest increment occurred in *Heritiera* (241%), then *Bruguiera* and *Phoenix* (229 and 224% respectively) and lowest in *Excoecaria* (147%) (Fig. 3F). Similarly, EST was highest increased in *Phoenix* (287%), *Bruguiera* (257%) and *Heritiera* (241%), lowest in *Excoecaria* (154%) (Fig. 3G). ACP reached its maximum increment in *Bruguiera* (293%) and *Xylocarpus* (267%) and lower in *Excoecaria* (139%) (Fig. 3H).

3.5 Statistical analysis

Estimated total protein and four enzymes from two habitats were taken in account. A two-tailed bivariate correlation coefficient (Pearson coefficient) was calculated among the each parameter (Table 2). The analysis showed that in case of the relationship between Protein and SOD, all species in saline environment have inverse relationship (at 0.01% level) except of *Bruguiera,* wherein it was significant at 0.05% level. In PRO vs. PRX significant inverse relationship was observed only in *Bruguiera* (0.05%) and *Phoenix* (0.01%) whereas the other three plants (*Excoecaria, Heritiera* and *Xylocarpus*) showed no statistically significant relationship. Correlation between PRO and EST obtained a significant positive relationship at 0.01% level only in *Bruguiera* and *Excoecaria* in saline inhabitants and others did not show any relationship. The only inversely correlation was obtained in *Excoecaria* (saline plant) at 0.01% level, where as in case of other plants it showed no relationship.

Species / Treatment	*B. gymnorrhiza*		*E. agallocha*		*H. fomes*		*P. paludosa*		*X. granatum*		Treat-ment
	Ns	S	Ns	S	Ns	S	Ns	S	Ns	S	
PRO	0.186	-0.571**	0.380	-0.754*	0.045	-0.529*	-0.383	-0.731*	0.145	-0.705*	SOD
	0.110	0.301	0.308	-0.442	0.795**	-0.348	0.698*	0.187	0.555	-0.013	POX
	0.213	0.667*	0.603	-0.66*	0.468	0.099	0.554	-0.517	0.518	-0.363	EST
	-0.206	0.430	-0.510	-0.65*	0.192	-0.9**	-0.254	0.495	-0.407	-0.403	ACP

* Significant at 0.01%; ** Significant at 0.05%; Ns – non saline, S – saline.

Table 2. Correlations among the different enzymes and total proteins in the plants of two habitats.

4. Discussion

Five typical mangroves (*Bruguiera gymnorrhiza, Excoecaria agallocha, Heritiera fomes, Phoenix paludosa* and *Xylocarpus granatum*) from *in situ* grown where salinity level of the substrate was quiet high (15 - 27 ppt) and *ex situ* (mesophytic) habitat (salinity level was 1.8 - 2 ppt) were investigated with respect to their comparative rates of net photosynthesis, stomatal conductance, and expression of two antioxidative enzymes, both qualitative and quantitative estimation.

Among the five investigated taxa *B. gymnorrhiza, E. agallocha* and *P. paludosa* had optimum PAR requirements for maximum photosynthesis that were higher in Sundarbans than those in the mesophytic taxa, whereas, the peak photosynthetic rates were higher in the mesophytic soil. But *H. fomes* and *X. granatum* showed the reverse phenomenon, where at comparatively low PAR the highest net photosynthetic rate occurred. Krauss and Allen (2003) pointed out that *B. sexangula* prefers low salinity combined with low light intensity. Cheesman and Lovelock (2004) experimentally showed that in *Rhizophora mangle* under low salinity conditions, net CO_2 exchange and photosynthetic electron transport becomes light saturated at less than 500 μmol $m^{-2}s^{-1}$. In Sundarbans however, despite tidal influence, high salinity makes the substrate physiologically dry. In order to check desiccation and xylem embolism, mangrove leaves reduce the rate of water efflux (Nandy and Ghose, 2001) that may enhance the tendency to elevate the leaf temperature with subsequent decline in photosynthesis. The present observation revealed that in all five species, stomatal conductance was reduced ranged by 25% to 52% under salinity stress that effectively limited

CO_2 influx. Although reduced stomatal conductance imposed by high salinity restricts CO_2 diffusion, it may elevate the CO_2 partial pressure across the stomata that are utilized by mangrove leaves to maintain a consistently moderate rate of photosynthesis throughout the day, thus avoiding CO_2 starvation and photoinhibition. This result is in accordance with Cowan (1982) and Nandy and Ghose (2001). Naidoo et al. (2002) also measured the optimum PAR for highest photosynthesis in *B. gymnorrhiza* at Durban Bay site that is similar (around 1000 μmol $m^{-2}s^{-1}$) to the present data. The opposite phenomenon occurred in *H. fomes* and *X. granatum* can be explained as less affinity of these species towards high salinity, irradiance and temperature of the Sundarbans forest. Theoretically, high photosynthetic efficiency can increase water use efficiency as more carbon is assimilated per unit water transpired. In mangroves, a positive correlation was reported between photosynthesis and stomatal conductance - an important determinant of water use efficiency (Nandy et al., 2005, Dasgupta et al., 2011). The effect of salinity stress on the photosynthetic enzyme activities is postulated to be a secondary effect mediated by the reduced CO_2 partial pressure in the leaves caused by the stomatal closure (Lowler and Cornic, 2002; Meloni et al., 2003; DeRidder et al., 2007). Salt stress depletes the water potential of soil which in turns causes water stress in plants. This phenomenon triggers the signal for stomatal closure and consequently makes CO_2 deficit in the leaf cells. Tausz et al. (2004) reported that a breakage of proper coordination between CO_2 assimilation and chloroplast photo function electron transport chains ultimately leads to the transfer of electrons in an alternating electron acceptor like O_2.. The excess electron reduces molecular oxygen to produce ROS. During normal growth and development this pathway monitors the level of ROS, produced by aerobic metabolism, and controls the expression and activity of ROS-scavenging pathways. The basic ROS cycle may also perform fine metabolic tuning, e.g., suppression of photosynthesis, for reducing the production rate of ROS. There are many potential sources of ROS in plants. Some are reactions of normal aerobic metabolism, such as photosynthesis and respiration, while others belong to pathways enhanced during abiotic stresses, such as photorespiration. In the present work, though *H. fomes* and *X. granatum* showed relatively higher rate of net photosynthesis using less PAR in saline environment than those of the other three species, production of PRX and SOD were relatively low. Hence from the present work it can be postulated that these two species probably less ability to successful management of salt-stress than those of other three species investigated in the present day salinity level at Sundarbans. This study also reveals that in all the mangroves grown in non-saline soil, an increased rate of assimilation is coupled with increased stomatal conductance.

All the five investigated mangrove taxa from fresh water habitat showed an increase amount of total leaf protein than that of their saline habitat replicas. It was noted that the percent of increment varied in a wide range from 5% to 36%, in which the highest increment occurred in *Excoecaria* and *Phoenix* while lowest in *Heritiera* and *Xylocarpus* (6.05 and 5.7% respectively). This occurred probably as salinity imposed plants are adversely affected in their growth and metabolism due to osmotic effect of salt, nutritional imbalance, accumulation of incompatible toxic ions. The decreased protein content in saline environment might be due to enhance activity of protease (Parida et al., 2002). The present result was well accord to Rajesh et al. (1999), where they experimentally reported that in *Ceriops*, the total leaf protein decreased under higher concentration of saline treatment. Raymond et al. (1994) opined that stress induced protein degradation may be essential which provide amino acids for synthesis of new proteins suited for growth or survival under the modified condition. Mansour (2000) reported that protein biosynthesis decline under alt stress condition, while cells preferentially synthesis some specific stress proteins. Stress induced proteins accumulated in the cell which might be synthesized *de novo* in response to salt or might be present constitutively at low level (Pareek et al., 1997). Reports from earlier works also confirm that enhanced cellular accumulation of ROS due to salinity stress can expedite the destruction of membrane lipids, proteins and nucleic acids (Hernandez et al., 1999, 2000; Mansour et al., 2005; Ben-Amor et al., 2007; Eyidogan and Oz, 2007). In the present investigation, it was evident that the degradation of proteins in *Heritiera* and *Xylocarpus* from saline habitat was lower amount than the other three taxa investigated might be pointed to the synthesis of lesser amount of compatible amino acids in salt habitat. Parida et al. (2002) reported that the total soluble leaf proteins decreased in *Bruguiera parviflora* under NaCl treatment. This decrease might have the outcome of adverse effect of NaCl treatment resulted synthesis of certain low molecular weight proteins which are yet to be elucidated.

Among the various antioxidant enzymes, qualitative and quantitative estimation of two antioxidant enzymes (PRX and SOD) and two other important (hydrolyzing) enzymes (EST and ACP) from saline and fresh water grown plants were made. The results revealed that in most of the cases number of isoforms, band intensity and enzyme expression were higher in salt stressed plants than those of their mesophytic counterparts. It has been proved that during electron transport in the mitochondria and chloroplasts some leakage of electrons occurs and these leaked electrons react with O_2 during aerobic metabolism to produce Reactive Oxygen Species (ROS) such as superoxide (O_2^-), hydrogen peroxide (H_2O_2) and the hydroxyl radical (^-OH) (Halliwell and Gutteridge, 1985). These cytotoxic ROS may seriously affect the normal metabolism through oxidative damage of lipids, proteins and nuclic acids (Fridivich, 1989). During photosynthesis, the internal O_2 level become high and chloroplast are prone to generate ROS at that time (Foyer and Mullineaux, 1994). Plants synthesize a number of antioxidative enzymes to counteract these ROS, especially SOD converts O_2^- into H_2O_2 and PRX catalyze H_2O_2 (Asada, 1994). In salinity imposed plants the balance between the production of ROS and the scavenging activity of the antioxidants becomes disrupted which ultimately results in oxidative damage. Plants with high levels of antioxidants, either constitutive or induced, have been reported to provide sufficient resistance against oxidative damage (Parida et al., 2004; Jithesh et al., 2006). The present work resulted that both PRX and SOD expressions were high in saline plants and the increment were ranged between 139 - 257% in case of PRX and 147 - 241 in SOD. The present result was substantiated with the earlier works (Cheeseman, 1997; Takemura et al., 2000). In both the cases the increments were lower in *Heritiera* (139% in PRX, 147% in SOD) and *Xylocarpus* (142% in PRX, 166% in SOD) than that of the other three species of saline habitat. Parida et al. (2004) opined that high salt concentration enhanced the accumulation of free amino acids and polyphenols. Thus, NaCl stress not only imposes alterations in antioxidative metabolism, but also accumulation of osmolytes as adaptive measures. The numbers of isoforms were also increased in case of PRX and SOD in saline habitat plants. In *Bruguiera* (saline) the highest numbers of isoforms were expressed in case of PRX, but it was unchanged in case of *Heritiera* and *Xylocarous* (three isoforms in each habitat). This might be due to the relatively less suitability of those plants in the saline environment. SOD showed the excess isoforms in all saline plants than their fresh water counterpart. Therefore, it is evident that the salt imposed production of toxic ROS is mostly regulated by up regulation of antioxidative enzymes like PRX and SOD. Sahu & Mishra (1987) reported changes in enzymatic activity of peroxidase during senescence of rice leaves when submitted to salt stress. They observed that NaCl increased peroxidase activity which could be related to regulation of membrane permeability, cell wall formation and oxidation of accumulated substances due to salt stress. It was also proved that peroxidases are enzymes related to polymer synthesis in cell wall (Bowles, 1990), as well as with prevention of oxidation of membrane lipids (Kalir et al., 1984).

Biosynthesis of Esterase (EST) revealed that in all five species it is in higher amount in the *in-situ* taxa investigated. The fresh water grown plants synthesized esterase enzyme with less number of isoforms except Excoecaria, where the numbers of isoforms were same (3), but band intensity was more in saline plants. Highest number of isoforms occurred in Heritiera (saline - 4; in non-saline - 2) and Xylocarpus (saline - 5; in non-saline - 1). Still the percentage of increment was lower in the above two taxa than the other three from saline habitat (123 and 156% respectively), the other three species ranged between 241 - 287% of esterase increment. This result supplemented by Hassanein (1999), where he experimentally proved that nine different esterase isoenzymes were detected in embryos of seeds germinated in 105 mM NaCl, whereas only five of them were detected in the embryos of untreated seeds. Pectins are major components of the primary plant cell wall. They can be both methylesterified and acetylesterified and de-esterification occurs by specific esterases (Cécile et al., 2006). Al-Hakimi and Hamada (2001) reported that the contents of cellulose, lignin of either shoots or roots, pectin of root and soluble sugars of shoots were lowered with the rise of NaCl concentration. Hence, esterases play a major role to counteract the salt induced imbalance in cell wall formation.

Acid Phosphatases (ACP) are a group of enzymes that catalyze the hydrolysis of a variety of phosphate esters. These enzymes are widely distributed in plants and are related to phosphate supply and metabolism from a vast array of phosphate esters which are essential for normal growth and development of plant organs (Olczak et al., 2000). The present work revealed that amount of increment in saline grown plant occurred ranging from 139 - 293%. It may be due to fact that under conditions of stress, growth is restricted and delivery of phosphate is impaired, thus resulting in the activation of the cellular phosphatases that release soluble phosphate from its insoluble compounds inside or outside of the cells thereby modulates osmotic adjustment by free phosphate uptake mechanism (Fincher, 1989). Jain et al. (2004) also demonstrated that the endosperm acid and alkaline phosphatase activities were significantly higher after salt treatment, than that of the control in pearl millet. Olmos and Hellin (1997) observed that acid phosphatases are known to act under salt and water stress by maintaining a certain level of inorganic phosphate which can be co-transported with H^+ along a gradient of proton motive force. Hence, the plants in which the ACP increments were observed lower might be less suited in higher salt environment.

The present investigation revealed that a significant inverse correlation obtained between the concentration of the antioxidative enzymes, peroxidase and SOD with total protein in the case of *Bruguiera gymnorrhiza, Excoecaria agallocha* and *phoenix paludosa* in saline habitat. This elevation in the antioxidant enzyme concentration level may have taken place to scavenge more number of free radicals that are produced during stress (Davies, 2000, Dasgupta et al., 2010) and the decrease in protein concentration might be the result of formation of more compatible osmolytes to restore more negative water potential in cell sap. Both these phenomenon might provide some combat force to the plants against salinity stress. On the other hand no such statistical significant relationship between antioxidant enzymes and total protein concentration was found in case of *Heritiera fomes* and *Xylocarpus granatum.* This relationship, as discussed above may provide some important clue towards the proper salt management mechanism for sustainable existence in the hostile environment. Therefore, the absence of it might be one of the reasons towards less adaptibility for the plants in present situation. Though there are scopes yet to elucidate in detail regarding the significance of increment of these enzymes in salt imposed plants, the present work might provide the base-line information and a system necessary to conduct future research in relation to the genetic basis of salt tolerance.

5. References

Alim, A. 1979. Instruction manual for plantations in coastal areas. In: White, K. L. (Ed). Research considerations in coastal afforestation. Food and Agricultural Organization, UNDP/FAO Project BDG/72/005. Chittgong, Bangladesh: Forest Research Institute: 65–75.

Al-Hakimi, AMA and Hamada , AM. 2001. Counteraction of Salinity Stress on Wheat Plants by Grain Soaking in Ascorbic Acid, Thiamin or Sodium Salicylate. Biologia Plantarum. 44(2): 253-261.

Alscher, RG., Donahue, JL and Cramer CL. 2002 Reactive oxygen species and antioxidants: Relationships in green cells. Physiol. Plant., 100: 224-233.

Amirjani, MR. 2010. Effect of salinity stress on growth, mineral composition, proline content, antioxidant enzymes of soybean. Am. J. Plant Physiol. 5(6): 350-360.

Asada, K. 1994. Production and action of active oxygen species in photosynthetic tissues. In: Foyer, CH and Mullineaux, PM (eds.), Causes of Photooxidative Stress and Amelioration of Defense Systems in Plants. CRC Press, Boca Raton, Ann. Arbor, London, Tokyo, pp. 77–104.

Balen, B., Krsnik-Rasol M and Zadro, I. 2004. Eserase activity and isozymes in relation to morphogenesis in *Mammillaria gracillis* Pfeiff. tissue culture. Acta Bot. Croat. 63(2): 83–91.

Ball, MC. 1988. Ecophysiology of mangroves. Trees 2:129-142.

Banarjee, LK. 1999. Mangroves of Orissa coast and their ecology. Bishen Singh Mohendra Pal singh, Dehra Dun, pp 41.

Ben-Amor, N., Jimenez, A., Megdiche, W., Lundqvist, M., Silvilla, F and Abdelly, C. 2007. Kinetics of the antioxidant response to salinity in the helophyte *Cackile maritime* J. Integr. Plant Biol., 49: 982-992.

Bowles, DJ. 1990. Defense-related proteins in higher plants. Annual Review of Biochemistry. 59: 873-907.

Bray, EA. 1993. Molecular responses to water deficit. Plant Physiol. 103:1035–1040.

Chaffey, DR., Miller, FR and Sandom, KH. 1985. A forest inventory of the Sundarbans, Bangladesh. Land Resources Development Centre, Surrey. Pp. 196.

Chaves, MM., Flexas, J and Pinheiro, C. 2009. Photosynthesis under drought and salt stress: regulation mechanisms from whole plant to cell. Annals of Botany 103(4):551-560.

Cécile T, Francoise, L and Pierre, VC. 2006. Polymorphism and modulation of cell wall esterase enzyme activities in the chicory root during the growing season. J. Exp. Botany. 57(1): 81-89.

Cheeseman, JM. 1994. Depressions of photosynthesis in mangrove canopies. In: Baker NR, Bower JR (Eds.) Photoinhibition of photosynthesis: Molecular Mechanisms to the Field; Bios Scientific Publishers, Oxford. p 377-389.

Cheeseman, JM., Clough, BF., Carter, DR., Lovelock, CE., Eong, OJ and Sein, RG. 1991. The analysis of photosynthetic performance in leaves under field conditions: A case study using *Bruguiera* mangroves. Photosynthesis Research 29: 11-22.

Cheeseman, JM., Herendeen, LB., Cheeseman, AT and Clough, BF. 1997. Photosynthesis and photoprotection in mangroves under field conditions. Plant Cell and Environ. 20: 579–588.

Cheesman, JM and Lovelock, CE. 2004. Photosynthetic characteristics of dwarf and fringe *Rhizophora mangle* L. in Belizean mangrove. Plant Cell and Environment 27(6): 769-780.

Choudhury, JK. 1996. Mangrove forest management. Mangrove rehabilitation and management project in Sulawesi. p. 297.

Cowan, IR. 1982. Regulation of water use in relation to carbon gain in higher plants. In: Water Relations and Carbon Assimilation. Physiological Plant Ecology. II, Springer-Verlag, Berlin. p 589-614.

Curtis, SJ. 1993. Working plan for the forests of the Sundarbans Division for the period from 1st April 1931 to 31st March 1961, Calcutta, India. Bengal Government Press. 175 p.

Das, S and Mukherjee, KK. 1997. Morphological and biochemical investigations on Ipomea seedlings and their species interrelationships. Ann. Bot. 79: 565–571.

Das, S. 1999. An adaptive feature of some mangroves of Sundarbans, West Bengal. J. Plant Biol. 42(2): 109–116.

Dasgupta, N., Nandy, P., Tewari, C and Das, S. 2010. Salinity imposed changes of some isozymes and total leaf protein expression in five mangroves from two different habitats. *Journal of Plant Interactions* 5(3): 211-221.

Dasgupta, N., Nandy, P and Das, S. 2011. Photosynthesis and antioxidative enzyme activities in five Indian mangroves with respect to their adaptability. *Acta Physiologiae Plantarum* 33: 803-810.

Davies, KJA. 2000. Oxidative stress, antioxidant defenses and damage removal, repair and replacement systems. IUBMB Life 50: 279-289.

DeRidder, BP and Salvucci, M. 2007. Modulation pf Rubisco activase gene expression during heat stress in cotton (*Gossyoium hirsutum* L.) involves post-transcriptional mechanisms. Plant Sci., 172, 246-252.

Desingh, R and Kanagaraj, G. 2007. Influence of salinity strss on photosynthesis and antioxidative systems in two cotton varieties. Gen. Appl. Plant Physiol. 33(3-4):221-234.

Eyidogen, F and Oz, MT. 2007. Effect of salinity on antioxidant responses on chickpea seedlings. Acta. Physiol. Plant., 29: 485-493.

Fincher, GB. 1989. Molecular and cellular biology association with endosperm mobilization in germinating cereal grains. Annual Rev. Plant Physiol. Plant Mol. Biol., 40: 305–46.

Flowers, TJ., Troke, PF and Yeo, AR. 1977. The mechanism of salt tolerance in halophytes. Annu. Rev. Plant Physiol. 28: 89–121.

Food and Agricultural Organization (FAO). 2003. New global mangrove estimate. Available from: http://www.fao.org/documents/show_cdr.asp?url_file=/docrep/007/j1533e/j1533e52.htm.

Foyer, CH and Mullineaux, PM. 1994. Causes of photooxidative stress and amelioration of defense systems in plants. CRC Press, Boca Raton, Ann Arbor, London, Tokyo, pp. 343–364.

Frenkel, C. 1972. Involvement of Peroxidase and Indole-3-acetic Acid Oxidase Isozymes from Pear, Tomato, and Blueberry Fruit in Ripening. Plant Physiol. 49: 757-763.

Fridivich, I. 1989. Biological effects of the superoxide radical. Arch. Biochem. Biophys. 247: 1–11.

Greenway, H and Munns, R. 1980. Mechanisms of salt tolerance in nonhalophytes. Annu Rev Plant Physiol. 31: 149-190.

Halliwell, B and Gutteridge, JMC. 1985. Free Radicals in Biology and Medicine. Clarendon Press, Oxford.

Hasegawa, PM., Bressan, RA., Zhu, JK and Bohnert, HJ. 2000. Plant cellular and molecular responses to high salinity. Annu. Rev. Plant Physiol. Plant Mol. Biol. 51: 463–499.

Hassanein AM. 1999. Alterations in Protein and Esterase Patterns of Peanut in Response to Salinity Stress.Biologia Plantarum. 42(2): 241-248.

Hernandez, JA., Campillo, A., Jimenez, A., Alarcon, JJ and Sevilla, F. 1999. Response of antioxidant systems and leaf water relations to NaCl stress in pea plants. New Phytol., 141: 241-251.

Hernandez, JA., Jimenez, A., Mullineaux, P and Sevilla, F. 2000. Tolerance of pea (*Pisum sativum* L.) to long term salt stress is associated with induction of antioxidant defenses. Plant Cell Environ., 23: 853-862.

Hoagarth, PJ. 1999. The Biology of mangroves. Academic Press. London.

Hurkman, WJ., Fornia, CS and Tanaka, CK. 1989. A comparison of the effects of salt on polypeptides and translatable mRNAs in roots of a salt-tolerant and salt-sensitive cultivar of barley. Plant Physiol. 90:1444–1456.

Huttová, J., Tamás, L and Mistrik, I. 2002. Aluminium induced acid phosphatase activity in roots of Al-sensitive and Al-tolerant barley varieties. Roslinná Výroba 48(12): 556–559.

Jain A, Sharmai, AD and Singh, K. 2004. Plant Growth Hormones and Salt Stress-Mediated Changes in Acid and Alkaline Phosphatase Activities in the Pearl Millet Seeds. Int. J. Agri. Biol., 6(6): 960-963.

Kalir, A., Omri, G and Poljakoff-Matber, A. 1984. Peroxidase and catalase activity in leaves of Halimione portulacoides exposed to salinity. Physiologia Plantarum. 62: 238-244.

Karim, A. 1988. Environmental factors and the distribution of mangroves in Sundarbans with special reference to *Heritiera fomes* Buch. Ham. Ph. D Thesis, University of Calcutta, Calcutta, India.

Karim, A. 1994. Vegetation. Mangroves of the Sundarbans. Vol. 2. Bangladesh (eds. Z. Hussain and G. Acharya), pp. 43-74, IUCN, Bangkok, Thailand.

Kathiresan, K and Moorthy, P. 1993. Influence of different irradiance on growth and photosynthetic characteristics in seedlings of *Rhizophora* species. *Photosynthetica* 29: 143-146.

Kathiresan, K and Moorthy, P. 1994. Photosynthetic responses of *Rhizophora apiculata* Blume seedlings to a long-chain aliphatic alcohol. Aquat. Bot. 47:191-193.

Keith, H., Emmanuelle, V., Hélène, B and Charistian, A. 1983. Superoxide dismutase assay using alkaline dimethylsulfoxide as superoxide anion-generating system. Analyt. Biochem. 135: 280–287.

Kishor, PBK, Hong, Z., Miao, G., Hu, CA and Verma, PS. 1995. Overexpression of pyrroline-5-carboxylate synthetase increases proline production and confers osmotolerance in transgenic plants. Plant Physiol. 108: 1387-1394.

Krauss, KW and Allen, JA. 2003. Influence of salinity and shade on seedling photosynthesis and growth of two mangrove species *Rhizophora mangle* and *Bruguiera sexangula*, introduced to Hawaii. Aquat. Bot. 77(4): 311-324.

Laemmli, UK. 1970. Cleavage of structural proteins during the assembly of the head of bacteriophage T4. Nature 227:680–685.

Lawlor, DW and Cornic, G. 2002. Photosynthetic carbon assimilation and associated metabolism in relation to water deficits in higher plants. Plant Cell Environ. 25, 275-294.

Li, Y. 2009. Effect of NaCl stress on antioxidative enzymes of glycine soja sieb. Pak. J. Biol. Sci., 12: 510-513.

Lowrey, OH., Rosebrough, N., Farr, AL and Randall, RJ. 1951. Protein measurements with folin phenol reagent. J. Biol. Chem. 193:265-275.

Macfarlane, GR and Burchett, MD. 2001. Photosynthetic pigments and peroxidase activity as indicators of heavy metal stress in the grey mangrove, *Avicennia marina* (Forsk.) Vierh. Marine Pollution Bulletin 42(3): 233–240.

Mader, M. 1976. Die Localization der Peroxidase Iso-. enzymgruppe G, in der Zellwand von Tabak-Geweben. Planta 131: 11–15.

Mansour, MMF. 2000. Nitrogen containing compound and adaptation of plants to salinity stress. Biol. Plantarum 43(3): 491-500.

Mansour, MMF., Salama, KHA., Ali, FZM and Abou Hadid, AF. 2005. Cell and plant responses to NaCl in *Zea mays* L. cultivars differing in salt tolerance. Gen. Applied Plant Physiol., 31: 29-41.

Meloni, DA., Oliva, MA., Martinez, CA and Cambraia, J. 2003. Photosynthesis and activity of superoxide dismutase, peroxidase and glutathione reductase in cotton under salt stress. Environ. Exp. Bot. 49, 69-76.

Naidoo, G., Tuffers, AV and von Willert, DJ. 2002. Changes in gas exchange and chlorophyll fluorescence characteristics of two mangroves and a mangrove associate in response to salinity in the natural environment. Trees 16(2-3): 140-146.

Nandy (Datta), P and Ghose, M. 2001. Photosynthesis and water-use efficiency of some mangroves of Sundarbans, India. J. Plant Biol. 44: 213-219.

Nandy (Datta), P., Das, S., Ghose, M and Spooner-Hart, R. 2007. Effects of Salinity on Photosynthesis, Leaf Anatomy, Ion Accumulation and Photosynthetic Nitrogen Use Efficiency in Five Indian Mangroves. Wetl. Ecol. and Manage. 15: 347–357.

Nandy (Datta), P., Das, S and Ghose, M. 2005. Relation of leaf micromorphology with photosynthesis and water efflux in some Indian mangroves. Acta Botanica Croatica 64 (2): 331-340.

Naskar, KR and Guha Bakshi, DN. 1983. A brief review on some less familiar plants of the Sundarbans India. J. Econ. and Taxonomic Bot. 4(3): 699–712.

Olczak, M., Kobialka, M and Watorek WX. 2000. Characterization of diphosphonucleotide phosphatase/phosphodiesterase from yellow lupin (Lupinus luteus) seeds. Biochimica et Biophysica Acta. 1478(2): 239-247.

Pareek, A., Singla, SL and Grover, A. 1997. Salt responsive proteins/genes in crop plants. *In*: Jaiwal PK, Singh RP, Gulati A (eds.): Strategies for improving salt tolerance in higher plants. Pp. 365-391. Oxford and IBH Publishing Co. New Delhi, India.

Parida, A., Das, AB and Das, P. 2002. NaCl stress causes changes in photosynthetic pigments, proteins and other metabolic components in the leaves of a true mangrove, *Bruguiera parviflora*, in hydroponic cultures. J. Plant Biol. 45:28–36.

Parida, AK., Das, AB and Mohanty, P. 2004a. Defense potentials to NaCl in a mangrove, *Bruguiera parviflora*: Differential changes of isoforms of some antioxidative enzymes. J. Plant Physiol. 161(5): 531–542.

Parida, AK., Das, AB and Mohanty, P. 2004b. Investigations on the antioxidative defence responses to NaCl stress in a mangrove, *Bruguiera parviflora*: Differential regulations of isoforms of some antioxidative enzymes. Plant Growth Regulation. 42(3): 213–226.

Parida, AK., Das, AB and Mitra, B. 2004. Effect of salt and growth, ion accumulation, photosynthesis and leaf anatomy of the mangrove, *Bruguiera parviflora.* Trees 18: 167-174.

Prochazkova, D and Wilhelmova, N. 2007. Leaf senescence and activities of the antioxidant enzymes. Biol. Plantarum. 51:401-406.

Rajesh, A., Arumugam, R and Venkatesalu, V. 1999. Response to *ceriops roxburghiana* to NaCl stress. Biol. Pantarum 42(1): 143-148.

Raymond, P., Broquisse, R., Chevalior, C., Couee, I., Dieuaide, M., James, F., Just, D and Pradet A. 1994. Proteolysis and proteolytic activities in the acclimation to stress: the case of sugar starvation in maize root tips. *In*: Cherry JH (ed.) Biochemical and cellular mechanisms of stress tolerance in plants. Pp. 325-334. Springer-Verlag, Berlin, Germany.

Sahu, AC and Mishra, D. 1987.Changes in some enzyme activities during excised rice leaf senescence under NaCl - stress. Biochemie and Physiologic der Pflanzen.182: 501-505.

Shannon, LM., Key, E and Lew, JY. 1966. Peroxidase isozyme from Horseraddish root. I. Isolation and physical properties. J.Biol. Chem. 249(9): 2166–2172.

Stonier, T and Yang, Hsin-Mei. 1973. Studies on Auxin Protectors: XI. Inhibition of Peroxidase-Catalyzed Oxidation of Glutathione by Auxin Protectors and *o* Dihydroxyphenols. Plant Physiol. 51: 391–395.

Swire-Clark, GA and Marcotte, WR. Jr. 1999. The wheat LEA protein Em functions as an osmoprotective molecule in Saccharomyces cerevisiae. Plant Mol. Biol. 39:117–128.

Takemura, T., Hanagata, N., Sugihara, K., Baba, S., Karube, I and Dubinsky, Z. 2000. Physiological and biochemical responses to salt stress in the mangrove, *Bruguiera gymnorrhiza*. Aquat. Bot. 68: 15–28.

Tausz, M., Sircelj, H and Grill, D. 2004. The glutathione system as a stress marker in plant ecophysiology: is a stress response concept valid? J. Exp. Bot. 55: 1955-1962.

Traczynski, MC., Jensen, RG and Bohnert, HJ. 1993. Stress protection of transgenic tobacco by production of the osmolyte msnnitol. Science, 259: 508-510.

Upadhyay, V.P., Rajiv Ranjan and Singh, JS. 2002. The Human Mangrove Conflicts-The way out. Current Science. 83: 1328-1336.

UPOV. 1997. Adopted report of the Technical committee. Thirty third Session, Geneva, October 16–18, 1996. Union for Protection of New Varieties of Plants TC/33/11.

Walter, H. 1971. Ecology of tropical and subtropical vegetation. Oliver and Boyed, Edinburgh, Scotland.

Xu, D., Duan, X., Wang, B., Hong, B., Ho, TDH and Wu, R. 1996. Expression of a late embryogenesis abundant protein gene, HVA1, from barley confers tolerance to water deficit and salt stress in transgenic rice. Plant Physiol. 110:249–257.

Xu, XY., Abo, M., Okubo, A and Yamazaki, S. 2001. Salt-stress responsive membrane proteins in *Rhodobacter sphaeroides* f. sp *denitrificans* IL106. J Biosc Bioeng. 91:228–230.

Ye, Y., Tam Nora, FY., Wong, YS and Lu, CY. 2003. Growth and physiological responses of two mangrove species (*Bruguiera gymnorrhiza* and *Kandelia candel*) to water logging. Envrion. Exp. Bot. 49(3): 209–221.

Zeidler, M. 2000. Electrophoretic analysis of plant. Acta Univ.Palacki. Olomuc. Fac. Rerum Nat. Biol. 38: 7–16.

Zimmermann, MH. 1983. Xylem structure and the ascent of sap, Springer-Verlag, Berlin.

17

Submergence Tolerance of Rice Species, *Oryza glaberrima* Steudel

Jun-Ichi Sakagami
Japan International Research Center for Agricultural Sciences
Japan

1. Introduction

Oryza glaberrima, an African monocarpic annual rice derived from *Oryza barthii,* is grown in traditional rice producing wetland areas of West Africa. *Oryza sativa,* an Asian rice that varies from annual to perennial, is derived from *Oryza rufipogon* (Sakagami et al., 1999). Genotypes of *O. glaberrima* are inherently lower yielding than those of *O. sativa* and are therefore cultivated in fewer areas (Linares, 2002). However, because they grow adequately in unstable environments such as those with water stress, they appear to tolerate severe environmental stress. Flooding imposes severe selection pressure on plants, principally because excess water in the plant surroundings can deprive them of certain basic needs, notably of oxygen and of carbon dioxide and light for photosynthesis. It is a major abiotic influence on species' distribution and agricultural productivity world-wide. Strong submergence-induced elongation is a widespread escape mechanism that helps submerged plants regain or retain contact with the aerial environment on which they depend (Arber, 1920). This mechanism enables plants to resume anaerobic metabolism and photosynthetic fixation of CO_2 by raising their shoots above water. Escape strategies based on elongation by stem or leaves are prominent characteristics of deep-water and floating rice. However, rapid elongation by leaves of young plants in response to short-term submergence flash flood (for up to 2 weeks) adversely affects tolerance by depleting carbohydrates that would otherwise support survival during and after submergence (Chaturvedi et al., 1995; Setter & Laureles, 1996; Kawano et al., 2002; Ram et al., 2002: Jackson & Ram, 2003; Joho et al., 2008). The submergence tolerance gene, *Sub-1A,* depresses shoot elongation under short-term submergence to ensure survival. Submergence-tolerant rice varieties tend to accumulate more starch in their stem section than susceptible varieties do. They experience less carbohydrate depletion after submergence (Karin et al., 1982; Emes et al., 1988). To improve the circumstances of tolerant plants to survive under flooding conditions is a major constraint for sustainable agriculture in unstable environments undergoing climate change. Consequently, in this chapter, we describe physiological mechanisms related to photosynthesis on submergence tolerance for rice species that are widely cultivated in West Africa.

2. Physiological mechanism on flooding tolerance in rice species

Kawano (2009) showed that suppression of underwater elongation brought about by the mutated form of *Sub-1A* in *O. sativa* is beneficial for the endurance of complete submergence. Consequently, non-shoot-elongation-cultivars during submergence show tolerance to short-term submergence, so-called flash flooding, for a few days or weeks. Sakagami et al. (2009) emphasized that this trait is inappropriate when selecting and breeding cultivars of *O. sativa* or *O. glaberrima* in cultivated rice for resilience to longer term submergence. Under these circumstances, a vigorous ethylene-mediated underwater elongation response by leaves is necessary to return leaves to air contact and full photosynthetic activity for long-term complete submergence.

2.1 Anaerobic metabolism in submerged rice plants

The rate of gas exchange is very slow in water because of small diffusion coefficients for gases (oxygen, 0.201 cm^{-2} s^{-1} in air; 2.1×10^{-5} cm^{-2} s^{-1} in water) (Armstrong, 1979). When water becomes stagnant, the oxygen concentration becomes especially low at night because of the nighttime respiration of algae. Rice plants increase the rate of alcoholic fermentation under low oxygen environments. However, alcoholic fermentation produces only two molecules of ATP per glucose molecule, which is not efficient when compared with aerobic respiration, through which 32 molecules of ATP are produced per glucose molecule. Therefore, rice cannot survive in a low oxygen environment for a long period because of the shortage of carbohydrates in the rice plants for use in energy production. Furthermore, photosynthesis is limited by low irradiance when the plant is submerged. It is necessary to improve photosynthetic capacity and the effective use of photosynthetic products as well as to survive under water.

Strategy	Quiscence	Escape	
Submergence tolerance	Slowing of ethylen-promoted leaf elongation to conserve energy	Rapid leaf elongation	Rapid internodal or stem elongation to resume anaerobic metabolism and photosynthesis
Ecological adaptation	Flash floods less than two weeks	Short-term submergence (Shallow-water)	Long-term submergence (Deep-water)
Gene expression(e.g)	*SUB1A*	*SUB1C*	*SNORKEL1, 2*
Carbohydrate consumption	Low (limited by *Sub1A*)	High	High

Table 1. Strategy by submergence tolerance of rice

Rice has adapted to submergence-prone environments through the use of two strategies (Table 1): submergence tolerance to flash floods where a rapid increase in water level causes partial to complete submergence for up to 2 weeks, and shoot elongation to short to long term submergence. *Sub1A* gene in *O. sativa* reportedly confers submergence tolerance to flash foods through a quiescence strategy in which cell elongation and carbohydrate metabolism in young seedlings is repressed during submergence (Fukao et al., 2006). This strategy is a predominant tolerance mechanism that is driven by adjustment of metabolism.

A strategy with shoot elongation shows two different mechanisms: rapid shoot elongation in shallow floods in a short-term submergence and internodal or stem elongation in deep water in long-term submergence. Based on our analysis, most *O. glaberrima* varieties adapt well when floods are deeper and when they entail long-term submergence (Fig. 1). These mechanisms for plant survival under submergence are affected by the conservation of energy and carbohydrate accumulation (Perata et al., 2007).

Fig. 1. Growing rice of *O. glaberrima* along the Niger River in Niger

2.2 Submergence tolerance with elongation for deep water

Rapid shoot elongation for young seedlings is usually disadvantageous in conditions of short-term submergence with deep water conditions because lodging usually occurs once floodwaters recede. This water regime adapts well, using submergence tolerance with a quiescence strategy. By tolerance, cell elongation and carbohydrate metabolism are repressed. Furthermore, fast shoot elongation can restore contact between the leaves and air, but it can also result in death if carbohydrate reserves are depleted before emergence in leaves above the water surface. Leaf elongation during submergence is controlled by the interaction of at least three plant hormones: ethylene, GA, and ABA (Kende et al., 1998). Accumulated ethylene is probably the primary signal which triggers the plant to start a cascade of reactions leading to enhanced cell elongation (Voesenek et al., 2006) because ethylene is accumulated in rice plants during submergence because of the fact that gas diffusion is 10^4-fold slower in solution than in air (Armstrong, 1979). The cascade model was proposed from the study of stem elongation in deepwater rice (Kende et al., 1998).

2.2.1 Submergence escapes mechanism with shoot elongation

Rapid elongation of the leaves and leaf sheath is advantageous for rainfed lowland varieties because it enables them to avoid submergence stress when moderate flooding occurs during the early vegetative stage. Deepwater rice is often characterized as floating rice. Nevertheless, the differences in characteristics of floating rice and deepwater rice remain unclear. In fact, the physiological mechanisms of growth differ between the two. Some rice plants can survive and stand without floating in water at 1 m water depth. In this chapter, such rice plants that stand without floating in water are designated as deepwater rice to distinguish them from floating rice. In general, the plant height of deepwater rice reaches 140–180 cm in the absence of submergence (Catling, 1992), but the abilities of deepwater rice shoots to extend are varied. Deepwater rice can maintain an aerobic metabolism during submergence via development of its canopy above water because of the elongation of its internodes and because of its long leaves. Deepwater rice's ability to elongate in a single day is less than that of floating rice. However, deepwater rice can adapt to submergence under conditions in which the water level increases 5 cm per day (Catling, 1992). However, this type of tall plant architecture often causes lodging after the water recedes.

2.2.2 Internode elongation

Setter et al. (1988) demonstrated that the adverse effects are caused mainly by reduced photosynthesis capacity because of CO_2 starvation in the shoot organs during submergence. Furthermore, they suggested a relation between ethylene concentration, leaf chlorosis and leaf elongation. Partial submergence treatment to deep water rice never affects carbohydrate and sugar contents in newly developed leaves under the water compared to the control (Setter et al., 1987). Elongation with floating ability is the most important morphological feature of deepwater rice. In particular, internode elongation is a more important mechanism for increasing shoot length. Internode elongation is related closely to plant hormones. Submergence lowers the O_2 level in rice internodes. Then low O_2 levels simulate ethylene synthesis. Ethylene accumulation occurs in the submerged internodes. Then high internodal ethylene concentration increases the sensitivity of tissues to gibberelic acid or increases the concentration of physiologically active gibberellins, thereby leading to commonly observed growth responses (Rose-John & Kende, 1985). Deepwater rice differs in its ability to accumulate carbohydrate contents within the cultivar's carbohydrate content, which does not correlate with the total internode length or plant length (Vergara et al., 1975).

3. Flooding response of *O. glaberrima*

O. glaberrima, a monocarpic annual derived from *O. barthii* (Sakagami et al., 1999), is grown in traditional rice production in the wetlands of West Africa. It is highly adapted to deepwater inundation in countries such as Gambia, Guinea, Mali, Niger, Senegal, and Sierra Leone in West Africa (Inouye et al., 1989). The first gene pool of *O. glaberrima* was inferred as an inland delta of the Niger River because of the high gene diversity among species. In Guinea, for example, coastal or lowland areas are heavily affected by submergence during the rainy season. Rice plants are often partially or completely submerged for more than a month. Such prolonged submergence often triggers crop failures. Guinea's farmers prefer to cultivate *O. glaberrima* fields with prolonged submergence because of such advantageous traits as those explained above. Cultivars of *O. glaberrima* are roughly divisible into two ecotypes: upland and lowland. However, it might be that *O. glaberrima* is a valuable rice species for flooding conditions in all cases. Tolerance of other abiotic and biotic stress such as drought (Maji et al., 2010), rice yellow mottle virus (Thiemele et al., 2010), African rice gall midge (Nwilene et al., 2009), and iron toxicity (Majerus et al., 2007) has been found in some cultivars of *O. glaberrima*. However, it is vulnerable to NaCl salinity (Awala et al., 2010), grain shattering (Koffi, 1980), and lodging (Dingkuhn, 1998). It is reasonable to presume that the indigenous cultivated species of African rice can provide useful genes for improvement of tolerance to major stress in Africa.

3.1 Responses to short-term submergence "flash flood"

The flooding response of *O. glaberrima* should be discussed thoroughly, but it is not clear from Futakuchi's report (2001) whether shoot elongation contributes to flooding tolerance in different water regimes or not. To elucidate the physiological responses of young rice plants to short-term submergence stress, so-called flash flooding, under rainfed conditions for *O. glaberrima* by comparison with several genotypes for lowland adapted, deepwater adapted shoot elongated escape and *Sub1* of *O. sativa*, 30-day-old seedlings were submerged completely for 10 d at 45 cm water depth at 13 d after transplantation in a lowland field (Joho et al., 2008). In fact, *O. glaberrima* showed higher shoot elongation ability during submergence than any genotype of *O. sativa* that we tested. However, *O. glaberrima* lodged easily after the end of submergence because of longer and more rapid shoot elongation during submergence. Therefore, it triggered a decrease in its survival rate (Fig. 2).

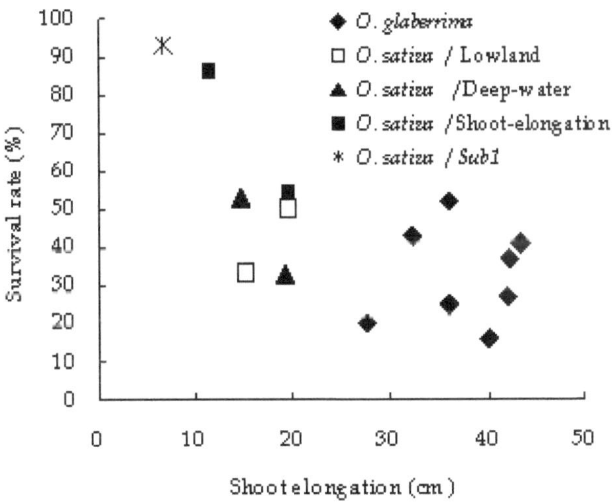

Fig. 2. Effect of shoot clongation during submergence on survival rate after desubmergence. Survival rate is observed at 14 d after desubmergence.

The submergence-tolerant genotype (*Sub1*) of *O. sativa* maintained the dry matter weight of the leaf blade during submergence through the inhibition of shoot elongation using the quiescence strategy, thereby attaining a survival rate of 93%. The escape strategy for *O. glaberrima* is therefore the effective usage of stored carbohydrates for shoot elongation in a severely photosynthesis-limited environment. However, failure to regain contact with air and the oxygen, carbon dioxide, and light it supplies invariably gives rise to severe carbohydrate depletion. Therefore, this escape strategy carries a high risk for young rice plants (Kawano et al., 2009). We reported that *O. glaberrima* is susceptible to short-term submergence, although it might adapt to prolonged flooding because of improved restoration of aerial

Genotype	Non-submergence		Complete-submergence	
	Shoot elongation ($cm\ d^{-1}$)	Shoot biomass increase ($g\ d^{-1}$)	Shoot elongation ($cm\ d^{-1}$)	Shoot biomass increase ($g\ d^{-1}$)
O. sativa L.				
BA8A	1.38	0.41	1.78	0.04
Balante	0.94	0.34	1.17	0.02
Banjoulou	0.95	0.32	1.23	0.02
Cinquant-deux	1.25	0.55	2.12	0.16
CK20	1.45	0.37	1.58	0.03
CK211	1.32	0.40	1.82	0.04
CK4	1.31	0.33	1.49	0.03
CK41	1.52	0.41	1.71	0.04
Danta rouge	1.54	0.48	2.43	0.07
EH-IA-CHIU	1.16	0.42	1.39	0.02
FR13A*	1.19	0.38	Death	Death
Gallale Blanc	0.65	0.24	1.89	0.12
Haïra koreye	1.43	0.45	2.15	0.05
IR49830-7-1-2-2	0.97	0.40	1.40	0.02
IR62293-2B-18-2-2-1-3-2-3	1.59	0.47	1.61	0.04
IR67520-B-14-1-3-2-2*	1.01	0.35	Death	Death
IR70027-8-2-2-3-2*	1.50	0.46	Death	Death
IR71700-247-1-1-2	1.19	0.42	1.65	0.02
IR73018-21-2-B-2-B*	0.95	0.35	Death	Death
IR73020-19-2-B-3-2B*	1.32	0.29	Death	Death
Kaolac	1.23	0.36	1.60	0.03
Kaorin	0.99	0.42	1.55	0.02
Köticondre	1.03	0.49	1.42	0.03
Marsal	1.09	0.45	1.72	0.05
N 22	1.26	0.43	1.69	0.05
N'ckrome	0.97	0.46	1.70	0.04
NIK 1A	1.17	0.53	1.30	0.02
Nylon	1.18	0.32	1.58	0.02
Protocolo	1.47	0.41	2.10	0.04
Reymont	1.03	0.22	1.21	0.01
ROK21	1.22	0.20	1.60	0.03
SHAI-KUH	1.08	0.38	1.29	0.03
Vandana	1.46	0.41	1.65	0.03
WAR1(ROK22)	1.26	0.32	1.58	0.03
Wonsongg orgle	0.88	0.26	1.53	0.02
O. glaberrima Steud.				
Aawba	1.14	0.32	1.84	0.08
Bakin Iri	1.06	0.51	1.87	0.12
CG14	0.97	0.40	1.67	0.07
Dam Iri	1.25	0.57	2.02	0.16
Dembou bourawana blanc	0.75	0.18	1.99	0.09
Djéifata noir	0.96	0.36	2.25	0.11
Djingua noir	0.76	0.35	1.91	0.14
Douboutou II	0.91	0.36	1.89	0.07
Gbagaye	1.10	0.34	1.95	0.06
Gbobaye	1.13	0.46	1.87	0.06
Kossa barkaneye	1.26	0.44	1.96	0.11
Mala Noir II	0.99	0.31	2.08	0.13
Mala Noir III	0.89	0.32	2.24	0.16
Mogo	1.28	0.47	2.24	0.14
Mokori	1.05	0.36	2.25	0.10
Pegnesso	0.71	0.23	2.06	0.07
RAM23	0.70	0.44	2.13	0.14
Salifore	1.39	0.49	1.80	0.08
Saligbeli	1.23	0.38	2.00	0.10
Salikutaforé	1.14	0.30	1.68	0.04
Samandényi	1.08	0.20	1.62	0.07
Sukéré	1.42	0.42	1.90	0.15
Tierka banc	0.91	0.31	2.06	0.13
Tombobökéri II	1.25	0.30	1.95	0.11
W0492	1.87	0.49	1.87	0.06
Wana thireye	1.25	0.42	2.06	0.13
Yélé 1A	0.63	0.32	1.96	0.17
Average(±SE)				
O. sativa L.(n=30)	1.20±0.04	0.39±0.02	1.63±0.05	0.04±0.01
O. glaberrima Steud.(n=27)	1.08±0.05	0.37±0.02	1.97±0.03	0.10±0.01
O. sativa x *O. glaberrima*	NS	NS	**	**

*Genotypes are charactatised by *Sub1*

Table 2. Effect of submergence to shoot elongation and biomass in the field experiment photosynthesis and survival rate through shoot elongation ability. Enhancement of shoot elongation during submergence in water that is too deep to permit re-emergence by small seedlings represents a futile escape strategy that is used at the expense of existing dry matter in circumstances where underwater photosynthetic carbon fixation is negligible. Consequently, it compromises survival or recovery growth once floodwater levels recede and plants are exposed again to the aerial environment. Consequently, shoot elongation capability to revert to anaerobic growth condition is vital for long-term flood survival.

3.2 Responses to long-term submergence "deep water"

Various lines of 35 *O. sativa* and 27 *O. glaberrima*, including some classified as short-term submergence tolerant, were compared for submergence tolerance in field and pot experiments to long-term submergence tolerant varieties in other words, deepwater varieties (Sakagami et al., 2009). Plants were submerged completely for 31 d in a field experiment, and partially or completely for 37 d in a pot experiment in a growth chamber. Leaf elongation and growth in shoot biomass during complete submergence in the field were significantly greater in *O. glaberrima* than in *O. sativa* (Table 2).

Submergence-tolerant cultivars of *O. sativa* were unable to survive prolonged complete submergence for 31–37 d, which indicates that the mechanism of suppressed leaf elongation that confers increased survival of short-term submergence is inadequate for surviving long periods underwater. The *O. sativa* deepwater-adapted cultivar 'Nylon' and the 'Yele1A' cultivar of *O. glaberrima* succeeded in emerging above the floodwater. The photosynthetic rate was higher in deeply submerged plants than in non-submerged plants. The photosynthetic rate at 37 d after submergence in partial and complete submergence was closely related to the net assimilation rate during submergence (Fig. 3), which caused greatly increased shoot length, shoot biomass and leaf area, in association with an increased net assimilation rate compared with the lowland-adapted *O. sativa* 'Banjoulou'.

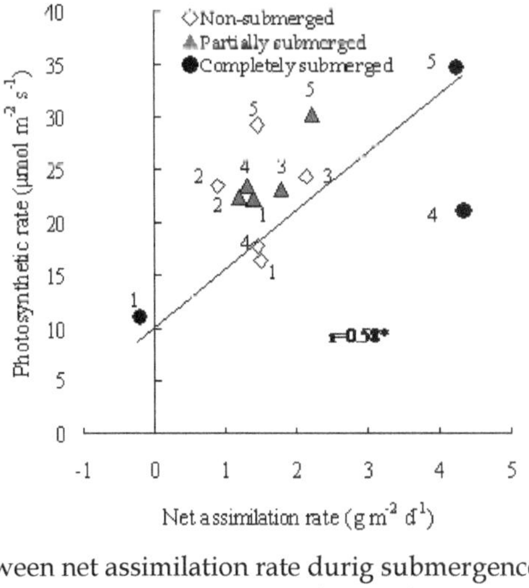

Fig. 3. Relathinship between net assimilation rate durig submergence and photosynthetic rate after 37 d submergence in a pot experiment. The number next each symbol indicate the cultivars: 1, Banjoulou; 2, IR71700; 3, IR73020; 4, Nylon; 5, Yele1A. Net assumilation rate indicates the increase of dry weight per unit area during 37 d submergence.

The superior tolerance of deepwater *O. sativa* and *O. glaberrima* genotypes to prolonged complete submergence appears to be attributable to their greater photosynthetic capacity developed by leaves that had newly emerged above the floodwater. Vigorous upward leaf elongation during prolonged submergence is therefore critical for ensuring shoot emergence from water, as are leaf area extension above the water surface and a subsequent strong increase in shoot biomass.

Actually, 'Yele1A' had an especially large capacity for shoot elongation when submerged. Watarai and Inoue (1998) noted that high internodal elongation contributes to shoot elongation using *O. glaberrima* under flooding regimes. Faster shoot elongation of *O. glaberrima* genotypes underwater is mainly caused by leaf elongation, but not internodal elongation. Consequently, internode and leaf elongation underwater share certain similarities in *O. glaberrima*, both presumably being stimulated by ethylene.

3.3 Unique physiological mechanism to complete submergence of "Saligbeli"

Lodging, plant height, and dry matter accumulation for 99 cultivars in *O. sativa*, *O. glaberrima*, and interspecific hybridization progenies (IHP) were measured when 12-day-old seedlings were submerged for 7 days in pots and in fields. Upland rice (*O. sativa*) showed greater shoot elongation, greater reduction in dry matter accumulation during submergence, and higher lodging, which indicate low flash flood tolerance. The physiological traits of most *O. glaberrima* and upland rice (*O. sativa*) for resistance against flash flooding were opposite those of submergence-tolerant cultivars, as evidenced from the results of a principal component analysis (Fig. 4). Axis I is the first principal component.

$$Y=-0.403942x_1 + 0.434866x_2 + 0.329416x_3 - 0.271996x_4 \quad (1)$$

Axis II is the second principal component.

$$Y=-0.068947x_1 - 0.080874x_2 + 0.618871x_3 - 0.772613x_4 \quad (2)$$

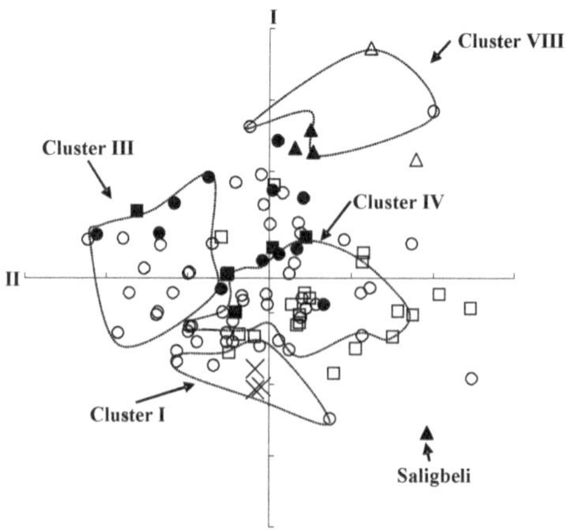

Fig. 4. Principal component analysis of physiological traits linked to submergence. (●)Upland *sativa*, (○)Lowland *sativa*, (▲)Upland *glaberrima*, (△)Lowland *glaberrima*, (■)Upland IHP, (□)Lowland IHP, (×) Submergence tolerance(*Sub1*)

x_1, x_2, x_3 and x_4 in (1), (2) represent in dry matter accumulation after desubmergence, lodging score, shoot elongation and increase in dry matter accumulation during submergence respectively. It is accounted for 74.0% of the total number of genotypes with the first and second principal components.

In Cluster I, III, and VIII, the main genotypes belonging to each cluster group were classified on the principal component analysis. Cluster I, including submergence tolerance genotype, and Cluster VIII, including *O. glaberrima*, were positioned in opposite regions.

The physiological response of Saligbeli cultivar differed from those of other *O. glaberrima* genotypes in terms of submergence tolerance. Saligbeli was found by the author in coastal regions in Guinea. Saligbeli exhibited enhanced shoot elongation with increased dry matter accumulation after the end of submergence, as was found also for the submergence-tolerant cultivar in a pot experiment (Table 3). The difference between pot and field experiments might be attributable to different characteristics of the submergence environment, such as turbidity. These features of Saligbeli were apparently a unique means to cope with submergence. These experiments revealed that enhancement of shoot-elongation during submergence are accomplished using dry matter of leaves that had developed before submergence.

Species	Geniotype	Shoot elongation (cm)$^{1)}$	Increase of DMW (mg plant^{-1}) During submergence (7d)	After desubmergence (14d)	Ratio of DMA $^{2)}$	Lodging score$^{3)}$
O. glaberroma	Aawba	16.6	-6.4	-18.2	0.18	3
	Saligbeli	**12.1**	**13**	**59.2**	**0.99**	**2**
	Samandenyi	15.1	-3.2	-4.0	0.34	6
	Sedou Bayebeli	13.2	-2.0	-19.9	0.12	6
	CG14	19.5	2.6	0	0.30	6
	DouboutouII	34.6	-8.6	-21.4	0.07	6
O. sativa (Sub1)	IR70027-8-2-2-3-2	3.1	4.0	31.0	0.99	1
	IR73020-19-2-B-3-2B	1.2	6.0	23.2	0.98	1
	IR49830-7-1-2-2	4.6	2.4	26.8	0.84	1

1) Increase of plant height during submergence, 2) Ratio of dry matter accumulation (DMA) was determined by dividing the submergence in the control, and 3) Score 7 is the highest and 1 is the lowest in lodging degree after desubmergence.

Table 3. Physiological traits linked to submergence tolerance in *O. glaberrima* and *O. sativa* of *Sub1*

4. Conclusion

Submerged rice is in an anaerobic environment because of the 10^4-fold slower gas diffusion underwater than in air. Furthermore, levels of oxygen, and carbon dioxide and light for photosynthesis drastically differ according to the floodwater period, depth, temperature, and turbidity. African rice, *O. glaberrima* can lodge readily under aerobic conditions after desubmergence because of weakening of the shoot base, which causes rapid leaf elongation and which increases plant mortality through photosynthetic products accumulated before submergence is exhausted under short-term submergence with the rapid increase of water level: so-called flash flooding. However, cultivars of *O. glaberrima* adapt to long-term complete submergence apparently because of their greater photosynthetic capacity developed by leaves that have newly emerged above floodwaters through rapid shoot elongation. The Saligbeli cultivar of *O. glaberrima*, with its unique physiological mechanisms, is apparently well-adapted to both conditions for short and prolonged submergence. It therefore holds promise as a selecting and breeding rice genotype for use in different flood-prone regions in Africa.

5. Acknowledgment

I am very grateful to O. Ito for continued support and stimulating discussion. I thank N. Kawano, Y. Joho, and A. Cisse for support with several experiments in Guinea and Japan.

6. References

Armstrong, W. (1979). Aeration in higher plants. In *Advances in Botanical Research* Vol.7, Woolhouse, H.W.W. (Ed.), pp. 225–332, Academic Press, ISBN 9780120059072, London, UK

Arber, A. (1920) *Water Plants: a study of aquatic angiosperms.* Cambridge University Press. ISBN 1108017320, Cambridge UK

Awala, S.K.; Nanhapo, P.I.; Sakagami, J-I.; Kanyomeka, L. & Iijima, M. (2010). Differential Salinity Tolerance among *Oryza glaberrima, Oryza sativa* and Their Interspecies Including NERICA. *Plant Production Science* Vol.13, No.1, (January 2010), pp. 3–10, ISSN 1343-943X

Catling, D. (1992). *Rice in deep water*, Palgrave Macmillan, ISBN 0333549791, London, UK.

Chaturvedi, G.S.; Mishra, C.H.; Singh, O.N.; Pandey, C.B.; Yadav, V.P.; Singh, A.K.; Dwivedi, J.L.; Singh, B.B. & Singh, R.K. (1995). Physiological basis and screening for tolerance for flash flooding. In: *Rainfed lowland rice: agricultural research for high-risk environments*, Ingram K.T., (Ed.) pp. 79–96, International Rice Research Institute, ISBN 971-22-0067-1, Los Banos, Philippines

Dingkuhn, M.; Jones, M.P.; Johnson, D.E. & Sow, A. (1998). Growth and yield potential of *Oryza sativa* and *O. glaberrima* upland rice cultivars and their interspecific progenies. *Field Crops Research Vol.*57, No.1, (May 1998), pp. 57–69, ISSN 0378-4290

Emes, M.J.; Wilkins, C.P.; Smith, P.A.; Kupkanchanakul, K.; Hawker, K.; Charlton, W.A. & Cutter, E.G. (1988). Starch utilization by deepwater rice during submergence. *Proceedings of the 1987 International Deepwater Rice Workshop*, pp. 319–326, ISBN 971-104-187-1, Bangkok, Thailand, October 1987.

Fukao, T.; Xu, K.; Ronald, P.C. & Bailey-Serres, J. (2006). A Variable Cluster of Ethylene Response Factor-Like Genes Regulates Metabolic and Developmental Acclimation Responses to Submergence in Rice. *Plant Cell* Vol.18, No.8, (August 2006), pp. 2021–2034, ISSN 1040-4651

Futakuchi, K.; Jones, M.P. & Ishii, R. (2001). Physiological and morphological mechanisms of submergence resistance in African rice (*Oryza glaberrima* Steud.). *Japanese Journal of Tropical Agriculture* Vol.45, No.1, (March 2001), pp. 8–14, ISSN 0021-5260

Inouye, J.; Hakoda, H. & Ng, N.Q. (1989). Preliminary studies on some ecological characteristics of African deep water rice (*Oryza glaberrima* Steud.). *Japanese Journal of Tropical Agriculture* Vol.33, No.3, (September 1989), pp. 158–163, ISSN 0021-5260

Jackson, M.B. & Ram, P.C. (2003). Physiological and molecular basis of susceptibility and tolerance of rice plants to complete submergence. *Annals of Botany* Vol.91, No.2, (January 2003), pp. 227–241, ISSN 0305-7364

Joho, Y.; Omsasa, K.; Kawano, N. & Sakagami, J-I. (2008). Growth responses of seedlings in *Oryza glaberrima* Steud. to short-term submergence in Guinea, West Africa. *Japan Agricultural Research Quarterly* Vol.42, No.3, (July 2008), pp. 157–162, ISSN 0021-3551

Karin, S.; Vergara, B.S. & Mazaredo, A.M. (1982). Anatomical and morphological studies of rice varieties tolerant of and susceptible to complete submergence at seedling stage. *Proceedings of the 1981 International Deepwater Rice Workshop*, pp. 287–291, ISBN 971-104-017-4, Bangkok, Thailand, 2–6 November 1981 (Krin et al. 1982)

Kawano, N.; Ella, E.; Ito, O.; Yamauchi, Y. & Tanaka, K. (2002). Comparison of adaptability to flash flood between rice cultivars differing in flash flood tolerance. *Soil Science and Plant Nutrition* Vol.48, No.5, (October 2002), pp. 659–66, ISSN 0038-0768

Kawano, N.; Ito, O. & Sakagami, J-I. (2009). Morphological and physiological responses of rice seedlings to complete submergence (flash flooding). *Annals of Botany* Vol.103, No.2, (January 2009), pp. 161–169, ISSN 0305-7364

Kende, H.; Van der Knaap, E. & Cho, H-T. (1998). Deepwater rice: a model plant to study stem elongation. *Plant Physiology* Vol.118, No.4, (December 1998), pp. 1105–1110, ISSN 0032-0889

Koffi, G., 1980. Collection and conservation of existing rice species and varieties in Africa. *Agronomie Tropicale* Vol 34, pp. 228-237

Linares, O.F. (2002). African rice (*Oryza glaberrima*) : history and future. *Proceedings of the National Academy of Sciences of the United States of America*, Vol.99, No.25, (December 2002), pp. 16360–16365, ISSN 0027-8424

Majerus, V.; Bertin, P. & Lutts, S. (2007) Effects of iron toxicity on osmotic potential, osmolytes and polyamines concentrations in Africa rice (*Oryza glaberrima* Steud.). *Plant Science* Vol.173, No.2, (August 2007), pp. 96–105, ISSN 0168-9452

Nwilene, F.E.; Traore, A.K.; Asidi, A.N.; Sere, Y.; Onasanya, A. & Abo, M.E. (2009). New records of insect vectors of rice yellow mottle virus (RYMV) in Cote d'Ivoire, West Africa. *Journal of Entomology* Vol.6, No.4, (October-December 2009), pp. 189–197, ISSN 1812-5670

Perata, P. & Voesenek, L.A. (2007) Submergence tolerance in rice requires Sub1A, an ethylene-response-factor-like gene. *Trends in Plant Science* Vol.12, No.2, (February 2007), pp. 43–46, ISSN 1360-1385

Ram, P.C.; Singh, B.B.; Singh, A.K.; Ram, P.; Singh, P.N.; Singh, H.P.; Boamfa, I.; Harren, F.; Santosa, E.; Jackson, M.B.; Setter, T.L.; Reuss, J.; Wade, L.J.; Pal Singh, L. & and Singh, R.K. (2002). Submergence tolerance in rainfed lowland rice: physiological basis and prospects for cultivar improvement through marker-aided breeding. *Field Crops Research* Vol.76, Special issue 2-3, (July 2002), pp. 131–152, ISSN 0378-4290

Rose-John, S. & Kende, H. (1985). Short-term growth response of deep-water rice to submergence and ethylene. *Plant Science* Vol.38, No.2, (March 1985), pp. 129–134, ISSN 0168-9452

Sakagami, J-I.; Isoda, A.; Nojima, H. & Takasaki, Y. (1999). Growth and survival rate after maturity in *Oryza sativa* L. and *O. glaberrima* Steud. *Japanese Journal of Crop Science*, Vol.68, No.2, (June 1999), pp. 257–265, ISSN 0011-1848

Sakagami, J-I.; Joho, Y. & Ito, O. (2009). Contrasting physiological responses by cultivars of *Oryza sativa* and *O. glaberrima* to prolonged submergence. *Annals of Botany* Vol.103, No.2, (January 2009), pp.171–180, ISSN 0305-7364

Setter, T.L.; Kupkanchanakul, T.; Kupkanchanakul, K.; Bhekasut, P.; Wiengweera, A. & Greenway, H. (1987). Concentrations of CO^2 and O^2 in floodwater and in internodal lacunae of floating rice growing at 1±2 m water depths. *Plant, Cell and Environment* Vol.10, No.9, (December 1987), pp. 767–776, ISSN 0140-7791

Setter, T.L.; Jackson, M.B.; Waters, I.; Wallace, I. & Greenway, H. (1988). Floodwater carbon dioxide and ethylene concentrations as factors in chlorosis development and reduced growth of completely submerged rice. *Proceedings of the 1987 International Deepwater Rice Workshop*, pp. 301–310, ISBN 971-104-187-1, Bangkok, Thailand, October 1987.

Setter, T.L. & Laureles, E.V. (1996). The beneficial effect of reduced elongation growth on submergence tolerance of rice. *Journal of Experimental Botany* Vol.47, No.10, (October 1996), pp. 1551–1559, ISSN 0022-0957

Thiemele, D.; Boisnard, A.; Ndjiondjop, M.N.; Cheron, S.; Sere, Y.; Ake, S.; Ghesquiere, A. & Albar, L. (2010). Identification of a second major resistance gene to Rice yellow mottle virus, RYMV2, in the African cultivated rice species, *O. Glaberrima*. *Theoretical and Applied Genetics* Vol.121, No.1, (June 2010), pp. 169–179, ISSN 0040-5752

Vergara, B.S., A. Mazaredo, S.K. De Datta & W. Abilay (1975). Plant age and internode elongation in floating rice varieties. Proceedings of the 1974 International Seminar on Deepwater Rice., pp178-183. Bangladesh Rice Research Institute, Dacca

Voesenek, L.A.C.J.; Colmer, T.D.; Pierik, R.; Millenaar, F.F. & Peeters, A.J.M. (2006). Tansley review. How plants cope with complete submergence. *New Phytologist* Vol.170, No.2, (April 2006), pp. 213–226, ISSN 0028-646X

Watarai, M. & Inouye, J. (1998). Internode elongation under different rising water conditions in African floating rice (*Oryza. glaberrima* Steud.). *Journal of the Faculty of Agriculture, Kyushu University* Vol.42, No.3-4, (March 1998), pp. 301–307, ISSN 0023-6152

18

Functional Analysis of *LHCB1* in *Arabidopsis* Growth, Development and Photosynthetic Capacity

M. Aghdasi, S. Fattahi and H. R. Sadeghipour
Dept. of Plant Physiology, Faculty of Science, Golestan University, Gorgan, Iran

1. Introduction

The light-harvesting chlorophyll a/b-binding proteins of photosystem II (LHCII) are the major components of the photosynthetic machinery in plants which contain more than 60% of plant chlorophyll (Peter and Thornber 1991). LHCII has four related roles in plant photosynthesis i.e. collecting and transferring excitation energy to the reaction centers of photosystem II (PS II) and photosystem I (PS I) to promote photosynthetic electron transport (Ruban et al., 1999, Van Amerongen and Dekker, 2003), organization of the plant photosynthetic system by maintaining the tight appression of thylakoid membranes in chloroplast grana (Allen and Forsberg, 2001), distribution of excitation energy between PS II and PSI by reversible phosphorylation at its N-terminal side (Allen and Forsberg, 2001, Kargul and Barber 2008), and protection of photosynthetic system from excess energy under light saturated conditions (Horton et al 1996 and 2008).

The LHCII proteins can be grouped into six subfamilies (LHCB1-6) which are encoded by *LHC* gene family (Jansson 1999). CP29, CP26 and CP24 are the minor proteins that are encoded by *LHCB4, LHCB5* and *LHCB6* genes, respectively. LHCB1, LHCB2 and LHCB3 are the major pigment-binding proteins which are encoded by *LHCB1, LHCB2* and *LHCB3* genes, respectively (Ruban et al 1999, Lucinski and Jackowski 2006). LHCB1, LHCB2 and LHCB3 polypeptides each with about 232 amino acid residues are similar in sequence, structure and function (Standfuss and Ku¨hlbrandt 2004). LHCB1-3 precursors are synthesized in cytoplasm and following transport into chloroplasts inserted into thylakoid membranes (Li et al., 2000). LHCB1 and LHCB2 are the most abundant proteins in the light harvesting antenna complex (Ruban et al., 1999). The N-terminal domain in both LHCB1 and LHCB2 lies on the stromal side where it is involved with adhesion of granal membranes and photo-regulated by reversible phosphorylation of its threonine residues (Boekema et al 1999, Anderson 2000).

The composition and structure of LHCII complex is regulated by different factors. For example light intensity can change the amount of light-harvesting complex components (Anderson et al., 1986, Bailey et al., 2001). Meanwhile it has been reported that the expression of *LHCB1* can be down regulated by accumulation of sugars such as glucose, sucrose and trehalose (Vinti et al., 2005, Aghdasi et al., 2009).

The antisense suppression of *Arabidopsis LHCB1* also leads to *LHCB2* suppression. These plants have reduced state transitions and capacity for feedback de-excitation important to adapt to changes in light intensity (Anderson et al., 2003). Over expression of *LHCB1-2* from pea in tobacco plants led to increased grana stacking and photosynthetic capacity at low irradiance. The transgenic plants also displayed increased cell volume, larger leaves, increased biomass and increased seed weight, and greater leaf number per plant at flowering, when grown under low irradiance levels (Labata et al., 2004).
So far, the function and importance of *LHCB1* alone in *Arabidopsis* growth, development and photosynthetic capacity have not been understood very well. In the current study, we screened Leclere and Bartel collection to identify mutants in *LHCB1* (Leclere and Bartel 2001). This led to the identification of one mutant in At1g29920 gene, *lhcb1*, with pale green phenotype. Characterization of the *lhcb1* mutant was achieved through its comparison with the wild type (WT) plants when both grown under normal and low irradiances. Furthermore, the over-expression of *Arabidopsis LHCB1* was carried out to confirm the function of the encoded protein in growth and development.

2. Methods and materials

2.1 Plant materials and growth conditions:

The *Arabidopsis thaliana* wild type (WT) plants ecotype Columbia-0 (COL-0), transgenic lines and *lhcb1* mutant seeds were planted in compost and watered twice per week. Plants were grown in controlled growth chamber under normal (150 µmol photon m^{-2} s^{-1}) and low (70 µmol photon m^{-2} s^{-1}) irradiances and a 25 °C day/ 20 °C night temperature regime.

2.2 Screening for *lhcb1* mutant

The collection of *Arabidopsis* 35S-cDNA lines described by LeClere and Bartel (2001) was used in this study. Seeds from 331 pools from this collection were screened. They were surface sterilized by the gas method sterilization (Clough and Bent, 1998). Sterilized seeds were plated on ½ Murashige and Skoog (MS) medium solified with 0.8 % agar (Murashige and Skoog, 1962). Seeds were stratified in darkness at 4 °C for 2 days, before transferring to growth chamber at 25 °C. A pale green mutant was characterized from this collection. The mutant plants were transferred to soil to generate second seed generation (S2). Seeds from S2 generation were grown on medium with 12.5 mg/L PPT (Phosphinotrice). Growth on PPT, allows the segregation of the T-DNA insertion carrying the CaMV promoter driven cDNA expression cassette. After 14 days, seedlings were screened for segregation of T-DNA inserted on 12.5 mg/L PPT. Seedlings resistant to PPT were transferred to soil along with WT plants. Upon flowering of the plants, crosses were carried out with the WT plants. The individual silliques were collected in one bag after ripening. To do seed re-screening, they were sown separately from each silique on ½ MS medium supplemented by 12.5 mg/L PPT.

2.3 DNA extraction and PCR analysis

Three small leaves were frozen in liquid nitrogen, pulverized with glass beads for 2 minutes at 2800 rpm in a Dismembranator (Braun, Germany), and then DNA was extracted using the Pure Gene DNA isolation kit (Amersham Pharmaciabiotec, England) according to the manufacturer's protocols. To determine the presence of the 35S cDNA fragments in the pale green *lhcb1* plants, PCR was performed with primers 35S-F (CGCACAATCCCACTATCCTTCGCAAG) Nos-R (GATAATCATCGCAAGACCGGAACAGG) primers. A mixture of *Taq* and PFU enzymes at unit ratio of 50:2 was used. After denaturation for 2 minutes at 94 °C, DNA was amplified with 35 cycles (30 sec 94 °C, 30 sec 56 °C and 2 min 72 °C). PCR was completed with a final step at 72 °C for 5 minutes. An aliquot from the PCR product was run on agarose gel and the remaining was cleaned using DNA purification kit (Amersham Biosciences, England).

2.4 RNA extraction and cDNA synthesis

Total RNA was extracted from 10 days old *Arabidopsis* plants. Whole plant material was snap frozen in liquid nitrogen, pulverized with glass beads for 2 minutes at 2800 rpm in a dismembrator (Braun, Melsungen, Germany). Total RNA was isolated with the RNeasy plant mini kit (QIAGEN USA, Valencia, CA). RNA concentration and purity were determined by measuring the absorbance at 260 nm. To remove any possible contamination by genomic DNA, 10 ng of RNA was treated with 2 U of DNAse I (DNA-free, Ambion, Austin, USA). The absence of DNA was analyzed by performing a PCR reaction (40 cycles, similar to the real-time PCR program) on the DNaseI-treated RNA using *Taq* DNA polymerase. Reverse transcriptase PCR (RT-PCR) experiments were performed using 1 ng of total extracted RNA and used for first-strand cDNA synthesis with 60 U of M-MLV reverse transcriptase (Promega, Madison, WI), 0.5 µg of odT16v (custom oligo from Invitrogen, Carlsbad, CA) and 0.5 µg of random hexamer (Invitrogen, USA). PCR was performed with forward and reverse primers (5′-ctcaacaatggctctctcct-3′ and 5′- aacccaagaactgaaaatccaa-3′). Amplification conditions were performed as initial DNA denaturation at 94°C for 2 minutes followed by 35 cycles of 1 minute denaturation at 94°C , 30 second annealing at 56 °C and 2 minutes of extension at 72 °C with a final extension time at 72°C for 10 minutes. An aliquot of the PCR product was run on an agarose gel (1%) and the remaining PCR product was cleaned using a DNA purification kit (Amersham Biosciences, England).

2.5 Cloning cDNA fragments into pGEM-T Easy vector

The resulting cDNA fragments from the previous steps were ligated into the pGEM-T Easy vector. For this purpose, cDNA was concentrated to 3 µl (25 ng) and was then added to 5 µl of 2X ligation buffer, 1 µl of T4 Ligase and 1 µl of pGEM-T easy vector. The ligation mixture was incubated over night at room temperature. An aliquot (100 µl) from the competent *E. coli* were taken from the -80 °C freezer and thawed on ice for 20 min. The over- night ligation mixture was added to the cells and the mixture was left on the ice for 20 minutes. Heat shock was applied for 50 sec at 42 °C, followed by a 5 min cooling period on ice. One ml of lysogeny broth (LB) medium was added and cells were incubated at 37 °C for 1 h. The LB plates contained 50 µg/ml of Ampicillin for selection. Isopropyl-β-D-1-hiogalactopyranoside (IPTG) and XGalactopyranoside (X-Gal) were added for screening of blue and white colonies. To check colonies containing the plasmid with the ligated fragments, restriction enzyme analysis was performed. Plasmids were isolated from 5 colonies using a plasmid miniprep kit (Sigma, USA). In the digestion mixture, 2 µl of plasmid, 1 µl of 10 X buffer, 6 µl of milli-Q water and 1 µl of *Eco*R1 were used. Samples were digested at 37 °C for 1.5 h. The obtained fragments were analyzed by agarose gel electrophoresis.

2.6 Sequence analysis

DNA sequencing was carried out at the sequencing facility in Wageningen University. Sequences obtained from analysis with forward and reverse primers (T7: 5´ atttaggtgacactatag 3´ and SP6: 5´ taatacgactcactataggg 3´) were aligned and the PCR fragment structures reconstructed by BLAST (Basic Local Alignement Search Tool) searches in TAIR (http://www.arabidopsis.org/Blast/).

2.7 cDNA over-expression constructs and re-transformation into Col-0

Full length cDNA were isolated and purified from the pGEM-T easy vector clones and over-expressed in wild type (WT) *Arabidopsis* plants. The CaMV 35S expression cassette was isolated by digestion with *Eco*RV from the pUC-18 vector. The cassette was filled with Klenow and dNTP and subsequently ligated into pBin19 (*Hin*dIII/*Eco*RI) to yield pBin-35S. Purified fragments were cloned into the pBin-35S expression cassette, resulting in pBin35S/cDNA/NOS. The plasmids containing *LHCB1* gene was digested with *Xba*I and *Xmn*I restriction enzymes. The construct was introduced by electroporation into *Agrobacterium tumefaciens* containing the pGV2260 plasmid. The resulting bacteria were used to transform *Arabidopsis* by floral dip method (Clough and Bent, 1998). Transgenic seedlings were selected on half MS media containing 50 mg/L of Kanamycin. Transgenic seedlings were grown in soil medium under either normal or low irradiances for further phenotypic characterization.

2.8 Expression analysis of the *LHCB1* gene

Quantitative-PCR (Q-PCR) analysis was performed to determine the expression level of *LHCB1* gene. Total RNA was extracted from *Arabidopsis* seedlings as described above. Following treatment of RNA with DNAase I, cDNA was synthesized using the M-MLV reverse transcriptase system (Promega, Madison, WI). Q-PCR was carried out by ABI-prism 7700 Sequence Detection System (PEApplied Biosystems, Foster City, CA). For each reaction, 12.5 µl of green PCR Master Mix (Applied Biosystems, UK) and 2.5 µl of gene-specific primers were used. Each experiment was repeated 3 times. Relative gene expression was based on the comparative Ct method (User Bulletin No. 2: ABI PRISM 7700 sequence detection system, 1997) using *AtACTIN2* as the calibrator reference (5´-ATGTCTCTTACAATTTCCCG-3´ and 5´- CAACAGAGAGAAGATGACT- 3´). The Q-PCR data were normalized against *AtACTIN2*.

2.9 Pigment content and Florescence measurements

Chlorophylls a, b and total chlorophyll were determined spectrophotometrically as described by Jeffery and Humphery (1975). In brief, 100 mg of fresh rosette leaves from 3-week-old *Arabidopsis* plants were grounded in liquid nitrogen and extracted with 80% (v/v) acetone. Absorbance was then measured at 647, 652 and 664 nm. The concentrations of chlorophylls a, b and total chlorophyll were then calculated.

Anthocyanin content was determined using the protocol of Mita *et al.* (1967). Frozen and homogenized leaves (20 mg) were extracted for 1 day at 4 °C in 1 ml of 1% (v/v) hydrochloric acid in methanol. The mixture was centrifuged at 23,000 ×*g* for 15 minutes and the absorbance of the supernatant was measured at 530 and 657 nm. Relative anthocyanin concentrations were calculated using the formula [$A530-(1/4 \times A657)$]. The relative anthocyanin content was defined as the product of relative anthocyanin concentration and the extract volume. One anthocyanin unit equals to one absorbance unit [$A530-(1/4 \times A657)$] in 1 ml of the extraction solution.

Chlorophyll *a* florescence was measured with OPTI-Sciences OS-30 fluorometer (Walz, Effeltrich, Germany). The *Arabidopsis* plants were adapted in the dark 15 minutes before measurement. F_0 (the initial fluorescence level of PSII reaction center) was measured in the presence of a 10 µmol photons m^{-2} s^{-1} measuring beam. The maximum fluorescence level in the dark adapted state (F_m) was determined by using a 0.8 s saturating irradiance pulse. The fluorescence parameter F_v/F_m was calculated using the DualPAM software

2.10 Chloroplast isolation and determination of Hill reaction rate

The rate of Hill reaction in the chloroplast preparations of WT, *lhcb1* and *LHCB1* over-expressed plants was measured according to Trebst (1972). Leaves (0.25 g) were homogenized in a cold mortar in a buffer consisting of 20 mM Tris-HCl (pH 7.5), 0.3 M sucrose, 10 mM EDTA and 5 mM MgCl2 and chloroplast were isolated. The rate of Hill reaction in the illuminated chloroplast preparations was determined spectrophotometrically by recording the decrease in absorbance at 600 nm due to Dichlorophenol indo phenol (DCPIP) reduction. The rate of Hill reaction was expressed as the changes in absorbance per milligram chlorophyll per minute (ΔOD.min $^{-1}$. mg chl $^{-1}$).

2.11 Carbohydrate and protein determination

The soluble and insoluble sugars were determined spectrophotometrically by the phenol-sulfuric acid method (Kuchert, 1985). The leaf soluble and total proteins were determined according to methods of Bradford (1976) and Markwell (1988), respectively.

2.12 SDS-PAGE analysis of chloroplast proteins

The reducing SDS-PAGE of the chloroplast protein samples was carried out according to Fling and Gregerson (1986). For the SDS PAGE analysis of the chloroplast proteins, chloroplasts were isolated from leaves (1 g) as described above and finally suspended in 100 µl of the homogenization buffer. Then 5.0 ml n-Hexan: 2-propanol (3:2; V/V) was added and after a thorough mixing, it was centrifuged at 4000 *g* for 15 min. The lipid-pigment containing upper phase was discarded and the remaining pellet was re-extracted with another 5.0 ml of n-Hexane: 2-propanol mixture. The protein precipitate obtained after doing the second centrifugation, was washed with 5.0 ml acetone (80% V/V) and dried under a stream of nitrogen gas. The dried protein precipitates were dissolved in electrophoresis sample and following quantification by the Markwell (1988) method, aliquots corresponding to 50 µg protein were resolved on a 15% SDS-containing Acrylamide gel.

2.13 Statistical analyses

Data from all experiments were processed by statistical SAS package (version 9). The reported values were means of three replicates. Means were compared for significance using the Duncan's test.

3. Results

3.1 Isolation of the *lhcb1* mutant

After screening the LeClere and Bartel mutant seed collection, a pale green mutant was identified. The pale coloration was uniformly displayed by all leaves throughout the whole life of the mutant. The selected mutant was fully fertile (Fig. 1A).

Fig. 1. Characterization of the *Arabidopsis lhcb1* mutant. A: Phenotype of the *lhcb1* mutant, as compared to the wild type (WT) Columbia-0 ecotype. B: Q-PCR analysis of *LHCB1* expression level in WT and *lhcb1*mutant plants.

To ensure the presence of T-DNA insertions in the selected line, a secondary screening was carried out on seeds from the selected pale green plants of the primary screen. Segregation analysis on PPT (phosphinotrice) showed that the line was homozygous for the T-DNA insertion. Cosegregation of resistance to PPT (flanked to T-DNA) and the pale green phenotype confirmed that the phenotype of the mutant has cosegregated with T-DNA insertion (Data not shown). To find out if the selected line contains cDNA fragments, we performed PCR using a forward primer on the CaMV35S promoter and a reverse primer on the nopaline synthase poly-adenylation sequence (LeClere and Bartel 2002). PCR reactions in the selected line yielded one fragment only. Sequence analysis of the PCR product indicated that cDNA fragment was full length, with ATG and TGA, in sense orientation and it encodes LHCB1 (At1g29920).

To find out whether the pale green phenotype is dominant or recessive, backcrossing between the selected line and WT plants was carried out. Analysis of segregation of resistance to PPT revealed that PPT resistance segregates as a single locus.

The level of *LHCB1* expression was determined in seedlings from both WT and the selected mutant. The expression level of *LHCB1* in the selected mutant was significantly lower than that of the WT plant. This indicates that in the selected line, the pale green phenotype is due to co-suppression of *LHCB1* expression (Fig.1B). This mutant hereafter named *lhcb1* mutant.

3.2 Transformation of cDNA construct into WT plants

Transformation of the WT plants (Colombia-0 ecotype) with full length cDNA of *LHCB1* yielded 20 independent lines with resistance to the selection marker. Two independent transgenic lines were selected for mRNA level analysis (TR-1 and TR-2). The transcript levels of *LHCB1* in TR-1 and TR-2 lines were increased by approximately 53% and 47% respectively, compared to WT plants (Fig. 2).

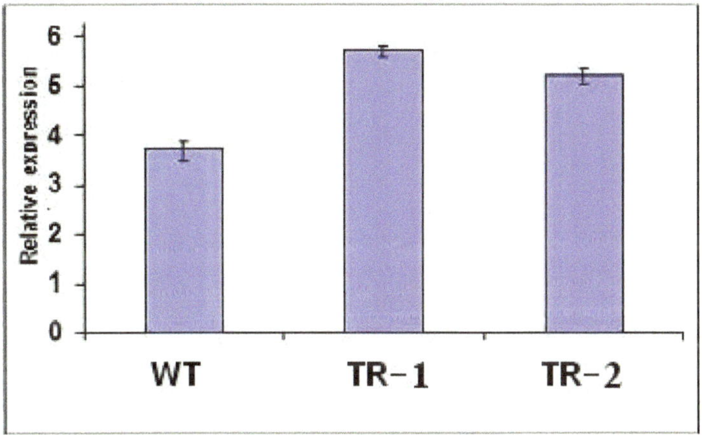

Fig. 2. *LHCB1* expression level in Transformed (TR) lines, as compared to WT plants.

3.3 Characterization of *lhcb1* mutant and WT plants

The *lhcb1* mutant plants displayed differences to the WT plants in growth, morphology, leaf area, dry and fresh weight when grown under normal and low irradiances. The *lhcb1* mutants showed pale green phenotype with smaller leaf area (Fig.1 and 3A). Dry and fresh weights were significantly lower in *lhcb1* mutant than that of the WT plants (Table 1). There was no significant difference in height between *lhcb1* mutant and WT plants under normal irradiances (Fig.1, Table 1). Plants grown under low irradiances were taller than those grown under normal irradiances. Under normal light conditions, the height of 4-week old *lhcb1* and WT plants were 13.66±3.16 and 15±1 cm respectively, while under low irradiances the height of *lhcb1* and WT plants were approximately 2 times more (Fig. 3B and Table 1). Relative to the WT plants, the *lhcb1* mutants were indifferent with respect to flowering time and fertility. The WT and the *lhcb1* mutant plants approached to flowering stage after 38 and

36 days, respectively under normal irradiances, however, this stage was shortened to 27 and 25 days, respectively under low irradiances. Carbohydrate analyses showed that both soluble and insoluble contents were similar in both *lhcb1* mutant and WT plants (Table 1).

Fig. 3. Morphological characteristics of *lhcb1* mutant, transformed (TR-1) and WT plants grown under normal and low irradiance conditions. A: leaf morphology of plants grown under normal irradiance conditions, B: TR-1 and WT plants grown under normal irradiance conditions, C: WT, TR-1 and *lhcb1*mutant plants grown under low irradiance conditions.

3.4 Reduced chlorophyll contents and photosynthesis capacity in the *lhcb1* mutant

As the *lhcb1* mutant plants were clearly pale green in color when compared to WT ones, the chlorophyll contents of them were compared following growth under normal and low irradiances. Under normal light conditions total chlorophyll content of 3-week-old WT leaves was 2.09±0.1 mg/g fresh weight, while total chlorophyll content in the *lhcb1* mutant was 1.02±0.9 mg/g fresh weight (52% less). Meanwhile under low irradiances, total chlorophyll content of 3-week-old WT leaves was 2.65±0.8 mg/g, but that of the *lhcb1* mutant was 1.38±0.15 mg/g fresh weight which is about 48% of the amount found in WT plants. There was an increase in the ratio of Chl *a/b* ratio in the *lhcb1* mutants compared to WT plants grown under any light conditions. The ratio of Chl *a/b* increased from 2.55 in the WT plants to 3.39 in the mutant plants under normal light conditions. Under low irradiances the figure rated to 1.81 in the WT plants which increased to 3.97 in the *lhcb1* mutant plants (Table 1).

Irradiance	Plant	height (cm)	number of days to flowering	dry weight (g)	fresh weight (g)	soluble sugar (µg/g)	insoluble sugar (µg/g)
Normal	WT	15.00±1	38	0.087±0.01	0.70±0.08	549.7±65.5	103.1±5.3
	lhcb1	13.66±3.16	36	0.035±0.01	0.50±0.15	576.9±557.1	107.3±38.9
	TR-1	31.33±2.08	25	0.14±0.04	1.10±0.26	700.2±690.1	115.4±83.1
Low	WT	31.22±3.44	27	0.09±0.01	0.91±0.12	3194±105.5	259.9±105
	lhcb1	26.66±2.45	25	0.04±0.01	0.47±0.65	2993±193	260.8±96.7
	TR-1	35.42±4.34	25	0.16±0.02	1.53±0.19	3625±169.7	3900±328.8

Table 1. Data on biomass, leaf total carbohydrate and total protein contents of *Arabidopsis* wild type (WT), *lhcb1* mutant and transformed (TR) plants grown under normal and low irradiance conditions.

LHCII functions as an auxiliary antenna for PSII. PSII and LHCII are close to each other in the stacked granal thylakoids. In this survey, PSII activity was analyzed by measuring the F_v/T_m value that is an indicator of the intrinsic efficiency of PSII. There was not any significant difference in fluorescence parameter (F_v/F_m) between WT and *lhcb1* mutant plants grown under normal and low irradiances. The F_v/F_m value was o.86±0.01 in WT and 0.84±0.03 in *lhcb1* mutant plants grown under normal irradiances (Table 1). These results revealed that PSII efficiency was not affected by the mutation in *LHCB1*.

We further examined the water oxidation capacity of the photosynthetic machinery of both WT and *lhcb1* mutant plants under normal irradiance condition. Measured as the rate of Hill reaction, the water oxidation capacity was significantly decreased in the *lhcb1* mutants compared to WT plants (Fig. 4).

3.5 The chloroplast protein composition of the *lhcb1* mutant and WT plants

Leaf materials from *lhcb1* mutant and WT plants were analyzed for total protein measurement and chloroplast protein composition. There was no significant difference in the total protein amount between *lhcb1* mutant and WT plants (Fig. 5A). Furthermore, chloroplasts were isolated from both the mutant and WT plants and their polypeptide compositions were analyzed by SDS-PAGE. The protein band patterns of both WT and *lhcb1* mutant chloroplasts were essentially similar. However, one protein band with a molecular mass of about 25 kDa was absent in the chloroplasts protein preparations of *lhcb1* mutant plants (Fig. 5B)

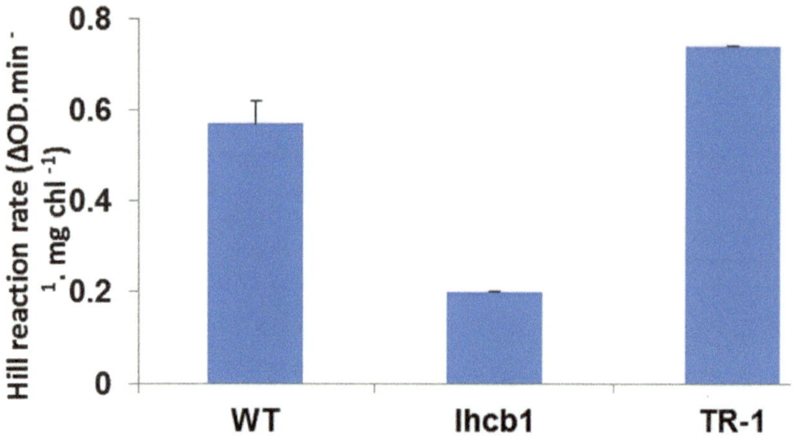

Fig. 4. The water oxidation capacity of WT, *lhcb1* mutant and TR-1 plants grown under normal irradiance conditions as measured by the rate of Hill reaction

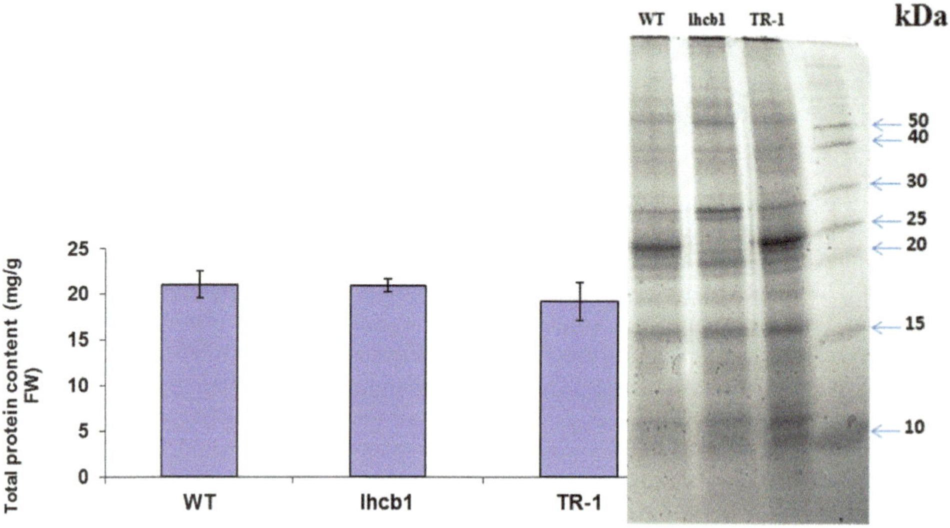

Fig. 5. Total protein content (A) and SDS-PAGE profile of WT, *lhcb1* mutant and transformed (TR) plants (B).

3.6 Transgenic plants display differences in vegetative morphology and growth under normal irradiances

Plants transformed with the full length cDNA of *LHCB1* displayed differences to the WT plants in growth, morphology, height, leaf area, dry and fresh weights, soluble and insoluble sugars when grown under normal irradiances (Fig. 3A, B and Table 1). Under normal irradiance conditions, transgenic plants (TR-1) had larger leaves than that of WT plants (Fig. 3A). Transgenic plants were generally taller (31.33±2.08 cm compared to 15±1 cm in WT), with increased dry weight (73% more than WT) and fresh weight (75% more than WT). TR-1 plants approached the flowering stage after 25 days under normal light conditions, while in the WT plants this figure extended to 38 days. The average of seed number per silique and the weight of 1000 seeds increased by over-expression of *LHCB1* (41.7±4.4 in TR-1 versus 32.8±2.84 in WT). The leaves of transgenic plants contained a significantly higher soluble and insoluble sugar contents compared to WT plants. Thus, the soluble and insoluble sugars were respectively 700.2±690.01 µg/g FW and 115.4±83.1 µg/g in transgenic plants whereas these figures were 549.7±65.5 µg/g and 103.1±53.1 µg/g in WT plants (Table 1).

Further analysis revealed that there were not any significant differences in Chl *a*, Chl *b*, ratio of Chl *a:b* and total Chl content among WT and Transgenic plants. Meanwhile the efficiency of PSII i.e. F_v/F_m value in transgenic plants was close to that in the WT plants (Table 2). But water oxidation capacity i.e. the rate of Hill reaction was significantly higher in transgenic plants compared to WT ones (Fig 4).

irradiance	Plant line	Chlorophyll content (mg/g FW)				F_v/F_m
		chl *a*	chl *b*	total chl	chl *a/b*	
Normal	WT	1.48± 0.11	0.58±0.08	2.09±0.19	2.55	0.86±0.01
	lhcb1	0.78±0.14	0.23±0.05	1.02±0.19	3.39	0.84±0.03
	TR-1	1.31±0.05	0.53±0.05	1.84±0.10	2.47	0.83±0.05
Low	WT	1.70± 0.01	0.94±0.09	2.65±0.08	1.81	0.84±0.03
	lhcb1	1.10±0.11	0.27±0.04	1.38±0.15	3.97	0.86±0.02
	TR-1	1.52±0.07	0.58±0.05	2.11±0.12	2.59	0.87±0.04

Table 2. Chlorophyll content (chl), Chlorophyll ratio (chl *a/b*) and photosynthetic parameter of *Arabidopsis* wild type (WT), *lhcb1* mutant and transformed (TR) plants grown under normal and low irradiance conditions.

There was not any significant difference in total protein content between transgenic and WT plants. Analysis of the chloroplastic proteins showed that most of the bands were similar in both transgenic and WT plants. However, a polypeptide with a molecular mass of about 25 kDa displayed relatively thicker band intensity in chloroplast protein preparations of the transgenic plants (Fig. 5B).

3.7 Characterization of transgenic plants grown under low irradiances

Further characterization of transgenic plants was achieved after planting them under low light conditions. The height of transgenic plants was close to that in WT ones. It was interesting that *lhcb1* mutants had also the same height as both WT and transgenic plants under low irridiances (Fig. 3C). There was not any significant difference in height between

TR-1 and WT plants (35.42±4.34 cm versus 31.22±3.44 cm). Flowering occurred after 25 and 27 days respectively, in transgenic and WT plants. Dry and fresh weights, soluble and insoluble sugar contents were obviously higher in transgenic plants than in WT ones (Table 1). There was an increase in total Chl content in transgenic and WT plants, compared to those grown under normal irradiances. Total chlorophyll in WT plants was fairly greater than in transgenic plants under low irradiances (2.65±0.8 mg/g FW versus 2.11±0.2 for transgenic plant). There was not any significant difference in the efficiency of PSII (F_v/F_m value) between transgenic and WT plants.

4. Discussion

A novel system designed to co-suppress or over-express cDNA in *Arabidopsis* was developed by LeClere and Bartel (2001). They constructed a binary vector containing a novel *Arabidopsis* cDNA library driven by the CaMV35S promoter. T-DNA in this vector contains a bar-gene cassette for PPT selection of the transgenic plants and a cassette with a randomly cloned cDNA inserted between CaM35S promoter and nopaline synthase (NOS) polyadenylation (polyA) sequences. This method has the advantage that the inserted cDNA can be amplified using PCR with primers in the promoter and polyA sequences (LeClere and Bartel 2001). The cDNA insertion could be responsible for the observed phenotype if both the phenotype and cDNA co-segregate as a dominant trait. Definitive confirmation for the correlation between cDNA and the phenotype in plants exhibiting dominant trait, could be obtained following the transformation of cDNA expression cassette into WT plants. Screening of 331 pools of T4 seeds from this collection displayed one pale green mutant. Sequence analysis of the amplified gene identified *LHCB1* cDNA fragment in this mutant. It was full length, with ATG and TAG, and in sense orientation. Q-PCR data revealed that this construct co-suppressed the endogenous *LHCB1* transcript.

Studies carried out so far to reveal the functional significance of LHCII protein-chlorophyll complexes in phenotypic alterations of plants, have suffered from segregating the specific role played by each individual polypeptides constituting the complex. Thus either the impacts of over-expression of *LHCB1-2* (Labate et al., 2004) or antisense cosupression of *LHCB1-2* (Andersson et al., 2003) has been described. However, in the present study some functional significance of LHCB1 protein was investigated by producing homozygous *lhcb1* mutants (which retained *LHCB2*) and *LHCB1* over-expressed plants. Furthermore, the *LHCB1* suppression / over-expression was accompanied with the corresponding decrease / increase of a polypeptide with a molecular mass of about 25 kDa on the SDS-PAGE gel which is very close to the reported mass range of this protein (Huber et al., 2001, Zolla et al., 2003). These clearly indicate that the genetic manipulations carried out on *Arabidopsis* are translated also at the protein level.

Silencing of *LHCB1* in *Arabidopsis* significantly reduced their chlorophyll content with respect to WT plants. It was evidenced by their pale green coloration and resulted in the increased ratio of Chl *a/b*. As the *lhcb1* mutants displayed significant reduction of biomass and leaf area with respect to WT plants, it can be said that loss of *LHCB1* has greatly compromised the efficiency of carbon assimilation. Reduced chlorophyll content associated with biomass decline has also been reported for *lhcb1-2* antisense plants (Andersson et al., 2003). These might partly be attributed to the significant reduction of water oxidation capacity of the mutant versus WT plants (Fig. 4). As for the *lhcb1-2* antisense plants

(Andersson et al., 2003), no significant alteration in the quantum efficiency of PSII occurred for the *lhcb1* mutants (Table 2). Regarding that the *lhcb1* mutants displayed increased ratio of Chl *a/b* very probably due to LHCB1 loss, it is expected that they observe reduced non-photochemical quenching and feedback de-excitation (Andersson et al., 2003) with respect to WT plants. This possibly makes them more susceptible to photoinhibitory conditions which ultimately reduce their fitness. Although far from higher plants, a mutant of *Chlamydomonas reinhardtii* which lacks a major polypeptide of LHCII, also suffers from nonphotochemical quenching and thus is prone to photoinhibition (Elrad et al., 2002). Considering that LHCB1 is a major target protein for phosphorylation / de-phosphorylation required for state transition (Lunde et al., 2000), its loss in *lhcb1* mutants might decreases the capacity for state transition, a feature which has been reported for *lhcb1-2* antisense plants (Andresson et al., 2003).

Many phenotypic characteristics of transgenic *Arabidopsis* TR-1 versus the WT plants under normal irradiances were similar to those reported for transgenic tobacco plants overexpressing pea *LHCB1-2* (Labate et al., 2004). Thus the *Arabidopsis* TR-1 plants exhibited taller stature, greater biomass, increased carbohydrate contents and larger seed size as compared to WT plants. Similar to transgenic tobacco plants, there were no differences in Chlorophyll content, Chl *a/b* ratio and the quantum efficiency of PSII with respect to WT plants. However, in contrast to the tobacco transgenic plants, the flowering time of *Arabidopsis* TR-1 plants was shortened compared to WT plants. This might be due to photoperiodic behavioral differences of the two species. The increased carbohydrate content of *Arabidopsis* TR-1 plants and their overall greater biomass might represent more efficient carbon assimilation. The non-photochemical quenching of the *Arabidopsis* TR-1 is expected to be greater than WT plants, a feature which has been reported for tobacco *LHCB1-2* transgenes (Labate et al., 2004). On the other hand, the lack of *LHCB1-2* in *Arabidopsis* mutants is associated with reduced nonphotochemical quenching (Anderson et al., 2003). Taking into account that the water oxidation capacity of *Arabidopsis* TR-1 plants (and thus their potential for NADPH generation) is also greater than WT plants, the idea of more efficient photosynthesis is justified. The increase in photosynthetic carbon assimilation efficiency of TR-1 plants might also be explained by the increase in chloroplast number and improved granal stacking (Labate et al., 2004). The increased seed number per silique of TR-1 plants might have been resulted from their photosynthetic superiority with respect to WT plants.

Under low irradiances, no significant differences occurred in stature of TR-1 plants and WT ones. The differences in flowering time between them were also abolished. Apparently under these conditions differences in water oxidation capacity of WT and *LHCB1* over-expressed plants does not play anymore role in carbon assimilation competence of the transgenic plants. Further insight on the functional significance of LHCB1 can be obtained by studying the physiological responses of *lhcb1* mutant and TR-1 plants under conditions which are known to limit photosynthesis to a great extent. Thus studies focusing on the behavior of these plants under various environmental stresses might be conductive in elucidating the specific role of various LHCII proteins on plant fitness.

5. References

Allen J.F, Forsberg J (2001) Molecular recognition in thylakoid structure and function. Trends Plant Sci 6: 317–326

Aghdasi M, Schluepmann H (2009) Cloning and expression analysis of two photosynthetic genes under trehalose treatment. International Journal of Biotechnology 7: 179-187

Anderson J.M (1986) Photoregulation of the composition, Function and structure of thylakoid membranes. Ann. Rev. Plant Physiol. Plant Mol. Biol. 37:93-136

Andersson B, Yang D.H, Paulsen H (2000) The N-terminal domain of the light-harvesting chlorophyll a/b-binding protein complex (LHCII) is essential for its acclimative proteolysis". *FEBS Lett.* 466 (2–3): 385–388

Anderson J, Wentworth M, Walters R, Howard C.H, Ruban A.V, Horton P, and Jansson S (2003) Absence of Lhcb1 and Lhcb2 proteins of the light-harvesting complex of photosystem II effects on photosynthesis, grana stacking and fitness. Plant J. 35: 350-361

Bailey S, Walters RG, Jansson S. Horton P (2001) Acclimation of *Arabidopsis* thaliana to the light environment. Planta J. 213: 794-801

Boekerma E.J, Van Breemen J.F.L, Van Room H, Dekker J.P (1999) Arrangement of photosystem II supercomplexes in crystalline macrodomainss within the thylakoid membrane of green plant chloroplasts. J. Mol. Biol. 301: 1123-1133

Bradford MM (1976) A rapid and sensitive method for the quantitation of microgram quantities of protein utilizing the principle of protein-dye binding. Anal. Biochem. 72:248-254

Clough S.J, Bent A.F (1998) Floral dip: a simplified method for Agrebacterium-mediated transformation of Arabidopsis thaliana. Plant J. 16:734-745

Elard D, Niyogi K.K, Grossman A (2002) A major light-harvesting polypeptide of photosystem II functions in termal dissipation. Plant Cell. 14: 1801-1816

Fling S, Gregerson D (1986) Peptide and protein molecular weight determination by electrophoresis using a high-molarity tris buffer system without urea. Analytical Biochemistry. 155:83-88

Jansson S (1999) A guide to the Lhc genes and their relatives in Arabidopsis. Trends Plant Sci 4: 236–240

Jeffery S, Humphrey G.F (1975) New spectrophotometric aquations determining chlorophyll a, b, c1 and c2 in higher plants, algae and phytoplankton. Plant aphysiol. 167:191-194 (1975).

Horton P, Ruban A.V, Rees D, Noctor G, Pascal A.A, Young A.J (1991) control of the light harvesting function of chloroplast membranes by aggregation of LHCII chlorophyll-protein complex. FEBS letters 292: 1-4

Horton P, Ruban A.V, Walters R.G (1996) Regulation of light harvesting in green plants. Annu. Rev. Plant Physiol. Plant Mol. Biol. 47:655-684

Horton P, Johnson MP, Perez-Bueno ML, Kiss AZ, Ruban AV (2008) Does the dynamic structure and macro-organization of Photosystem II in higher plant grana membrane regulate light harvesting states? FEBS Journal. 275:1069-1079

Huber C.G, Tipperio A.M, Zolla L (2001) Isoforms of Photosystem II antenna proteins in different plant species reveald by liquid chromatography-electroscopy ionization mass spectroscopy. J. of Biological Chemistry. 49: 45755-45761

Kar M, Mishra D (1976) Catalase, Peroxidase and Polyphenol oxidase activities during rice leaf senescence. Plant Physiol 57: 315-319

Kargul J, Barber J (2008) photosynthetic acclimation: structural reorganization of light harvesting antenna-roleof redox-dependent phosphorylation of major and minor chlorophyll *a/b* binding proyeins.275: 1056-1068

Ku¨hlbrandt W, Wang D.N, Fujiyoshi Y (1994) Atomic model of plant light-harvesting complex by electron crystallography. Nature 367:614–621

Labate M, Kenton K.o, Zdenka W. K, Costa Pinto L, Real M, Romano M.R, Barja P.R, Granell A, Friso G, Van Wijk K, Brugnoli E, Labate C.A (2004) Constitutive expression of pea Lhcb 1-2 in tobacco affects plant development, morphology and photosynthetic capacity. Plant Plant Mol Biol. 55: 701-714

Lass M.H (1984) Experiments in Plant Physiology. George Allen & Unwin Publishers, London

Li X.P, Bjo¨rkman O, Shih C, Grossman A.R, Rosenquist M, Jansson S, Niyogi K.K (2000) A pigment-binding protein essential for regulation of photosynthetic light harvesting. Nature 403:391–395

Lucinski R, Jackowski G (2006) The structure, function and degradation of pigment-binding proteins of photosystem II. Acta Biochemica polonica. 53: 693-708

Lunde C, Jensen P.E, Haldrup A, Knoetzel J, Scheller H.V (2000) The PSI-H subunit of photosystem I is essential for state transition in plant photosynthesis. Nature. 408: 613-615

Markwell M.A.K, Hass S.M, Tolbert NE, and Bieber LL (1988) Protein determination in membrane and lipoprotein sample: Manual and automated procedure. Methods in Enzymology. 72:296-303

Mita S, Murano N, Akaike Nakamura K (1997) Mutants of *Arabidopsis* thaliana with pleiotropic effects on the expression of the gene for beta-amylase and on the accumulation of anthocyanin that is inducible by sugars. Plant J. 11:841-851

Nilsson A, Stys D, Drakenberg T, Spangfort M.D, Forsen S, Allen J.F (1997) Phosphorylation controls the three- dimensional structure of plant light harvesting complex II. J Biol Chem 272, 18350–18357.

Peter G.F, Thornber J.P (1991) Biochemical composition and organization of higher plant photosystem II light-harvesting pigmentproteins. J Biol Chem 266: 16745–16754

Ruban A.V, Lee P.J, Wentworth M, Young A.J, Horton P (1999) Determination of the stoichiometry and strength of binding of xanthophylls to the photosystem II light harvesting complexes. J Biol Chem 274: 10458–10465

Standfuss J, Ku¨hlbrandt W (2004) The three isoforms of the light harvesting complex II: spectroscopic features, trimer formation, and functional roles. J Biol Chem 279: 36884–36891

Trebst A (1972) Measurement of Hill reaction and photoreduction . Methods Enzymol. 24: 146-165

Van Amerongen H, Dekker J.P (2003) Light-harvesting in photosystem II. In Light-Harvesting Antennas in Photosynthesis 219–251. Kluwer Academic Publishers, Dordrecht

Zhang M, Senoura T, Yang X, Chao Y, Nishizawa K (2011) *Lhcb2* gene expression analysis in two ecotype of *Sedum alfredii* subjected to Zn/Cd treatments with functional analysis of *SaLhcb2* isolated from Zn/Cd hyperaccumulator. Biotechnol letter (in press)

Zolla L, Timperio A, Walcher W, Huber C.G (2003) Proteomics of light-harvesting proteins in dfferent plant species. Analysis and comparison by liquid chromatography-electroscopy ionization mass spectrometry. Plant Physiol. 131: 198-214

19

Development and Application of Molecular Markers to Breed Common Bean (*Phaseolus vulgaris* L.) for Resistance to Common Bacterial Blight (CBB) – Current Status and Future Directions

Kangfu Yu, Chun Shi, and BaiLing Zhang
Agriculture and Agri-Food Canada, Canada

1. Introduction

Common bean (*Phaseolus vulgaris* L.) is grown and consumed principally in developing countries in Latin America, Africa, and Asia. It is a major source of dietary protein that complements carbohydrate-rich sources such as rice, maize, and cassava. It is also a rich source of dietary fibres, minerals and certain vitamins (Gepts et al.2008). Common bacterial blight (CBB), incited by *Xanthomonas axonopodis* pv. *Phaseoli* (Smith) Dye (*Xap*), is one of the most destructive bacterial diseases of common bean. CBB is a seed-transmitted disease that is a major yield-limiting factor of common bean production worldwide. CBB can reduce seed quality through staining and browning, which render the bean seed unacceptable for the food processing industry (Yu et al., 2000a). Currently, CBB outbreaks are managed through the use of expensive pathogen free seeds reproduced in certain locations (Scott & Micheals, 1992) and seed treatment with antibiotics such as Streptomycin or via foliar spraying with copper-based compounds (e.g. Kocide™) that are not only costly, but partially effective, and have serious long-term consequences on human and animal health (Forbes & Bretag, 1991; Fininsa, 2003). The exploitation of natural resistance to CBB is the only effective and environmentally sound approach to control this disease in bean production. Sources of genetic resistance to CBB have been identified in common bean and its related species, tepary bean (*P. aculifolius*) and runner bean (*P. coccineus*), but most of them are inherited as quantitative trait loci (QTL) and vary in their levels of genetic effects and their expressions are influenced by environmental conditions (Kelly et al.2003; Miklas et al. 2006). However, dominant gene controlling CBB resistance in common bean was also reported (Zapata et al. 2010). Conventional breeding for resistance to CBB is further aggravated by the pathogen variability, linkage of resistance with undesirable traits (Liu et al. 2008), and different genes conditioning resistance in different plant organs, including leaves, pods, and seeds (Jung et al. 1997; Liu et al., 2009; Aggour et al., 1989; Mutlu et al., 2008; López et al., 2006; Mkandawire et al., 2004; Zapata, 1997). The advent of DNA-based molecular marker (MM) technology has provided an efficient selection tool to breeders in plant breeding (Tanksley et al. 1989). Molecular marker can be defined as a gene or a piece of DNA that is located on a chromosome and can be used as a point of reference (Weber & Wricke, 1994). Since DNA-based MM is phenotypic and environmental neutral, and can be accurately and automatically analyzed with little quantity of DNA (nanogram) in a laboratory at any time for any plant tissue, they can reduce the breeding cost and improve the selection efficiency. Significant progresses on the development and application of MMs to breed bean for CBB resistance have been made in recent years. In this chapter, we will review the current status on the development and application of MMs for CBB breeding in common bean and discuss the future prospects of research in this area.

2. Genetic resistance sources of the CBB resistance gene(s)/or QTL

CBB resistance bean plants can somehow decrease or stop the movement of the *Xap* pathogen through vascular tissues, which can abate the accumulation of bacterial population in leaves or barricade internal seed infection (Goodwin et al., 1995; Aggour & Coyne,1989). Genetic resistance to CBB is relatively low in common bean in comparison with its related species, scarlet runner bean and tepary bean (Singh & Munoz, 1999; Coyne and Schuster, 1973; Yoshii et al., 1978; Mohan, 1982).

Although the genetics of CBB resistance in *P. coccineus* are largely unknown, low to moderate levels of resistance from *P. coccineus* were transferred to common bean (Table 1; Freytag et al., 1982; Park & Dhanvantari, 1987; Miklas et al., 1994; Miklas et al., 1999; Singh & Munoz, 1999). The CBB resistance source for the bean line XR-235-1-1 released by Freytag et al., (1982) was from scarlet runner (*P. coccineus*) bean line PI273667 (Ethiopia). Two minor CBB resistance QTL, derived from PI273667, were identified in the XR-235-1-1 bean line and together they explained about 27% of the phenotypic variation for CBB resistance (Yu et al., 1998). Four CBB resistance lines, C1, C2, C3 and C4 were developed by inter-specific crosses between Common bean and *P. coccineus* (Park and Dhanvantari, 1987). The CBB resistance source for the multiple disease resistance line TARS VCI-4B released by Miklas et al. (1994) was from two *P. coccineus* PI lines, PI311950 and PI311977 (Mexico). Miklas et al. (1999) released four CBB resistant bean lines, ICB-3, ICB-6, ICB-8 and ICB-10, which derived their CBB resistance either from *P. coccineus* or from Great Northern varieties. Singh and Munoz (1999) screened 166 promising germplasm accessions of four *Phaseolous* species including 55 Scarlet Runner beans and found that moderate resistance present in *P. coccineus*. Except the two CBB resistance QTL in XR-235-1-1 was mapped to linkage group A (chromosome 7) and F (chromosome 8) (Yu et al., 1998), the rest of the CBB resistance genes / or QTL from *P. coccineus* are unknown.

Because tepary bean has the highest level of resistance to CBB (Arnaud-Santana et al., 1993; Coyne and Schuster, 1983; Mohan, 1982; Schuster et al., 1983; Singh and Munoz, 1999; Zapata et al., 1985), efforts have been made successfully to transfer the genetic factors controlling CBB resistance from tepary bean into common bean through inter-specific hybridizations (Thomas and Waines, 1984; McElroy, 1985; Parker, 1985; Scott & Michaels, 1992; Singh & Munoz, 1999). The XAN lines, XAN-159, XAN-160 and XAN-161 were developed at CIAT (Centro International de Agricultura Tropical) through inter-specific crosses between *P. vugaris* and *P. acutifolius* accession PI319443 (Thomas and Waines, 1984; Jung et al., 1997). PI319443 showed quantitative inheritance, predominately additive effects, and partial dominance for CBB resistance (McElory, 1985). Based on the CBB resistance data, McElory (1985) hypothesized that one prominent major gene and two other minor genes controlling CBB resistance in PI319443 (McElory, 1985), and the XAN-159, derived its CBB resistance from PI319443, is likely to obtain most of the genes. Parker (1985) transferred CBB resistance from tepary bean into *P. vulgaris* by hybridizing PI440795 (*P. acutifolius*) to 'ICA Pijao' (*P. vulgaris*) and crossing the F_1 progeny to 'Ex Rico 23' (*P. vulgaris*). CBB resistance in both OAC-88-1 line and OAC- Rex Cultivar are originated from PI 440795 (Scott & Michaels, 1992; Michaels et al., 2006). The VAX lines, VAX1, VAX2, VAX3, VAX4, VAX5 and VAX6, were developed from inter-specific cross between the common bean cultivar ICA Pijao and the tepary bean accession G40001 in 1989 by embryo rescue (Mejia-Jimenez et al., 1994; Singh & Munoz, 1999). By far, tepary bean PI319443, PI44079, and G40001 (Table 1) are sources of the major CBB resistance gene(s) /or QTL that have been incorporated into common bean breeding programs (Yu et al., 2000a; Yu et al., 2004; Liu et al., 2008; Bai et al., 1997; Tar'an et al., 2001; Jung et al., 1997; Miklas et al., 2000; Singh & Munoz, 1999; Pedraza et al., 1997).

Miklas et al. 2003, however, revealed that the major CBB resistance gene(s)/or QTL present in the great northern cultivars GN# 1 and GN#1 Selection 27 (GN#1Sel 27) was actually derived from common bean cultivar, Montana No. 5 that was released as a cultivar in 1947 from a selection out of the common great northern landrace (Sutton & Coyne, 2007). CBB resistance gene(s) / or QTL were also identified from the Mesoamerica bean line BAC 6 (Jung et al., 1996). A dominant gene conferring resistance to CBB was recently found in the small white bean line PR0313-58 (Zapata et al. 2010). Four CBB resistance QTLs were also reported in the Middle American breeding line BAT93 (Nodari et al., 1993b). Table 1 summarized the major CBB resistance sources commonly used by common bean breeders.

3. Genomic mapping of bean with MMs

A genetic map is a graphic representation of the arrangement of a gene or a MM on a chromosome, which can be used to locate and identify the gene or group of genes that determines a particular inherited trait. If a linkage between a MM and a trait is established, the MM can be used to carry out indirect selection of the trait in a plant breeding program. The first comprehensive genetic map of bean was developed in 1992 with a seed and flower color marker (P), nine seed protein, nine isozyme and 224 restriction fragment length polymorphism (RFLP) marker loci. Eleven linkage groups (LGs), corresponding to the *P. vulgaris* L. (n = 11), were established in the backcross population derived from the cross between the Mesoamerican breeding line XR-235-1-1 and the Andean cultivar, Calima. The 11 linkage groups covered 960 centiMorgans (cM) of the bean genome (Vallejos et al., 1992). The second bean genetic map was constructed in an F_2 population from the cross of BAT93/Jalo EEP 558 (BJ) with 152 MMs, and 143 of the 152 MMs were assigned to 15 LGs, which covered 827 cM of the bean genome (Nodari et al., 1993a). Adam-Blondon et al. (1994) developed the third bean genetic map with 51 RFLP, 100 random amplified polymorphic DNA (RAPD), and 2 sequence characterized amplified region (SCAR) loci in a backcross population derived from the Ms8EO2 x Corel cross. Twelve LGs were formed in this map that covered 567.5 cM of the bean genome.

Because there are different breeding objectives in bean, which include crop quality, disease and pest resistance, tolerance to abiotic stresses, domestication syndrome, etc. (Gepts et al., 1998, 1999), no single mapping population would segregate for all the economic traits of interest. Thus, genes for these traits have been mapped to different segregating populations. In addition, each map was constructed with most of markers that are not common among them. Therefore, there is a need to compare among these maps to determine the linkage

Origin	cv. or lines	Gene/QTL number	MM	Explained phenotypic variation (R^2)	Reference
Resistance source from common bean (*P. vulgaris*)					
Montana No. 5	GN#1		SAP6	35%	Miklas et al. 2003
	GN#1Sel 27				
BAC6		3	BC409	12%(first trifoliate leaves); 13%(later-developed trifoliate leaves); 17%(pods)	Jung et al. 1996, 1999
			K19		
BAT 93		4		75% (all QTLs)	Nodari et al. 1993b
Belneb RR-1		4	W10	44% (first trifoliate leaves); 41% (pods)	Ariyarathne et al., 1999
	PR0313-58	1	SAP6	55%	Zapata et al. 2010
Resistance source from tepary bean (*P. acutifollus*)					
PI 440795	OAC-88-1	2	R7313	81% (all QTLs)	Bai et al. 1996
	OAC-Rex	3	PV-ccct-001	42.2%(One QTL)	Tar'an et al. 2001
PI 319443	XAN 159	4	BC420 SU91	28% (leaves); 11% (pods); 18% (seeds) 17%	Jung et al. 1997 Mutlu et al. 2005a
	HR 45	2	BC420	25-52%	Liu et al. 2008
	HR 67	2	BC420	70%	Yu et al. 2004
G40001	VAX 1				Singh & Munoz, 1999
	VAX 2				
	VAX 3				
	VAX 4				
	VAX 5				
	VAX 6				
Resistance source from scarlet runner bean (*P. coocineus*)					
PI 273667	XR235-1-1	2		27%	Yu et al. 1998
PI 311950 PI 311977	TARS VCI-4B				Miklas et al. 1994
PI 165421	C1				Park & Dhanvantari, 1987
	C2				
	C3				
	C4				

Table 1. The genetic origins of the major CBB resistance gene(s) /or QTL of common bean (*P. vulgaris*) derived from the *Phaseolus* species. The major QTL and the associated MM present in the genotypes are also shown if they are available.

relations for the different genes and MMs located on the different maps. A deliberate effort was therefore initiated to make a consensus map by determining the segregation in a core mapping population of a few markers selected from each linkage group (LG) on the different maps (Freyre et al., 1998; Kelly et al., 2003). The recombinant inbred line (RIL) population of the cross between the Middle American genotype BAT93 and Andean genotype Jalo EEP558 (BJ) was chosen to develop the integrated core map because it displayed high level of polymorphisms for RFLP and other MMs (Nodari et al., 1992; McClean et al., 2002), and segregated for multiple host–microorganism interactions (Nodari et al., 1993b; Geffroy et al., 1999). Two or more markers from the Florida map (Vallejos et al., 1992), the Paris map (Adam-Blondon et al., 1994), the Davis map (Nodari et al., 1993a) and the Nebraska–Wisconsin maps (Jung et al., 1996, 1997–1999) were mapped in the BJ population. This approach allowed researchers to correlate and align the different LGs of these maps and to establish the collinear relationships of the different LGs among the maps with more than two markers being mapped per LG (Freyre et al., 1998; Gepts, 1999). Although no detailed correlation (i.e., over short distances of the order of 10–15 cM or less) among the LGs of the different maps was provided, the information developed thus far allows at least a rough comparison of the location for genes or QTL placed on independent maps (Kelly et al., 2003). The total map length of the consensus or core map is approximately 1200 cM (Gepts, 1999) and the average relationship between genetic and physical distances is approximately 500 kb/cM, which was verified around the *Phs* (phaseolin) locus on LG B7 (Llaca and Gepts, 1996). The core map is composed of some 550 markers, including RFLP, RAPD, SCAR, allozyme, and seed protein markers; when considering markers from the other, correlated maps, at least 1000 markers have been mapped which would average one marker per 1-2 cM (Freyre et al., 1998; Gepts, 1999; Kelly et al., 2003, Miklas et al., 2006).

Another development in the genetic mapping of the bean genome includes the development of simple sequence repeat (SSR) or microsatellite markers (Yu et al., 1999, 2000b; Blair et al., 2003; Melotto et al., 2005; Hanai et al., 2007). The first preliminary study of the existence of microsatellite markers in the bean genome was conducted by Yu et al. (1999, 2000b), who were able to map 15 of the SSR markers onto the core linkage map. One hundred fifty SSR markers derived from either gene-coding or anonymous genomic-sequences were developed by Blair et al. (2003) and 100 of them were mapped to an F_9 RIL population of the DOR364 x G19833 cross and the BJ core mapping population (Freyre et al., 1998). These loci are integrated with the 15 microsatellite markers previously mapped to seven chromosomes on the BJ population (Yu et al., 2000b). Therefore these studies bring to a total of 115 microsatellite loci on the common bean genetic map, and provides coverage for every chromosome in the genome with from five to 20 SSR markers each (Blair et al., 2003). Two hundreds forty-three SSR sequences were identified from 5,255 bean expressed sequence tag (EST) sequences (Melotto et al., 2005) and 471 SSRs were found from 714 genomic sequences in a microsatellite-enriched genomic library (Hanai et al., 2007). These SSR markers will be invaluable for fine mapping the bean genome and tagging genes of economic interest.

Single nucleotide polymorphisms (SNPs) discovery and marker development is the current focus of several genetic mapping efforts. SNPs are by far the most common form of DNA polymorphism in a genome (Hyten et al., 2010). It was estimated that common bean genome has one SNP per 88 bp (Gaitán-Solís et al., 2008). With an estimated bean genome size of 588 Mb (Liu et al., 2010), over 6 millions of SNPs would present in the entire bean genome. Three hundreds eighteen SNPs were detected from 5,255 bean EST sequences (Melotto et al., 2005). With EST sequence information from two bean genotypes, the Mesoamerican cultivar Negro Jamapa 81 and the Andean cultivar G19833 (Ramirez et al., 2005) , over 1,800 SNPs and indels were detected (McConnell et al., 2010). PCR primers were designed to amplify the BAT93 and Jalo EEP558 fragments and 534 useful fragments with an average size of 500bp were sequenced. Of the 534 gene fragment, 395 (74%) were polymorphic between BAT93 and Jalo EEP558. Three hundreds of the 395 gene fragments (most of them have at least one SNP) were mapped on the BJ core map population (McConnell et al., 2010). The genetic map (LOD 2.0) composed of the 300 gene-based markers, 103 core map markers and 24 other markers covered 1545.5 cM of the bean genome (McConnell et al., 2010). Using a modified deep sequencing of reduced representation library (RRL) (Van Tassell et al., 2008) approach, named as Muli-tier RRL, Hyten et al (2010) discovered 3487 SNPs from the parental lines, BAT 93 and Jalo EEP 558, which was used to develop the bean core map RIL population (Freyre et al., 1998). Because of the abundance of SNPs present in the bean genome, it is expected that SNPs will soon become the most commonly used MMs in bean mapping and breeding programs. Because the SNPs discovered so far are mainly from bean genotypes between, rather than within, the bean gene pools, one limitation of using these SNPs in common bean breeding programs, however, is the lack of polymorphisms (about 10%) of these SNPs between parental lines within each of the bean market classess (Shi et al., 2011b). Therefore, discovering SNPs within gene pool is necessary for routine application of SNPs in bean improvement.

Bacterial artificial chromosome (BAC) libraries are important resources for physical mapping and the development of molecular markers (Yu, 2011). The Development of SSR markers from BAC-end sequences is very cost-effective and offers genome-wide coverage as all repeat types are systematically sampled in the randomly selected BACs (Cho et al., 2004). BAC libraries were constructed for 10 common bean genotypes (Table 2). The first common bean BAC library was developed by Vanhouten & MacKenzie (1999) with the Sprite snap bean-derived genotype for physical mapping of the nuclear fertility restorer *Fr* locus. In 2006, four BAC libraries, three for common bean genotypes BAT93, G21245, G02771, and one for lima bean, cv Henderson (*P. lunatus*), were developed to study the evolution of the arcelin–phytohemagglutinin–a-amylase inhibitor (APA) multigene family (Kami et al., 2006) The four BAC libraries have a range of 9-20 fold genome coverage that should make them useful genetic resources for studying common bean and lima bean. The BAT 93 BAC library has been used successfully for cytogenetic studies of bean chromosomes (Fonseca, et al. 2010; Pedrosa-Harand et al. 2009). BAC libraries were also developed for common bean genotypes G19833 (Schlueter et al., 2008), G12949 (Galasso et al., 2009), HR45 (Liu et al., 2010), G02333 (Melotto et al., 2003), HR67 and OAC-Rex (Gepts et al., 2008).

The G19833 BAC library was used for BAC end sequence analysis to develop BAC derived SSR markers and for physical mapping of the common bean genome (Fonseca, et al., 2010; Córdoba et al., 2010). Liu et al. (2010) used the HR45 BAC library to physically map the major CBB resistance QTL of common bean to the end of chromosome 6. Currently, the OAC-Rex BAC library is being used to sequence the whole genome of the CBB resistance cultivar, OAC-Rex (Pauls, et al. personal communication) and whole genomic sequencing of G19833 is also under way (McClean et al. personal communication). However, the potential utilization of the bean BAC libraries for MM development has not been fully explored in bean.

P. vulgaris L.	Genotype	Vector	R-site[a]	No. of clones	A.I-size[b] kb	Cov[c].	Ref.
(Common bean)	Sprite	pECSBC4	RcoRI	33,792	100	5.3	Vanhouten, & MacKenzie.1999
	Bat 93	pIndigoBac5	HindIII	110,592	125	21.7	Kami et al. 2006
	G21245	pIndigoBac5	HindIII	55296	105	9.2	Kami et al. 2006
	G02771	pIndigoBac5	HindIII	55296	139	12.1	Kami et al. 2006
	G12949	pIndigoBac5	HindIII	30720	135	6.5	Pedrosa-Harand et al. 2009
	G19833	pIndigoBac 536	HindIII	55296	145	12.6	Schlueter et al. 2008
	HR45	pIndigoBac5	HindIII	33,024	107	5.5	Liu et al. 2010
	HR67	BIBAC2	BamHI	22,560	300	10.6	Gepts et al. 2008
	OAC-Rex	BIBAC2	BamHI	31,776	150	7.5	Gepts et al. 2008
	G02333	pBeloBac11	HindIII	24,960	125	4.9	Galasso, et al. 2009

Table 2. BAC libraries developed for common bean genotypes. [a] R-site is restriction cutting site for cloning; [b] A.I.-size is the average insert size of the BAC library in kilo bases; [c] Cov is the times of genome equivalent coverage.

Cytological analysis of the common bean chromosomes has long been hampered by the small size and overall similarity of its (2n=22) chromosomes. Unambiguous identification of each bean chromosome was not possible until the double fluorescent in situ hybridization (FISH) with 45S and 5S rRNA probes, followed by 4'-6-diamidino-2-phenylindole (DAPI) counterstaining techniques were developed (Moscone et al., 1999). The first common bean mitotic chromosome nomenclature was proposed by Moscone et al. (1999). Chromosomes were characterized with respect to size, morphology, heterochromatin content and distribution of rDNA genes by fluorescent in situ hybridization (FISH) and assigned numbers from 1 to 11, based on size from largest to smallest, using the European cultivar 'Wax' as a reference (Pedrosa et al. 2003). Because the numbering of linkage groups according to Freyre et al. (1998), B1 to B11, has been widely used by the bean community, it was agreed during the Phaseomics III meeting in 2004 that chromosomes should be reassigned numbers based on the linkage group nomenclature (Pedorsa et al., 2008). The integration of the common bean LGs and the chromosome map was a breakthrough in common bean genetics. The standard nomenclature can be found in the BIC website (http://www.css.msu.edu/bic/Genetics.cfm). In two recent studies, a cytogenetic map of common bean was built by FISH of BACs selected with markers mapping to the 11 linkage groups, plus 2 plasmids for 5S and 45S ribosomal DNA and one bacteriophage (Pedrosa et al., 2009; Fonseca et al., 2010). It was found that about 50% of the bean genome is heterochromatic and that genes and repetitive sequences on the bean chromosomes are intermingled in the euchromation and heterochromatin.

4. Molecular tagging of CBB resistance gene(s) /or QTL and application of MMs in bean breeding

The positioning of MMs linked to CBB on the bean genetic maps has revolutionized our understanding of resistance to CBB (Kelly et al., 2003). Nodari et al. (1993) used 150 restriction fragment length polymorphism (RFLP) markers and 70 F_2-derived F_3 families of the BAT 93 X Jalo EEP558 cross to map the genetic factors controlling CBB resistance present in BAT 93. They identified 4 putative QTLs located on 4 different linkage groups. These four QTLs explained 75% of the phenotypic variation for CBB resistance. Random amplified polymorphic DNA (RAPD) markers were used by Jung et al. (1996) to tag the QTL in a recombinant inbred line (RIL) population derived from the cross between BAC 6 x HT 7719. The CBB resistance of this RIL population was from the BAC 6 Mesoamerican bean line. RAPD marker BC409.1250 was found to associate with a major QTL conditioning CBB resistance to the *Xap* EK-11 strain in first trifoliate leaves (account for 12% of the phenotypic variation), later-developed trifoliate leaves (13% of the phenotypic variation), and pods (17% of the phenotypic variation). Another QTL conditioning CBB resistance to the *Xap* Epif-IV strain in the later-developed trifoliate leaves (10% of the phenotypic variation) was found to link with RAPD marker U16.600-H11.650. Other QTL were also detected but they were of less significance (account for less than 10% of the phenotypic variation). MMs associated with QTL of less effect are usually not very effective in MAS. The two genomic regions that enclose the CBB resistance QTL were confirmed later with 3 additional mapping populations and 1 additional *Xap* strain (Jung et al., 1999). It is interesting that the BC 409.1250 RAPD marker was significantly associated with CBB resistance to all 3 *Xap* strains (EK-11, Epif-IV, and DR-7) in all the 4 populations {BAC 6 x HT 7719 (BH), Venezuela x BAC6 (BV), Belneb RR-1 x A55 (BA), and PC50 x BAC6 (PB)}. The BC 409.1250 marker, however, was not linked to any of the other markers used in their experiments (Jung et al., 1996, 1999).

The other genomic region tagged by U16.600-H11.650 in the BH population (Jung et al., 1996) was also confirmed in 3 of the 4 populations (BH, BA, and PB). Two RAPD markers, K19.450 and AG15.660, were mapped to the same location across the 4 populations (Jung et al., 1999). The BC409.1250 RAPD marker was converted into a sequence characterised amplified region (SCAR) marker (Paran & Michelmore, 1993) for reliable PCR amplification in MAS of bean breeding programs (Jung et al., 1999; Ariyarathne et al., 1999). The two genomic regions tagged, respectively, by BC409.1250 and K19.450 (Jung et al.,1996, 1999) was later mapped to the same linkage group (LG) B10 by Ariyarathne et al. (1999). Another RAPD marker, W10.550, located in the same LG was also identified, which is tightly associated with the CBB resistance QTL in first trifoliate leaf and pod. This QTL explained up to 44% and 41% of the phenotypic variation in first trifoliate leaf and pod, respectively (Ariyarathne et al., 1999). These results provideed strong evidence that more than one QTL are located on the bean linakge group 10 (Chromosome 10). The SCAR marker, derived from the BC409.1250 RAPD marker, was named as BAC 6 SCAR marker, which is one of the commonly used MM in MAS of bean for CBB resistance (http://www.css.msu.edu/bic/PDF/SCAR_Markers_2010.pdf).

By using selective genotyping and bulked segregant analysis (Michelmore et al., 1991), Miklas et al. (1996) detected two genomic regions affecting CBB resistance with an $F_{5:7}$ RIL population derived from the Dorado x XAN 176 cross. One genomic region defined by RAPD marker AP6.820 on LG PR14 explained 60%, 10% and 30 of the phenotypic variation in greenhouse-leaf, field-leaf and greenhouse-pod reactions, respectively. The AP6.680 RAPD marker was mapped to the end of LG 10 by Miklas et al. (2000a) and converted to the SCAR marker SAP6.680 (Miklas et al., 2003) which was positioned between the W10.550 and the BAC 6.1250 markers on LG10 in the BAC 6/HT 7719 population (Ariyaranthe et al., 1999). So there is a good chance that both the SAP 6.680 and the BAC 6.1250 markers are actually linked to the same QTL (Miklas, personal communication). The other genomic region on LG PR7 or B7 defined by A19.600 explained 12% and 40% of the phenotypic variation in greenhouse-leaf and field-leaf reaction. The occurrence of common genomic region affecting CBB resistance in different tissues and under different environmental conditions indicate that the expression of CBB resistance may be conditioned by more than one genetic factors clustered together or by one genetic factor with pleiotropic effect.

Jung et al. (1997) identified four genomic regions present in XAN 159, a bean line was developed by inter-specific crossing between tapary bean PI319443 and common bean, conditioning CBB resistance at first trifoliate leaves, pods and seeds. One genomic region on their LG 5 defined by BC420.900 RAPD marker was found to contain a major QTL conferring CBB resistance to *Xap* strain DR-7 and EK-11 in leaf (accounted for 28% of the phenotypic variation), pod (11%) and seed (18%). The CBB resistance in XAN159 was successfully transferred into common bean by Park & Dhanvantari (1994) and Park et al. (2007) to develop the HR45 and HR67 bean germplasm, which are highly resistant to CBB in both field and greenhouse test (Yu et al., 2000a; Liu et al., 2008). The major QTL of XAN159 origin linked with BC420.900 RAPD marker was confirmed by Yu et al. (2000a) in a RIL population derived from the HR 67 x W1744d cross. This QTL explained 62% of the phenotypic variation in CBB reaction tested in the greenhouse environment. The BC420.900 RAPD marker was also converted into a SCAR marker, which is one of the most commonly used markers for MAS in breeding bean for CBB resistance (Yu et al., 2000a; Liu et al., 2008; Kelly et al., 2003; Miklas et al., 2006; Fourie and Herselman 2002, Park and Yu 2004; Ibarra-Perez and Kelly 2005; Liu et al. 2005; Mutlu et al.2005b). An SSR marker in the bean nitrate reductase gene was also found to have tight linkage with the BC 420.900 marker and CBB resistance in a RIL population derived from the HR67 x OAC-95-4 cross (Yu et al., 2004). The same RIL population was also used by Liu et al. (2008) to develop several sequence tagged site (STS) markers that are tightly linked to this major QTL. They also confirmed that the BC420.900 QTL is located on LG B6 (chromosome 6) rather than LG B7 reported by Yu et al. (2004). One negative aspect of this QTL is its linkage with the *V* gene, which conditions purple flower color and dark seed colors (Jung et al., 1997; Miklas et al., 2006; Kelly et al., 2003; Liu et al., 2008). This linkage drag is unfavourable for breeding CBB resistance pinto, cranberry, and red kidney beans If XAN159, HR67 or HR45 were used as the CBB resistance source (Liu et al., 2008; Miklas et al., 2006). Bai et al. (1997) detected tight association between the RAPD marker BC73.700 and a major CBB resistance QTL in OAC-88-1, a breeding line developed by the inter-specific cross between tapary bean PI440795 x common bean 'ICA Pijao' (Scott & Michaels, 1992). This QTL could explain 45% of the phenotypic variation and believed to be located on LG8 (Chromosome 8) (Kelly et al., 2003) because of its linkage with the SU91.700 SCAR marker that is tightly linked to a major QTL in XAN159, which derived its CBB resistance from tapary bean PI 319443 (Pedraza et al., 1997; Kelly et al., 2003; Liu et al., 2009). The BC73.700 RAPD marker was also converted into a SCAR marker for efficient use in MAS (Beattie et al., 1998). The exact map location of the SU91 is

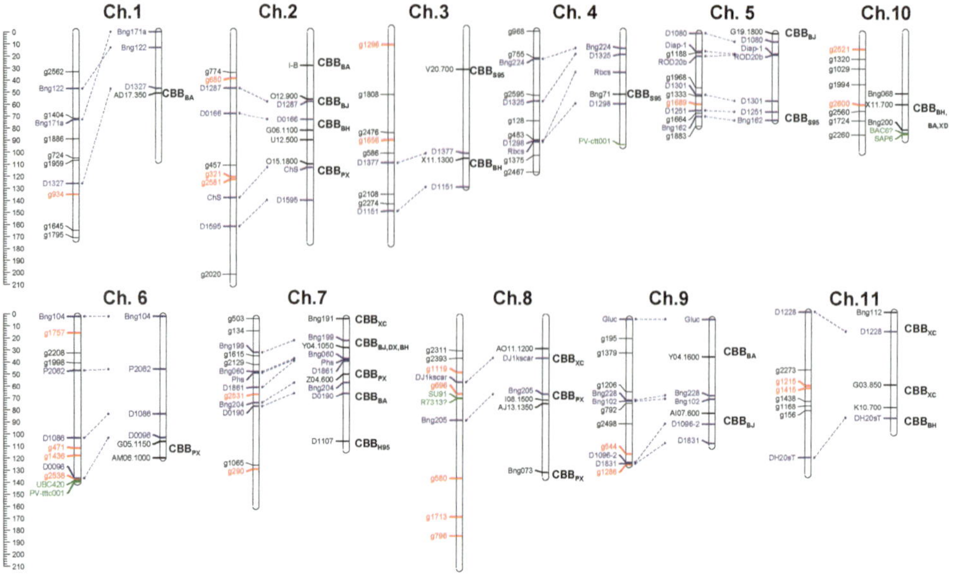

Fig. 1. The chromosome distribution of the identified CBB resistance gene (s) /or QTLs and the associated molecular markers, resemble the maps presented by Shi et al. (2011b). For each linkage group, the map on the left is reproduced from McClean (2007) map {(http://www.comparative-legumes.org/); McConnell et al. 2010}, the map on the right is reproduced from the bean core map {(http://www.comparative-legumes.org/); Freyre et al. 1998}, adopted from Miklas et al (2006) . Both maps are integrated by shared markers except for linkage group B10 (Shi et al. 2011b). In McClean (2007) map, only molecular markers used in association study (in black, Shi et al. 2011b), shared markers (in blue, Freyre et al. 1998), and the Breeder-friendly SCAR and SSR markers (in green) are shown. The markers in red were found significantly ($P \leq 0.05$) associated with CBB resistance (Shi et al. 2011b). In Freyre (1998) map, loci placed on the left side of each chromosome were shared markers in blue and molecular markers closest to previous identified CBB-QTLs. To the right of each linkage group are previously identified CBB-QTLs in different populations (Miklas et al. 2006). Symbols in subscript represent the source population of the QTL: BA Belneb-RR-1/A55, BJ BAT93/JaloEEP558, BH BAC6/HT7719, DX DOR364/XAN176, H95 HR67/OAC95, PX PC50/XAN159, S95 Seaforth/OAC95 and XC XR-235-1-1/Calima. Marker UBC420, PV-ttc001, PV-ctt001, SU91, SPA6 and QTL locations are approximate because most were not directly mapped in the BAT93/JaloEEP558 population. The total distance of each linkage group is expressed in cM (Kosambi mapping function).

not determined so far. Recent studies, however, found that SU91.700 SCAR marker appears to have two similar PCR fragments co-migrating on agrose gel and shows distorted segregation in some mapping populations (Xie et al., unpublished results). Interestingly, the two QTL associated with BC73.700 and SU91.700 were derived from tapary bean PI 440795 and PI 319433, respectively. This suggests that the two QTL may be the same or of the same origion. Tar'an et al. (2001) identified three QTL from OAC-95-4, which is a sibling breeding line of the OAC-88-1 and the experimental name of the OAC-Rex cultivar (Michaels et al., 2006). One

Major QTL	Chromosome Location	Linked MMs	Primer sequences	Size (bp)	Reference
PV-cttt001	4 or 5 ?	PV-cttt001 SSR	F- gagggtgtttcactattgtccctgc R- ttcatggatggtggaggaacag	152	Tar'an et al. 2001 Kelli et al. Unpublished
BC 420	6	BC420 SCAR	F- gcagggttcgaagacacaxtgg R- gcagggttcgcccaataacg	900	Yu et al. 2000a
		PV-tttc001 SSR	F- tttacgcaccgcagcaccac R- tggactcatagaggcgcagaaag	161	Yu et al. 2004
		STS 183	F- cctatgtacttcttgagggagac R- agaagcccagggacttggat	142	Liu et al. 2008
		STS 333	F- cataagatgaatggttcttgac R- ccatttggtgagattcactt	274	Liu et al. 2008
		GTM 1	F- ccacctgccacatagacctt R- tctcgagaagggcagaggta	459	Yu et al. Unpublished
		GTM 2	F- cgagactcgtgtgctctctg R- acgaaggttgattcccagtg	519	Yu et al. Unpublished
SU 91	8	SU91 SCAR	F- ccacatcggttaacatgagt R- ccacatcggtgtcaacgtga	700	Pedraza et al. 1997
		GTM 3	F- atggtggagacgagatgacc R- tccgacattgaaaccagttg	425	Yu et al. Unpublished
		GTM 4	F- ggcgacggcttctttgac R- tccaaagaccaaagggtgag	464	Yu et al. Unpublished
R 7313	8	R7313 SCAR	F- attgttatcgtcgacacg R-aatatttctgatcacacgag	700	Bai et al. 1997 Beattie et al. 1998
SPA6	10	SPA6 SCAR	F-gtcacgtctccttaatagta R-gtcacgtctcaataggcaaa	820	Miklas et al. 2003
BAC6	10	BAC6 SCAR	F- taggcggcggcgcacgttttg R-taggcggcggaagtggcggtg	1250	Jung et al. 1999

Table 3. Commonly used MMs associated with the major CBB resistance gene(s) /or QTL and their primer sequences

major QTL associated with the SSR marker *PHVPVPK-1* was mapped on LG G5 (corresponding to B5 or chromosome 5), which explained over 42% of the phenotypic variation. The SSR marker *PHVPVPK-1* linked to this major CBB resistance locus in the study of Tar'an et al. (2001) is equivalent to the PV-ctt001 reported by Yu et al. (2000b) that was positioned on LG B4 (chromosome 4) of the 'BAT93' – 'Jalo EEP558' core map (Yu et al.

2000b). The mapping position discrepancy may be because the marker is located at the end of a linkage group in both maps, a position known to be difficult to map accurately, especially in a small population. Differences in the genetic background between the two mapping populations may also contribute to the discrepancy in marker location (Tar'an et al., 2001). Recent studies, however, indicate again that PV-ctt001 is likely located on B4 (chromosome 4) rather than B5 (chromosome 5) and the QTL associated with PV-ctt001 is not significant in the absence of BC 420 (Kelli et al., unpublished results). The other major QTL was positioned on LG G2 (corresponding to B4 or chromosome 4), which was tagged by*BNG71DraI* and explained 36% of the phenotypic variation for CBB resistance.

The SCAR markers, BC420, SU91 and SAP6 linked with the 3 major QTL of *P. acutifolius* (BC420 and SU91) and *P.vulgaris* (SAP6) origins on B6, B8, and B10, respectively, have been used for MAS of CBB resistance (Table 3, Mutlu et al., 2005b; Yu et al., 2000a) and to validate QTL presence in resistant lines selected by phenotypic selection (see review by Miklas et al., 2006, Fouie & Herselman, 2002). Because of the differences in molecular weights among the 3 SCAR markers, they can be multiplexed in one PCR reaction (Miklas et al., 2000) to speed up MAS for combined resistance to CBB (Miklas et al., 2006). The SU91 SCAR marker associated QTL tentatively positioned on LG 8 or chromosome 8 has been used the most often for MAS (Miklas et al 2005, Mutlu et al 2005b, Hou et al. 2010, Navabi et al., unpublished results; Yu et al unpublished results). Epistatic interactions, however, between BC420 and SU91 CBB resistance QTL were detected (Vandemark et al., 2008), which would complicate the use of MMs in MAS. Epistatic interactions between other CBB resistance QTL were also reported in several other studies (Jung et al.1997; Tar'an et al., 1998; O'Boyle et al., 2007,).

5. Cloning CBB resistance gene(s) / or QTL

Current technical progress in the area of molecular biology and genomics have made the cloning of QTL [i.e. the identification of the DNA sequences (coding or non-coding) responsible for QTL] possible. To date, most plant QTL have been cloned using a positional cloning approach following identification in experimental crosses. In some cases, an association between sequence variation at a candidate gene and a phenotype has been established by analysing existing genetic accessions (Salvi & Tuberosa, 2005). A literature survey shows that although about 150 research papers reporting original QTL data are published yearly (average of 2000-2004, considering Arabidopsis, soybean, rice, sorghum, maize, barley and wheat), only a handful of studies have reported the cloning of QTL (Salvi & Tuberosa, 2005). In common bean, there is only one report on efforts to clone the major CBB resistance QTL by Yu et al. (2010). Two approaches were used by Yu's group to clone the major CBB resistance QTL associated with the BC 420 SCAR and the SU 91 SCAR markers present in the CBB resistance line HR45. Namely, map-based position cloning and candidate gene approaches. With the first approach, they first mapped the BC420 QTL to chromosome 6 at a distal region (Figure 3A, Yu et al., 2000a, Liu et al., 2008,). A BAC library was then developed to physically map the QTL to a genomic region of about 750 kb with 6 BAC clones (Figure 3 B, Liu et al., 2010). One BAC clone, 4k7, containing the BC420 marker was sequenced. The sequenced 4k7 BAC was assembled into a 90kb single contig for functional annotation. Since the BAC was selected by marker BC420, the entire sequence of BC420 was fully recovered from this BAC sequence. Twenty-one genes were *ab initio* predicted by FGENESH using Medicago gene model, including 11 from sense chain and 10 from anti-sense chain (Figure 3C). Although no homology to any previously identified common bean genes was found, six of the putative genes were supported by common bean ESTs and three of them were supported by runner bean ESTs. The expression of 6 putative genes with supported bean ESTs was assessed and verified by real time RT-PCR. For each putative gene, one or two primer pairs were designed and tested in the contrasting NILs (Near Isogenic Lines) (Vandemark et al. 2008). Fifty-seven percent (8 of 14) of the primer pairs were polymorphic. Seven of them are dominant markers present in the NILs harboring the BC420-QTL, but one is a co-dominant marker. Based on the simple repetitive elements found in the BAC sequences, seven SSR markers were designed and tested in the contrasting NILs. Three of them turned out to be polymorphic, including two dominant and one co-dominant markers. Overall, eleven new markers have been developed in association with CBB resistance in HR45. Another BAC clone harbouring the SU91 SCAR marker was also identified from the HR45 BAC library and sequenced. A single contig of 58kb was assembled. Sixteen genes were also predicted by FGENESH using Medicago gene model (Figure 2).

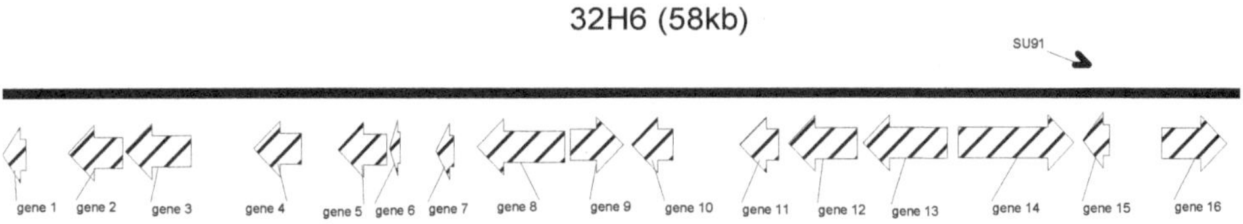

Fig. 2. Annotated genes that are present in the SU91 BAC clone.

In parallel, cDNA-amplified fragment length polymorphism (AFLP) technique was used to identify the candidate genes (CG) that are differentially expressed in the leaves of HR45 sampled at different time-periods after inoculation (Shi et al., 2011a). Selective amplifications with 34 primer combinations allowed the visualization of 2,448 transcript-derived fragments (TDFs) in infected leaves; 10.6% of them were differentially expressed. Seventy-seven differentially expressed TDFs (DE-TDFs) were cloned and sequenced. 50.6% (39 of 77) of the DE-TDFs representing modulated bean transcripts were not previously reported in any EST database then. The expression patterns of 10 representative DE-TDFs were further confirmed by real-time RT-PCR. BLAST analysis suggested that 40% (31 of 77) of the DE-TDFs were homologous to the genes related to metabolism, photosynthesis, and cellular transport, whereas 28% (22 of 77) of the DE-TDFs showed homology to the genes involved in defence response, response to stimulus, enzyme regulation, and transcription regulation. Thus, the 22 pathogenesis-related DE-TDFs were selected as functional candidate genes (FCGs) in association with CBB resistance. Meanwhile, six of the FCGs were *in silico* mapped to the distal region of the chromosome 6 (the genomic region of the CBB resistance QTL linked to BC420 in HR45) and were chosen as positional candidate genes (PCGs) for comparative mapping. Comparing the CGs found from map-based cloning to the CGs derived from cDNA-AFLP, none of them is overlapped. This indicates that gene expression studies may characterize the downstream transcriptional cascade of the QTL. The PCGs could be the genes for CBB resistance, whereas the FCGs genes that map to other locations may be involved in the molecular responses related to the QTL (Shi et al., 2011a).

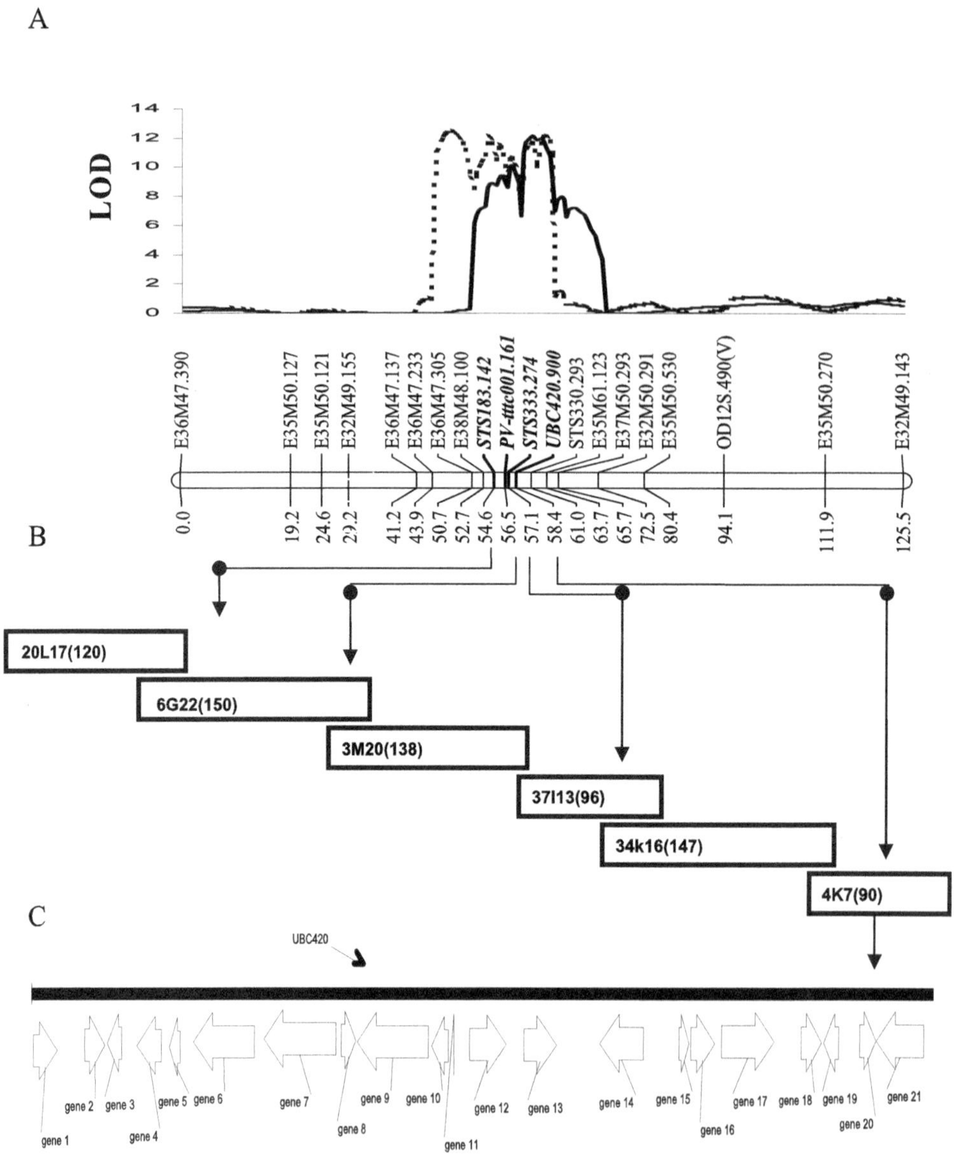

Fig. 3. Genetic, physical and sequence map of the BC420 QTL. A) A genetic map showing the map location of the BC420 QTL in the HR 67 X OAC-95-4 recombinant inbred line population (Liu et al. 2008);B) A physical map showing the contig that contains the BC420 SCAR marker (Liu et al. 2010); C) Annotated genes present in the BAC clone 4k7 (Yu et al. unpublished).

6. Summary and future prospects

Significant progresses have been made in introducing CBB resistance gene(s)/or QTL into common bean from its related species, in dissecting genetic factors underlying CBB resistance, and in mapping CBB resistance QTL on bean LG or chromosomes. Molecular markers linked to several major CBB resistance QTL have been developed and used in marker assisted breeding of common bean for CBB resistance. However, all of the QTL mapping studies reviewed so far have been based on the analysis of populations derived from bi-parental crosses that segregated for trait(s) of interest. To date, at least 24 different CBB resistance QTL have been positioned across all eleven LG or chromosomes of common bean (Figure 1, Kelly, et al., 2003; Miklas et al., 2006; Shi et al., 2011b). Because these QTL were mapped in a number of different bi-parental populations, the MMs associated with each of the major QTL only segregate in one or a few of the mapping populations (Shi et al., 2011b). As a result, co-alignment of the QTL associated MMs from the different studies could not be done accurately and several questions remain unanswered. For example, what is the map location of SU91 QTL relative to the R7313 QTL on LG 8 or Chromosome 8? Are they the same or different? What is the map location of SPA6 QTL relative to BAC6 QTL on LG10 or chromosome 10? Are the same or different? Thus markers linked to these QTLs are not immediately available for use in other bean breeding programs. Validation of QTL effects in other genetic backgrounds is necessary prior to widespread application of the QTL linked markers for MAS (Shi et al. 2011b).

Alternatively, association mapping is a new QTL mapping approach that can be done in natural populations, cultivars released over years, and/or the breeding materials within a breeding program (Oraguzie et al., 2007). These types of populations or a subset of them may represent a smaller set of the available genetic diversity within a breeding program. Collections of these breeding lines may provide great potential for applied association mapping experiments because they are routinely evaluated in the breeding programs and regional trials to assess their local adaptation or response to biotic and/or abiotic stresses (Oraguzie et al., 2007).

Association mapping is increasingly being used to identify marker-QTL linkage associations using plant materials routinely developed in breeding programs. Compared with conventional QTL mapping approach, association mapping using breeding populations should be a more practical strategy for cultivar development, considering that markers linked to a major QTL can immediately be applied for MAS, once new QTL are identified. For instance, in soybean (*Glycine max* L. Merr.) two markers, Satt114 and Satt239, were found to be associated with iron deficiency chlorosis loci using advance breeding lines (Wang et al., 2008). In rice (*Oryza sativa* L.), microsatellite markers associated with yield and its components were identified in a variety trial, and many of them were located in regions where QTL had previously been identified (Agrama et al., 2007). Association mapping studies have also been used to investigate the genetic diversity within crop species. High levels of LD (Linkage Disequilibrium) (pairwise LD: 56%; average $r^2 = 0.1$) was found in common bean (Monica et al., 2009). Much higher LD was observed in domesticated populations (pairwise LD: 57.3%; average $r^2 = 0.18$) compared to wild populations (pairwise LD: 31.5%; average $r^2 = 0.08$) (Monica et al., 2009). In the presence of high LD, lower marker density is required for a target region with greater potential for detecting markers strongly associated with the target gene polymorphism, even if distant physically (Shi et al., 2011b). Thus, whole-genome-scan association study is feasible for bean domestic populations (Monica et al., 2009). In association mapping, where unlike conventional QTL mapping, populations of un-structurally related individuals are employed, it is important to consider population structure and kinship among individuals, because false associations may be detected due to the confounding effects of population admixture (Oraguzie et al., 2007).This may indeed be the case for populations sampled from large collections, breeding materials, or from released cultivars. Therefore, it is necessary to apply appropriate statistical methods that account for population structure and kinship among individuals. A Mixed Linear Model (MLM) approach has been developed to account for multiple levels of relatedness simultaneously as determined by kinship estimates based on a set of random genetic markers (Yu et al., 2006). This model has been proven useful in genome-wide association studies to control the biases that may be caused by population structure and relatedness in other species e.g., maize (*Zea mays* L.) (Yu et al., 2006), rice (Wang et al., 2008). Another issue for association mapping is reliability, an issue of particular concern when the goal is to discover marker/trait associations that have broad application. Shi et al. (2011b) conducted the first association mapping study in common bean for CBB resistance. Using CBB resistance data collected in a CBB field nursery from 395 of the 469 dry bean lines of different market classes representing plant materials routinely developed in a bean breeding program in Ontario, Canada and 132 SNPs evenly distributed across the bean genome, significant associations between CBB resistance data, collected at 14 and 21 days after inoculation, and 14 MMs were detected. Among the 14 MMs, previously identified BC420 and SU91 SCARS were confirmed for their association with CBB resistance (Shi et al., 2011b). The rest of the markers were SNPs, which were co-localized with or close to the CBB-QTLs identified previously in bi-parental QTL mapping studies. Given the abundance of SNPs exits in the common bean genome, the possibility of automation for the analysis of SNPs, and the efficiency of association mapping approach, it is expected that the use of association mapping with SNPs will be the method of choice for mapping CBB resistance gene(s) /or QTL in the future.

Despite of the progresses that have been made for understanding CBB resistance in common bean, there are still a number of challenges lying ahead, such as 1) what is the molecular mechanisms underlying the epistatic interactions between CBB resistance QTL, such as the BC 420 and SU91 QTL? 2) The epistatic interaction between CBB resistance gene(s)/or QTL would make the application of MMs to assist in breeding CBB resistance bean and the cloning of the QTL more complex, and 3) the recalcitrant nature of common bean to transformation make the validation of CGs coding for CBB resistance more difficult to accomplish. So it is an urgent task for bean researchers to develop novel approaches, such as virus induced gene silencing (VIGS) (Zhang et al., 2010) for validating the functions of CGs in common bean.

7. Acknowledgement

The authors would like to thank the financial supports from the Ontario White Bean Producers' Marketing Board, The Ontario Coloured Bean Growers' Association, and the Agriculture and Agri-Food Canada.

8. References

Adam-Blondon, A., Se´vignac, M., & Dron, M. (1994). A genetic map of common bean to localize specific resistance genes against anthracnose. *Genome*, 37, pp. 915-924.

Aggour, A.R., & Coyne, D.P. (1989). Heritability, phenotypic correlations, and associations of the common blight disease reactions in beans. *J. Am. Soc. Hortic. Sci.*, 114, pp. 828–833.

Agrama, H.A., Eizenga, G.C., & Yan, W. (2007). Association mapping of yield and its components in rice cultivars. *Molecular Breeding*, 19, pp. 341-356.

Ariyarathne, H.M.,Coyne, D.P., Jung, G., Skroch, P.W., Vidaver, A.K., Steadman, J.R., Miklas, P.N., & Bassett, M.J. (1999). Molecular mapping of disease resistance genes for halo blight, common bacterial blight, and bean common mosaic virus in a segregating population of common bean. *J Am Soc Hort Sci*, 124, pp. 654–662.

Arnaud-Santana, E.T., Mmbaga, T., Coyne, D.P., & Steadman. J.R. (1993). Sources of resistance to common bacterial blight and rust in elite *Phaseolus vulgaris* L. germplasm. *HortScience*, 28, pp. 644–646.

Arnaud-Santana, E., Coyne, D.P., Eskridge, K.M., & Vidaver, A.K. (1994). Inheritance, low correlations of leaf, pod, and seed reactions to common blight disease in common beans, and implications for selection. *J. Am. Soc. Hortic. Sci.*, 119, pp. 116–121.

Bai, Y., Michaels, T.E., & Pauls, K.P. (1997). Identification of RAPD markers linked to common bacterial blight resistance genes in *Phaseolus vulgaris* L. Genome, 40, pp. 544–551.

Beattie, A., Michaels, T.E., & Paul, K.P. (1998). An efficient reliable method to screen for common bacterial blight (CBB) resistance in *Phaseolus vulgaris* L. *Annu. Rept. Bean Improv. Coop.*, 41, pp. 53-54.

Blair, M. W., Pedraza, F., Buendia, H. F., Gaitan-Solis, E., Beebe, S. E., Gepts, P. & Tohme, J. 2003. Development of a genome-wide anchored microsatellite map for common bean (*Phaseolus vulgaris* L.). *Theor. Appl. Genet.*, 107, pp. 1362-1374.

Cho, S.H., Chen, W.D., & Muehlbauer, F.J. (2004). Pathotype-specific genetic factors in chickpea (*Cicer arietinum* L.) for quantitative resistance to ascochyta blight. *Theoretical and Applied Genetics*, 109, pp. 733-739.

Córdoba, J. M., Chavarro, C., & Schlueter, J. A. (2010). Integration of physical and genetic maps of common bean through BAC-derived microsatellite markers. *BMC Genomics*, 11, pp. 436-445.

Coyne, D.P., & Schuster, M.L. (1973). *Phaseolus* germplasm tolerant to common blight bacterium (*Xanthomonas phaseoli*). *Plant Dis. Rep.*, 57, pp. 111-114.

Coyne, D.P., & Schuster, M.L. (1983). Genetics of and breeding for resistance to bacterial pathogens in vegetable crops. *HortScience*, 18, pp. 30–36.

Fininsa, C. (2003). Relationship between common bacterial blight severity and bean yield loss in pure stand and bean-maize intercropping systems. *International Journal of Pest Management*, 49, pp. 177-185.

Forbes, C.J., & Bretag, T. W. (1991). Efficacy of foliar applied streptomycin for the control of bacterial blight of peas. *Australasian Plant Pathology*, 20, pp. 115-118.

Fonseca, A., Ferreira, J., dos Santos, T. R. B. Mosiolek, M., Bellucci E., Kami, J., Gepts, P., Geffroy, V., Schweizer, D., dos Santos, K. G. B., & Pedrosa-Harand, A. (2010). Cytogenetic map of common bean (*Phaseolus vulgaris* L.). *Chromosome Research*, 18, pp. 487–502.

Fourie, D., & Herselman, L. (2002). Breeding for common blight resistance in dry beans in South Africa. *Annu Rep Bean Improv Coo,* 45, pp. 50–51.

Freyre, R., Skroch, P. W., Geffroy, V., Adam-Blondon, A. F., Shirmohamadali, A., Johnson, W.C., Llaca, V., Nodari, R.O., Pereira, P.A., Tsai, S. M., Tohme, J., Dron, M., Nienhuis, J., Vallejos, C. E., & Gepts, P. (1998). Towards an integrated Linkage map of common bean. 4. Development of a core linkage map and alignment of RFLP maps. *Theoret. Appl. Genet.,* 97, pp. 847–856.

Freytag, G. F., Bassett, M. J., & Zapata, M. (1982). Registration of XR-235-1-1 bean germplasm. *Crop Sci.,* 22, pp. 1268-1269.

Gaitán-Solís, E., Choi, I.Y., Quigley, C., Cregan, P., & Tohme, J. (2008). Single nucleotide polymorphisms in common bean: their discovery and genotyping using a multiplex detection system. *The Plant Genome,* 1, pp. 125-134.

Galasso, I., Piergiovanni, A.R., Lioi, L., Campion, B., Bollini, R., & Sparvoli, F. (2009). Genome organization of Bowman–Birk inhibitor in common bean (*Phaseolus vulgaris* L.). Mol. Breeding, 23, pp. 617–624.

Geffroy, V., Sicard, D., de Oliveira, J.C.F., Se´vignac, M., Cohen, S., Gepts, P., Neema, C., Langin, T., & Dron, M. (1999). Identification of an ancestral resistance gene cluster involved in the coevolution process between *Phaseolus vulgaris* and its fungal pathogen *Colletotrichum lindemuthianum. Mol. Plant–Microb. Interact.,* 12, pp. 774–782.

Gepts, P. (1998). Origin and evolution of common bean, past event and recent trends. *HortScience,* 33, pp. 1124–1130.

Gepts, P. (1999). Development of an integrated linkage map. Common Bean Improvement in the Twenty-First Century. In: *Developments in Plant Breeding,* Singh, S.P. (eds), pp.53–91, Kluwer Academic Publishers, Netherlands.

Gepts, P., Aragão, F.J.L., de Barros, E., Blair, M.W., Brondani, R., Broughton, W., Galasso, I., Hernández, G., Kami, J., Lariguet, P., McClean, P., Melotto, M., Miklas, P., Pauls, P., Pedrosa-Harand, A., Porch, T., Sánchez, F., Sparvoli, F., & Yu, K. (2008). Genomics of Phaseolus Beans, a Major Source of Dietary Protein and Micronutrients in the Tropics. In *Genomics of Tropical Crop Plants,* Moore, P.H., & Ming, R. (eds.), pp. 113-140, Springer, Germany.

Goodwin, P.H., Sopher, C.R., & Michaels, T.E. (1995). Multiplication of *Xanthomonas* pv. *phaseoli* and intercellular enzyme activities in resistant and susceptible beans. *Phytopathology,* 143, pp. 11–15.

Hanai, L.R., de Campos, T., Camargo, L.E.A., Benchimol, L.L., de Souza, A.P., Melotto, M., Carbonell, S.A.M., Chioratto, A.F., Consoli, L., Formighieri, E.F., Siqueira, M.V.B.M., Tsai, S.M., & Vieira, M.L.C. (2007). Development, characterization, and comparative analysis of polymorphism at common bean SSR loci isolated from genic and genomic sources. *Genome,* 50, pp. 266-277.

Hyten, D.L., Song, Q., Fickus, E.W., Quigley, C.V., Lim, J., Choi, I., Hwang, E., Pastor-Corrales, M., & Cregan, P.B. (2010). High-throughput SNP discovery and assay development in common bean. *BMC Genomics,* 11, pp. 475-482.

Ibarra-Perez, F. J., & Kelly, J. D. (2005). Molecular markers used to validate reaction of elite bean breeding lines to common bacterial blight. *Annu. Rep. Bean Improv. Coop.,* 48, pp. 98-99.

Jung, G., Skroch, P.W., Coyne, D.P., Nienhuis, J., Arnaud-Santana, E., Ariyarathne, H.M., Kaeppler, S.M., & Bassett, M.J. (1997). Molecular-markers-based genetic analysis of tepary bean derived common bacterial blight resistance in different developmental stage of common bean. *J. Am. Soc. Hort. Sci.*, 122, pp. 329–337.

Jung, G., Coyne, D.P., Skroch, P., Nienhuis, J., Arnaud-Santana, E., Bokosi, J., Ariyarathne, H., Steadman, J., Beaver, J. & Kaeppler, S. (1996). Molecular markers associated with plant architecture and resistance to common blight, web blight, and rust in common beans. *J Am Soc Hort Sci*, 121, pp. 794–803.

Jung, G., Coyne, D.P., Bokosi, J.M., Steadman, J.R., & Nienhuis, J. (1998). Mapping genes for specific and adult plant resistance to rust and abaxial leaf pubescence and their genetic relationship using random amplified polymorphic DNA (RAPD) markers in common bean. *J. Am. Soc. Hort. Sci.*, 123, pp. 859–863.

Jung, G., Skroch, P.W., Nienhuis, J., Coyne, D.P., Arnaud-Santana, E., Ariyarathne, H.M., & Marita, J.M. (1999). Confirmation of QTL associated with common bacterial blight resistance in four different genetic backgrounds in common bean. *Crop Sci.*, 39, pp. 1448–1455.

Kami, J., Poncet, V., Geffroy, V., & Gepts, P. (2006). Development of four phylogenetically-arrayed BAC libraries and sequence of the APA locus in *Phaseolus vulgaris*. *Theor. Appl. Genet.*, 112, pp. 987–998.

Kelly, J.D., Gepts, P., Miklas, P.N., & Coyne, D.P. (2003). Tagging and mapping of genes and QTL and molecular marker-assisted selection for traits of economic importance in bean and cowpea. *Field Crops Res.*, 82, pp. 135–154.

Liu, S., Yu, K., & Park, S.J. (2008). Development of STS markers and QTL validation for common bacterial blight resistance in common bean. *Plant Breeding*, 127, pp. 62-68.

Liu, S., Yu, K., & Park, S.J. (2009). Marker-assisted breeding for resistance to common bacterial blight in common bean. In: *Plant Breeding*, Huttunen, N., & Sinisalo, T. (eds), pp. 211-226. Nova Scientifc Publisher, New York, U.S.A.

Liu, S., Yu, K., Park, S. J., Conner, R. L., Balasubramanian, P., Milndel, H., & Kichn, F.A. (2005). Marker-assisted selection of common beans for multiple disease resistance. *Annu. Rep. Bean Improv. Coop.*, 18, pp. 82-83.

Liu, S., Yu, K., Huffner, M., Park, S.J., Banik, M., Pauls, K.P., & Crosby, W. (2010). Construction of a BAC library and a physical map of a major QTL for CBB resistance of common bean (*Phaseolus vulgaris* L.). *Genetica*, 138, pp. 709-716.

Llaca, V., & Gepts, P. (1996). Pulsed field gel electrophoresis analysis of the phaseolin locus region in *Phaseolus vulgaris*. *Genome*, 39, pp. 722–729.

López, R., Asensio, C., & Gilbertson, R.L. (2006). Phenotypic and genetic diversity in trains of common blight bacteria (*Xanthomonas campestris* pv. *phaseoli* and *X. campestris* pv. *phaseoli* var. *fuscans)* in a secondary center of diversity of the common bean host suggests multiple introduction events. *Phytopathology*, 96, pp.1204–1213.

McClean, P., Lee, R., Otto, C., Gepts, P., & Bassett, M. (2002). Molecular and phenotypic mapping of genes controlling seed coat pattern and color in common bean (*Phaseolus vulgaris* L.). *J. Hered.*, 93, pp.148–152.

McElroy, J.B. (1985). Breeding for dry beans, *P. vulgaris* L., for common bacterial blight resistance derived from *Phaseolus acutifolius* A. Gray. Ph. D. diss., Cornell Univ., Ithaca, NY.

McConnell, M., Mamidi, S., Lee, R., Chikara, S., Rossi, M., Papa, R., & McClean, P. (2010). Syntenic relationships among legumes revealed using a gene-based genetic linkage map of common bean (*Phaseolus vulgaris* L.). *Theoretical and Applied Genetics*, 121, pp. 1103-1116.

Mejía-Jiménez, A., Munz, C., Jacobsen, H.J., Roca, W.M., & Singh, S.P. (1994). Interspecific hybridization between common and tepary beans: Increased hybrid embryo growth, fertility, and efficiency of hybridization through recurrent and congruity backcrossing. *Theor. Appl. Genet.*, 88, pp. 324–331.

Melotto, M., Fransisco, C., & Camargo, L.E.A. (2003). Towards cloning the *Co-4*^{2} locus using a bean BAC library. Ann. Rep. Bean Improv. Coop, 46, pp.51–52.

Melotto, M., Monteiro-Vitorello, C.B., Bruschi, A.G., & Camargo, L.E.A. (2005). Comparative bioinformatic analysis of genes expressed in common bean (*Phaseolus vulgaris* L.) seedlings. *Genome*, 48, pp. 562-570.

Michaels, T. E., Smith, T. H., Larsen, J., Beattie, A. D., & Pauls, K. P. (2006). OAC Rex common bean. *Can. J. Plant Sci.*, 86, pp. 733–736.

Miklas, P.N., Beaver, J.S., Grafton, K.F., & Freytag, G.F. (1994). Registration of TARS VCI-4B multiple disease resistant dry bean germplasm. *Crop Sci.*, 34, pp.1415.

Miklas, P.N., Coyne, D.P., Grafton, K.F., Mutlu, N., Reiser, J., Lindgren, D.T., & Singh, S.P. (2003). A major QTL for common bacterial blight resistance derives from the common bean great northern landrace cultivar Montana No. 5. *Euphytica*, 131, pp.137–146.

Miklas, P.N., Johnson, E., Stone, V., Beaver, J.S., Montoya, C., & Zapata, M. (1996). Selective mapping of QTL conditioning disease resistance in common bean. *Crop Sci.*, 36, pp.1344–1351.

Miklas, P.N., Kelly, J.D., Beebe, S.E., & Blair, M.W. (2006). Common bean breeding for resistance against biotic and abiotic stresses: from classical to MAS breeding. *Euphytica*, 147, pp. 105–131.

Miklas, P.N., Larsen, K.M., Terpstra, K., Hauf, D.C., Grafton, K.F., & Kelly, J.D. (2007). QTL analysis of ICA Bunsi-derived resistance to white mold in a pinto x navy bean cross. *Crop Sci.*, 47, pp. 174–179.

Miklas, P.N., Stone, V., Daly, M.J., Stavely, J.R., Steadman, J.R., Bassett, M.J., Delorme, R., & Beaver, J.S. (2000a). Bacterial, fungal, and viral disease resistance loci mapped in a recombinant inbred common bean population ('Dorado'/XAN 176). *J Am. Soc. Hort. Sci.*, 125, pp. 476–481.

Miklas, P.N., Smith, J.R., Riley, R., Grafton, K.F., Singh, S.P., Jung, G., & Coyne, D.P. (2000b). Marker-assisted breeding for pyramided resistance to common bacterial blight in common bean. *Annu. Rep. Bean Improv. Coop*, 43, pp. 39–40.

Miklas, P.N., Zapata, M., Beaver, J.S., & Grafton, K.F. (1999). Registration of four dry bean germplasm resistant to common bacterial blight: ICB-3, ICB-6, ICB-8, and ICB-10. *Crop Sci.*, 39, pp. 594.

Mkandawire, A.B.C., Mabagala, R.B., Guzman, P., Gepts, P., & Gilbertson, R.L. (2004). Genetic diversity and pathogenic variation of common blight bacteria (*Xanthomonas campestris* pv. *phaseoli* and *X. campestris* pv. *phaseoli* var. *fuscans*) suggests co-evolution with the common bean. *Phytopathology*, 94, pp.593–603.

Mohan, S.T. (1982). Evaluation of *Phaseolus coccineus* Lam. Germplasm for resistance to common bacterial blight of bean. *Turrialba*, 32, pp. 489–490.

Monica, R., Elena, B., Elisa, B., Laura, N., Domenico, R., Giovanna, A., & Roberto, P. (2009). Linkage disequilibrium and population structure in wild and domesticated populations of *Phaseolus vulgaris* L. *Evolutionary Applications,* 2, pp. 504-522.

Moscone, E.A., Klein, F., Lambrou, M., Fuchs, J., & Schweizer, D. (1999). Quantitative karyotyping and dual-color FISH mapping of 5S and 18S–25S rDNA probes in the cultivated *Phaseolus* species (*Leguminosae*). *Genome,* 42, pp. 1224–1233.

Mutlu, N., Miklas, P., Reiser. J. & Coyne, D. (2005a). Backcross breeding for improved resistance to common bacterial blight in pinto bean (*Phaseolus vulgaris* L). *Plant Breed,* 124, pp. 282-287.

Mutlu, N., Miklas, P.N., Steadman, J.R., Vidaver, A.V., Lindgren, D., Reiser, J., & Pastor-Corrales M.A. (2005b). Registration of pinto bean germplasm line ABCP-8 with resistance to common bacterial blight. *Crop Sci.*, 45, pp. 806–807.

Mutlu, N., Vidaver, A.K., Coyne, D.P., Steadman, J.R., Lambrecht, P.A., & Reiser J. (2008). Differential pathogenicity of *Xanthomonas campestris* pv. *phaseoli* and *X. fuscans* subsp. *fuscans*) stains on bean genotypes with common blight resistance. *Plant Dis.*, 92, pp. 546–554.

Nodari, R.O., Koinange, E.M.K., Kelly, J.D., & Gepts, P. (1992). Towards an integrated linkage map of common bean. I. Development of genomic DNA probes and levels of restriction fragment length polymorphism. *Theoret. Appl.Genet.*, 84, pp. 186–192.

Nodari, R.O., Tsai, S.M., Gilbertson, R.L., & Gepts, P. (1993a). Towards an integrated linkage map of common bean. II. Development of an RFLP-based linkage map. *Theoret. Appl. Genet.*, 85, pp.513–520.

Nodari, R.O., Tsai, S.M., Guzman, P., Gilbertson, R.L., & Gepts, P. (1993b). Towards an integrated linkage map of common bean. III. Mapping genetic factors controlling host–bacteria interactions. *Genetics,* 134, pp. 341–350.

O'Boyle, P. D., Kelly, J . D., & Kirk, W. W. (2007). Use of marker assisted selection to breed for resistance to common bacterial blight in common bean. *J. Am. Soc. Hort. Sci.*, 132, pp. 381-386.

Oraguzie, N.C., Rikkerink, E.H.A., Gardiner, S.E., & Silva, H.N.D. (2007). *Association mapping in plants*, Springer, ISBN 9780387358444, New York.

Park, S.J., & Dhanvantari, B.N. (1987). Transfer of common blight (*Xanthomonas campestris* pv. *phaseoli*) resistance from *Phaseolus coccineus* Lam. to *P. vulgaris* L. through interspecific hybridization. *Can. J. Plant Sci.*, 67, pp. 685–695.

Park, S.J., & Dhanvantari, B.N. (1994). Registration of common bean blight-resistant germplasm, 'HR45'. *Crop Sci.*, 34, pp. 548.

Park, S.J., & Yu, K. (2004). Molecular marker-assisted selection techniques for gene pyramiding of multiple disease resistance in common bean: a plant breeder prospective. *Annu. Rep. Bean Improv. Coop,* 47, pp. 73-74.

Park, S. J., Yu, K., Liu, S., & Rupert, T. (2007). The release of white bean HR67. *Annu. Rep. Bean Improv. Coop.*, 50, pp. 221-222.

Parker, J.P.K. (1985). Interspecific transfer of common bacterial blight resistance from *Phaseolus acutifolius* A. Gray to *P. vulgaris* L. M.S. Thesis, Univ. of Guelph, Guelph, Canada.

Pedraza, F., Gallego, G., Beebe, S., & Tohme, J. (1997). Marcadores SCAR y RAPD para la resistencia a la bacteriosis comun (CBB). In : *Taller de mejoramiento de frijol para el siglo XXI: Bases parauna estrategia para America Latina (In Spanish)*, Singh, S.P., & Voysest, O. (ed.), pp. 130-134, CIAT, Cali, Colombia.

Pedrosa-Harand, A., Kami, J., Gepts, P., Geffroy. V., & Schweizer, D. (2009). Cytogenetic mapping of common bean chromosomes reveals a less compartmentalized small-genome plant species. *Chromosome Research*, 17, pp. 405–417.

Pedrosa-Harand,A., Porch, T., & Gepts, P. (2008). Standard nomenclature for common bean chromosomes and linkage groups. *Annu. Rep. Bean Improv. Coop.*, 51, pp. 106-107.

Pedrosa-harand, A., Vallejos, C.E., Bachmair, A., & Schweizer, D. (2003). Integration of common bean (*Phaseolus vulgaris* L.) linkage and chromosomal maps. *Theor. Appl. Genet.*, 106, pp. 205–212.

Ramirez, M., Graham, M.A., Blanco-Lopez, L., Silvente, S., Medrano-Soto, A., Blair, M.W., Hernandez, G., Vance, C.P., & Lara, M. (2005). Sequencing and analysis of common bean ESTs. Building a foundation for functional genomics. *Plant Physiol.*, 137, pp. 1211–1227.

Salvi, S., & Tuberosa, R. (2005). To clone or not to clone plant QTLs: present and future challenges. *Trends in Plant Science*, 10, pp. 297-304.

Schlueter, J.A., Goicoechea, J.L., Collura, K., Gill, N., Lin, J., Yu, Y., Kudrna, D., Zuccolo, A., Vallejos, C.E., Muñoz-Torres, M., Blair, M.W., Tohme, J., Tomkins, J., McClean, P., Wing, R.A., & Jackson, S.A. (2008). BAC-end sequence analysis and a draft physical map of the common bean (*Phaseolus vulgaris* L.) genome. *Tropical Plant Biol.*, 1, pp. 40–48.

Schuster, M.L., Coyne, D.P., Behre, T., & Leyna, H. (1983). Sources of *Phaseolus* species resistance and leaf and pod differential reactions to common blight. *HortScience*, 18, pp. 901–903.

Scott, M.E., & Michaels, T.E. (1992). *Xanthomonas* resistance of *Phaseolus* interspecific cross selections confirmed by field performance. *HortScience*, 27, pp. 348-350.

Shi, C., Chaudhary, S., Yu, K., Park, S.J., Navabi, A., & McClean, P.E. (2011a). Identification of candidate genes associated with CBB resistance in common bean HR45 (*Phaseolus vulgaris* L.) using cDNA-AFLP. *Molecular Biology Reports*, 38, pp. 75-81.

Shi, C., Navabi, A, & Yu, K. (2011b). Association mapping of common bacterial blight resistance QTL in Ontario bean breeding populations. *BMC Plant Biology*, 11:52.

Singh, S.P., & Munoz, C.G. (1999). Resistance to common bacterial blight among *Phaseolus* species and common bean improvement. *Crop Sci.*, 39, pp. 80-89.

Sutton, L.A., & Coyne, D.P. (2007). Vegetable cultivar descriptions for North America: bean-dry, lists 1-26 combined, Accessed on 10 June 2011, Available from: <http://cuke.hort.ncsu.edu/wehner/vegcult/beandry.html>.

Tanksley, S.D., Young, N.D., Paterson, A.H., & Bonierbale, M.W. (1989). RFLP mapping in plant breeding: new tools for an old science. *Biotechnology*, 7, pp. 257-264.

Tar'an, B., Michaels, T.E., & Pauls, K.P. (1998). Stability of the association of molecular markers and common bacterial blight resistance in common bean (*Phaseolus vulgaris* L.). *Plant Breeding*, 117, pp. 553-558.

Tar'an, B., Michaels, T.E., & Pauls, K.P. (2001). Mapping genetic factors affecting the reaction to *Xanthomonas axonopodis* pv. *phaseoli* in *Phaseolus vulgaris* L. under field conditions. *Genome*, 44, pp. 1046-1056.

Thomas, C.V., & Waines, J.G. (1984). Fertile backcross and allotetraploid plants from crosses between tepary beans and common beans. *J. Hered.*, 75, pp. 93–98.

Vallejos, E.C., Sakiyama, N.S., & Chase, C.D. (1992). A molecular marker-based linkage map of *Phaseolus vulgaris* L.. *Genetics*, 131, pp. 733–740.

Vandemark, G.J., Fourie, D., & Miklas, P.N. (2008). Genotyping with real-time PCR reveals recessive epistasis between independent QTL conferring resistance to common bacterial blight in dry bean. *Theor. Appl. Genet.*, 117, pp. 513-522.

Vanhouten, W., & MacKenzie, S. (1999). Construction and characterization of a common bean bacterial artificial chromosome library. Plant Molecular Biology, 40, pp. 977-983.

Van Tassell, C.P., Smith, T.P., Matukumalli, L.K., Taylor, J.F., Schnabel, R.D., Lawley, C.T., Haudenschild, C.D., Moore, S.S., Warren, W.C., & Sonstegard, T.S. (2008). SNP discovery and allele frequency estimation by deep sequencing of reduced representation libraries. *Nat. Methods*, 5, pp. 247-252.

Weber, W. E., & Wricke, G. (1994). *Genetic markers in plant breeding. Plant Breeding Suppl. 16*, Parey, Berlin.

Yoshii, K., Galvez, G.E., & Alvarez, G. (1978). Screening bean germplasm for tolerance to common blight caused by *Xanthomonas phaseoli* and the importance of pathogenic variation to varietal improvement. *Plant Dis. Rep.*, 62, pp.343-347.

Yu, J., Pressoir, G., Briggs, W.H., Vroh, B.I., Yamasaki, M., Doebley, J.F., McMullen, M.D., Gaut, B.S., Nielsen, D.M., Holland, J.B., Kresovich, S., & Buckler, E. S. (2006). A unified mixed-model method for association mapping that accounts for multiple levels of relatedness. *Nature Genetics*, 38, pp. 203-208.

Yu, K. (2011). Bacterial Chromosome Libraries of pulse crops - characteristics and applications. *Journal of Biomedicine and Biotechnology*, in press.

Yu, K., Haffner M., & Park S.J. (2006). Construction and characterization of a common bean BAC library. *Annu. Rep. Bean Improv. Coop*, 49, pp. 78–79.

Yu, K., Park, S.J., & Poysa, V. (1999). Abundance and variation of microsatellite DNA sequences in beans (*Phaseolus* and *Vigna*). *Genome*, 42, pp. 27–34.

Yu, K., Park, S.J., & Poysa, V. (2000a). Marker-assisted selection of common beans for resistance to common bacterial blight: efficacy and economics. *Plant Breed*, 119, pp. 411–415.

Yu, K., Park, S.J., Poysa, V., & Gepts, P. (2000b). Integration of simple sequence repeat (SSR) markers into a molecular linkage map of common bean (*Phaseolus vulgaris* L.). *J. Hered.*, 91, pp. 429–434.

Yu, K., Park, S.J., Zhang, B., Haffner, M., & Poysa, V. (2004). An SSR marker in the nitrate reductase gene of common bean is tightly linked to a major gene conferring resistance to common bacterial blight. *Euphytica*, 138, pp. 89–95.

Yu, K., Shi, M.C., Liu, S., Chaudhary, S., Park, S.J., Navabi, A., Pauls, K.P., McClean, P., Miklas, P.N., & Fourie, D. (2010). Cloning the major CBB resistance QTL of common bean through map-based cloning and gene profiling approaches - current status and future prospects. *Annu. Rep. Bean Improv. Coop*, 53, pp. 27-28.

Yu, Z.H., Stall, R.E., & Vallejos, C.E. (1998). Detection of genes for resistance to common bacterial blight of beans. *Crop Sci.*, 38, pp. 1290–1296.

Wang, J., McClean, P.E., Lee, R., Goos, R.J., & Helms, T. (2008). Association mapping of iron deficiency chlorosis loci in soybean (*Glycine max* L. Merr.) advanced breeding lines. *Theoretical and Applied Genetics*, 116, pp. 777-787.

Zapata, M., Beaver, J.S., & Porch, T.G. (2010). Dominant gene for common bean resistance to common bacterial blight caused by *Xanthomonas axonopodis* pv. *Phaseoli*. *Euphytica,* 179, pp. 373–382.

Zapata, M., Freytag, G.F., & Wilkinson, R.E. (1985). Evaluation for bacterial blight resistance in beans. *Phytopathology,* 5, pp. 1032 - 1039.

Zapata, M. (1997). Identification of *Xanthomonas campestris* pv. *Phaseoli* races in *Phaseolus vulgaris* leaves. *Agron. Mesoam.*, 8, pp. 44 - 52.

Zhang, C., Bradshaw, J. D. , Whitham, S. A., & Hill, J. H. (2010). The development of an efficient multipurpose bean pod mottle virus viral vector set for foreign gene expression and RNA silencing. *Plant Physiology,* 153, pp. 52–65.

20

Analysis of Rural Ecosystem in Japan Using Stable Isotope Ratio

Atsushi Mori
Renewable Resources Engineering Division, National Institute for Rural Engineering
Japan

1. Introduction

1.1 Rural ecosystem in Japan

Rice is one of the most important cereal crops especially in large part of Asia and a part of Africa. It is split up into two groups. One is upland rice, grown on dry soil. Most rice is grown in flooded rice paddy fields and most of them in Japan are irrigated well. Rice is considered to come to Japan about three thousand years ago from southern China, analyzing Carbon 14 of unearthed rice seed from paddy field remains.

Paddy fields are, so to speak, fabrication of floodplain area. Some areas near river are flooded in rainy season and change into shallow puddles in dry season. Wild rice plant probably has spread in such wet lands. Ancestor reclaimed and cultivated paddy fields and irrigated them after a period of primitive harvest of wild rice.

Paddy fields are also special and sacred place where spirit lives and give us fertility. The spirit is said to live in the deep mountain during winter and come down to communities when spring comes. Farmers welcome the spirit with honor. We can see their soul by devoting a sprig of camellia to the spirit at the inlet of the paddy field when irrigation water inflowed into the paddy field at the beginning of cultivation. Paddy fields have affected traditional human society and life in Japan.

Contribution and influence of paddy fields is not exclusive not only to human beings. The artificial wet lands provide some wild lives good spawning, growing, feeding site. **Photo 1** is scene before puddling, preparation for rice planting. It shows water network between canal and paddy fields. Water level is lower than photo's level in non-irrigation period. So paddy fields are dry in non-irrigation period. A weir which is set at the beginning of irrigation period makes water level high as **Photo 1** and fishes such as loach can swim up into paddy fields in order to spawn. Frogs also benefit from paddy field and canals. Tokyo Daruma Pond Frog (*Rana porosa porosa*) (**Photo 2**) spawns in paddy fields and grows up near canals after metamorphosis. They have been using the new habitats efficiently.

Mosaic land use in rural area is characteristic in Japan. **Photo 3** is typical land use in hilly and mountainous rural area. Of course, there some farming types in Japan, including dairy husbandry, large scale upland crop or rice farming. But hilly and mountainous rural area account for 40 percents of cultivated acreage in Japan. Lower area in hilly and mountainous rural area is used for rice farming, being irrigated from canals. Dry field locates in upper area.

Photo 1. Mt. *Yakeishi* reflected on the surface of the newly irrigated paddy fields, Iwate Prefecture, Tohoku district

Photo 2. *Satoyama* and paddy fields that were poured with irrigation water before puddling

Photo 3. Tokyou Daruma Pond Frog in a paddy field

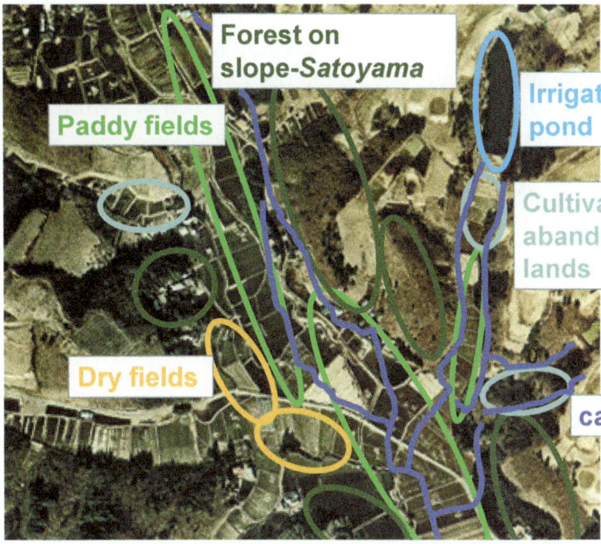

By National Land Information web mapping system-Ministry of Land, Infrastructure, transport and Tourism

Photo 4. Typical land use in hilly and mountainous rural area

Village forest called *Satoyama* in Japanese had a role of wood fuel and compost production. Underbrush in *Satoyama* had been cut well not to transit into overplanting several decades ago. Fuel revolution and developing pressure altered *Satoyama* into golf course or run-down bush. *Satoyama* was also precious habitat for wild lives similarly as with paddy field and canal are comfortable habitat for fishes and frogs.

Mosaic structure help many wild lives adapted to these various environments. Japanese Brown Frog (*Rana Japonica*) move to *Satoyama* after metamorphosis, feeding and overwinter there. Both paddy fields and *Satoyama* are necessary for their life cycle. So, plural mosaic land use aggregate (we can recognize them as each ecosystem)

Photo 5. Drainage canal separated from dual-purpose canal, irrigation and drainage

Thus, rural ecosystem in Japan is influenced by paddy fields that account for half of farming land total. Ecosystem affected by agriculture, especially by paddy fields is called Secondary Nature. A word of secondary may include an inferior nuance. However the term Secondary Nature never has such a nuance. A lot of living lives are seen in a rural area. A number of rare species also survive in rural area with human activities.

Photo 6. Drop work is constructed for moderation of channel slope, opposing swimming up of fishes

The rural ecosystem in Japan has deteriorated during recent years because of land improvement that permitted agricultural production to be efficiently, chemical materials input into paddy fields and development in rural area. Land improvement as a term must be less familiar with general readers. I would introduce farm land consolidation that is main construction project in land improvement to readers. Traditional paddy fields in Japan were so narrow and cultivators had some fields from place to place. Because the same channel had been used for irrigation and drainage, ground water level was so high that farmers had not been able to bring agricultural machines to paddy fields. For, soil bearing capacity was low before independent setting of irrigation and drainage. Dual-purpose canal was separated into irrigation channel and drainage canal by farm land consolidation. As underground drainages are constructed in many consolidation projects, moisture of soil become dry. Japanese Brown Frog that has a habit of coming to paddy fields to spawn from *Satoyama* in early spring, non-irrigated period, lose a chance to spawn in dry paddy fields. Canal is also renovated from soil to concrete by farm land consolidation. Concrete canal has superiority from the point of agriculture, that is, water management, maintenance. On the other hand, there is inferiority from the aspect of environment. Water velocity becomes large, covering by concrete. Considerable aquatic animals can't swim against velocity after repair. We can see some water drop facilities that adjust gradient and water velocity in complete canal. Even though they are needed to keep canal performance, they prevent fishes from swimming up to upstream. If small animals, for example Japanese Brown Frog that comes from *Satoyama* to spawn, drop into concrete canal, they never escape from concrete canal because they don't have suckers with their fingers like Japanese Tree Frog. Thus, land improvement consisted in such as agricultural canal renovation influenced

habitat of aquatic animals so much, although it gives us a lot of contribution. Technology has both sides, convenience for our ordinary life and harmfulness for ecosystem.
It is only one decade ago that eco-friendly construction seems to be needed in Japanese rural area. Deterioration and damage of ecosystem were the background of revision of Land Improvement Act in 2001. The revision included that land improvement project must attend to harmony with the environment.
However, biology of rural wild lives and how biodiversity could be conserved are not well known. It's even less of a surprise that how we shall do to attend to the environment concretely. Does the facility built in order to conserve wild animals in a canal work well actually? Is the effect of the facility enough to keep the ecosystem? We have to investigate not only technical methods for water facilities but also basic biology of rural living organism and characteristic of ecosystem

2. Biological analysis using stable isotope ratio

Some elements have isotopes that don't emit radiation and are called stable isotopes. Carbon-13 and Nitrogen-15 are two such isotopes often used to analyze a food web. These isotopes inevitably exist in nature, and also are present in the bodies of organisms. Stable isotope ratios for carbon and nitrogen are expressed as $\delta^{13}C$ and $\delta^{15}N$, respectively.
Stable isotope ratio is defined as following formula.

$$\delta = \frac{R_{sample} - R_{standard}}{R_{standard}} \times 1000(‰)$$

δ : Carbon stable isotope ratio: $\delta^{13}C$

Nitrogen stable isotope ratio: $\delta^{15}N$

$R_{standard}$: in case of C; Carbon stable isotope ratio of Belemnites in PD-stratum, South Carolina

In case of N; Nitrogen stable isotope ratio in Nitrogen gas in the air

R_{sample} : Stable isotope ratio of sample, $^{13}C/^{12}C$, $^{15}N/^{14}N$

Plants are classified into 3 groups by the photosynthesis mechanism, namely C_3, C_4 and CAM plants. Scarce plant species in the paddy fields of Japan belong to the CAM plant group. A considerable number of plants in the paddy area are C_3 plants, and all trees in forests, *Satoyama*, belong to this group.
Sugarcane, corn, eulalia, crabgrass, and giant foxtail belong to C_4 plants. C_4 plants can often be found in the periphery of a paddy field. They are usually members of Gramineae that are natives of tropical regions. The $\delta^{13}C$ values for C_4 plants are approximately -13‰, with a range from -15‰ to -10‰. The $\delta^{13}C$ values of organisms are considerably lower than C_4 plants, approximately -27‰, with a range from -35‰ to -25‰.
The values for algae in an environment that abounds in CO_2 are between C_3 plants and C_4 plants, at approximately -20‰ as shown by previous investigations.
The type of plants present as producers in a food web can be predicted by analysis of $\delta^{13}C$, because $\delta^{13}C$ changes little under prey-predator interactions. Therefore, when the $\delta^{13}C$ of an animal is approximately -30‰, the animal is considered to be in a food web derived from C_3 plants (**Fig. 1**).

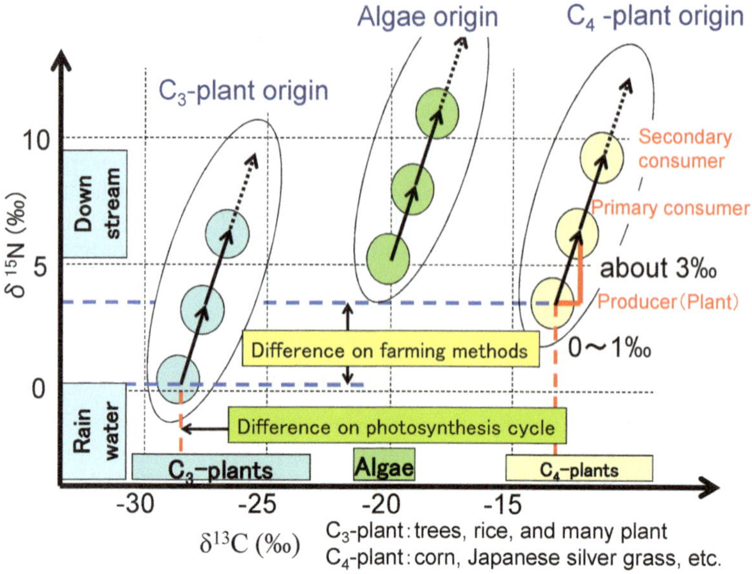

Fig. 1. C-N map representing food web

We can see several characteristics in Japanese rural vegetation. One is mosaic botanical distributions. They are paddy fields, fields, grass lands, orchard, forests between mountain and arable area that is called *Satoyama* in Japanese language. It is not uncommon that these distributions are found in even $1km^2$. Only C_3 plants are found in *Satoyama*. On the other hand, significant proportion of plants near paddy fields belong to C_4 plants though rice crop itself is C_3 plants. They often grow in the terrestrial area like land temporarily fallow.

Algae are also important producer in rural ecosystem. A lot of aquatic animals feed on algae. Difference of $\delta^{13}C$ between detritus and algae. Almost detritus in a agricultural drainage canal origins from litter produced in *Satoyama* and paddy fields, where only C_3-type plants grow. The difference still remains at animal level. That is, $\delta^{13}C$ of animals fed detritus is lower than that of animals fed on algae.

3. Actual investigations

1. Analysis of food web

Using the characteristics of carbon and nitrogen stable isotope ratio described previously, we can analyze food-web. I would like to offer you the analysis example of which sample is collected in rural area. The investigate area was Isawa plain that locates in Iwate prefecture, Tohoku district.

The samples were dredged from two aquatic areas, a drainage canal and a small irrigation pond. The canal had been constructed eco-friendly, of which bottom hadn't lined with concrete. Stones, earth and sand had been deposited to a substantial extent. The pond located on the boundary between plain that and a hill. A lot of leaves have fallen into the pond of which water was brownish-red, probably because of the elution of organic element from leaves.

Fig. 2 shows C-N map (food-web) in the canal. $\delta^{13}C$ of samples that is the most important value to distinguish food-web was approximately from -25‰ to -20‰. Attached matter that has fed insects, fishes though insects was consisted with detritus and algae according to the microscopic visualization. The value of attached matter confirmed it because the value located between $\delta^{13}C$ of C_3 plants and that of algae. Of course, we can't afford to ignore the existence of C_4 plats. A basin of the canal was occupied by paddy fields where rice that belongs to C_3 plants grows. The top of food web was Amur Minnow (*Phoxinus logowskii steindachneri*). Their stomachs were filled with caddice-worms.

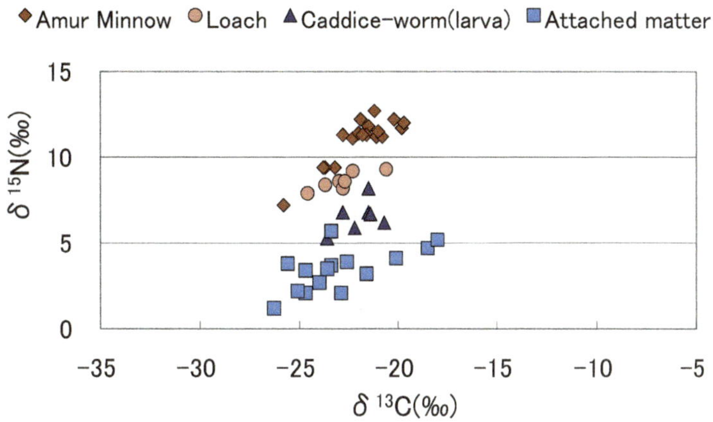

Fig. 2. C-N map of collected matters from a canal

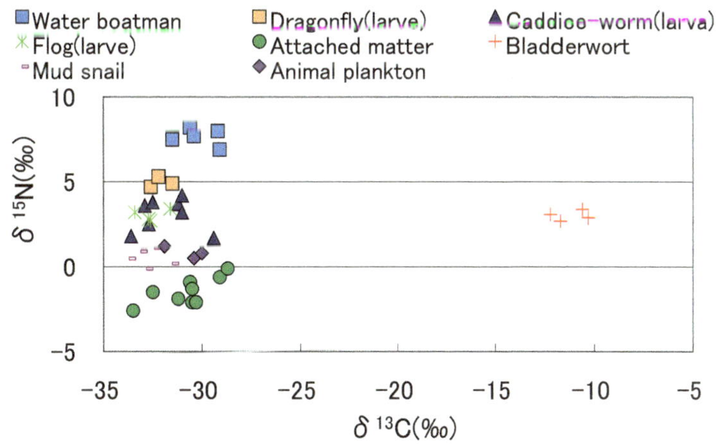

Fig. 3. C-N map of collected matters from a pond

How was the food web in the pond? **Fig. 3** indicated quite lower values than that of the canal. The food web depended on C_3 plant apparently. The top of the pyramid should be small insets, water boatman (*Notonecta triguttata*) that suck fluids of small fishes, tadpoles, insects, etc. Such a curious food web like this sometimes appears in an aquatic area where no fish live. Bladderwort is insectivorous plant,

2. Analysis of Food source

A loach is one of the most familiar freshwater fish for Japanese. It had been swimming everywhere, so catching it had been common play for children. It has been important food, especially for certain rural area even today. Loaches have been decreasing like other fishes. Revamping of canal much affected their life, because they favor sandy bottom that disappeared after construction.

To increase them is becoming popular in these days, as they are also the feed for white stork (*Ciconia boyciana*, Critically Endangered) and Japanese crested ibis (*Nipponia Nippon*, Extinct in the Wild). Loaches that are their feed to release to wild also attract attention. To invest feed of loaches is an essential problem besides environmental improvement.

When irrigation water is poured into paddy fields in spring, loaches begin to swim up into paddy fields, where they spawn. Young fishes swim down to canals in early summer after feeding and growing there. Sampling and analyzing were continued from June to September to grasp the change of $\delta^{13}C$ of fishes bodies. Sample loaches were caught in a paddy field in Kaminokawa town, Tochigi prefecture. C-N map of the loaches is shown as **Fig. 4.**

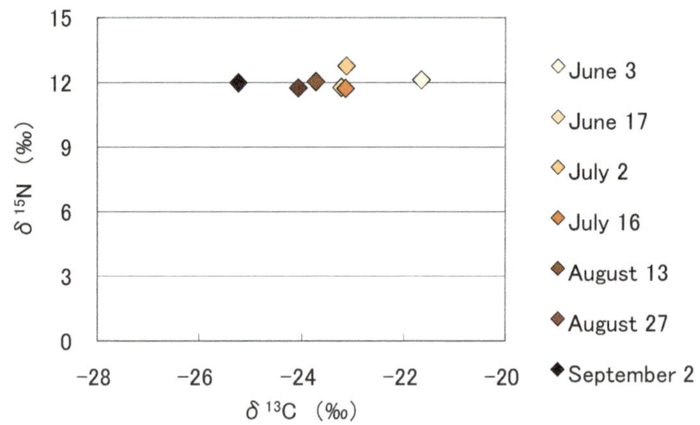

Fig. 4. Seasonal C-N map of loach

Along with the season, stable carbon isotope ratio was decreasing and decreasing. This phenomenon declared that food source of loaches has changed from terrestrial matter to aquatic matter. Along with the season, stable carbon isotope ratio was decreasing and decreasing. Copepod in early phase, bloodworm in late phase may be main feed by consulting the past reference and benthic examination in paddy fields. Another was that trophic level hasn't changed, showing the fish was carnivorous even though loaches are said that they become to be omnivorous with growth.

Secondly, I show the analysis of food resource of shell. *Matsukasagai* shell (*Pronodularia japanensis*), a kind of unionid, is necessary for Japanese bitterling because it has a habit of spawning, inserting their ovipositors into the shells. Extinction of the shell means the same of the fish. The shells are described as they feed plankton, but detail is not well known. So I investigated them using stable isotope ratio.

They suck FPOM (Fine Particulate Organic Matter) with water, filtering and eating real feed. The rest of FPOM is excreted. **Fig. 5** shows the result of investigation including FPOM and their spoor in certain habitat, Nakagawa River basin in Tochigi prefecture, where Tokyo bitterling (*Tanakia tanago*) that was protected species lived. There was a pond in upper stream of the habitat. The habitat is described as Point A in this article. Carbon isotope ratio of *Matsukasagai* shell was low, -32.5‰. The value of spoor was -28.9‰, locating right between *Matsukasagai* shell and FPOM. Phenomenon kindred to it was seen in flesh water calm, *Corbicola* sp.. These represented that FPOM contained their feed and unwanted materials were excreted, of which carbon isotope ratio was higher than their feed.

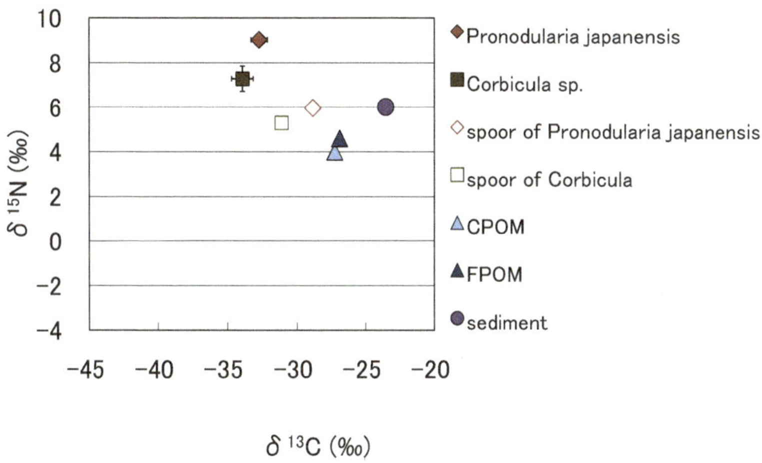

Fig. 5. C-N map in Point A

Mataukasagai shells were caught in Point B near Point A, too. A pond located upstream of point B just like Point A. The result was also similar to Point A, that is The value of spoor (-33.3‰) located right between *Matsukasagai* shell (-39.9‰) and FPOM (-27.0)(**Fig. 6**). The low value of -39.9‰ indicated notable information, considering the origin of the organic matter. C_3 plant and detritus derived from it never indicates such low value. Carbon isotope ratio of plankton may represented the value of -20‰ approximately in a circumstance of which gas exchange is well-balanced, however I suppose that quite little ecosystem in land water wouldn't not be balanced on CO_2 in practice, DOC (Dissolved Organic Carbon) that is resource of photosynthesis of algae including phytoplankton may lower than DOM derived from CO_2 in the air. This might be attributed recycling of DOC in the water, preparation and dissolution, value of $\delta^{13}C$ might decrease during recycling because of isotope fractionation.

Fig. 7 shows relationship between shell length and $\delta^{13}C$. $\delta^{13}C$ value of *Matsukasagai* shells increased with their growth. This suggested that feeding habitat would change as they became larger.

This hypothesis was confirmed to some extent by other analysis shown as Fig. 8. It is about 50km distance from Point A and B to Point C, where *Inversiunio jokohamensis* that is relative species of *Inversidens japanensis* lived. Basin area of Point C was different from that of former two points. Basin area of Omoigawa River in **Fig. 8** is quite distant from A and B even all of them locate in Tochigi Prefecture. Harakawa River locates in Iwate Prefecture, as described before, in a northern direction from Tochigi Prefecture.

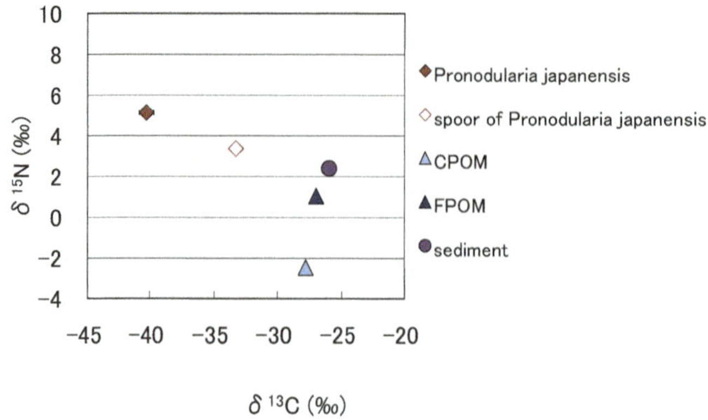

Fig. 6. C-N map in Point B

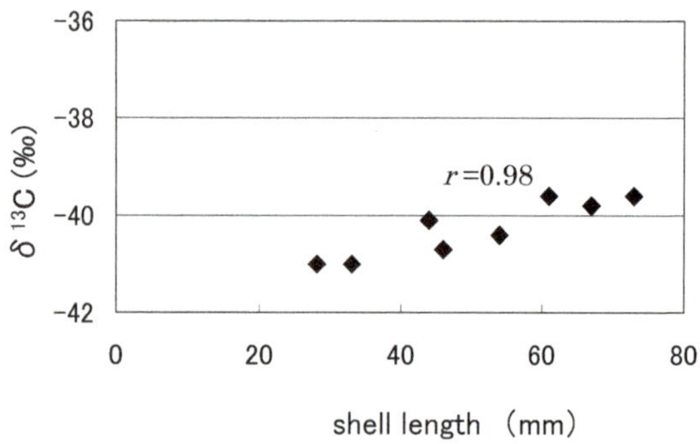

Fig. 7. Relationship between Shell length and $\delta^{13}C$

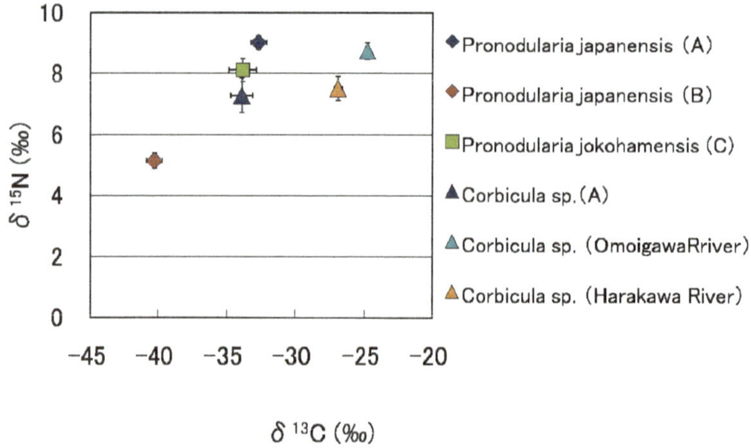

Fig. 8. C-N map of bivalve in various habitats

All of these three species are said to be plankton feeders. We can assert with probability that there are characteristics of value of DOC in each aquatic environments.

3. Analysis of animal moving

Using behaviors described previously, we can analyze not only food webs but also ecological characteristic of animals. Japanese brown frog (*Rana japonica*) is typical amphibian in rural area. Their populations are said to being decreasing rapidly after revamping of agricultural drainage canal. Prime cause is presumed that concrete agricultural drainage canal cut network between terrestrial area and aquatic area. According to qualitative knowledge in the past, adults of Japanese brown frog live in *Satoyama* and overwinter there. After coming out from hibernation, they move to paddy fields and lay spawns in paddy fields. *Satoyama* is important habitat for the frogs indeed. But I have also often seen them near paddy fields like land temporarily fallow during my field research. Are their wintering places only *Satoyama*? We need to know how they move to *Satoyama* quantitatively.

As some C_4-plants grow in lands temporarily fallow, $\delta^{13}C$ of frogs body will higher than value of C_3-plants if they feed in the land. However, catching frogs is too difficult because they move to paddy fields during nights and they have agility. So, I came up with usage of spawns.

Sampling was conducted in hilly and mountainous rural area, where the frogs have preference to live, in Tochigi prefecture (**Fig.9**). The investigated paddy field A borders *Satoyama* on two sides, fallow lands on two sides. Field B borders *Satoyma* on one side, a fallow land on one side, and other paddy fields on two sides. Over 100 egg masses were skimmed up from each field, using a dipper, and dropped back after being sampled about 20 eggs from every mass. The dedicated small tin cups to analyze were filled with eggs after being defatted by methanol and chloroform, crushed. Stable isotope ratios of the samples were analyzed with a mass spectrometer, Delta plus XP, manufactured by Thermo Finnigan.

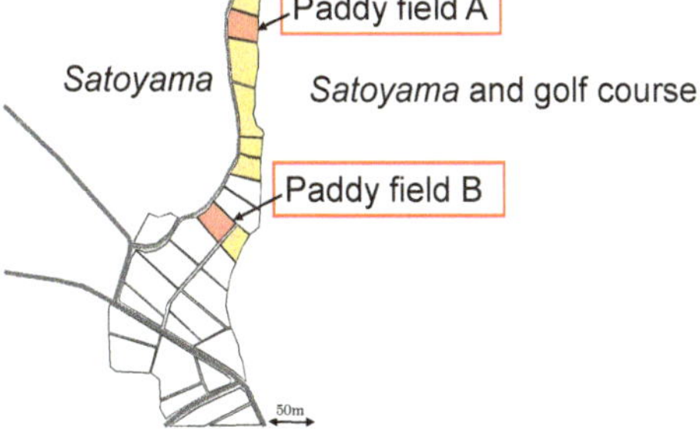

Fig. 9. Investigated fields

Fig. 10 shows the total result of analysis. The histogram has two peaks. One peak that of central value is minus 25 per mill indicates their mother frogs fed in the C_3 environment. We can see that another group exists, seeing higher value from-21 to -16 per mill values; their mother's foods included appreciably insects that had fed C_4 plants. The result says that most Japanese brown frogs move to *Satoyama* after metamorphosis or spawn, where they feed and winter.

Then, is there relationship between land use around the paddy field and moving? I estimated that $\delta^{13}C$ value of eggs in field A (**Fig.11**) might be higher than in field B (**Fig.12**) because the former had two borders to fallow lands where C_4 plants grew. But there wasn't considerable difference between A and B. That suggests that frogs in the population move to *Satoyama* at a constant rate.

What does the result mean? Agricultural drainage canals are usually routed through boundary area between paddy fields and *Satoyama*. Most frogs that come to paddy fields to spawn fall easily into a concrete agricultural drainage canal and wouldn't be able to escape from there. Conservation of the frogs must be needed when canal would be revamped not to let the population disappear. The problem is we don't prepare the effective methods actually. Slopes at the canal walls constructed to help escaping from the canal aren't used frequently. Only a few individuals that escaped from the canal used the slopes may not have so important meaning, for most of individuals died in the canal before spawn. Capping on the opening section of canal also is a good method. But it costs a lot and farmers often disagree to cap it because they won't use water to wash their farm tools. Some complain of anxiety that it may negatively affect to rise water temperature though rising water in while flowing through in the canal is nothing much.

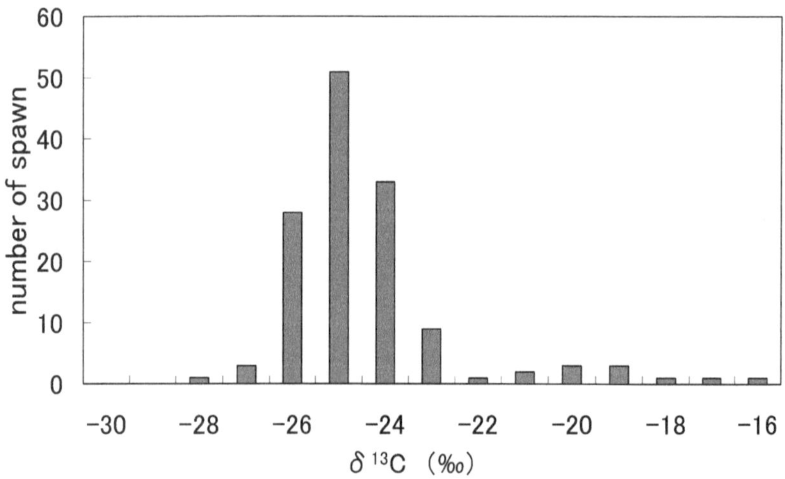

Fig. 10. Histogram of δ 13C distribution of spawns (entirety)

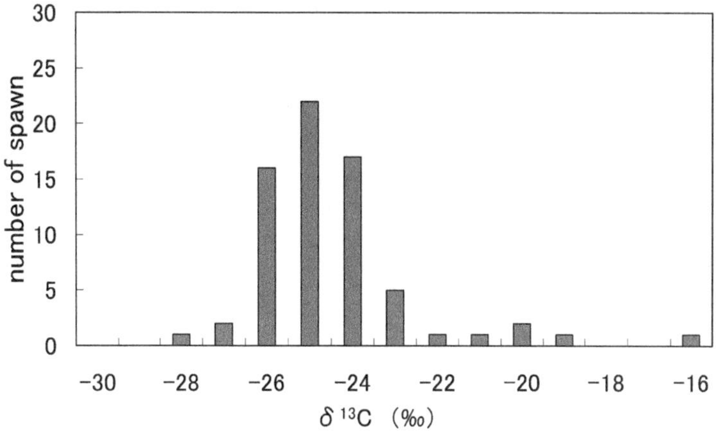

Fig. 11. Histogram of δ 13C distribution of spawns (Paddy field A)

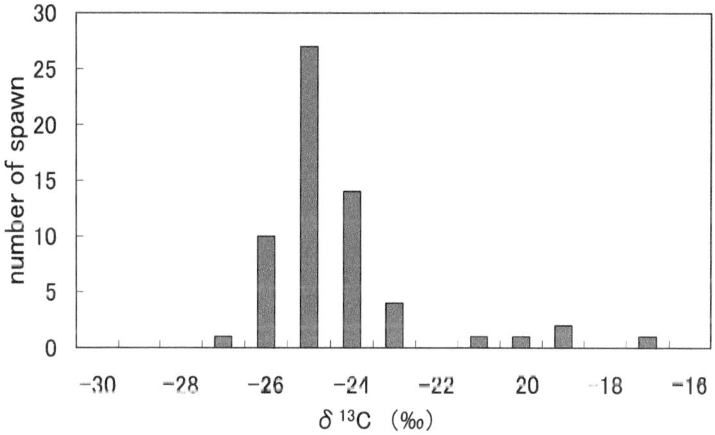

Fig. 12. Histogram of δ 13C distribution of spawns (paddy field B)

4. Analysis of ecosystem structure

Paddy field at hill-bottom or certain area with it is called *Yatsuda* in Japanese language. *Yatsuda* means the latter here. *Yatsuda* includes paddy fields and *Satoyama*, aquatic area and terrestrial area. And canals almost always run at the border area, where boundary area between Paddy fields and *Satoyama*. I considered that I could comprehend plural small food-webs near canals, which are virtually identical ecosystem for each land use there.

I focused on spiders for the food-web analysis because they located top of a micro tentative food-web. We can divide spiders into two groups, web-builder and hunting, from the point of their life types. Furthermore I separated hunting types into phytophilous hunting spiders and epigaeic hunting spiders because each type fed particular foods, considering their habitat.

I analyzed carbon stable isotope ratio ($\delta^{13}C$) of spiders (Arachnida) which were caught nearby farm ditches and a levee in a paddy field at hill-bottom (*Yatsuda*) with their life types. $\delta^{13}C$ of web-builder spiders (**Fig. 13**) was -26.0‰, which indicated that they have been greatly affected by insects emerged from the ditches that had depended on the organic matter supplied from village forest (*Satoyama*). $\delta^{13}C$ and its distribution of long-jawed water spiders (T*etragnatha maxillosa*) was approximately equal to silver vlei spiders (*Leucauge magnifica*). That suggested that both of spiders depended on the same food resources. Although $\delta^{13}C$ average of phytophilous hunting spiders (**Fig. 14**) was close to the web-building type, the standard deviation was higher than that. $\delta^{13}C$ of epigaeic hunting spiders was -22.2‰, which indicated that C_4 plants rose $\delta^{13}C$ value of the spiders (**Fig.15**). It was suggested that $\delta^{13}C$ of epigaeic hunting spiders lay in close relation with the rate of C_4 plants to the hole plants on the levee and approximately one third of carbon in their bodies was derived from C_4 plants. Diversity of material flow in creatures by carbon stable isotope ratio in the area of *Yatsuda* ditches was revealed, which is the eco-tone between *Satoyama* and a paddy field (**Fig. 16**). Animals in a narrow area shared foods and are compartmentalization, which will be recognized as diversity of ecosystems.

Canals in boundary area between *Satoyama* and paddy fields are often consolidated while land improve project is implemented. Of course, new canals after construction are of concrete. Complex and diversified ecosystem there will be transformed into homogenized one.

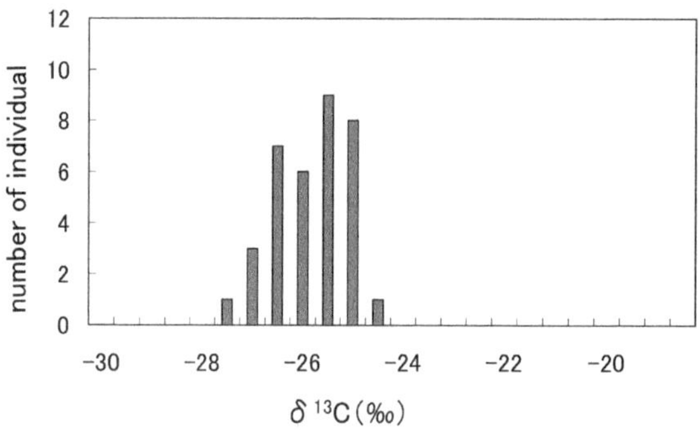

Fig. 13. Histogram of web-builder spiders

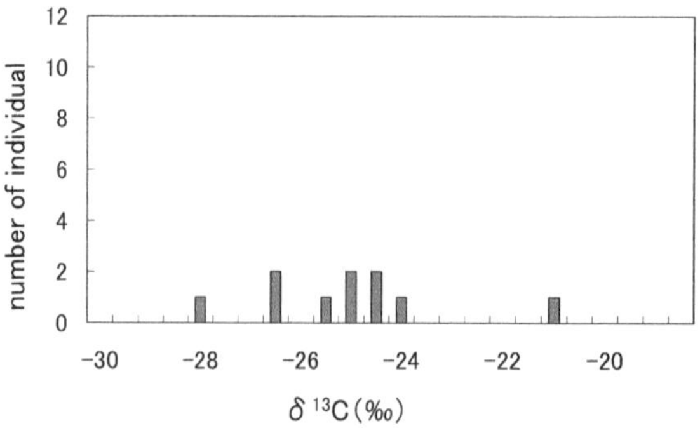

Fig. 14. Histogram of phytophilous hunting spiders

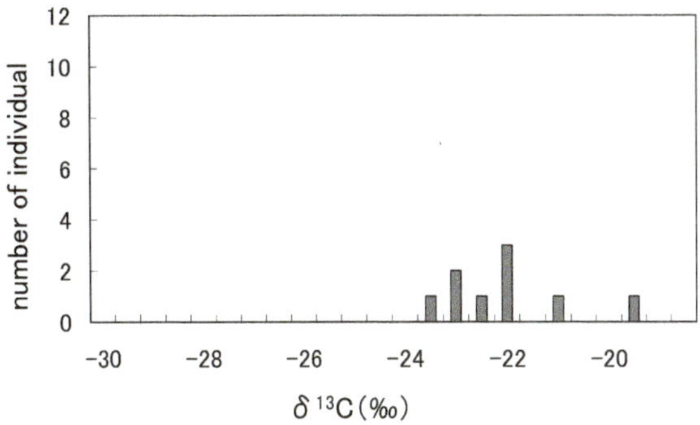

Fig. 15. Histogram of epigaeic hunting spiders

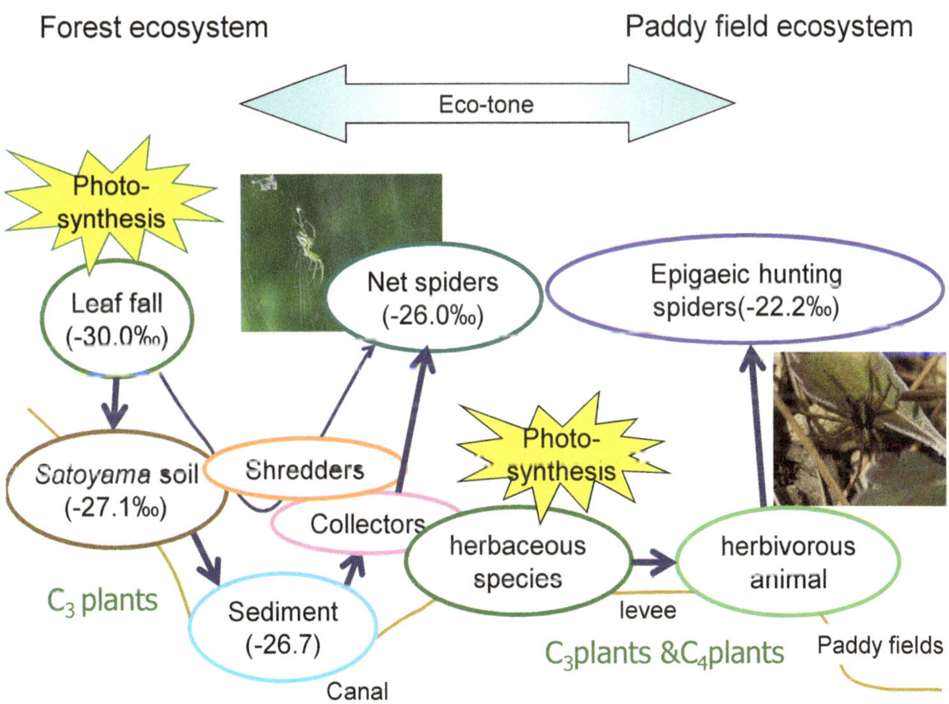

Fig. 16. Food web and carbon flow around canals in *Yatsuda* area

5. Analysis of population structure

May population of animals be homogeneous? Isn't there revolving of members in the population? Constitution of population is an important issue to sustain it as population with complements will still continue even if members would decrease. Stable isotope ratio gave me a vast amount of help.

Stable isotope ratio of animals has an interesting behavior that is caused by difference of turnover time for their parts. Turnover time of muscles and stomachs is short. On the other hand, that of bone is quite slower than these parts. If there is deference among parts, actually between bone and other parts, there is potential that the frog has fed in strange habitat in past times because carbon isotope ratio of previous food is presumed to remain in bone yet. Material of animal tells where he lived.

Carbon isotope ratio of the three parts of Tokyo Daruma Pond frog (*Rana porosa porosa*) that was caught at *Yatsuda* area was analyzed. The result is shown as **Fig. 17**. We notice that there is variability among the group and within certain individual internally. Specifically, vales of bone of individual number 5 and 19 are lower than that of muscle and stomach. This indicates that the frog has grown at another habitat and joined the population at a later time, remaining particular value of carbon on its bone. Overall values of number 6 and 10 are different from majority individual. This means they joined the group quite recently, thus carbon isotope ratio of the body hasn't changed into background value of catch area (habitat of the population) completely yet. Background value will be affected by weighted average of stable isotope ratio of producer. Number 6 might grow at the circumstances where considerable C_4 plants grow. On the contrary, number 10 might grow at the circumstances where all but C_3 plants grew and it fed insects that depended almost exclusively on C_3 plants.

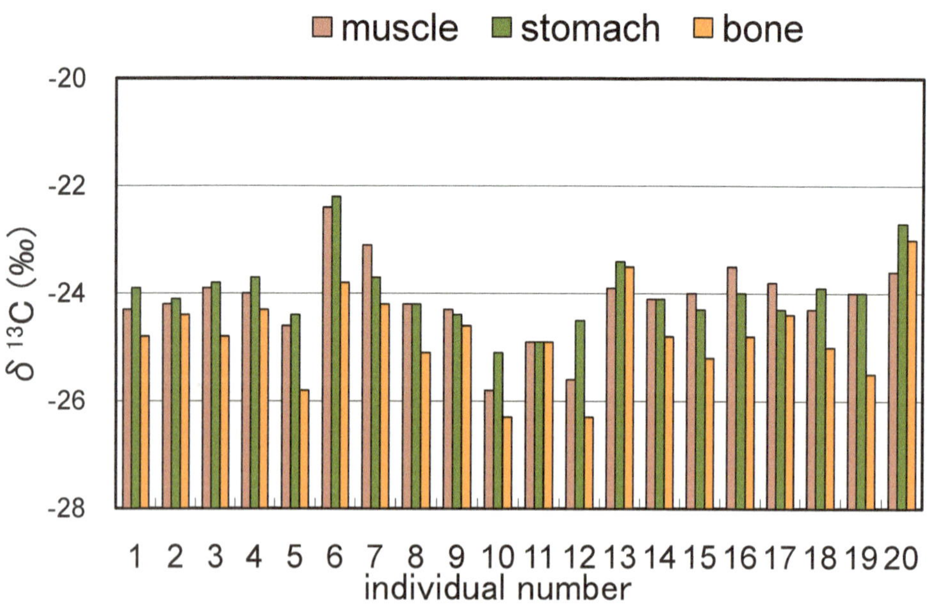

Fig. 17. Carbon isotope ratio of Tokyo Daruma Pond Frog for each parts

These truths educate us important things. Being consisted with individuals moved from plural habitats is also important obviously from the biology point of view, but how in-migrants joined the group includes internal issue for conservation. The species live near paddy fields and/or canals, into which frogs jump when they are attacked by predators. Some attacked frogs escaped into flowing canal will be washed away to downstream although most frogs will drift to near dropped point. Even the washed frogs try to land, but they won't be able to escape from there, if the canal will be constructed by concrete. Existence of in-migrants is the proof of dropping, being washed and landing of frogs.
Canals seem to have a role of pathway of aquatic animals not just a role of agricultural production. Breakpoints of network have harmful effects not only for swimming up of fishes but also accession of frogs from other populations. This is just my own supposition, in-migrants might have a role of gene transfer to certain population.

4. Ecosystem and photosynthesis

Stable isotope ratio has been providing a lot of relevant knowledge about rural ecosystem. To pursue the ecological phenomenon in rural area from the point of biology, it has contributed so much. And findings from the development of ecological engineering are also needed. I'm convinced that stable isotope ratio will help us to conserve rural ecosystem as an evaluation method of the engineering effects.
What thus far discussed depends on the difference between photosynthesis cycle, C_3 plants, C_4 plants and algae. Photosynthesis gives us not only organic matter that is vital for all of animals but also efficient method to investigate and to conserve ecosystem.

5. References

Mori Atsushi (2007): Behavior of Stable Isotope Ratios of Attached Matter in a Drainage Canal, Japan Agricultural Research Quarterly, 41(2), 133-138.

Mori Atsushi, Mizutani Masakazu, Matsuzawa Shinichi, (2007): Origin Estimation of Carbon of Spiders (Arachnida) by Carbon Stable Isotope Ratio, Irrigation, Drainage and Rural engineering Journal, 251, 565-571.[In Japanese]

Mori Atsushi, WATABE Keiji, Koizumi Noriyuki, Takemura Takeshi (2009): Organ Specific Carbon Stable Isotope Ratio of Tokyo Daruma Pond Frog, Irrigation, Drainage and Rural engineering Journal, 261, 93-94.[In Japanese]

Mori Atsushi, Mizutani Masakazu, Matsuzawa Shinichi(2006): What does Distribution of Carbon Isotope Ratio of the Animal Population Show?, 2006th Lecture meeting of The Japanese society of Irrigation, Drainage and Rural engineering.[In Japanese]

Mori Atsushi, Mizutani Masakazu, Shioyama Fusao, Nakakuki Genichi, Matsuzawa Shinichi (2007): Estimation of Loach Food in a Paddy Field by Stable Isotope Ratio Method, 2007th Lecture meeting of The Japanese society of Irrigation, Drainage and Rural engineering.[In Japanese]

Mori Atsushi, Watabe Keiji, Koizumi Noriyuki, Takemura Takeshi (2008): Estimation of Japanese Brown Frog Movement by Stable Isotope Ratio Method, 2008th Lecture meeting of The Japanese society of Irrigation, Drainage and Rural engineering.[In Japanese]

Mori Atsushi, Watabe Keiji, Yoshida Yutaka (2009): Analysis of the Feeding Habit of *Pronodularia japanensis* by Stable Isotope Ratio Method, 2009th Lecture meeting of The Japanese society of Irrigation, Drainage and Rural engineering.

Permissions

The contributors of this book come from diverse backgrounds, making this book a truly international effort. This book will bring forth new frontiers with its revolutionizing research information and detailed analysis of the nascent developments around the world.

We would like to thank Dr. Mohammad Mahdi Najafpour, for lending his expertise to make the book truly unique. He has played a crucial role in the development of this book. Without his invaluable contribution this book wouldn't have been possible. He has made vital efforts to compile up to date information on the varied aspects of this subject to make this book a valuable addition to the collection of many professionals and students.

This book was conceptualized with the vision of imparting up-to-date information and advanced data in this field. To ensure the same, a matchless editorial board was set up. Every individual on the board went through rigorous rounds of assessment to prove their worth. After which they invested a large part of their time researching and compiling the most relevant data for our readers. Conferences and sessions were held from time to time between the editorial board and the contributing authors to present the data in the most comprehensible form. The editorial team has worked tirelessly to provide valuable and valid information to help people across the globe.

Every chapter published in this book has been scrutinized by our experts. Their significance has been extensively debated. The topics covered herein carry significant findings which will fuel the growth of the discipline. They may even be implemented as practical applications or may be referred to as a beginning point for another development. Chapters in this book were first published by InTech; hereby published with permission under the Creative Commons Attribution License or equivalent.

The editorial board has been involved in producing this book since its inception. They have spent rigorous hours researching and exploring the diverse topics which have resulted in the successful publishing of this book. They have passed on their knowledge of decades through this book. To expedite this challenging task, the publisher supported the team at every step. A small team of assistant editors was also appointed to further simplify the editing procedure and attain best results for the readers.

Our editorial team has been hand-picked from every corner of the world. Their multi-ethnicity adds dynamic inputs to the discussions which result in innovative outcomes. These outcomes are then further discussed with the researchers and contributors who give their valuable feedback and opinion regarding the same. The feedback is then collaborated with the researches and they are edited in a comprehensive manner to aid the understanding of the subject.

Apart from the editorial board, the designing team has also invested a significant amount of their time in understanding the subject and creating the most relevant covers. They scrutinized every image to scout for the most suitable representation of the subject and create an appropriate cover for the book.

The publishing team has been involved in this book since its early stages. They were actively engaged in every process, be it collecting the data, connecting with the contributors or procuring relevant information. The team has been an ardent support to the editorial, designing and production team. Their endless efforts to recruit the best for this project, has resulted in the accomplishment of this book. They are a veteran in the field of academics and their pool of knowledge is as vast as their experience in printing. Their expertise and guidance has proved useful at every step. Their uncompromising quality standards have made this book an exceptional effort. Their encouragement from time to time has been an inspiration for everyone.

The publisher and the editorial board hope that this book will prove to be a valuable piece of knowledge for researchers, students, practitioners and scholars across the globe.

List of Contributors

Róbson Ricardo Teixeira, Jorge Luiz Pereira and Wagner Luiz Pereira
Federal University of Viçosa, Viçosa, Minas Gerais State, Brazil

Ben Dhiab Rym
Institut National des Sciences et Technologies de la Mer, Monastir, Tunisia

Gilberto Barbante Kerbauy, Cassia Ayumi Takahashi, Alejandra Matiz Lopez, Aline Tiemi Matsumura, Leonardo Hamachi, Lucas Macedo Félix, Paula Natália Pereira, Luciano Freschi and Helenice Mercier
São Paulo University, Brazil

José Gomes-Laranjo
Centre for the Research and Technology of Agro-Environmental and Biological Sciences (CITAB), Universidade de Trás-os-Montes e Alto Douro (UTAD), Vila Real, Portugal

Lia-T. Dinis, Ester Portela, Teresa Pinto, Changhe Zhang and Afonso Martins
Centre for the Research and Technology of Agro-Environmental and Biological Sciences (CITAB), Universidade, de Trás-os-Montes e Alto Douro (UTAD), Vila Real, Portugal

Luís Martins
Department of Forest & Landscape, Universidade de Trás-os-Montes e Alto Douro (UTAD) Apartado, Vila Real, Portugal

Marta Ciordia Ara, Isabel Feito Díaz and Juan Majada
(SERIDA), Consejería de Medio Rural y Pesca, Villaviciosa, Asturias, Spain

Francisco Peixoto
CECAV, University of Trás-os-Montes e Alto Douro, Vila Real, Portugal

S. Pereira Lorenzo and A.M. Ramos Cabrer
University of Santiago de Compostela, Department of Crop Production, Campus de Lugo, Spain

Rita Costa
INRB, I.P. L-INIA Quinta do Marquês, Oeiras, Portugal

Giedrė Samuolienė and Pavelas Duchovskis
Institute of Horticulture, Lithuanian Research Centre for Agriculture and Forestry, Kaunas, Lithuania

Natascha Oppelt
Kiel University, Germany

Fungyi Chow
University of Sao Paulo, Department of Botany, Institute of Bioscience, São Paulo, SP, Brazil

Katarína Kráľová and František Šeršeň
Faculty of Natural Sciences, Comenius University in Bratislava, Slovak Republic

Qingmin Han
Forestry and Forest Products Research Institute, Japan

Ronald Osinga
Wageningen University, Aquaculture & Fisheries, Wageningen, The Netherlands

Roberto Iglesias-Prieto and Susana Enríquez
Unidad Académica Puerto Morelos, Instituto de Ciencias del Mar y Limnología, Universidad Nacional Autónoma de México, Apdo, México

Miriam Zigman, Zvy Dubinsky and David Iluz
The Mina and Everard Goodman Faculty of Life Sciences, Bar-Ilan University, Ramat-Gan, Israel

David Iluz
The Department of Geography and Environment, Bar-Ilan University, Ramat-Gan, Israel

Germani Concenço
Embrapa Agropecuária Oeste, Dourados-MS, Brazil

Ignacio Aspiazú
Universidade Estadual de Montes Claros-MG, Brazil

Evander A. Ferreira
UFVJM, Diamantina-MG, Brazil

Leandro Galon
Universidade Federal do Pampa, Itaqui-RS, Brazil

Alexandre F. da Silva
Embrapa Milho e Sorgo, Sete Lagoas-MG, Brazil

Shin-ichi Shibata and Haruhiko Kimura
Division of Electrical Engineering and Computer Science, Graduate School of Natural Science & Technology, Kanazawa University, Japan

Takashi Oyabu
Graduate School of Strategic Management, Kanazawa Seiryo University, Japan

Ramón Díaz-Ruiz
Colegio de Postgraduados, Campus Puebla, México

Sauren Das
Agricultural and Ecological Research Unit, Indian Statistical Institute, India

Jun-Ichi Sakagami
Japan International Research Center for Agricultural Sciences, Japan

M. Aghdasi, S. Fattahi and H. R. Sadeghipour
Dept. of Plant Physiology, Faculty of Science, Golestan University, Gorgan, Iran

Kangfu Yu, Chun Shi, and BaiLing Zhang
Agriculture and Agri-Food Canada, Canada

Atsushi Mori
Renewable Resources Engineering Division, National Institute for Rural Engineering, Japan